Geoinformatics: Data to Knowledge

edited by

A. Krishna Sinha
Department of Geosciences
Virginia Polytechnic Institute and State University
Blacksburg, Virginia 24061
USA

THE GEOLOGICAL SOCIETY OF AMERICA

Special Paper 397

3300 Penrose Place, P.O. Box 9140 ▪ Boulder, Colorado 80301-9140 USA

2006

Published by The Geological Society of America, Inc.
3300 Penrose Place, P.O. Box 9140, Boulder, Colorado 80301-9140, USA
www.geosociety.org

Printed in U.S.A.

GSA Books Science Editor: Abhijit Basu

Library of Congress Cataloging-in-Publication Data

Geoinformatics: data to knowledge / edited by A. Krishna Sinha
p.cm. – (Special paper ; 397)
Includes bibliographical references.
ISBN 0-8137-2397-3
1. Geographic information systems. 2. Geodatabases. I. Sinha, A. Krishna, 1941- II. Special papers (Geological Society of America) ; 397

G70.212.G4735 2006
910.285--dc22

2005055215

Cover: Image courtesy of the National Aeronautics and Space Administration. Imagery developed by the Earth Observatory team (http://earthobservatory.nasa.gov/).

10 9 8 7 6 5 4 3 2 1

Contents

Preface v

1. ***Representing geoscientific knowledge in cyberinfrastructure: Some challenges, approaches, and implementations*** 1
Boyan Brodaric and Mark Gahegan

2. ***A community effort to construct a gravity database for the United States and an associated Web portal*** 21
G. Randy Keller, T.G. Hildenbrand, R. Kucks, M. Webring, Allen Briesacher, Kristine Rujawitz, A.M. Hittleman, D.R. Roman, Daniel Winester, R. Aldouri, John Seeley, Jorge Rasillo, Roberto Torres, William J. Hinze, Ann Gates, Vladik Kreinovich, and Leonardo Salayandia

3. ***The Georgia Basin Digital Library for Sustainable Development: A digital library approach to addressing the societal relevance of geoscientific knowledge*** 35
Rob Harrap, Sonia Talwar, Murray Journeay, Boyan Brodaric, Ryan Grant, Joost Van Ulden, and Shannon Denny

4. ***An integrated chronostratigraphic data system for the twenty-first century*** 53
Paul J. Sikora, James G. Ogg, Anthony Gary, Cinzia Cervato, Felix Gradstein, Brian T. Huber, Charles Marshall, Jeffrey A. Stein, and Bruce Wardlaw

5. ***A geochemical database for western North American volcanic and intrusive rocks (NAVDAT)*** 61
J. Douglas Walker, Todd D. Bowers, Ross A. Black, Allen F. Glazner, G. Lang Farmer, and Richard W. Carlson

6. ***Building the American Mineralogist Crystal Structure Database: A recipe for construction of a small Internet database*** 73
Hareesh Rajan, Hinako Uchida, Deborah L. Bryan, Ranjini Swaminathan, Robert T. Downs, and Michelle Hall-Wallace

7. ***Geoscience concept models*** 81
Stephen M. Richard

8. ***Managing scientific data: From data integration to scientific workflows*** 109
Bertram Ludäscher, Kai Lin, Shawn Bowers, Efrat Jaeger-Frank, Boyan Brodaric, and Chaitan Baru

9. ***Active tectonics, tectonic geomorphology, and fault system dynamics: How geoinformatics can help*** 131
J Ramón Arrowsmith

10. ***Fusion of Landsat ETM+ and radar data to enhance the extraction of surface and near-subsurface information*** 141
Hongjie Xie and G. Randy Keller

11. ***Data fusion in geophysics: Seismic tomography and crustal structure in Poland as an example*** 153
Matthew G. Averill, G. Randy Keller, Kate C. Miller, Piotr Sroda, Tiffni Bond, and Aaron Velasco

12. ***Schema to ontology for igneous rocks*** 169
A.K. Sinha, A. Zendel, B. Brodaric, C. Barnes, and Jihane Najdi

13. ***GeoNet: Use of grid technologies in geoinformatics for the transition zone between the Colorado Plateau and the Basin and Range province*** 183
Leonardo Salayandia, Yuan Huang, Ann Q. Gates, and Steven Roach

14. ***Development of ontologies for earth system science*** 195
Rob Raskin

15. ***The case for transforming the geological survey knowledge system: Where digital geoscience spatial models meet the semantic Grid*** 201
T.V. Loudon

16. ***The concept of geological uncertainties and new ways of their geomathematical evaluation*** 211
György Bárdossy and János Fodor

17. ***Application of fuzzy arithmetic and prior information to the assessment of the completeness of a mineral exploration program: A case study*** 217
György Bárdossy and János Fodor

18. ***The system integrator in distributed data systems: The OPeNDAP Data Connector (ODC)*** . . . 225
Peter Cornillon, John Chamberlain, and Daniel Holloway

19. ***Bird's-eye view of data mining in geosciences*** 235
Rahul Ramachandran, John Rushing, Xiang Li, Chandrika Kamath, Helen Conover, and Sara Graves

20. ***A new GIS-driven geophysical database for the southwestern United States*** 249
Jesse B. Yoburn, Matthew J. Fouch, J Ramón Arrowsmith, and G. Randy Keller

21. ***Designing a modular architecture for the structural geology ontology*** 269
Hassan A. Babaie, John S. Oldow, Abbed Babaei, Hans G. Avé Lallemant, and A. John Watkinson

Preface

The science of informatics in the broadest sense has been several thousand years in the making. The history of dealing with numbers probably originated within ancient communities through utilization of symbols representing numbers such as 1, 2, 3, and so on. The manipulation of numbers through additive, decimal, and exponent notations has yielded the binary system required for modern computations. With the emergence of large storage devices and high-speed processing of data, it has become possible to organize vast amounts of data as digital products with ontologic tags and concepts for smart queries. Coupling this computational capability with earth science data defines the emerging field of geoinformatics.

Several decades ago, digital archives of data were mostly geographical records. With the emergence of plate tectonics, the need to access other types of data (maps, rock, geophysical and fossil databases) became important. As a consequence, disciplinary and subdiscipline-oriented databases have developed rapidly, but are difficult to integrate due to issues of format and a limited understanding of the conceptual relationships between data.

Since geology was established as a science several centuries ago, observations led to conclusions that were integrative in concept (layering, thickness, type of rock, etc.) and clearly had profound implications for its birth. Subsequently, earth scientists have splintered into dozens of specialties, each individually unable to explain complex earth systems. As information about the earth and other planets become abundant, databases created to store information with rigid structures and complex formats pose integration challenges, and the emerging science of geoinformatics may act as the integrator.

Integration of data is also required to make geosciences a societally relevant science. For example, concerns about climate changes require an integrated understanding of stratigraphy, sea-level changes, fossil record, isotopes, and tectonics. Clearly, core research tasks of the emerging science of geoinformatics include enabling scientists to find, model, and visualize data in an open, Web-based environment

The emerging field of geoinformatics will enable a new generation of geoscientists to conduct research and enhance educational opportunities through easy access to both data and knowledge. It is also likely that integration will create disciplines and scientific thrusts that we are currently unable to envision. It is imperative for the earth science community to remember that, as our national economy moves from service-based to knowledge-based, geoinformatics can facilitate this transition for our science.

Twenty one research papers, co-authored by ninety-six researchers from both earth and computer sciences, have been included in this volume to provide the first-ever organized presentation of the science of informatics as it relates to geology. Readers will readily recognize the vast intellectual content represented by these papers as they seek to address the core research goals of geoinformatics. As editor, I am grateful to many of my colleagues for undertaking the difficult task of reviewing multidisciplinary research papers, and as such they deserve special acknowledgement. The reviewers included J.R.A. Giles, Clinton Smyth, Gary Raines, Ramon Arrowsmith, Matthew Fouch, Naveen Ashish, José Hurtado, James Spotila, Ross Black, John Ebel, John Hole, Atilla Aydin, Shuhab Khan, Rebecca Dodge, Mohamed Abdelsalam, Ruixin Yang, Vipin Kumar, Pusheng Zhang, Cyril Pshenichny, Susana Montes, S. Henley, Eric Christiansen, Cin-Ty Lee, Michael Gertz, V.J.S. Grauch, James Boxall, Nadine Schuurman, Carl Francis, David Mogk, Hassan Babaie, Nancy McMillan, Nathan Green, Peter Sadler, Karl Fleisch, Ian Jackson, W. Hatton, Boyan Brodaric, Andrew Curtis, and Chuck Meertens. Review of the preface by Margaret Banton is also gratefully acknowledged.

A. Krishna Sinha

Geological Society of America
Special Paper 397
2006

Representing geoscientific knowledge in cyberinfrastructure: Some challenges, approaches, and implementations

Boyan Brodaric*
Geological Survey of Canada, 234B-615 Booth St., Ottawa, Ontario K1A 0E9, Canada, and *GeoVISTA Center, Department of Geography, The Pennsylvania State University, University Park, Pennsylvania 16802, USA*

Mark Gahegan*
GeoVISTA Center, Department of Geography, The Pennsylvania State University, University Park, Pennsylvania 16802, USA

ABSTRACT

Networks of databases and tools are being constructed in the geosciences to facilitate geoscientific computing and to stimulate new research. These efforts are contributing to an emerging geoscientific cyberinfrastructure that has the potential to be an important new operational paradigm for the geosciences. However, limited research has been undertaken on the adaptation of emerging cybertechnology to geoscience knowledge and practice. Consequently, we examine the nature of geoscientific knowledge and practice and identify a series of challenges related to representing and using these in emerging computational networks. The first main challenge emphasizes enhanced representation of geoscientific knowledge via ontology development, and the second calls for enhanced tools for processing the knowledge within database networks. We also briefly restate a schema that addresses several of the representation challenges and outline its implementation by different government agencies in three database networks. The identification of challenges, their treatment in the schema, and the implementations all contribute in part to the conceptualization and realization of a cyberinfrastructure for the geosciences.

Keywords: ontology, cyberinfrastructure, geoscience knowledge, knowledge representation, geoscience reasoning.

1. INTRODUCTION

Geoscience[1] must leverage a broad suite of information in order to address problems that span many geographic scales and scientific disciplines. However, use of geoscientific information is inhibited by its fragmentation along agency and thematic boundaries: the information is resident in autonomous databases that for historical reasons are hosted in a variety of sectors, including government, industry, and academia, and that are segmented thematically as per stratigraphy, geochronology, and geophysics. Expected increases in both information volume and Web-based accessibility suggest such fragmentation will continue to rise, exacerbating difficulties in finding, integrating, and ultimately using the information. This leads to a scenario in which routine information handling and research are seriously impeded, particularly in light of the prevailing trend in all sectors toward large-scale, multidisciplinary projects

*E-mails: brodaric@NRcan.gc.ca, mng1@psu.edu.
[1]"Geoscience" and "Earth science" are used synonymously herein.

Brodaric, B., and Gahegan, M., 2006, Representing geoscientific knowledge in cyberinfrastructure: Some challenges, approaches, and implementations, *in* Sinha, A.K., ed., Geoinformatics: Data to Knowledge: Geological Society of America Special Paper 397, p. 1–20, doi: 10.1130/2006.2397(01). For permission to copy, contact editing@geosociety.org.

that are increasingly reliant on computational tools and digital information. In response, several initiatives are pursuing greater cohesion of information through database networks in which multiple-source databases are linked to provide networkwide access to information and associated tools (e.g., Geosciences Network [GEON], http://www.geongrid.org; Southern California Earthquake Center [SCEC], http://www.scec.org/cme/; Canadian Geoscience Knowledge Network [CGKN], http://www.cgkn.net; U.S. Geological Survey National Geological Map Database [NGMDB], http://ngmdb.usgs.gov/Info/).

Database network initiatives must overcome many thorny technological hurdles, including those associated with the following tasks: (1) registering information and tools with the network—the *resources registration* task; (2) finding relevant information and services across the network—the *resources discovery* task; (3) integrating the discovered information—the *information interoperability* task; (4) integrating the various software tools—the *services interoperability* task; (5) using the integrated information and tools for scientific work—the (possibly collaborative) *analysis and processing* task; and (6) disseminating the scientific results, derived from the collective information, for scientific or educational purposes—the *publishing* task.

In this paper we focus on information interoperability (task 3). We identify several challenges to geoscientific information interoperability, present an approach that addresses some of the challenges, and briefly discuss three implementations that demonstrate the approach. In section 2 we explore the needs of geoscientific interoperability and conclude with two main challenges: the first emphasizes enhanced representation of geoscientific knowledge, and the second stresses enhanced tools for processing the knowledge during data integration activities. In section 3 we elaborate these challenges by examining the nature of geoscientific knowledge, thereby identifying additional challenges in reasoning, operation, and representation. In section 4 we outline a schema useful for structuring aspects of geoscientific knowledge in database networks, in response to several of the representation challenges, and in section 5 we sketch out its implementation in three projects, two that deal with the problem of geological map information interoperability and a third that deals with the role of geoscience information in modeling sustainable development. We close with a brief summary in section 6.

2. INFORMATION INTEROPERABILITY

For our purposes information interoperability (henceforth "interoperability") refers to the ability of a human or machine agent to (1) receive unified and integrated results in response to a query of a database network, and (2) to update the contents of the network. An example of the former, querying, is a request to the network for all mapped geological formations located in a geographic region of North America that contain a certain lithology and age—without having to individually search each source database, such as those for geological maps, lithostratigraphic lexicons, and geochronologic databases, and without necessarily having to reconcile differences in format and content manually. An example of the latter, updating, involves adding or deleting information by posing a single request to the network that is then executed within the appropriate source databases.

Interoperability is difficult because of distribution, autonomy, and heterogeneity (Sheth and Larson, 1990). Distribution refers to how the information is spread amongst the multiple databases and their physical locations. Autonomy refers to the independence of each database within the network, including its control over local database design, response to external requests, and policies for sharing information. Heterogeneity refers to differences between databases that can be systemic, syntactic, schematic, and semantic (Sheth, 1999): i.e., the databases can be implemented on different computing platforms (systems), use different control languages and file formats (syntactic), be diversely structured (schematics), possess different content, have diverse concepts for the same content, and even have the same concepts for diverse content (semantics). To achieve interoperability, strategies for distribution, autonomy, and heterogeneity must be implemented: the databases need to be connected and the information appropriately distributed; the database stewards must agree how to share some information and functions and decide how to manage changes and transactions; and mechanisms must be adopted for unifying the information along the four dimensions of heterogeneity. Figure 1 illustrates reconciling information in these dimensions to enable exchange of meaning.

These requirements lead to a variety of possible database network architectures based on the level of coupling between the source databases and the overall database network (after Elmagarmid et al., 1999; Sheth and Larson, 1990). We concentrate our discussion of these architectures mainly on semantic heterogeneity and categorize the architectures according to the degree of overlap of concepts in the source database schema, including concepts in the database structure (i.e., in entities, relations, attributes, rules, functions) and in the vocabulary of the database contents (i.e., in attribute ranges, metadata).

1. Tightly coupled systems are characterized by low autonomy, information centralization, and low heterogeneity (high source-concept overlap), enabling the implementation of a common database schema across the various source databases (Hull, 1997); e.g., a data warehouse in which information from the source databases is duplicated (i.e., materialized) and where a common schema is used by all.

2. Loosely coupled systems are characterized by moderate autonomy, distribution, and heterogeneity amongst the sources. Partial source concept overlap enables the creation of a common, single global schema for the network, built on shared concepts and the union of nonoverlapping concepts. The global schema is thus somewhat independent of the source schema, requiring mappings to be developed between the global and various source schema to enable information integration. These mappings are typically implemented as logical views between the global schema and the source schema: e.g., the global schema can be defined as a global view over the source schema (GAV:

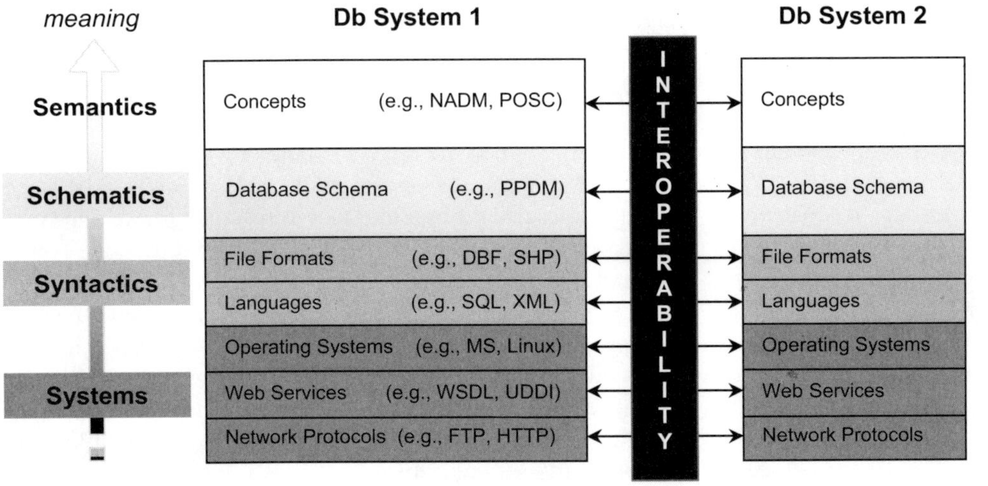

Figure 1. Interoperability levels increasing vertically for greater exchange of meaning between database systems; Public Petroleum Data Model Association (PPDM)—industry standard database schema for petroleum information (http://www.ppdm.org), North American Geologic Map Data Model Steering Committee (NADM)—evolving catalog of geological (map) concepts and related terminology (NADM, 2004), Petrotechnical Open Standards Consortium (POSC)—catalog of standard geological concepts for the petroleum industry (http://www.posc.org/), Web services—protocols for executing applications over the Web, languages—data definition, manipulation, and transport syntax.

global-as-view), the local source schema can be defined as views over the global schema (LAV: local-as-view; e.g., Ullman, 1997), or some hybrid (Calì et al., 2002). The global schema typically serves as a central conceptual portal into the database network, providing the illusion of a single unified data source.

3. Disjoint systems are highly autonomous, distributed, and semantically heterogeneous, precluding the creation of a global schema based primarily on source concepts. Interoperability is achieved by mapping source concepts onto more general knowledge, an *ontology*, that consists of general concepts, theories, models, processes, and other knowledge forms. The ontology can serve many roles in a disjoint system: (1) metadata about the network resources; (2) a global conceptual schema for the network that connects both data and tools; (3) a vocabulary for the network contents; (4) the foundation for a query interface that can act as a central portal into the network; (5) a representation of geoscientific knowledge. These roles are variously combined or compartmentalized within different implementations (Wache et al., 2001). When disjointedness is extreme, perhaps because very different types of information sources are regularly added and deleted from the network, mapping to a common domain ontology is unachievable and information might be integrated via *peer-to-peer* translations between sources (e.g., Fensel et al., 2002; Gribble et al., 2001; Panti et al., 2002).

As implied by Table 1, the boundaries between these architectures are fuzzy, reflecting a gradient in the overlap of source concepts.

The marked prevalence of semantic heterogeneity in geoscientific databases suggests the adoption of disjoint architectures for geoscientific database networks and consequently the need to develop a shared ontology from the conventions of geoscience. This semantic focus would supplement existing syntactic (e.g., Cox, 2001; Ramachandran et al., 2004) and schematic (e.g., Breunig, 1999; Gartner et al., 2001) interoperability orientations in geoscience.

A computer-based ontology (henceforth "ontology") is defined as an explicit specification of shared concepts and theories (Gruber, 1993), one that represents the intended meaning of a vocabulary (Guarino, 1998). This contrasts with philosophical ontology defined as the study of a priori concepts. (Smith, 1998a, 1998b; Guarino, 1998). Ontologies can in general be seen as having two purposes, enabling computer operations as

TABLE 1. CHARACTERIZATION OF ARCHITECTURES FOR DATABASE NETWORKS

	Autonomy	Distribution	Heterogeneity	Common Schema		
Tight	low	central	low	√	specific	GAV
Loose	↓	↓	↓	√	↓	LAV
				√		Ontology
Disjoint	high	distributed	high	—	general —	peer-to-peer

Note: See text for details.

well as aiding human comprehension (Fensel et al., 2002). In the former, an ontology can provide formal and explicit meaning for information, facilitating computer-based information processing; in the latter, an ontology can stimulate informal and possibly implicit meaning in humans to enable understanding of information and theory.

Ontologies are often categorized according to content or representational structure. Content-based approaches commonly distinguish between ontologies that apply across domains and those that pertain within domains; e.g., *formal* ontologies represent general a priori concepts such as "relation," whereas *material* ontologies represent domain concepts such as "fault"—geoscience ontology is therefore fundamentally material (Smith, 1998a). Structure-based categorizations distinguish ontologies according to mode of representation and the nature and degree of connectivity between entities in the ontology. For example, linguistic, axiomatized, and prototypical ontologies are differentiated in Sowa (2000, p. 493–496). Linguistic ontologies consist of loosely structured standard terms and their natural language definitions, such as a list of terms in a glossary (e.g., Jackson, 1997), or they are exemplified by more structured entities such thesauri that include term relations, e.g., "synonym." Most notably, linguistic ontologies are difficult to compute with but eminently comprehensible to humans. In contrast, axiomatized ontologies are quite computable as they use a logical language to represent the concepts and relations that underlie terms (e.g., Karp, 2000; NADM, 2004), but the computation is at present limited to classification issues such as verifying the logical consistency of taxonomic relations. In contrast to both linguistic and axiomatized ontologies, concept meanings in prototype ontologies are not defined a priori but are derived from prototypical examples, e.g., a geological formation described in terms of its type locality or localities. Prototypical concept definitions may not therefore be exact or crisp but may instead be approximate summaries of formative evidence. From a structural viewpoint, material geoscience ontologies may thus be linguistic or a *mixed* ontology containing axiomitized as well as prototypical concepts.

The relationship between a linguistic and a mixed ontology in a database network involves connecting terms and axiomatic-prototypical representations of term meanings. For example, synonymy, where multiple terms have a common meaning, can be viewed as the mapping of multiple linguistic terms onto a single axiomatic and/or prototypical concept (e.g., the common concept for "strike" and "azimuth"), and polysemy, where a single term has multiple meanings, can be viewed as the mapping of multiple axiomatic and/or prototypical concepts onto a single term (e.g., a geological unit name that refers to both a formation and a member).

A disjoint architecture for geoscience could then benefit from (1) a mixed material ontology, to bridge disjoint source database concepts and terms as well as link to prototypical evidence, and (2) a linguistic ontology that catalogs standard terms and maps them onto concepts in the axiomatic ontology (Klein, 2002). This architecture suggests at least four types of mappings, as shown in Figure 2: (1) between the source databases and the mixed ontology to deal with synonymy and polysemy in the database contents and schema (dashed lines, right); (2) between concepts and the prototypical evidence in the source databases, to document the origin of concepts (solid line, right); (3) between the mixed ontology and the linguistic ontology to identify standard vocabulary for concepts (solid lines, middle); and (4) between the ontologies and the user interface to specify requests and communicate data in standard terminology to people, tools, and other query sources (dashed lines, left). Moreover, the number of mappings implemented will increase in a pluralistic system in which multiple linguistic and mixed ontologies can be supported. The justification for a pluralistic approach rests in the notion that the task of science is to evolve knowledge, implying that ontologies will likely evolve through time and potentially conflict, following scientific progress and debate.

Consequently, a primary challenge to the successful deployment of geoscientific database networks is the development of reference or foundational ontologies for geoscience and the frameworks that host them. While geoscientific database schema have been widely implemented and even standardized in some agencies (e.g., Bain and Giles, 1997; Public Petroleum Data Model Association—www.ppdm.org), the need to formulate general geoscientific knowledge into widely applicable formalisms such as ontologies has been recognized (Loudon, 2000; Pennanen and Kortman, 2002; Pshenichny, 2002), and efforts are in early stages of construction (e.g., NADM, 2004; Pshenichny et al., 2003; Raskin and Pan, 2005) or are industry specific but not fully axiomitized (e.g., Petrotechnical Open Standards Consortium—www.posc.org). A database schema and ontology can be differentiated on the basis of purpose as well as conceptual and structural completeness: database schema (conceptual, logical, and physical) are geared to the specific requirements of a suite of data, particular tasks, and system constraints and are not typically fully axiomatized, whereas ontologies are application independent and possess broader, more general, and axiomitized content, including concepts and other knowledge forms discussed in section 3. The main technological difficulty in such ontology development is providing an ontological framework that can host multiple ontologies, including their regional and personal variations, and that can enable their scientific evolution. This is quite apart from the social challenges inherent in developing the content of such ontologies.

A second technological challenge to the successful deployment of geoscientific database networks is the incorporation of mathematical, geospatial, and temporal operators in information integration. Ontology-based disjoint systems (as well as those loosely coupled) are typically implemented using mediators and wrappers (Domenig and Dittrich, 1999; Wiederhold, 1996): mediators act as coordinators by distributing requests to source-specific wrappers that (1) repackage the requests for consumption by the source database and (2) translate query results from the sources into a shared schema (or ontology) prior to returning them to the mediator for amalgamation. Communication between mediators and wrappers is usually via mark-up languages (e.g., XML, RDF). For geoscientific interoperability the translation process may be especially

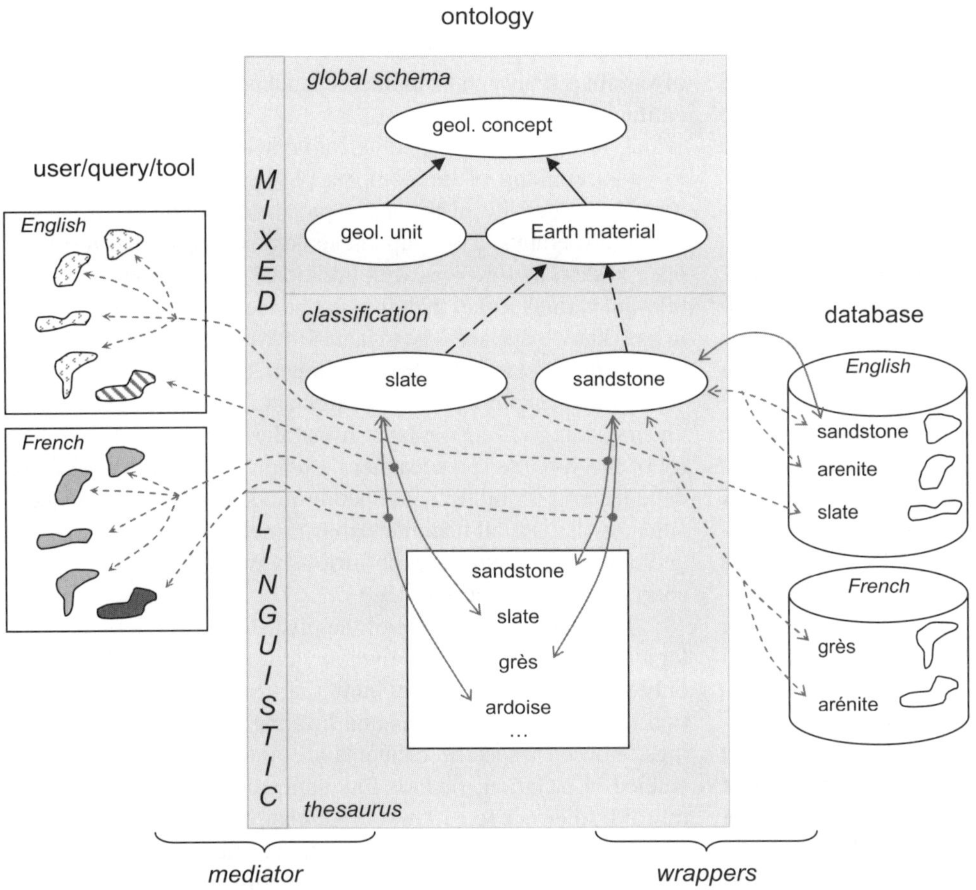

Figure 2. Viewing integrated data from French and English databases via mappings (gray lines) between databases (right disks), a multilingual linguistic ontology (lower middle rectangle), a mixed axiomatized-prototypical ontology (middle ovals), and two user/query/tool interfaces (left rectangles). A wrapper over each database usually executes the database to mixed ontology mapping, whereas a mediator usually executes the user to ontology mapping; mediators and wrappers can both execute mixed to linguistic ontology mappings.

complex because geoscientific features and disjoint concepts can be related via equations, temporal and geospatial relations such as proximity, adjacency, and containment, via analytic functions such as interpolation, via simulation functions such as modeling, and also via participation within processes. Moreover, complex data integration processes may also need human intervention at difficult stages. Existing database interoperability mechanisms thus need to incorporate geospatial, temporal, and process-oriented operations (Koch et al., 2000) and potentially be amenable to human guidance. Some of the needs related to automated data integration have been met in part by the development of wrappers and interchange formats for geospatial information (e.g., OGC Web Mapping Service [WMS], Web Feature Service [WFS]: http://www.opengis.org), but these are distinct from those created for nongeospatial information. Bridging this divide then additionally involves creating nongeospatial wrappers and a blended ontology that combines geoscientific, temporal, geospatial, and process ontologies so that the separately wrapped information can be integrated. It further involves developing mechanisms for recognizing and/or prompting human intervention and directing all these developments at geoscience. Current approaches to these issues are not directed at geoscience, in that they are either schematic, where (geo)scientific schema are extended with general space-time concepts (e.g., Ram et al., 1999; Renolen, 2000), or semantic, where geospatial ontology is linked to thematic and function concepts (Fonseca et al., 2002; Kuhn, 2001) or is an element of a top-level ontology framework for structuring features and processes (Grenon and Smith, 2004; Gangemi et al., 2003).

In summary, we have identified two broad technological challenges to information interoperability in geoscientific database networks that employ disjoint architectures.

Challenge 1: Developing the framework and contents for geoscientific ontology, including axiomatized, linguistic, and prototypical elements, and linkage to geospatial, temporal, process, and ontology.

Challenge 2: Developing tools to leverage geoscientific ontology in database networks.

Challenge 2.1: Developing mediators and wrappers that incorporate geoscientific, geospatial, temporal, and process-oriented operations in data integration activities and that utilize and/or prompt human intervention.

3. GEOSCIENTIFIC KNOWLEDGE

Knowledge, including scientific knowledge, can be categorized along several dimensions including type, quality (de Jong and Ferguson-Hessler, 1996), and mind centered versus experience centered (Reynolds et al., 1996). However, any categorization of knowledge is necessarily aligned with some philosophical worldview that makes various commitments, minimally, to (1) the objects we have knowledge of (philosophic ontology), and (2) a theory of knowledge—its nature, sources, limits, and methods (epistemology; Raper, 1999, 2000, chapter 1). For example, scientific realists hold that increasingly universal knowledge about an objective biophysical reality is obtainable via scientific method, while constructivists argue that knowledge is fundamentally relative to the social forces experienced by the individual, and middle positions suggest that approximate and incremental knowledge of biophysical reality is methodologically obtainable by individuals participating within an affecting social milieu (Gould, 2000). These commitments have not only philosophical relevance but also have critical impact on the representation of knowledge. A strong realist stance implies that knowledge can be directly identified and represented, that a single ontology for geoscience is obtainable and representable, and that automated machine integration of data is thus potentially feasible; a strong constructivist stance implies that knowledge is relative, that cognitive and social contexts should be represented, and that satisfactory data integration must be performed ultimately by humans but will be nonetheless inherently imperfect; lastly, intermediate positions emphasize the approximate and evolving nature of knowledge, implying that multiple ontologies and geoscientific interpretations of varying certainty should be represented, and that machine-based and/or human-based data integration may be required depending on certainties and complexities in the knowledge, methods, and data. Underpinning these positions are differing sentiments about how the knowledge acquisition and evaluation processes (epistemology) impact the identified knowledge (ontology). Computational ontologists have proposed staged ontologies that in lower "material" levels might vary within domains according to human conceptualization and action, reflecting constructivist and/or intermediate philosophical positions but tend toward stability at upper "formal" levels and the realist position (Guarino, 1998; Smith, 1998b). Recent commentary among geoscientists, who are situated mainly at the material level, adopts the intermediate position in which the unique characteristics of geoscience variably bind its knowledge results to the discovery process (Baker, 1996, 1999; Frodeman, 1995, 2003). Investigating these characteristics and their relation to both the forms of knowledge and to geoscientific process and cognition provides insight into how such knowledge should be represented at the material levels.

3.1. Unique Characteristics of Geoscientific Knowledge

Geoscientific knowledge (debatably) differs from knowledge in other sciences in terms of complexity, incompleteness, indeterminacy, singularity, explanation, and historicity.

1. Complexity: Geological knowledge is complex because its subject, the physical Earth system, is complex, both in terms of heterogeneity in the composition of materials and in terms of the multiple processes interacting with those materials at various spatial and temporal scales. There are therefore a large number of variables (concepts) and interdependencies involved in geoscientific knowledge.

2. Incompleteness: Geological knowledge is incomplete due to undersampling of the complex physical Earth; the sampling itself is limited by instrument, spatial, temporal, or socio-economic constraints. This implies geological knowledge forms an open system that evolves with additional observation. As all the relevant variables may therefore not be identified, there exist gaps in both knowledge and data (Mann, 1993). Geoscience knowledge is therefore bootstrapped from an imperfect knowledge base.

3. Indeterminacy: Complexity and incompleteness imply understanding is fragmentary, in that the vastness and complexity of the variable space cannot be uniquely solved by available data, leading to multiple competing models to describe a single situation that are all underdetermined in varying degrees. Hence geological knowledge can be variously interpreted and is inherently uncertain (Schumm, 1991).

4. Singularity: Because of the preceding characteristics geological situations are on the whole quite differentiated and thus only partially transportable, inhibiting their complete reproduction in laboratory situations and limiting the generality of findings. However, specific elements are transferable and might be studied in isolation, such as foundational elements of the theoretical framework (e.g., laws; Frodeman, 1995).

5. Explanation: Geoscience is more often aimed at explaining a specific physical situation, by describing characteristics and interpreting them in relation to background concepts and theory (e.g., plate tectonics), than generalizing situations into empirical regularities, new concepts, or theories, or predicting future situations. The primary product is thus an explanatory model: a reconstruction of the state of Earth materials and processes, at some spatiotemporal location, and their causal interactions with historical materials and processes. Due to indeterminacy and singularity, the criteria for selecting the optimum model is its overall coherence with observations and background knowledge rather than experimental replication. Coherence is often conveyed via graphic and textual narratives such as maps and reports using scientific and rhetorical reasoning (Engelhardt and Zimmermann, 1982).

6. Historicity: The role of history in explanation is emphasized in two respects: (1) both the objects to be explained as well as their causes can occur in the past, providing geoscience with a largely causal imperative, and (2) explanations often involve longer time periods than in other sciences.

These insights lead us to supplement the first challenge as follows.

Challenge 1.1: Embracing fragmentation and plurality in geoscientific ontology to account for multiple partial explanations and knowledge evolution and developing integration mechanisms to combine the plural views or map between them. This may involve integrating diverse scientific explanations or mapping between a scientific explanation and other use of the information, e.g., societal or cross-disciplinary.

3.2. Forms and Types of Geoscientific Knowledge

Geoscience shares with other sciences some typical knowledge forms, including observables, such as human- or instrument-detected occurrences, and inferred unobservables such as undetected occurrences, concepts, hypotheses, laws, theories, and principles. Occurrences refer to features, events, and processes; concepts refer to abstractions of observable or unobservable objects, e.g., items in a classification scheme or taxonomy; hypotheses are conjectural or lawlike propositions that summarize relations between occurrences or concepts; theories are internally consistent but not necessarily complete collections of hypotheses; and principles are framework guidelines that motivate and direct geoscientific activity. This sequence of forms, from occurrences to principles, increases in generality and decreases in observability, following general conventions in philosophy, logic, and artificial intelligence or AI (Shrager and Langley, 1990). Although the listing is no doubt incomplete it suffices for our immediate purposes.

Table 2 provides examples of these forms within the categories of situational (S_k) and theoretical (T_k) knowledge that we have derived to reflect the explanatory and empirical emphasis of geoscience. Situational knowledge can be seen as knowledge about the spatiotemporal structure of a real-world situation and involves forms that explain a specific geospatial place or region, whereas theoretical knowledge and its forms are generalizations that apply to many geospatial situations. For example, the geological unit concepts found on maps and in lithostratigraphic lexicons are situational in that they summarize rock bodies located at specific geographical locations, occurring at specific geological times, and identified for specific purposes, whereas the more general concept of a lithostratigraphic unit is theoretical as it describes how geological, spatial, and temporal properties can be applied (instantiated) across geospatial regions and geological times (e.g., North American Commission on Stratigraphic Nomenclature, 1983). The same distinction can be made between a particular mineral deposit and a mineral deposit type and between a specific process history for a situation and a general process theory (e.g., plate tectonics). Situational and theoretical knowledge are mutually affective, in that situational knowledge provides the evidence to derive theoretical knowledge that is in turn referenced to explain situations.

In order to allow geoscientific ontology to be scoped and prioritized we further make use of the well-known cognitive science distinctions of implicit (I_k) versus explicit (E_k) knowledge (Goschke, 1997) and adapt the AI distinctions of foreground (F_k) versus background (B_k) knowledge (Flach and Kakas, 2000). For us, foreground knowledge refers to knowledge that is being newly developed in a situation—e.g., during a field campaign, laboratory study, or other research activity—and consists of new situational knowledge as well as new implicit knowledge such as intuitions. Foreground knowledge might be viewed as an explanation being formulated by a scientist. In contrast, background knowledge refers to previous knowledge in the domain. Background knowledge includes situational and theoretical forms that can be explicitly expressed, as well as prior nonexpressible implicit knowledge. Implicit knowledge may, in a limited sense, be conveyed via loose explicit structures such as narratives and via more formal structures that represent situations and contexts. The relationships between these knowledge categories are stated below in equations 1–3, and the distinctions are illustrated in Figure 3.

$$F_k = S_k + I_k . \quad (1)$$

$$B_k = E_k + I_k . \quad (2)$$

$$E_k = S_k + T_k . \quad (3)$$

The scope of geoscientific ontology can now be stated as consisting of explicit background knowledge (equation 3)—e.g., prior models, model types, concepts, hypotheses, laws, and occurrences. Note that this scope moves closer to capturing implicit human knowledge explicitly by unconventionally including situations (Brodaric and Gahegan, 2002). Moreover, it is intended to capture foundational geological knowledge that can subsequently be transformed to suit diverse scientific and societal purposes.

TABLE 2. SITUATIONAL AND THEORETICAL KNOWLEDGE, THEIR FORMS, AND SOME PROPERTIES

	Form	Situational Knowledge	Theoretical Knowledge
observable particular	*occurrence*	*feature: e.g., polygon X*	N/A
↓	*concept*	*feature type: e.g., X formation, Y formation*	*feature type: e.g., lithostratigraphic unit classification: e.g., rock classification*
	hypothesis	*e.g., X older than Y*	*law: e.g., superposition*
	theory	*model: e.g. a mineral deposit; e.g., a process history*	*model type: e.g., VMS deposit type; e.g., Wilson Cycle*
	principle	*N/A*	*e.g., uniformitarianism, catastrophism*
unobservable general		geospatial particular	spatial general

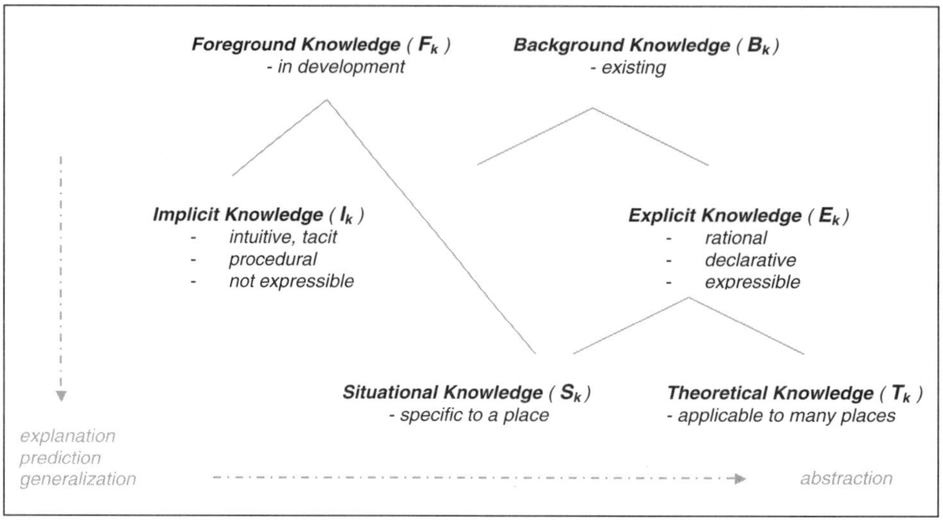

Figure 3. Types of geoscientific knowledge (bold text), their composition (solid lines), and associated reasoning (dashed lines).

The idea being that if foundational geoscientific knowledge, in all its partial and plural nature, is well represented it will not only stimulate new geoscientific research but also ease transformation to wider contexts within and beyond geoscience.

The scope also includes both prior observables and unobservables. However, because some observables and unobservables refer to occurrences stored in existing databases, the actual structure of a geoscience ontology intended for use with database networks can be refined further to exclude these occurrences and include *relations* to them instead, to avoid duplication of content between the ontology and the database.

There are four potential relations to be exploited; two are related to extension development and two to intension development (Brodaric and Gahegan, 2002). Extension development involves linking a database occurrence to the concept it exemplifies and corresponds to the database-mixed ontology mapping (dashed lines, right, Fig. 2). We identify two types of extension development: instantiation and classification. Instantiation refers to the assignment of concept properties to an occurrence, as well as occurrence creation via its inference from theory, whereas classification refers to the assignment of occurrences to the concept such that the occurrence gains membership in the class of examples denoting the concept. Conversely, intension development refers to the process and occurrences involved in generating a concept and corresponds to the mapping of a mixed ontology to the supporting, possibly prototypical, occurrences in databases (solid line, right, Fig. 2). We denote the two forms of intension development as clustering and categorization. Clustering refers to the development of unobservables by grouping (un)observables into regularities, whereas categorization refers to the development of unobservables conceptually distant from their inputs. The semantic distinctness of these four relations is justified by the fact that an occurrence could simultaneously participate in both intension and extension development, e.g., as an instance of one concept and evidence for another concept, such as when a lithology description (an observation) is reused in the definition of a specific geological formation. In terms of relevance to ontology structure, we view inclusion of the extension relations as optional because they mainly inform occurrence content and existence, but we view the intension relations as mandatory because they amplify concept meaning by informing concept genesis, and because they have scientific merit by recording the data and reasoning processes used to derive an unobservable, which facilitates scientific reexamination and testing.

Scoping geoscientific ontology as per equation 3 suggests ontology development might proceed bottom-up, beginning with a catalog of explicit situational knowledge *(S_k)* such as standard reference models and links to supporting premises (Brodaric and Gahegan, 2002). Alternatively, it might proceed top-down, with a catalog of explicit theoretical knowledge *(T_k)* that transcends situations and is generally shared within geoscience, e.g., standard concepts, such as a normal fault, as well as widely accepted theories, e.g., plate tectonics. Our experience indicates that any one approach in isolation soon ends up needing the other in practical applications. Thus we can add the following into the first challenge.

Challenge 1.2: Populating a mixed geoscientific ontology framework with explicit background knowledge consisting of reference models, theories, concepts, and prototypical occurrences, described formally as well as informally via narratives. What should not be understated here is the social engineering challenge of capturing a knowledge fragment for inclusion in such a framework. This involves on the one hand capturing individual knowledge, and on the other hand reaching consensus within a group and capturing group knowledge. A related issue is the ability to transform the captured knowledge within and between ontologies in the framework.

3.3. Reasoning with Geoscientific Knowledge

Understanding geoscientific knowledge that evolves and internally competes entails comprehending the factors leading to change or variety. These factors include not only the supporting evidence but also the reasoning process applied to it. Two modes of reasoning are suggested from the knowledge distinctions above: the development of foreground knowledge from the combination of background knowledge and ancillary foreground knowledge (equation 4), and the development of background knowledge from just foreground knowledge (equation 5):

$$\boldsymbol{F}_k\,\boldsymbol{B}_k \rightarrow \boldsymbol{F}_k \quad (4)$$

$$\boldsymbol{F}_k \rightarrow \boldsymbol{B}_k\,. \quad (5)$$

The first mode (equation 4) is associated with situational knowledge development and hence with the generation of explanations and predictions (Fig. 3). These coincide, respectively, with abductive and deductive forms of inference (Flach and Kakas, 2000). Due to the historical and causal nature of geoscience one of the $\boldsymbol{F}_k$ in equation 4 typically refers to a cause and the other to an effect. Reasoning forward from cause to effect (i.e., ***cause*** $\boldsymbol{B}_k \rightarrow$ ***effect***), using background knowledge, represents deduction and prediction, whereas reasoning backward from effect and background knowledge to cause (i.e., ***effect*** $\boldsymbol{B}_k \rightarrow$ ***cause***) represents abduction and explanation (see Engelhardt and Zimmermann, 1982). The common geoscience practice of developing models and process histories from observations is thus inherently abductive and explanatory; using such models to develop future situations is deductive and predictive. From a logical viewpoint, abductive explanations are less certain because effects have the potential to be satisfied by multiple causes (i.e., competing models, hypotheses, concepts, occurrences). The impact of this rule of logic is magnified in geoscience in that causes are often historical and unobservable, as detailed in section 3.1.

Deduction on the other hand generates logical consequences when theory is applied to a situation; it further possesses the falsification property whereby false consequences imply false premises (i.e., ***cause*** $\rightarrow$ (***effect*** $\boldsymbol{B}_k$)). Deduction is therefore superior to abduction from a logical viewpoint, but its certainty is also questioned when, from a pragmatic viewpoint, the theory and situation are often uncertain, incomplete, or indeterminate, allowing for variable predictive success.

The omission of $\boldsymbol{B}_k$ in equation 4 (i.e., $\boldsymbol{F}_k \rightarrow \boldsymbol{F}_k$ or ***(cause, effect)*** $\rightarrow \boldsymbol{F}_k$) represents generalization and induction, where empirical regularities in the form of unobservables are derived for the situation, e.g., the concept of unit X as derived from field observations. We denote abstraction to occur when generalizations transcend the local situation to induce background theoretical knowledge that may be applied to many situations (above in equation 5 and Fig. 3). Geoscientific generalization and abstraction may both be uncertain because, logically, inductions may be overturned with future contradictory evidence, and because, geologically, the underlying knowledge being generalized or abstracted may itself be uncertain, incomplete, or singular. The latter requires background knowledge, or the return of $\boldsymbol{B}_k$ to the logical form, to guide and validate the induction, leading to the claim that logical induction can be considered a weak form of abduction (Josephson and Josephson, 1994). This resonates with Peirce's later pragmatic account of inference (1987), which is not based on logical form but rather on scientific activities whereby abduction generally refers to the discovery of new knowledge, deduction to prediction generation, and induction to the process of evaluating (dis/confirming) predictions within a broader population (Sowa, 2000).

In the pragmatic view of abduction, an optimal knowledge form is selected, modified and/or generated as ***cause*** (in ***effect*** $\boldsymbol{B}_k \rightarrow$ ***cause***) and may consist of a causal state of affairs (i.e., undetected occurrences) or abstract unobservables (e.g., hypotheses, laws, theories, models). Abduction may therefore be aligned with instantiation, for occurrence generation, and with categorization, for generation of abstract unobservables—often from sparse knowledge, leading to its centrality in scientific insight and discovery. Though subsumed under pragmatic abduction, nonpragmatic induction can in contrast be viewed as generating abstract unobservables via pattern recognition over rich evidence and may correspondingly be aligned with clustering; it is however less novel as its products are stated in the language and concepts of the underlying evidence. The choice of which form to utilize or emphasize in scientific practice may then be dependent on the quantity and quality of the evidence at hand.

Challenge 1.3: Capturing the reasoning process used to determine explicit background knowledge and structuring geoscientific ontology accordingly.

Developing Geoscientific Ontologies

To demonstrate geoscience ontology development we provide an example in which logic is used to represent aspects of map unit concepts derived from situations. The example, though simplified, is cogent because many ontology tools are logic based (e.g., Web Ontology Language, or OWL; McGuinness and van Harmelen, 2003), can reason conceptual structure from evidence (e.g., Loom; MacGregor, 1991), and form the reasoning basis for the emerging cyberinfrastructure as conceived by designers of the Semantic Web (Burners-Lee et al., 2001). For the example assume the following:

1. three geological formation concepts, $\boldsymbol{c}_1$, $\boldsymbol{c}_2$, and $\boldsymbol{c}_3$;
2. two attributes for those concepts consisting of distinct lithologies $\{ \boldsymbol{L}_1 \ldots \boldsymbol{L}_n \}$ (e.g., granite) and ages $\{ \boldsymbol{A}_1 \ldots \boldsymbol{A}_m \}$ (e.g., Devonian);
3. several observables consisting of $< \boldsymbol{L}_i, \boldsymbol{A}_j >$ pairs, some of which belong to the map unit concepts as depicted in Figure 4.

The internal structure of concepts in logic-based ontologies is typically represented with necessary conditions, sufficient conditions, or necessary and sufficient conditions. Necessary conditions inform instantiation: they tell us what properties are known to be common (instantiated) in all instances of the concept; e.g.,

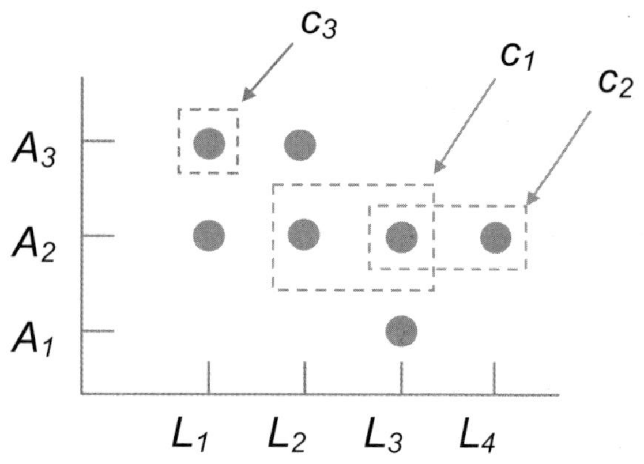

Figure 4. Map unit concept (c_i) representation using age and lithology $< A_m, L_n >$.

it may be that c_1 is characterized by age A_2 and a lithology L_2 throughout the unit [equations 6 and 7 below]. The negation of necessary conditions informs classification, by indicating which observables are not concept instances [equation 10 below]. Conversely, sufficient conditions also inform classification, by specifying which observables are concept instances or instances of concept properties [equations 8 and 9 below]; e.g., observable (L_2, A_2) is not an instance of map unit concept c_1 but rather is an instance of a complex property of c_1. Although sufficient conditions generally imply unique classification, e.g., observable (L_2, A_2) uniquely denotes c_1, assigning unique sufficient conditions to real concepts may in practice be difficult due to conceptual overlap, e.g., shared observable (L_3, A_2) belongs to c_1 and c_2 (Fig. 4). When the overlap is such that one concept wholly contains another, then the contained concept is a subclass of the other, e.g., granite is igneous; in such cases the necessary conditions for the container concept apply to the subclass, but their sufficient conditions will differ. Sufficient conditions also need not be necessary, e.g., it may be that (L_2, A_2) is not present throughout c_1, and necessary conditions need not be sufficient, e.g., if (L_3, A_2) is necessary for c_2 it cannot be uniquely sufficient because of overlap with c_1. Conditions that are both necessary and uniquely sufficient provide *definition*, as the concept is demarcated without overlap, e.g., c_3 [equation 11 below; Fig. 4].

Instantiation $c_1 \rightarrow A_2 \wedge L_2$ (6)

$c_2 \rightarrow A_2 \wedge L_4$ (7)

Classification $c_1 \leftarrow A_2 \wedge (L_2 \leq L \leq L_3)$ (8)

$c_2 \leftarrow A_2 \wedge (L_3 \leq L \leq L_4)$ (9)

$\neg c_1 \leftarrow \neg (A_2 \wedge L_2)$ (10)

Definition $c_3 \longleftrightarrow A_3 \wedge L_1$ (11)

Instantiation, classification, and definitional conditions facilitate reasoning in logic-based ontological systems not only about concepts and related instances but also about concept structure and conceptual consistency. These conditions can also illustrate the three primal forms of reasoning, deduction, induction, and abduction, and related extension and intension relations. For example, consider equation 8 to be a hypothesis about a geological unit concept in a particular model (represented by a map), and then several inferences, stated below, can be made: two of these inferences characterize intension development, indicating how unobservables might be obtained via clustering (equation 13) or categorization (equation 15), whereas the other two inferences characterize extension development, indicating how, in this case, occurrences might be classified (equation 12) or instantiated (equation 14). Note that as per convention, the complete left side denotes the *premise* for the right-side conclusion and the observables on the left side denote the *evidence*.

Deduction: $(A_2, L_2)\ (8) \rightarrow c_1$ (12)

Induction: $((A_2, L_2), c_1), ((A_2, L_3), c_1) \rightarrow (8)$ (13)

Abduction: $c_1, (8) \rightarrow \{ (A_2, L_2), (A_2, L_3) \}$ (14)

Abduction: c_1, (6–7, 9–11) $\rightarrow$ (8) (15)

In general, definitions are preferred by ontological systems for transacting computational tasks due to their representational simplicity and associated computational efficiency. But this preference is at odds with philosophers (Wittgenstein, 1953), linguists (Lakoff, 1987), and cognitive scientists (Smith and Medin, 1981) who suggest human concepts are not definitional; indeed, some cognitive research suggests that concepts can be contextual, varying with evidence, theory, actions, or motivations (Barsalou, 1999; Solomon et al., 1999). Evidence variance is apparent in geoscience where, for example, the age or lithology associated with a map unit varies within the unit from one geographic location to another and may gradually overlap with surrounding map units. This variance may reflect physical heterogeneity or diverse human interpretation. It implies nonetheless that some evidence will be more typical of the concept than others (e.g., for c_1, consider L_2 to be the dominant lithology and L_3 to be secondary, but where L_3 may play a more significant role in another concept such as c_2). Therefore the content and structure of concepts and instances may be variously shared (Brodaric and Hastings, 2002). An additional implication is that concepts will shift with changes to the contextual factors (Brodaric and Gahegan, 2001). Variance in evidence typicality and structure and concept evolution have representation and reasoning ramifications: originating inferences for some knowledge should be represented and augmented with contextual factors; representational schema for concepts and instances might be shared; and reasoning about concepts and their relations may involve statistical methods and logics that enable probable classification (e.g., multivalued, modal logic) and coexistence of conflicting concept versions (i.e., nonmonotonic logic).

By way of concluding this section we derive the following additional challenges.

Challenge 1.3.1: Representing the originating context for explicit knowledge, including premises (theory, evidence, and its typicality), inference type (deduction, induction, abduction), and metadata such as motivations, causal factors, and methods (calculation and simulation)

Challenge 1.3.2: Where possible, describing geological concepts with formal conditions.

Challenge 1.3.3: Representing geoscientific ontology with formal languages (e.g., OWL).

Challenge 1.3.4: Allowing concepts and instances to share representational schema.

Challenge 2.2: Developing probabilistic and revisionist reasoning tools to enable, for example, concepts, models, and theories to be identified, created, extended, presented, compared, and contrasted.

3.4. Operating with Geoscientific Knowledge

The scientific method has been studied by philosophers (e.g., Hempel and Oppenheim, 1988; Peirce, 1878; Popper, 1959) as well as geoscientists (e.g., Albritton, 1963; Harvey, 1969; Kitts, 1977; Martin, 1998; Schumm, 1991). Some of the results are empirically supported by psychologists (Feist and Gorman, 1998; Zimmerman, 2000) and implemented computationally (Langley, 2000; Shrager and Langley, 1990; Thagard, 1988) but with limited application to geoscience (e.g., Thagard and Nowak, 1990). The premise behind these computational efforts suggests that studies of the scientific method may provide a template for the computer-aided discovery and use of geoscientific knowledge (e.g., Gahegan and Brodaric, 2002)—in essence, that they may provide a geoscientific metaphor for using information and tools in cyberinfrastructure.

Initially, philosophic approaches to scientific method were geared to the development of theoretical knowledge but subsequent critiques resulted in better accommodation for aspects of situational and implicit knowledge. The early sequence can be summarized as consisting of: (1) neutral observation, (2) induction of problem statement or hypothesis, (3) deducing test criteria, (4) inductive confirmation of the hypothesis via further observation, and (5) communicating results (Hempel and Oppenheim, 1988). The sequence was amended by many including Popper (1959) who sought to mitigate its logical frailties by eliminating (1), because of theory-laden observation, relegating (2) to psychological processes and advocating theory falsification rather than confirmation in (3), because induction is susceptible to future contradiction. Subsequent responses emphasized the validity of abduction as an inference form and the role of implicit as well as background knowledge in observation and reasoning (Feyerband, 1975; Hanson, 1958; Kuhn, 1962). This reintroduces (1) and (2), recast in abductive terms, and leads to the notion that theories might be valid if their parts cohere overall in spite of some internal uncertainties, incompleteness, and conflicts. Geoscience method is often discussed in this recast framework (e.g., Schumm, 1991). The coherence principle implies potential nonmonotonicity, i.e., where new observations could trigger modification or invalidation of existing knowledge. We portray this possibility as a cycle in Figure 5, whereby new foreground knowledge could at any point stimulate further knowledge and cause a cascade of changes to existing knowledge forms and products (ovals), or a shift to alternate activities (boxes) and related inference forms (flags). We see this as relevant to situational knowledge, as it is particularly susceptible to reinterpretation. The knowledge cycle also suggests that any knowledge can be reused as input occurrences to begin the cycle again at a more abstract level. It further suggests a knowledge-driven view of presentation, in which knowledge is transformed to suit a particular medium for communication purposes, e.g., a map as a visual presentation of explicit knowledge.

Challenge 2.3: Developing a metaphor and tools grounded in geoscientific method to facilitate human use of geoscientific knowledge in a database network.

Challenge 2.4: Enabling knowledge-rich presentation by linking formal knowledge representation means, such as ontologies, with presentation forms, such as maps.

4. STRUCTURING GEOSCIENTIFIC KNOWLEDGE

Several challenges identified above impact the structure of geoscientific knowledge: the issue of plurality and fragmentation,

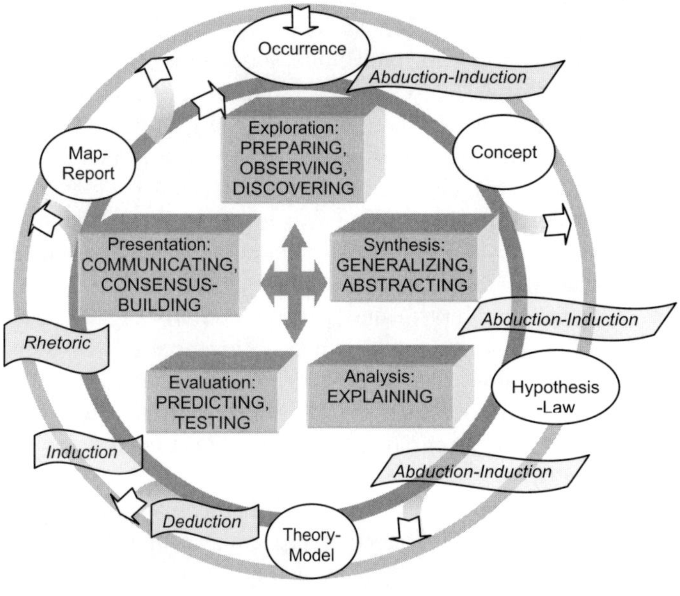

Figure 5. Geoscientific knowledge cycle (after Gahegan and Brodaric, 2002).

for representing multiple partial versions and competing knowledge (challenge 1.1); the need for context, for representing the factors used to derive some knowledge (challenge 1.3.1); structural and content overlap between concept and instance representations (challenge 1.3.3); and connecting presentation mechanisms with knowledge representations (challenge 2.4). These challenges point to the need for (1) an ontological framework and derived global schema that model how geoscientific knowledge is structured, and (2) schema content in the form of specific geoscientific concepts or theories. This complements conventional approaches to geological schema that emphasize specific database contents rather than general approaches to structuring geoscience knowledge: e.g., database schema (e.g., Gartner et al., 2001; Bain and Giles, 1997), domain ontologies (Raskin and Pan, 2005), and frameworks to explain geological occurrences on maps (Heyn et al., 2000; Voisard, 1999).

In this section we restate a global schema as an example of an approach that addresses some aspects of the four knowledge structuring challenges. The schema derives from an effort to create a standard North American geological map data model (Johnson et al., 1999) and has influenced its successor (NADM, 2004). The schema is geared to representing information associated with a geological map (challenge 2.4) from the viewpoint that a map denotes a visual presentation of a geoscientific model, one that may utilize several existing ontologies (challenge 1.1) and be derived from prototypical evidence gathered in the field and elsewhere (challenge 1.3.1, 1.3.3). Due to this map orientation the notion of a geological map is temporarily conflated with that of a geological model. As the details are relayed elsewhere (Brodaric and Hastings, 2002; Brodaric and Gahegan, 2002; Soller et al., 2002), we summarize its main elements below. Figure 6 depicts core components including symbols, concepts, occurrences, and descriptions.

1. Concept: refers to abstractions of observable or unobservable occurrences, such as those associated with geoscientific vocabulary: e.g., fault, unit, superposes, Precambrian. Concepts are often interrelated hierarchically, e.g., a map unit being a member of a formation, granite being an igneous rock, or semantically, e.g., unit A being correlated with unit B and overlying unit C.

2. Occurrence: typically refers to an observable feature, event, or process located in space and time; e.g., a measurement or observation or a map unit.

3. Symbol: refers to the elements comprising the cartographic display. Both concepts and occurrences can be symbolized; e.g., mines can be designated with a default symbol that can be overridden for a specific mine occurrence. Symbols may be assigned cartographic rules for scale dependencies via specific software.

4. Description: a numeric, textual and possibly spatial characterization of a concept or occurrence; e.g., a textual or geometric description of a map unit, or the numeric and/or textual details and location of a specific measurement or observation. Descriptions can be subclassed extensively, as shown in Figure 7.

The independence of descriptions in the schema is justified on the basis of graded typicality, in that the structure and content of a concept can be variously shared by its instances and evidence, which respectively represent the extension and intension development relations. Figure 7 illustrates shared structure in that the prototypical "granodiorite" occurrence in the inner loop shares lithology structure with the polygon instance of "X formation" in the outer loop. Content variance is also evident in that the prototypical "granodiorite" evidence for "X formation" in the inner loop contrasts with its instance description in the outer loop containing "monzogranite."

Table 3 categorizes evidence and instances based on the degree of similarity to concept descriptions. Of note is that high similarity indicates a potential "type locality" situation, in that the occurrence is very representative of the concept, e.g., the "granodiorite" site in the inner loop of Figure 7. The medium and low similarity situations serve to remind that normative definitions and generalizations are approximations for a complex empirical reality (Fig. 7, outer loop).

Figure 8 shows the grouping of symbols, concepts, occurrences, and descriptions into aggregations, which we call information models, and illustrates where some common geoscience information types fit into the schema.

1. **Symbolic (cartographic) model:** a collection of symbols representing a cartographic palette, such as a symbol standard for a mapping agency.
2. **Conceptual model:** a collection of concepts, their relations, and descriptions forming a specific vocabulary, taxonomy, classification scheme, or ontology.
3. **Occurrence model:** a collection of occurrences denoting a set of classified and described objects.
4. **Description model:** a collection of descriptions corresponding to a data set. In distributed data environments, a description model may refer to a remotely located data set, and the contained descriptions might then consist of pointers to the actual data. For example, a collection of spatial descriptions might be a geographic information system (GIS) layer or file, and a collection of nonspatial descriptions might be a database table.

These groupings enable the definition of a legend as a symbolized conceptual model and the definition of a map/geological model as a complex model composed of a legend and aggregations of occurrences. A geological model might then be defined by omitting the symbolic components. Note the schema does not at this point represent hypotheses or theories so its explanation capacity is limited. As illustrated in Figure 8, these definitions and resulting schema can serve as a knowledge layer over information. It also has the potential to be serve as an ontology repository that can facilitate information integration (as in section 2): e.g., a conceptual model can (partially) represent a mixed axiomatized-prototypical ontology, a specific description model can represent a linguistic ontology, an occurrence model can represent the instances being integrated, and their various relations represent the mappings between ontologies.

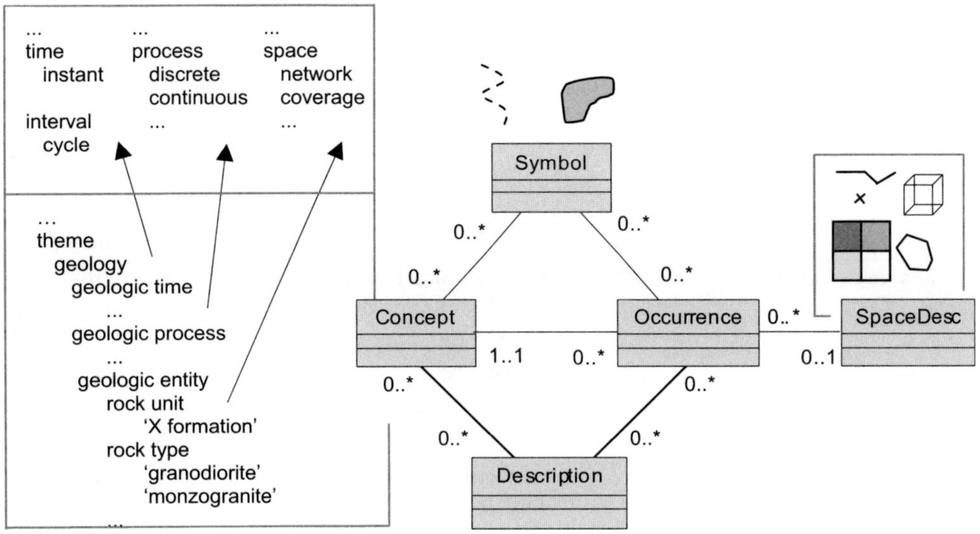

Figure 6. Core elements of the data model (from Brodaric and Hastings, 2002).

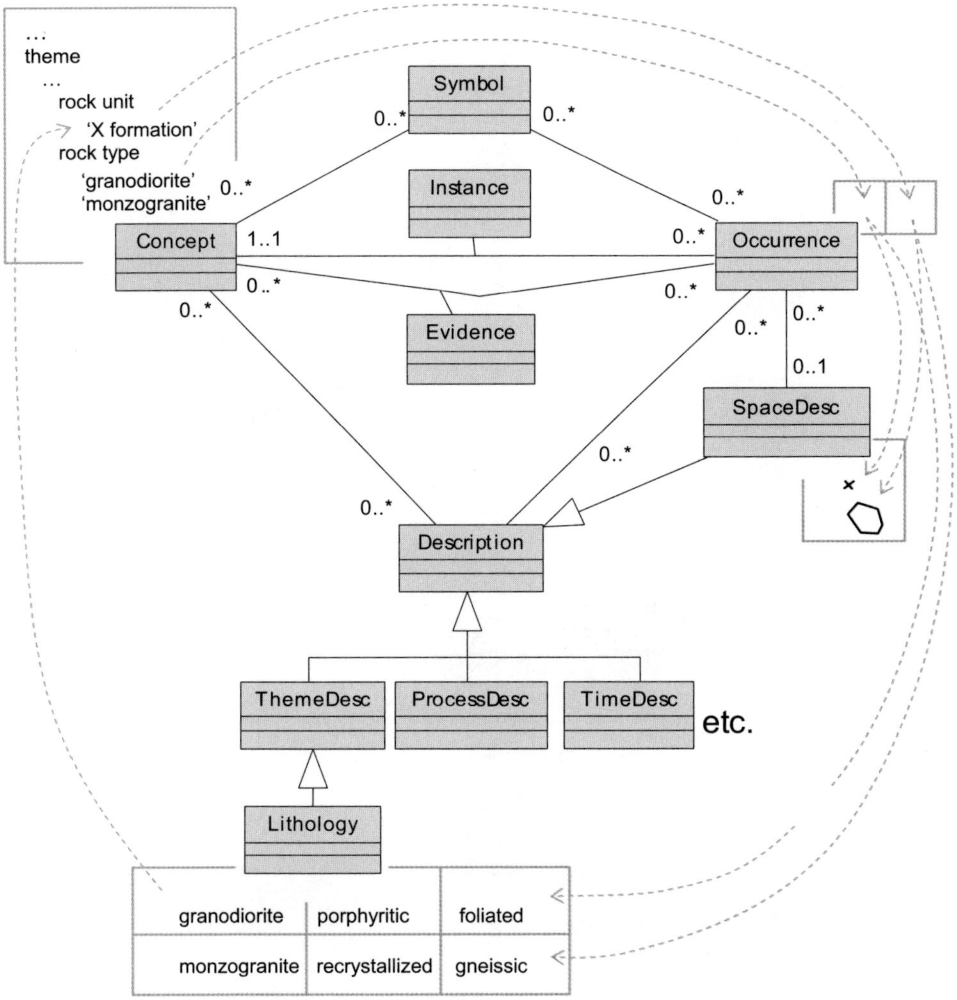

Figure 7. Concept instances, evidence, and shared descriptions (from Brodaric and Hastings, 2002).

TABLE 3. EVIDENCE AND INSTANCE TYPES GROUPED ACCORDING TO THEIR SIMILARITY TO THE CONTENT AND STRUCTURE OF CONCEPT DESCRIPTIONS

Shared Concept Description Structure	Shared Concept Description Content	Evidence/Instance Type
X	X	*N/A—very atypical*
X✓	X	*atypical-typical*
✓	✓	*prototypical*

5. APPLICATIONS

Variations of the schema described above are being implemented in three significant projects: the Canadian Geoscience Knowledge Network (CGKN; http://www.cgkn.net), U.S. Geological Survey National Geologic Map Database (NGMDB; Soller et al., 2002) project, and the Georgia Basin Digital Library (GBDL; Talwar et al., 2003). In each case the schema forms the basis for a conceptual database design that is extended into logical and physical designs for the project. The three implementations focus on different challenges: CGKN emphasizes the concept representation challenges (challenges 1.2, 1.3.1, and 1.3.3) by populating the schema with common geoscientific concepts and descriptions; NGMDB concentrates on the map representation challenge (challenge 3.2), using the schema to drive visualizations of geological information; and GBDL focuses on the plurality challenge (challenge 1.1) by representing the multiple perspectives on sustainable development in some communities.

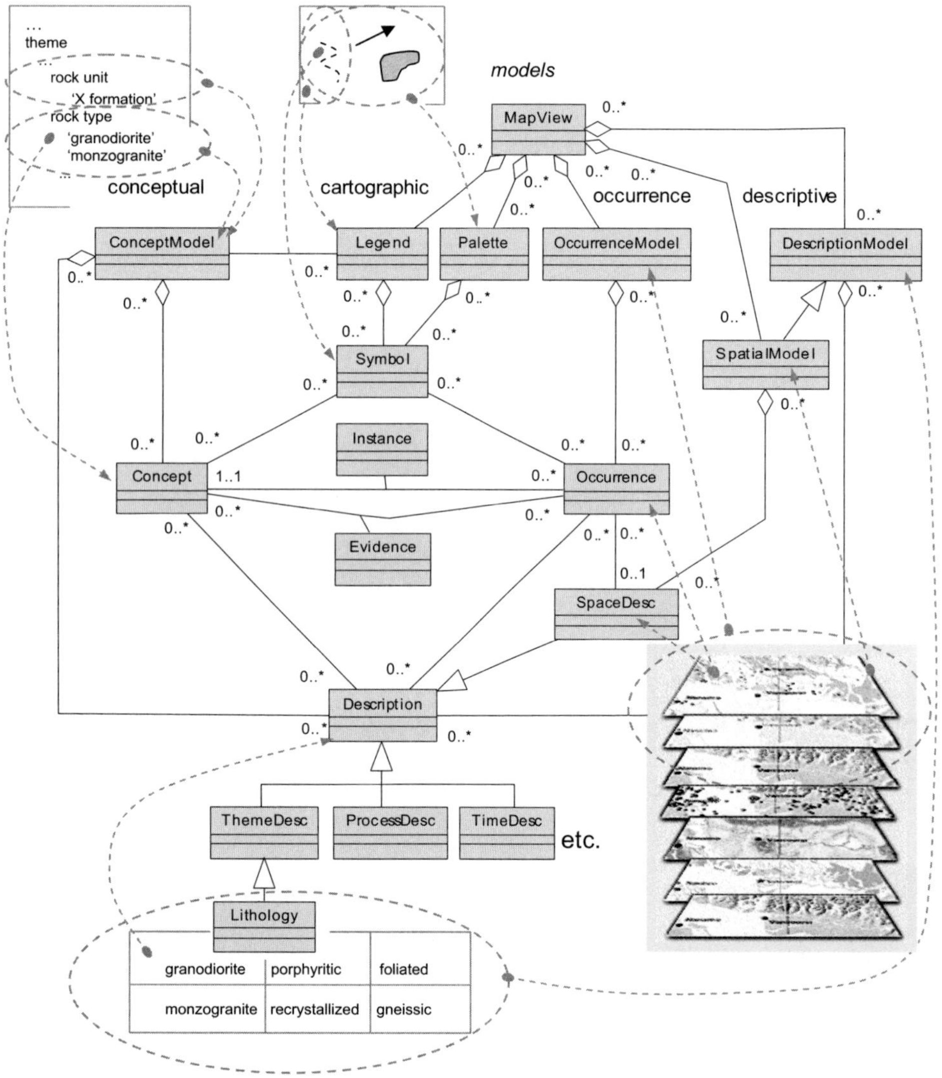

Figure 8. Populating the data model with conceptual, cartographic, occurrence, and descriptive data. Ovals denote collections of objects (after Brodaric and Hastings, 2002).

5.1. The Canadian Geoscience Knowledge Network

CGKN is a cooperative initiative of the Canadian government geoscience community that aims to provide a national Web-based portal to its information holdings. CGKN is being constructed as a loosely coupled network of heterogeneous databases distributed across the country. Interoperability in two subprojects, the national bedrock database and the surficial map database, is being developed through the use of a common source structure, derived from the schema above, and the use of evolving standards for geospatial Web services and scientific content (Davenport et al., 2002; Struik et al., 2002).

A major challenge faced by both subprojects is the acquisition from existing maps of national-scale map unit concepts and their descriptions. The schema fragment most heavily utilized by CGKN for this exercise involves the recursive concept and description relations: the map unit concepts possess many descriptive attributes including those for age and lithology, which themselves possess attributes. This recursion is partly exemplified in Figure 7 where "granodiorite" is both a concept with its own description and an attribute value for the description of "X formation." The subprojects have to date collected and entered several thousand map unit concepts and descriptions, as indicated in Table 4, as well as prototyped queries in the network (Boisvert et al., 2001). Specializing the schema for lithology descriptions is an ongoing major task.

TABLE 4. MAP UNIT CONCEPTS REPRESENTED IN A NATIONAL DATABASE

Concept	Subconcepts
Geological Province	2
Tectonic Terrane	17
Tectonic Assemblage	42
"Regional Unit"	99
(Supergroup, Supersuite)	2
"Regional Subunit"	159
Group, Suite, Complex	43
Formation, Lithodeme (+ informal equivalents)	161
Member (+ informal equivalents)	21
Total	2082

Note: From Davenport et al. (2002)

5.2. The U.S. Geological Survey National Geological Map Database

The NGMDB project is congressionally mandated to develop a database of geological map information for the United States, in cooperation with state and other agencies. NGMDB is being developed in three stages.

1. Metadata: a Web-based index of published maps, stratigraphic names, and current mapping projects (see http://ngmdb.usgs.gov/);
2. Spatial: a Web-based catalog of raster and vector map images; and
3. Nonspatial: a database of geological content underlying the spatial data.

The schema above was adapted for a prototype in NGMDB's third stage (Soller et al., 2002). The prototype emphasized using the schema as a framework for visualizing geological information in map form. Implementation in an object-oriented GIS (GE-SmallWorld™) demonstrated the schema is an effective means of hosting and visualizing multiple geological maps (models) from a single information source. Advanced display operations were added to objects in the schema to enable sophisticated visualization: e.g., **MapView→Display** displays a particular geological model (i.e., concepts, occurrences, and evidence) from the information source; **MapView→ReClass** reclassifies a map according to an attribute (e.g., dominant lithology, as in Fig. 9); and **Occurrence→Display** performs scale-sensitive display of occurrences: e.g., teeth appear on faults at more detailed scales, and certain

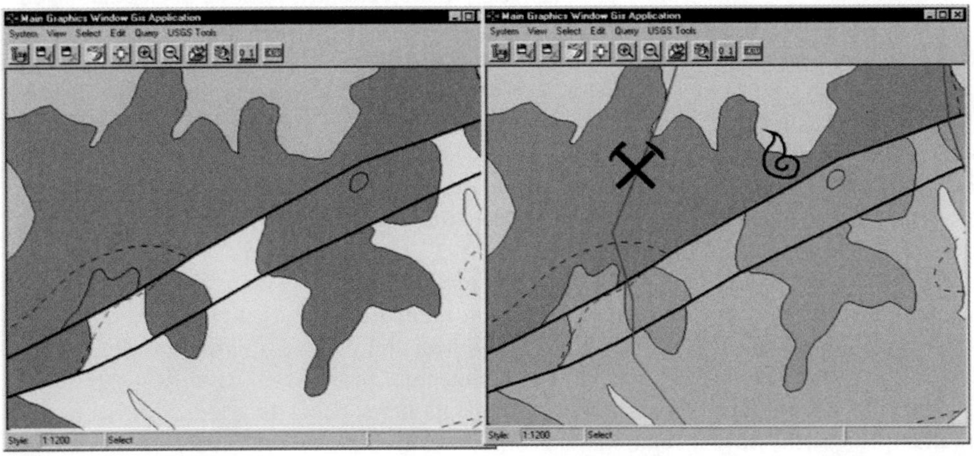

Figure 9. Reclassifying map units into rock types (from Brodaric and Hastings, 2002).

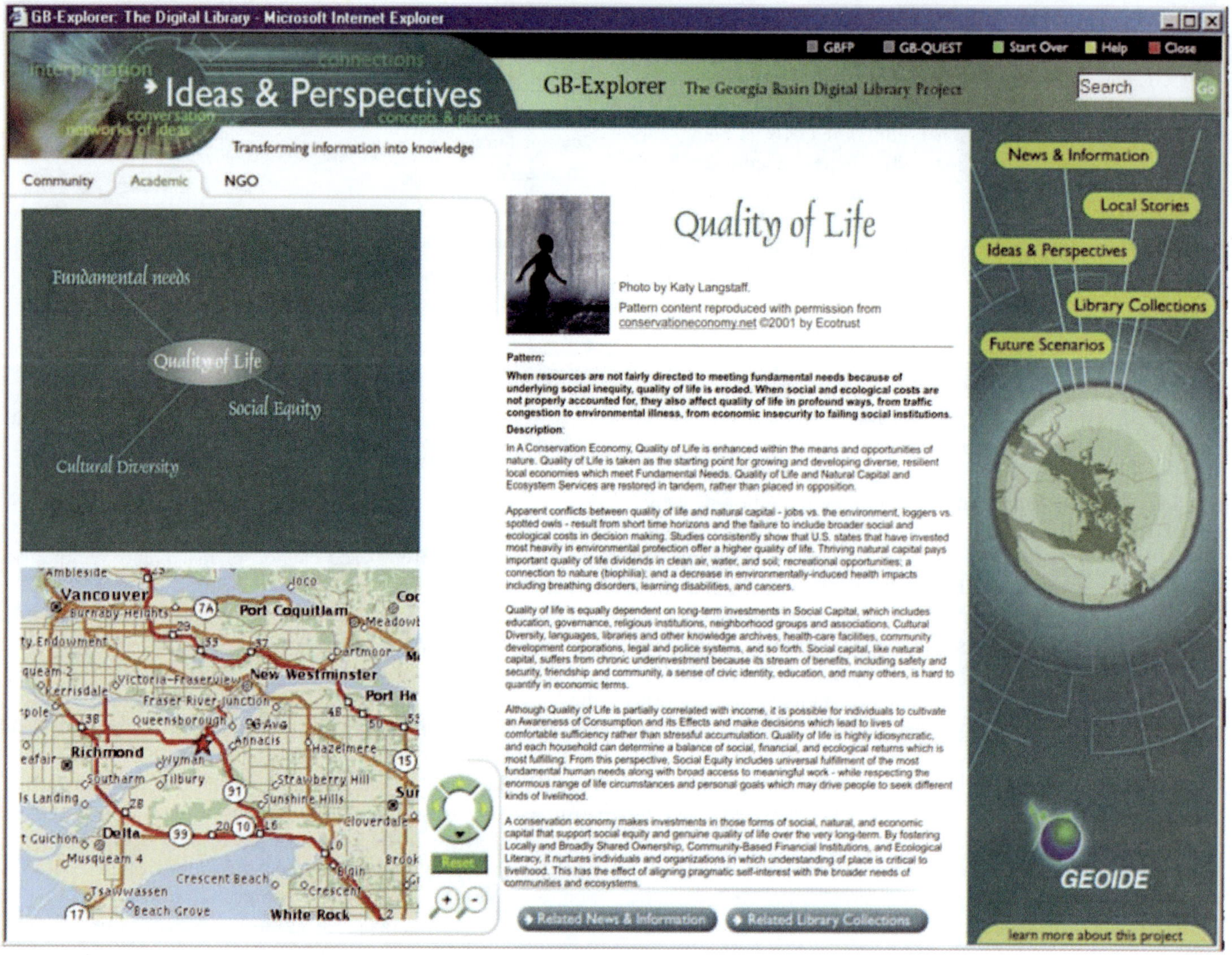

Figure 10. Exploring sustainable development concepts in Georgia Basin Digital Library.

map units are grouped together at regional scales through reclassification of the unit to a higher lithostratigraphic rank.

5.3. The Georgia Basin Digital Library

The GBDL aims to build awareness and understanding about sustainability amongst municipal and scientific communities in the Georgia Basin (southwestern British Columbia). It provides a Web-based mechanism to host and browse natural resource and socio-economic information intended for input into modeling tools that generate future urban scenarios. Such scenarios are useful to regional planners and policy makers during decision making. GBDL uses a variation of the schema described in section 4 to organize 64 natural resource geospatial data sets (i.e., spatial models), sustainability concepts, and multimedia descriptions (i.e., textual, image-based). It emphasizes multiple sustainability ontologies, including those reflecting community, academic, and other perspectives. We are currently working at associating sustainability and geoscience concepts in GBDL, to indicate how geoscience information can contribute to sustainability issues, and thus how it can conceptually feed into the modeling tools. Figure 10 displays the Ideas and Perspectives module where the sustainability ontologies can be explored by browsing a semantic network of relevant concepts (top left) whose selection triggers the display of related maps (bottom left) and multimedia descriptions (right).

The schema, populated with sustainability concepts and extended to contain their descriptions, figures prominently in the overall technological architecture of GBDL as illustrated in Figure 11. The architecture possesses three tiers: (1) the information tier implements the schema as a logical schema for a relational database system (SQL Server™); (2) the services tier

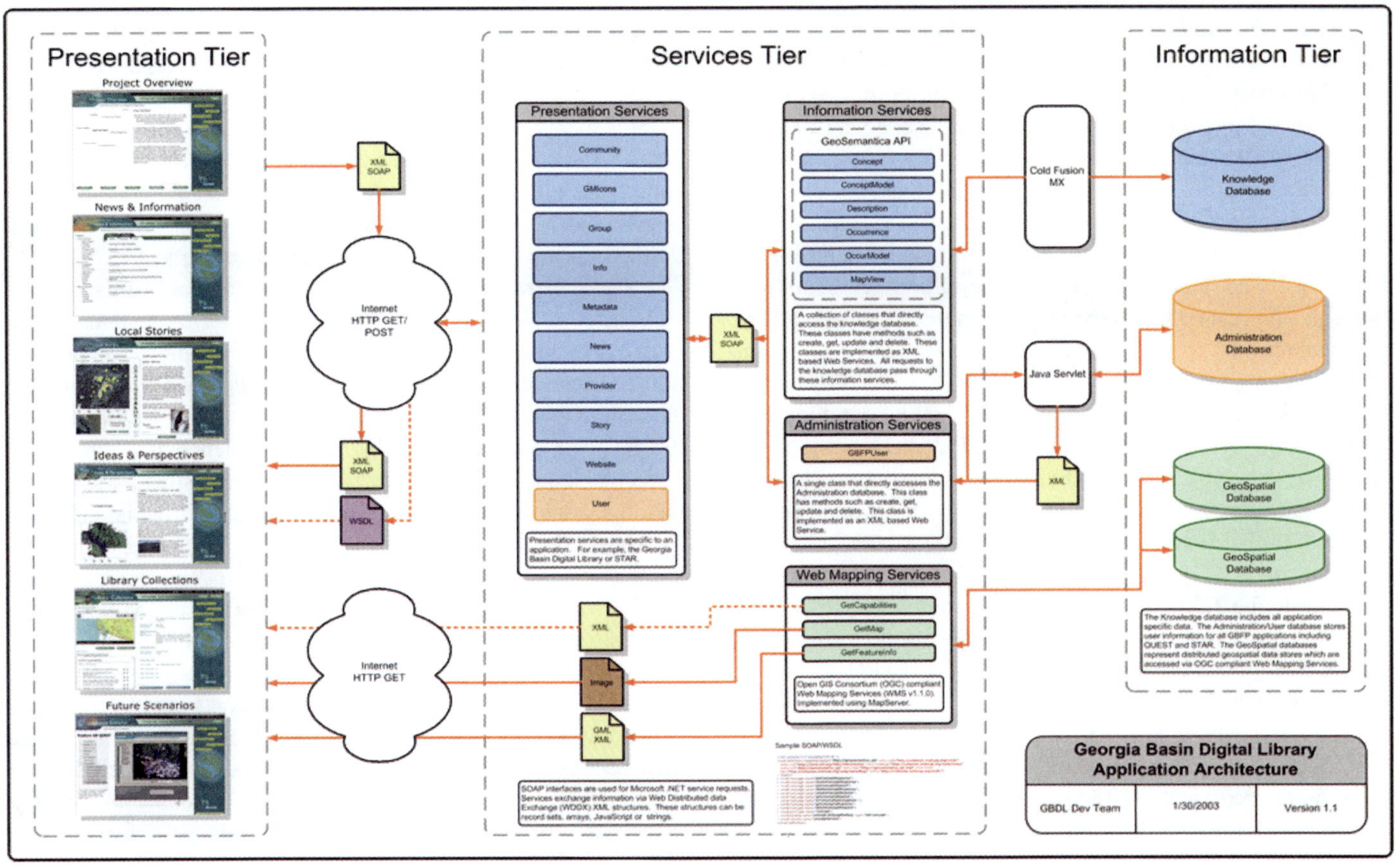

Figure 11. The three-tiered architecture of the Georgia Basin Digital Library.

provides on the one hand a Web service to access the sustainable development ontologies and descriptions, and on the other hand it provides standard Web services for using geospatial occurrences (i.e., via OGC, WMS, and WFS); and (3) the presentation tier provides a user interface to the library (information tier) and its functions (services tier), exclusively using the services tier to access the information.

6. CONCLUSIONS

The philosophical and cognitive aspects of the geoscientific process provide a basis for a knowledge representation framework for geoscientific information. We discuss several geoscience knowledge characteristics, forms, types, reasoning mechanisms, and activities and identify several challenges to representing geoscientific knowledge in cyberinfrastructure, reiterated and elaborated below.

Challenge 1: Developing the framework and contents for geoscientific ontology, including axiomitized, linguistic, and prototypical elements, and linkage to geospatial, temporal, and process ontology.

Challenge 1.1: Embracing plurality in geoscientific ontology to account for multiple explanations and knowledge evolution, and developing integration mechanisms to combine the plural views and map between them. This may involve integrating diverse scientific explanations or mapping between a scientific explanation and other use of the information, e.g., societal and cross-disciplinary. Rejecting plurality would contravene the aim of geoscientific research geared to generating revised or new data, explanations, and predictions.

Challenge 1.2: Populating a mixed geoscientific ontology framework with explicit background knowledge consisting of reference models, theories, concepts, and prototypical occurrences, described formally as well as informally via narratives. What should not be understated here is the social engineering challenge of capturing a knowledge fragment for inclusion in such a framework. This involves on the one hand capturing individual knowledge and on the other hand reaching consensus within a group and capturing group knowledge. A related issue is the ability to transform the captured knowledge within and between ontologies in the framework.

Challenge 1.3: Capturing the reasoning process used to determine explicit background knowledge and structuring geoscientific ontology accordingly.

Challenge 1.3.1: Representing the originating context for explicit knowledge, including premises (theory, evidence, and its typicality), inference type (deduction, induction, abduction), and metadata such as motivations, causal factors, and methods (calculation, simulation). Context includes the empirical basis for justifying and explaining some derived knowledge, a fundamental aspect of any science. Its exclusion from representation would reduce the scientific merit and use of an ontology.

Challenge 1.3.2: Describing geological concepts formally, where possible, with necessary and/or sufficient conditions, for computer-based processing. Precise definitions for geological concepts and related terms may soon be necessary for regulatory and legal reasons, as well as to advance automated computational processing of geoscientific information.

Challenge 1.3.3: Representing geoscientific ontology using formal languages such as RDF, XML, OWL, and others. This encompasses syntactical encoding of geoscientific ontology and may require in some cases potential language enhancements to deal with probable classification and prototypes, theory revision, and contexts.

Challenge 1.3.4: Allowing concepts and instances to share representational schema for prototypical concepts, to simplify schema design and aid in the representation of evidence.

Challenge 2: Developing tools to leverage geoscientific ontology in database networks.

Challenge 2.1: Developing mediators and wrappers that incorporate geoscientific, geospatial, temporal, and process-oriented operations in data integration activities and that utilize and/or prompt human intervention. These are required to enable scientific processing in the course of integrating information. Their omission would result in less sophisticated network behavior.

Challenge 2.2: Developing probabilistic and revisionist reasoning tools for operating on geoscientific knowledge, to enable, for example, concepts, models and theories to be identified, created, extended, presented, compared, and contrasted. As such tools will operate on geoscientific knowledge as well as data, they will ultimately provide the most concrete illustration of the new cyberenabled operational paradigm. Associated with this challenge are issues of operating with variable uncertainty.

Challenge 2.3: Developing a metaphor and tools grounded in geoscientific method to facilitate use of geoscientific knowledge in a database network. This involves the successful migration of conventional practices to cybertechnologies, which in turn depends on deriving user interfaces and operating frameworks from geoscientific process. At risk is decreased usability via an unstructured and unwieldy suite of information and tools.

Challenge 2.4: Enabling knowledge-rich presentation by linking formal knowledge representation means, such as ontologies, with presentation forms, such as maps. This will result in more meaningful documents in which scientific premises are embedded. At risk is understandability and potential gains from novel knowledge-enhanced visualization mechanisms.

Although our list of challenges is no doubt incomplete, it already represents some significant challenges to semantic interoperability in geoscience that we have encountered in both theory and practice. In our efforts to build geoscientific databases, connect, and interoperate with them, we focused herein on the first challenge—dealing with knowledge representation—and discuss a schema that serves as a foundation for (1) a high-level domain ontology representing shared geological concepts and other explicit knowledge, and (2) a conceptual database design framework for large geoscientific data repositories. We are also experimenting with using the schema as (3) a global schema and concept repository to be used during information integration in database networks.

The general nature of such a schema makes this type of approach particularly suitable for long-term databases that must cope with scientific discovery and evolution, for integration mechanisms that serve diversely structured and populated databases, and for geoscientific ontology where justification in the form of confirming and refuting evidence for constituent concepts should be recorded. Fundamentally, such an approach would seem to be required to accommodate a growing and shifting knowledge base that is characteristic of much of geoscientific knowledge. We are in the early stages of testing these ideas in the database networks described herein, and though initial results are positive, several of the challenges still need to be addressed: probably the most pressing is the development of geoscientific ontologies in which core geoscientific concepts are identified and structured and the development of tools that clearly demonstrate their utility. Addressing these and other outstanding challenges will provide a significant step forward in the conceptualization and construction of cyberinfrastructure for the geosciences.

ACKNOWLEDGMENTS

We gratefully acknowledge the Geological Survey of Canada and the U.S. Geological Survey for supporting this work and thank the reviewers for their careful and thoughtful comments, many of which led to improvements in the paper. This research was also supported in part by the U.S. National Science Foundation (NSF) via grants BCS-9978052 (HERO), ITR (BCS)-0219025, and ITR (EAR)-0225673 (GEON).

REFERENCES CITED

Albritton, C.C., Jr., editor, 1963, The Fabric of Geology: Reading, Massachusetts, Addison-Wesley, 372 p.

Bain, K.A., and Giles, J.R.A., 1997, A standard model for storage of geologic map data: Computers & Geosciences, v. 23, no. 6, p. 613–620, doi: 10.1016/S0098-3004(97)00043-5.

Baker, V., 1996, Hypotheses and geomorphological reasoning, *in* Rhoads, B.L., and Thorn, C.E., eds., The Scientific Nature of Geomorphology: New York, Wiley, p. 57–86.

Baker, V., 1999, Geosemiosis: Geological Society of America Bulletin, v. 111, no. 5, p. 633–645, doi: 10.1130/0016-7606(1999)111<0633:G>2.3.CO;2.

Barsalou, L.W., 1999, Perceptual symbol systems: The Behavioral and Brain Sciences, v. 22, p. 577–660, doi: 10.1017/S0140525X99002149.

Boisvert, E., Morin, A., Lauziere, K., and Lebel, D., 2001, Using the proposed North American Data Model in a distributed database environment, *in* Soller, D., ed., Digital Mapping Techniques '01—Workshop Proceedings: U.S. Geological Survey Open-File Report 01-223, p. 35–44, http://pubs.usgs.gov/of/2001/of01-223/boisvert.html.

Breunig, M., 1999, An approach to the spatial data and systems integration for a 3D geo-information system: Computers & Geosciences, v. 25, no. 1, p. 39–48, doi: 10.1016/S0098-3004(98)00104-6.

Brodaric, B., and Gahegan, M., 2001, Learning geoscience categories in-situ: Implications for geographic knowledge representation, *in* Proceedings of the Ninth ACM International Symposium on Advances in GIS, Atlanta, GA, Nov. 9–10, 2001: New York, ACM Press, p. 130–135.

Brodaric, B., and Gahegan, M., 2002, Distinguishing instances and evidence of geographical concepts for geospatial database design, *in* Egenhofer, M., and Mark, D., eds., Geographic Information Science—2nd International Conference, GIScience 2002, LNCS 2478: Berlin, Springer, p. 22–37.

Brodaric, B., and Hastings, J., 2002. An object model for geologic map information, *in* Richardson, D., and van Oosterom, P., eds., Advances in Spatial Data Handling, 10th International Symposium on Spatial Data Handling: New York, Springer, p. 55–68.

Burners-Lee, T., Handler, J., and Lassila, O., 2001, The Semantic Web: Scientific American, May 2001, p. 35–43.

Calì, A., Calvanese, D., De Giacomo, G., and Lenzerini, M., 2002, On the expressive power of data integration systems, *in* Proceedings of the 21st International Conference on Conceptual Modeling (ER 2002), LCNS: 2503, Berlin, Springer, p. 338–350.

Cox, S., 2001, Geology data transfer using XML, *in* Proceedings of the International Association for Mathematical Geology Annual Meeting, Cancun, Mexico, Sept. 6–12 (CD-ROM).

Davenport, P., Boisvert, E., Quat, M., Okulitch, A., Brodaric, B., Colman-Sadd, S., Nolan, L., Struik, B., MacIntyre, D., Tzeng, P., Scott, D., Gilbert, C., Abbott, G., Bassan, A., Journeay, M., Francis, J., and Houlahan, T., 2002, A scalable, digital map database of bedrock geology for Canada: A progress report, *in* Soller, D.R., 2002, Digital Mapping Techniques '02—Workshop Proceedings: U.S. Geological Survey Open-File Report 02-370, 214 p.

de Jong, T., and Ferguson-Hessler, M.G.M., 1996, Types and qualities of knowledge: Educational Psychologist, v. 31, no. 2, p. 105–113.

Domenig, R., and Dittrich, K., 1999, An overview and classification of mediated query systems: ACM SIGMOD Record, v. 28, no. 3, p. 63–72.

Elmagarmid, A., Rusinkiewicz, M., and Sheth, A., eds., 1999, Management of Heterogeneous and Autonomous Database Systems: San Francisco, California, Morgan Kaufmann, 413 p.

Engelhardt, W., and Zimmermann, J., 1982, Theory of Earth Science: New York, Cambridge, 381 p.

Feist, G.J., and Gorman, M.E., 1998, The psychology of science: Review and integration of a nascent discipline: Review of General Psychology, v. 2, no. 1, p. 3–47, doi: 10.1037//1089-2680.2.1.3.

Fensel, D., Staab, S., Studer, R., and van Harmelen, F., 2002, Peer-2-Peer enabled Semantic Web for Knowledge Management, *in* Davies, N.J., Fensel, D., and van Harmelen, F., eds., Ontology-based Knowledge Management: Exploiting the Semantic Web: London, UK, Wiley.

Feyerband, P., 1975, Against Method: London, Verso, 296 p.

Flach, P., and Kakas, A., eds., 2000, Abduction and Induction: Essays on their Relation and Integration: Boston, Kluwer, 309 p.

Fonseca, F., Egenhofer, M., Davis, C., and Camara, G., 2002, Semantic Granularity in ontology-driven GIS: Annals of Math and Artificial Intelligence, v. 36, no. 1–2, p. 121–151.

Frodeman, R., 1995, Geological reasoning: Geology as an interpretive and historical science: Geological Society of America Bulletin, v. 107, no. 8, p. 960–968, doi: 10.1130/0016-7606(1995)107<0960:GRGAAI>2.3.CO;2.

Frodeman, R., 2003, Geo-logic: Breaking ground between philosophy and the earth sciences: Albany, State University of New York Press, 184 p.

Gahegan, M., and Brodaric, B., 2002, Computational and visual support for geographical knowledge construction: Filling in the gaps between exploration and explanation, *in* Richardson, D., and van Oosterom, P., eds., Advances in Spatial Data Handling, 10th International Symposium on Spatial Data Handling: New York, Springer-Verlag, p. 11–26.

Gangemi, A., Guarino, N., Masolo, C., and Oltramari, A., 2003, Sweetening WordNet with DOLCE: AI Magazine, v. 24, no. 3, p. 13–24.

Gartner, H., Bergmann, A., and Schmidt, J., 2001, Object-oriented modeling of data sources as a tool for integration of heterogeneous geoscientific information: Computers & Geosciences, v. 27, p. 975–985, doi: 10.1016/S0098-3004(00)00135-7.

Goschke, T., 1997, Implicit learning and unconscious knowledge: Mental representation, computational mechanisms, and brain structures, *in* Lamberts K., and Shanks, D., eds., Knowledge, Concepts, and Categories: Cambridge, Massachusetts, MIT Press, p. 247–334.

Gould, S.J., 2000, De-constructing the science wars by reconstructing an old mold: Science, v. 287, p. 253–261, doi: 10.1126/science.287.5451.253.

Guarino, N., 1998, Formal Ontology and Information Systems, *in* Guarino, N., ed., Formal Ontology in Information Systems: Amsterdam, IOS Press, p. 3–15.

Grenon, P., and Smith, B., 2004, SNAP and SPAN: Towards Dynamic Spatial Ontology, Spatial Cognition and Computation, v. 4, p. 1.

Gribble, S., Halevy, A., Zachary, I., Rodrig, M., and Suciu, D., 2001, What can peer-to-peer do for databases and vice versa?, *in* Proceedings of the Fourth International Workshop on the Web and Databases (WebDB 2001), Santa Barbara, California, May 2001.

Gruber, T.R., 1993, A translation approach to portable ontology specifications: Knowledge Acquisition, v. 5, p. 199–220, doi: 10.1006/knac.1993.1008.

Hanson, N.R., 1958, Patterns of Discovery: An Inquiry into the Conceptual Foundations of Science: Cambridge, Cambridge University Press, 240 p.

Harvey, D., 1969, Explanation in Geography: London, Edward Arnold, 521 p.

Hempel, C.C., and Oppenheim, P., 1988, Studies in the Logic of Explanation, *in* J. Pitt, ed., Theories of Explanation: New York, Oxford University Press, p. 9–50.

Heyn, G., Kuebler, S., Richter, B., Skala, W., and Voisard, A., 2000, The Geo-hyp Project: Representing knowledge in geologic hypermaps: GEO-Informations systeme, v. 13, no. 13, p. 24–28.

Hull, R., 1997, Managing semantic heterogeneity in databases: A theoretical perspective, *in* Proceedings of the ACM SIGACT-SIGMOD-SIGART Symposium on Principles of Database Systems (PODS): New York, ACM Press, p. 51–61.

Jackson, J.A., 1997, Glossary of Geology, 4th ed.: Alexandria, Virginia, American Geological Institute, 769 p.

Johnson, B.R., Brodaric, B., Raines, G.L., Hastings, J.T., and Wahl, R., 1999, Digital geological map data model, v4.3: U.S. Geological Survey, http://www.nadm-geo.org/prd/Model43a.pdf.

Josephson, J.R., and Josephson, S.G., 1994, Abductive inference: Computation, philosophy, technology: New York, Cambridge University Press, 306 p.

Karp, P.D., 2000, An ontology for biological function based on molecular interactions: Bioinformatics (Oxford, England), v. 16, no. 3, p. 269–285, doi: 10.1093/bioinformatics/16.3.269.

Kitts, D.B., 1977, The Structure of Geology: Dallas, Texas, Southern Methodist University Press, 180 p.

Klein, M., 2002, Combining and relating ontologies: Knowledge-Based Systems, v. 15, no. 5-6, p. 265–273, doi: 10.1016/S0950-7051(01)00162-9.

Koch, C., Petta, P., Le Goff, J., and McClatchey, R., 2000, On information integration in large scientific collaborations: Technical Report TR-2000-28, Austrian Research Institute for Artificial Intelligence.

Kuhn, T.S., 1962, The Structure of Scientific Revolutions: Chicago, University of Chicago Press, 172 p.

Kuhn, W., 2001, Ontologies in support of activities in geographical space: International Journal of Information Science, v. 15, no. 7, p. 613–631, doi: 10.1080/13658810110061180.

Lakoff, G., 1987, Women, Fire and Dangerous Things: Chicago, University of Chicago Press, 614 p.

Langley, P., 2000, The computational support of scientific discovery: International Journal of Human-Computer Studies, v. 53, p. 393–410, doi: 10.1006/ijhc.2000.0396.

Loudon, T.V., 2000, Geoscience after IT: Part L. Adjusting the emerging information system to new technology: Computers and Geosciences, v. 26, no. 3-1, p. A109-A121.

MacGregor, R., 1991, Inside the loom classifier: SIGART Bulletin, v. 3, no. 2, p. 70–76.

Mann, C.J., 1993, Uncertainty in Geology, *in* Davis, J.C., and Herzfeld, U.C., eds., Computers in Geology—25 Years of Progress: New York, Oxford University Press, 298 p.

Martin, R.E., 1998, One Long Experiment: New York, Columbia University Press, 262 p.

McGuinness, D., and van Harmelen, F., eds., 2003, OWL Web Ontology Language Overview, http://www.w3.org/TR/owl-features/#s1.

North American Commission on Stratigraphic Nomenclature, 1983, North American Stratigraphic Code: American Association of Petroleum Geologists Bulletin, v. 67, p. 841–875.

North American Geologic Map Data Model Steering Committee (NADM), 2004, NADM Conceptual Model 1.0, A Conceptual Model for Geologic Map Information, http://geology.usgs.gov/dm/steering/teams/design/.

Panti, M., Penserini, L., and Spalazzi, L., 2002, A pure P2P approach to information integration, *in,* Technical Report, 2002-02-19, University of Ancona, Italy, Istituto di Informatica.

Peirce, C.S., 1878, Deduction, induction and hypothesis: Popular Science Monthly, v. 13, p. 470–782.
Pennanen, M., and Kortman, C., 2002, Semantic web—Towards more intelligent geoscience web applications, *in* Bayer, U., Burger, H., and Skala, W., eds., 8th Annual Conference of the International Association for Mathematical Geology, 15–20 September 2002, Berlin, Germany, v. 2, Terra nostra. Schriften der Alfred-Wegener-Stiftung 04/2002, p. 205–210.
Popper, K.R., 1959, The Logic of Scientific Discovery: New York, Basic Books, 479 p.
Pshenichny, C.A., 2002, Investigation of geologic reasoning as a new objective of geoscience and geohazard assessment: IAMG Newsletter, no. 65, p. 11.
Pshenichny, C.A., Moukhachov, V. P., and Khrabrykh, Z.V., 2003, Logical assessment of observational knowledge in volcanology: Journal of Volcanology and Geothermal Research, v. 128, no. 1–3, p. 287–298.
Ram, S., Park, J., and Ball, G., 1999, Semantic Model Support for Geographic Information Systems: IEEE Computer, v. 32, no. 5, p. 74–81.
Ramachandran, R., Graves, S., Conover, H., and Moe, K., 2004, Earth Science Markup Language (ESML): a solution for scientific data-application interoperability problem, Computers and Geosciences, v. 30, p. 117 124.
Raper, J., 1999, Spatial representation: The scientist's perspective, *in* Longely, P.A., Goodchild, M.F., Maquire, D.J., and Rhind, D.W., eds., Geographical Information Systems: Principles and Technical Issues: New York, Wiley, p. 61–70.
Raper, J., 2000, Multidimensional Geographic Information Science: London, Taylor & Francis, 300 p.
Raskin, R.G., and Pan, M.J., 2005, Knowledge representation in the semantic web for Earth and environmental terminology (SWEET): Computers & Geosciences, v. 31, p. 1119–1125.
Renolen, A., 2000, Modeling the real world: Conceptual modeling in spatiotemporal information system design: Transactions in GIS, v. 4, no. 1, p. 23–42, doi: 10.1111/1467-9671.00036.
Reynolds, R.E., Sinatra, G.M., and Jetton, T.L., 1996, Views of knowledge acquisition and representation: A continuum from experience centered to mind centered: Educational Pschologist, v. 31, no. 2, p. 93–104.
Schumm, S.A., 1991, To Interpret the Earth: Ten Ways to be Wrong: New York, Cambridge University Press, 133 p.
Sheth, A., and Larson, J., 1990, Federated Database Systems for Managing Distributed Heterogeneous, and Autonomous Databases: ACM Computing Surveys, v. 22, no. 3, p. 183–236, doi: 10.1145/96602.96604.
Sheth, A.P., 1999, Changing focus on interoperability in information systems: From system, syntax, structure to semantics, *in* Goodchild, M., Egenhofer, M., Fegeas, R, and Kottman, C., eds., Interoperating Geographic Information Systems: Dordrecht, Kluwer, p. 5–29.
Shrager, J., and Langley, P., eds., 1990, Computational Models of Scientific Discovery and Theory Formation: San Mateo, California, Morgan Kaufman, 498 p.
Smith, B., 1998a, Basic Concepts of Formal Ontology, *in* Guarino, N., ed., Formal Ontology in Information Systems: Amsterdam, IOS Press, p. 19–28.
Smith, B., 1998b, An introduction to Ontology, *in* Peuquet, D., Smith, B., and Brogaard, B., eds., The Ontology of Fields: Santa Barbara, California, National Center for Geographic Information and Analysis, p. 10–14.
Smith, E.E., and Medin, D.L., 1981, Categories and Concepts: Cambridge, Massachusetts, Harvard University Press, 203 p.
Soller, D.R., Brodaric, B., Hastings, J.T., Wahl, R., and Weisenfluh, G.A., 2002, The central Kentucky prototype: An object-oriented geologic map data model for the National Geologic Map Database: U.S. Geological Survey Open-File Report 02–202, version 1.0, http://pubs.usgs.gov/of/2002/of02-202/.
Solomon, K.O., Medin, D.L., and Lynch, E., 1999, Concepts do more than categorize: Trends in Cognitive Sciences, v. 3, no. 3, p. 99–104, doi: 10.1016/S1364-6613(99)01288-7.
Sowa, J.F., 2000, Knowledge Representation: Logical, Philosophical, and Computational Foundations: New York, Brooks/Cole, 594 p.
Struik, L.C., Quat, M.B., Davenport, P.H., and Okulitch, A.V., 2002, A preliminary scheme for multihierarchical rock classification for use with thematic computer-based query systems: Geological Survey of Canada, Current Research 2002-D10, 9 p.
Talwar, S., Journeay, M., Harrap, R., Brodaric, B., van Ulden, J., Grant, R., and Denny, S., 2003, The Georgia Basin Digital Library Project: A framework for a place-based exploration of sustainability issues: Geomatica, v. 57, no. 2, p. 155–164.
Thagard, P., 1988, Computational philosophy of science: Cambridge, Massachusetts, MIT Press, 240 p.
Thagard, P., and Nowak, G., 1990, The Structure of the Geological Revolution, *in* Shrager, J., and Langley, P., eds., Computational Models of Scientific Discovery and Theory Formation: Palo Alto, California, Morgan Kaufmann, p. 27–71.
Ullman, J.D., 1997, Information integration using logical views: Proceedings of the International Conference on Database Theory, v. 19–40, p. 1997.
Voisard, A., 1999, Abduction and deduction in geologic hypermaps, *in* Güting, R.H., Papadias, D., and Lochovsky, F., eds., Advances in spatial databases: 6th International Symposium, SSD'99, Hong Kong, China, July 20–23, 1999, Lecture Notes in Computer Science, no. 1651: Berlin, Springer, p. 311–329.
Wache, H., Vögele, T., Visser, U., Stuckenschmidt, H., Schuster, G., Neumann, H., and Hübner, S., 2001, Ontology-Based Integration of Information: A Survey of Existing Approaches, *in* Stuckenschmidt, H., ed., IJCAI-01 Workshop, Seattle, Washington, USA, April 4–5, 2001: Ontologies and Information Sharing, p. 108–117.
Wiederhold, G., 1996, Intelligent Integration of Information: Boston, Massachusetts, Kluwer, July 1996, 216 p.
Wittgenstein, L., 1953, Philosophical Investigations: Oxford, Blackwell, 232 p.
Zimmerman, C., 2000, The development of scientific reasoning skills: Developmental Review, v. 20, p. 99–149, doi: 10.1006/drev.1999.0497.

Manuscript Accepted by the Society 18 June 2005.

Printed in the USA

Geological Society of America
Special Paper 397
2006

A community effort to construct a gravity database for the United States and an associated Web portal

G. Randy Keller
Department of Geological Sciences, University of Texas, El Paso, Texas 79968, USA

T.G. Hildenbrand
R. Kucks
M. Webring
U.S. Geological Survey, Denver, Colorado, USA, and Menlo Park, California, USA

Allen Briesacher
Kristine Rujawitz
National Geospatial-Intelligence Agency, St. Louis, Missouri, USA

A.M. Hittleman
National Oceanic and Atmospheric Administration/National Geophysical Data Center, Boulder, Colorado, USA

Daniel R. Roman
Daniel Winester
National Oceanic and Atmospheric Administration/National Geodetic Survey, Silver Spring, Maryland, and Colorado, USA

R. Aldouri
John Seeley
Jorge Rasillo
Roberto Torres
Pan American Center for Earth and Environmental Science, University of Texas, El Paso, Texas 79968, USA

William J. Hinze
Purdue University, West Lafayette, Indiana, USA

Ann Gates
Vladik Kreinovich
Leonardo Salayandia
Department of Computer Science, University of Texas, El Paso, Texas 79968, USA

ABSTRACT

Potential field data (gravity and magnetic measurements) are both useful and cost-effective tools for many geologic investigations. Significant amounts of these data are

Keller, G.R., Hildenbrand, T.G., Kucks, R., Webring, M., Briesacher, A., Rujawitz, K., Hittleman, A.M., Roman, D.R., Winester, D., Aldouri, R., Seeley, J., Rasillo, J., Torres, R., Hinze, W.J., Gates, A., Kreinovich, V., and Salayandia, L., 2006, A community effort to construct a gravity database for the United States and an associated Web portal, *in* Sinha, A.K., ed., Geoinformatics: Data to Knowledge: Geological Society of America Special Paper 397, p. 21–34, doi: 10.1130/2006.2397(02). For permission to copy, contact editing@geosociety.org.

traditionally in the public domain. A new magnetic database for North America was released in 2002, and as a result, a cooperative effort between government agencies, industry, and universities to compile an upgraded digital gravity anomaly database, grid, and map for the conterminous United States was initiated and is the subject of this paper. This database is being crafted into a data system that is accessible through a Web portal. This data system features the database, software tools, and convenient access. The Web portal will enhance the quality and quantity of data contributed to the gravity database that will be a shared community resource. The system's totally digital nature ensures that it will be flexible so that it can grow and evolve as new data, processing procedures, and modeling and visualization tools become available. Another goal of this Web-based data system is facilitation of the efforts of researchers and students who wish to collect data from regions currently not represented adequately in the database. The primary goal of upgrading the United States gravity database and this data system is to provide more reliable data that support societal and scientific investigations of national importance. An additional motivation is the international intent to compile an enhanced North American gravity database, which is critical to understanding regional geologic features, the tectonic evolution of the continent, and other issues that cross national boundaries.

Keywords: gravity, geoinformatics, database, geophysics, data processing.

INTRODUCTION

Measurements of Earth's gravity field represent a cost-effective tool to aid a broad range of geoscientific investigations. Relatively local variations (anomalies with wavelengths <500 km) in the gravity field that are due to structures and compositional variations in the lithosphere are of primary interest to geologists and applied geophysicists, and over the years, studies of gravity anomalies have advanced our understanding of both large- and small-scale geological processes. As reviewed by Keller (1988), the results of many important gravity studies were published in the *Geological Society of America Bulletin* during its first 100 years of existence.

The use of gravity data has benefited from the fact that the scientific community working with these data has been traditionally small and well networked with a history of international cooperation and sharing of data. This has been particularly true in North America where government agencies, industry, academia, and professional societies have worked together to produce products such as the Bouguer gravity anomaly map of the United States (Society of Exploration Geophysicists, 1982) and the corresponding map for North America (Committee for the Gravity Anomaly Map of North America, 1987) that have led to the release of grids and large databases of point data. The availability of these data and technical developments such as GPS (Global Positioning System) and GIS (Geographic Information Systems) have produced a rising recognition of the utility of gravity data and thus have increased the demand for and use of gravity data by nonspecialists. In addition, these technical developments have made the interpretation and collection of high-quality gravity data much easier for specialists and nonspecialists alike.

One purpose of this paper is to describe a concerted and continuing effort, spurred by the Geoinformatics Initiative (www.geoinformatics.info), whose goal is a substantially upgraded digital gravity anomaly database for North America. For the conterminous United States, this effort is spearheaded by the geophysics groups at the National Geospatial-Intelligence Agency (NGA), the U.S. Geological Survey (USGS), the National Oceanic and Atmospheric Administration (NOAA), and academia represented by the University of Texas at El Paso (UTEP). From the beginning, it was clear that our effort should begin with the conterminous United States, and a broad cross section of the gravity community was invited to a workshop held in 2002. The resulting report (Hildenbrand et al., 2002) set up new standards that have been refined via community input and widely circulated (e.g., Hildenbrand et al., 2003; Hinze et al., 2005). This compilation effort has resulted in significant additions to and major refinement of the public domain portion of the United States database that is compiled and maintained by the NGA (formerly known as NIMA, National Image and Mapping Agency) and released publicly via NOAA's National Geophysical Data Center in 1999 (Fig. 1). Although our initial focus is the conterminous United States, the broader goal is improvement of the North American gravity database that was released over 15 years ago (Committee for the Gravity Anomaly Map of North America, 1987). This new database is critical in understanding the tectonic evolution of the continent and elucidating regional geologic features, particularly those crossing national boundaries.

Another purpose of this paper is to describe the evolving data system that we have constructed. In using the term *data system,* we mean that our ultimate goal is far more than a high-quality database. Our goal is an interlinked virtual framework where users can obtain and contribute data, receive help in regard to data collection and reduction procedures, find tools to manipulate the data, and find tutorials to aid them in understanding and utilizing the data. This data system is intended to serve the purposes of a diverse user community while being simple to access and flexible enough to meet the range and changing needs of users. The initial version of the Web portal to this system is available at http://paces.geo.utep.edu. This community effort will ultimately be seamlessly available via the Web sites of all of the partner institutions and organizations.

DATA REDUCTION

Once gravity measurements are made in the field, the goal is to "correct" (adjust) the readings for the known variations in gravity due to effects such as elevation and Earth's shape and rotation so that geologically meaningful anomalies can be calculated, mapped, and modeled. This process is called data reduction and amounts to modeling of known effects so that they can be removed to reveal anomalies. Gravity and geodesy are intimately related, so there are many elements of gravity data reduction that depend on geodetic considerations. Key considerations are the geodetic datum to which latitude and longitude are referenced and the geoid model or a model of Earth's shape (reference ellipsoid) to which elevation (height) is referenced (e.g., Li and Gotze, 2001). A detailed discussion of reduction procedures is beyond the scope of this paper. For the sake of completeness, however, a brief discussion of the fundamentals of this topic follows (see more detailed discussions in Blakely, 1995, and Telford et al., 1990). The specific corrections and formulae that we have implemented in our data system are presented in Hinze et al. (2005).

By international agreement, an International Gravity Formula has been derived that predicts the value of gravity at sea level for any location on Earth (e.g., Morelli, 1974). We denote this value as G_t, and as a function of latitude, it is dependent on the rotation of Earth and its elliptical shape. This formula is updated from time to time as more precise geodetic information becomes available.

Spring-based relative gravity meters measure differences in gravity precisely (routine sensitivity is 1 part in 10^7 or better). The usual value stated for the gravitational attraction of Earth is 980 cm/s^2, where 1 cm/s^2 is a unit called a Gal (in honor of Galileo). Thus, gravity variations are regularly measured with

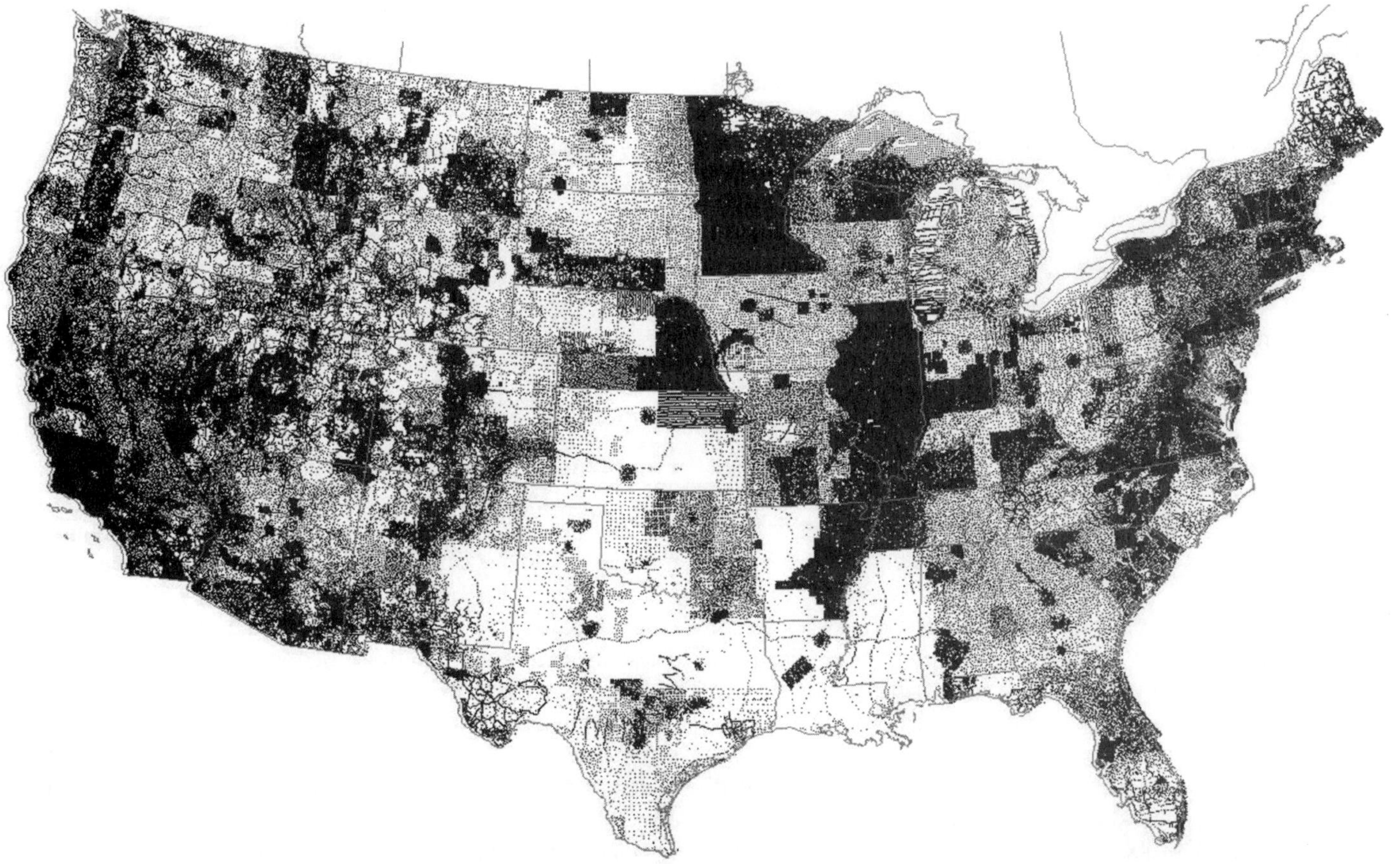

Figure 1. Map of the distribution of data points in the most recent (1994) National Geospatial-Intelligence Agency data set that is released via the National Geophysical Data Center. A total of 839,881 points are in this version of the database.

precisions of at least 0.1 milligal (mGal). Although one can conduct a local survey that delineates relative gravity anomalies that are useful, the best practice is to tie every survey to a base station where the absolute value of gravity (G_{base}) has been determined to a high accuracy. In order to facilitate this practice, our data system includes up-to-date information on the base station network in the United States. After corrections for small time variations in the measured values due to Earth tides and to drift within the workings of the gravity meter, a value for the total observed gravity at each field measurement site (G_{obs}) can be easily determined:

$G_{obs} = G_{base}$ + [(gravity meter reading at the station – gravity meter reading at the base station) * calibration constant for the meter].

The simplest version of a gravity anomaly is the difference between theoretical (G_t) and observed gravity (G_{obs}). However, elevation effects render this anomaly almost meaningless, and thus the application of an elevation correction is a key element in gravity reduction.

The first element of the elevation or height correction is to adjust for the Free Air effect. This effect is simply based on gravity decreasing with increasing height due to the location being at a greater distance from the center of mass of Earth as predicted by Newton's Law of Gravitation. Using some approximations for values such as the radius of Earth, a widely employed first-order approximation of the height variation (Free Air effect) in mGal is 0.3086 h, where h is the height of the gravity station in meters above a datum that can be chosen by the user. This datum has traditionally been sea level, and h has been the elevation relative to the geoid. However, the North American gravity database committee has recommended that ellipsoidal heights be used in the future. The value of 0.3086 h is called the Free Air correction. The expression for the Free Air effect reveals that we must know the height with an accuracy of a few tens of cm to make this correction with a precision equivalent to the gravity measurement itself. Today, we conduct many high-precision gravity surveys that require equally high-precision determinations of heights.

The application of the Free Air correction alone produces a Free Air Anomaly (FAA), and the symbolic equation to calculate this anomaly is:

$$\text{FAA} = G_{obs} - G_t + \text{Free Air correction.}$$

However, we also must take into account the mass above the datum that makes gravity greater at the point of the reading than the value expected at the datum. This mass is taken into account with the Bouguer correction, which is based on a mathematically straightforward volume integral and an estimate of the density of the rocks above the datum (normally 2,670 kg/m^3). It is important to keep in mind that, although the Bouguer correction accounts for the mass above the datum, the point of measurement is unchanged. The result is the Simple Bouguer Anomaly (SBA) that is defined as:

$$\text{SBA} = G_{obs} - G_t + \text{Free Air correction} - \text{Bouguer correction.}$$

In areas with significant topographic relief, it is important to correct for the effects of this topography by making a Terrain correction. When corrected for terrain, the result is the Complete Bouguer Anomaly (CBA).

Computation of Terrain Corrections

Although conceptually a simple volume integral, the calculation of the terrain correction is time consuming. Modern digital elevation models can be employed to calculate this correction except for topography very near (hundreds of meters) the location of the gravity measurement, which is called the inner zone. An important feature of the database is that terrain corrections have been calculated for all of the stations. This major undertaking was done using an improved algorithm that was written by one of the authors (Webring). The terrain correction numerical integration was performed using cylindrical segments with a top that may have a general slope in the radial direction. Eight terrain resolutions (see listing below) are used, ranging from 2 arc-second to 5 arc-minute intervals. Each terrain resolution is assigned a starting and stopping radii based on using ninety segments per cylinder and enough subcylinders to maintain approximately square segments. The listing below shows the inner radius of the zone in kilometers, the associated grid resolution, and the Hammer zone designation where appropriate:

Radius of zone (km)	Grid size (sec or min)	Hammer (1939) zone
0.895	2 s	G
1.529	4 s	H
2.615	8 s	I and J
6.653	15 s	K
9.903	30 s	L and M
21.944	1 min	
57.0	3 min	
117.0	5 min	

Note: Outermost radius = 166.7

Computation of Isostatic Residual Anomaly

Bouguer gravity anomaly maps have been used traditionally to provide a geologic picture of the subsurface over land areas (Simpson et al., 1986). At regional scales, Bouguer gravity maps display broad anomalies inversely correlated with regional topography. These anomalies are usually caused by deep-seated masses that isostatically support the topographic loads. To remove the effects of these topographic roots, an isostatic regional field is computed in a manner similar to the terrain model using the Airy-Heiskanen model. A modified version of the approach of Jachens and Roberts (1981) that assumes local compensation is being considered to calculate the isostatic correction of varying depths to a hypothetical crust-mantle boundary caused by varying loads of topography above the geoid. For each gravity station in the national database, topography will be modeled assuming a density of 2,670 kg/m^3 and using 3 min elements of topography to 166.7 km plus an interpolated value (Karki et al., 1961) beyond 166.7 km.

CONSTRUCTION OF THE DATABASE

Sources of Data

The National Geospatial-Intelligence Agency (NGA) has maintained a global database of gravity values for many years. Some of these data are not available publicly, but a large number of readings (about one million in the United States) were available in the latest release from the NGA via the National Geophysical Data Center (NGDC). Most investigators around the world who regularly gather and use gravity data have interacted with this agency. Thus, it was determined that the latest (1999) version of the NGA comprehensive database for the United States (Fig. 1) would be used as the starting point for our database and would be supplemented by a significant quantity of new data. An extensive search was undertaken to identify and acquire existing (legacy) data that have not been sent to the NGA or released to the public, and ~400,000 new gravity stations in the United States have already been identified and included in the new database (Fig. 2) that is presently accessible.

Because a major source of data will be totally new measurements, the Web portal has been designed to facilitate the entry and online reduction of new data. Contributed data will be (1) specifically recognized with the contributor's name and organization and cited as a contribution to the database, (2) processed in the same manner as the other data, and (3) freely available ultimately upon request by any user. Following the model of other public geophysical databases such as the massive seismic database available through the Incorporated Research Institutions for Seismology (IRIS) (www.iris.edu), provision will be made for data to be held confidential for up two years while the provider conducts his research.

Data and Metadata Formats

A recommendation was made to develop a new ASCII format for the gravity data to meet the objective of the database. The data are being stored in a relational database so the issue is ultimately not in the selection of a format but rather which data table entries are required and which ones are optional. The data

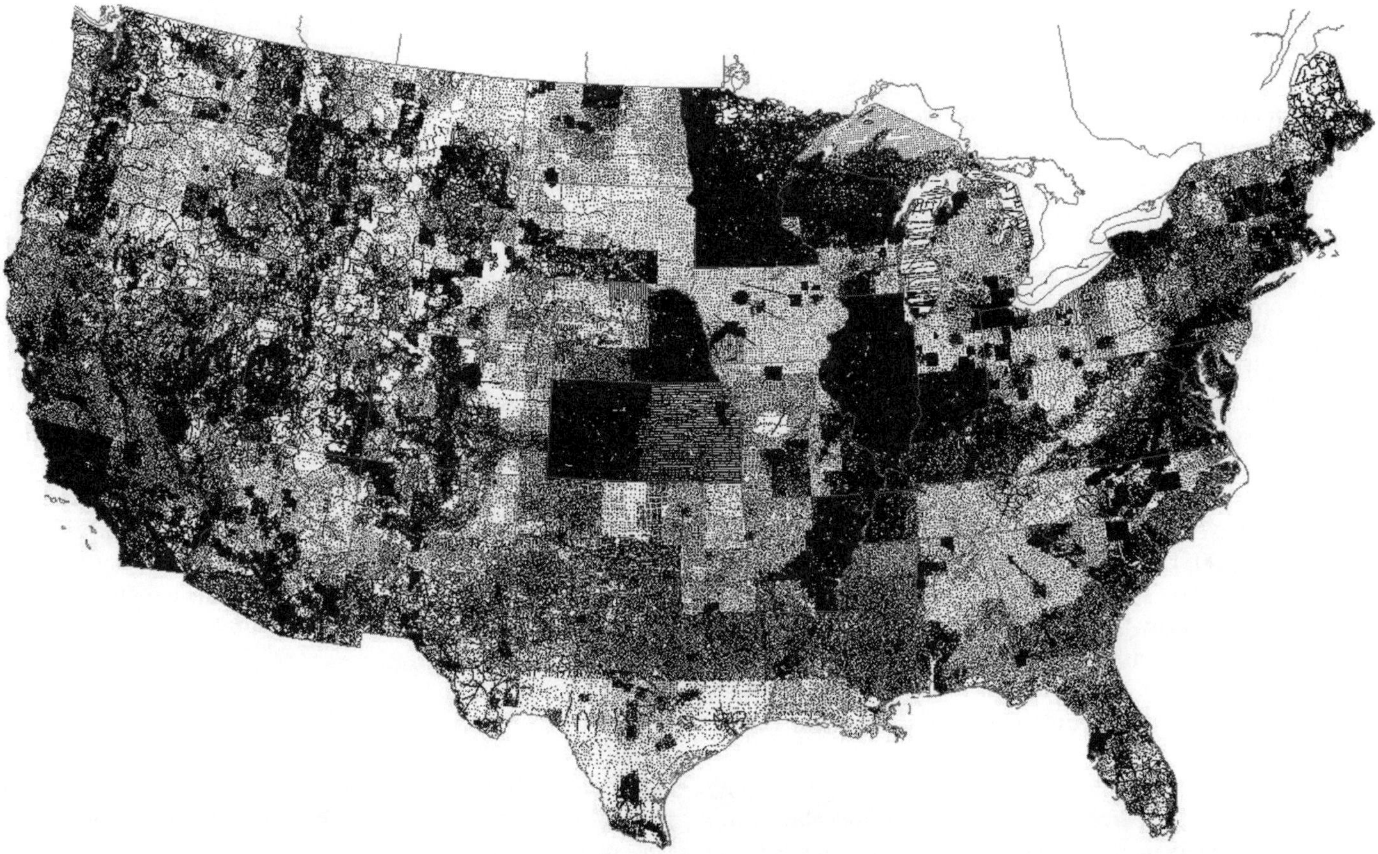

Figure 2. Map of the distribution of data points in the updated data set that is currently available on the Web portal. A total of 1,238,881 points are in this version of the database.

that the user downloads via the portal are in the form of simple ASCII files that are readily entered into a wide variety of software. New entries into the database include accuracies of values expected in a modern database (e.g., geographical coordinates in decimal degrees to 7 places, elevation in meters to 3 decimal places, observed gravity in mGals to 4 places). Most data sets will not attain this high accuracy, but accommodating high-resolution data increases the usability of the database especially in future efforts to track changes in gravity. Thus, those who make additions to the data set are being asked to include an estimated accuracy for their data.

We can now easily store much more information about each data point than in the past, and the system is being structured so that a significant amount of metadata will be requested for new data entries. For the near term, metadata residing with the NGA will be incorporated in the Web portal to convey important information on the various data sources and associated data now in the NGA's files. It will be possible to store information provided by data contributors such as a report describing the survey and descriptions of new base station locations as metadata. In addition to metadata and principal facts (latitude, longitude, elevation, observed gravity) and their accuracy, contributors of new data will be asked to enter information such as a brief description of the survey procedures (e.g., nature of GPS or other techniques to determine elevations), description of gravity meter used, approach used to determine inner zone terrain corrections, geodetic datum used, and base stations employed.

Standardization of Horizontal, Vertical, and Observed Gravity Datums

The default horizontal datum is currently NAD27, but as discussed in Hinze et al. (2005), we are in the process of converting to the International Terrestrial Reference Frame (ITRF) in conjunction with the 1980 Geodetic Reference System (GRS80) ellipsoid. This approach is essentially equivalent to using WGS84 (NAD83; NIMA, 2000). We will continue to include NAD27 coordinates in the data tables, and we will include software for conversion to other datums if there is a significant demand. The default vertical datum is currently the NGVD29 height above mean sea level on topographical maps, but we are in the process of conversion to NAVD88 heights (orthometric). As discussed by Hinze et al. (2005), we also plan to include heights measured relative to the 1980 Geodetic Reference System (GRS80) ellipsoid. Observed gravity will be tied to the International Gravity Standardization Network 1971 (IGSN71) (Morelli, 1974).

Eliminating Bad Points, Duplicates, and Redundant Data

Our goal is to provide a database to the scientific community that is as accurate and free of duplicate data as feasible. Since there is a growing demand for gravity data, we have moved with a sense of urgency to produce a much-improved database. However, the database building process is being designed to be evolutionary in nature so that over time, the product will be constantly expanded and upgraded.

Sophisticated algorithms have been developed to aid in the cleaning process to identify questionable or duplicate data (e.g., Torres et al., 2004). We recognize that the needs of the community can change, and all data obtained will be stored so that anything done during the cleaning process can be recovered at a later date. Since the search for anomalies is the object of most gravity surveys, one has to be careful when deciding to expunge a data point. In addition, we are mindful that several factors complicate the process of selecting data points that are bad, erroneous, redundant, or duplicates (see definitions below).

Revisions of gravity and geodetic datums and of reference ellipsoids have sometimes resulted in confusion about base station values, reduction formulae, and even coordinates of stations. These effects can be corrected when there is enough metadata available to identify them.

The real problem with many gravity readings, particularly older ones, is the accuracy of the elevation value. Our algorithms jointly search for suspicious values of observed gravity and elevation. In theory, we can use digital elevation models (DEM) to identify and even correct elevation values, but most DEMs presently available have at best 3 m vertical accuracy. The use of GPS has revolutionized the collection of gravity data, but great care must be exercised to insure that the resulting elevation values are properly tied to traditional geodetic datums.

Duplicate values can corrupt the results of statistical data processing and analysis. Thus as recommended by the Federal Geographic Data Committee (1998), we consider detecting and eliminating duplicates to be an important part of assuring and improving the quality of geospatial data. This issue is particularly important in the case of gravity measurements because over the years many groups have established databases by starting with the NGA compilation. They have taken the original data (principal facts) and reformatted them, probably making conversions between feet and meters and changing the way latitude and longitude are stored (e.g., decimal degrees versus degrees, minutes, and seconds). As new data are collected, they are simply merged with the original data. When a third party requests some data, the combined database is used, and the whole process starts over. This practice clearly produces many duplicate points in these derivative databases. Due to the conversions that are made, these duplications do not have the exact x, y, and z coordinates as the original data point. Thus, the search algorithm could not be built on the definition of duplicate being equal.

Definitions:

Bad—A data point that is associated with an erroneous value for a principal fact that cannot be corrected. Such points were very rare in the existing database (a small fraction of a percent), but they create unacceptable artifacts (bull's-eye anomalies) on maps created from a database.

Erroneous—A reading that has a correctable value for a principal fact such as observed gravity, which is incorrect due to the use of an incorrect base station value or elevation values that confuse feet and meters.

Redundant—A reading taken exactly (e.g., at a benchmark) where a reading already in the database base was taken. If precise enough and time tagged, these data can be used to detect temporal changes in gravity.

Duplicate—A reading that is repeated in a database but has no extra information value. Such readings can either be the result of a particular value reading being present in two data sets that were used in the compilation of the database or of two readings of comparable accuracy (but not precise enough to be useful from a time variation perspective) being made at essentially the same location.

SOME COMPLICATIONS CONCERNING INTEGRITY

Changes in Gravity Formulae

As the international geodetic community updates models of the shape of Earth, the International Gravity Formula (IGF) is adjusted. As discussed above, we are using the latest version of this formula, which dates from 1980. The earlier version that dated from 1963 produced theoretical gravity values that were ~1.5 mGal different from those produced by the 1980 formula. This difference is fairly subtle if one is only compiling regional gravity data, so there is a potential for data reduced unknowingly with different versions of the IGF to be merged together.

Changes in Base Station Values

The IGF depends on empirical estimates of the average value of gravity at the equator, and ultimately gravity surveys are tied to a network of base stations. The International Gravity Standardization Network (IGSN71) is the international standard for base station values (Morelli, 1974). However, this standard may be adjusted in the future as it was in 1971. One needs to be aware that these changes can cause significant problems (~14 mGal in the case of 1971) and take some time to be adopted by the community. By having a living database, we can adopt a new standard and recalculate all anomaly values quickly, thus avoiding confusion.

Elevation/Height—A Limiting Factor

Another important consideration is that the accuracy of reduced gravity data (usually Bouguer anomaly values) is heavily dependent on the precision of the determination of the elevation/height of the point where the gravity measurement was made. Thus, a pitfall in gravity surveys is that highly accurate gravity measurements are often coupled with much less accurate elevation control. In general, considerably more effort is required to obtain good elevation control than is required to make the gravity measurement. Traditional surveying techniques can produce very precise elevation values but are very laborious when the desire is to cover large areas with limited budgets. Thus, many other approaches have been used to obtain elevations for gravity readings that include: (1) locating gravity readings at known elevation control points (benchmarks and other survey markers), which is highly restrictive because these points are usually sparse; (2) using high-quality barometers, which leads to considerable uncertainty (at least ±2 m); and (3) using elevations posted on topographic maps. However in recent years, precision GPS technology has revolutionized gravity surveying by making it possible to obtain high precision x, y, and z coordinates almost anywhere on Earth's surface. Unfortunately, the legacy data (which constitute the majority of the public domain data) contain a mixture of elevation control often without adequate metadata about the accuracy and techniques employed. Thus, caution must always be exercised when comparing data from surveys of different vintages. Since minute variations in gravity are associated with phenomena such as the withdrawal of fluids from the subsurface, monitoring changes in gravity is an emerging field of study. To facilitate such studies, the data system requests considerable metadata about the precision of the gravity measurements and positioning (x, y, and z) so that data from repeat high precision surveys can be validly compared with the original data.

SOFTWARE TOOLS

Like the database, the package of software tools in the data system is being designed to grow and evolve through time. However, a number of tools have been identified for initial implementation.

Data Reduction

Many different groups around the world routinely reduce gravity data and produce results that are consistent within <1 mGal, exclusive of terrain corrections. This fact is a testament to the existence of a considerable amount of standardization. However, the level of precision that is now typical in gravity surveys justifies that care be taken to ensure that reduction equations and parameters are applied with a high degree of consistency. We feel that to accomplish this goal, all data need to be reduced with the same software. The easiest way to accomplish this is to set up the data system by rereducing the legacy data from the principal facts (latitude, longitude, elevation, observed gravity) and then to reduce all new data online with the same software. The new Web portal incorporates this concept. There are details about data reduction that are a matter of personal preference, and the data system being created does not intend to dictate to the community. Thus, all users are welcome to either use the reduced data provided by the system or rereduce the data with their own scheme from the principal facts and their own software. Since we have obtained input from a many workers in the North American geophysical community as well as international colleagues, we hope that the vast majority of users will use the native software in the system in order to achieve standardization.

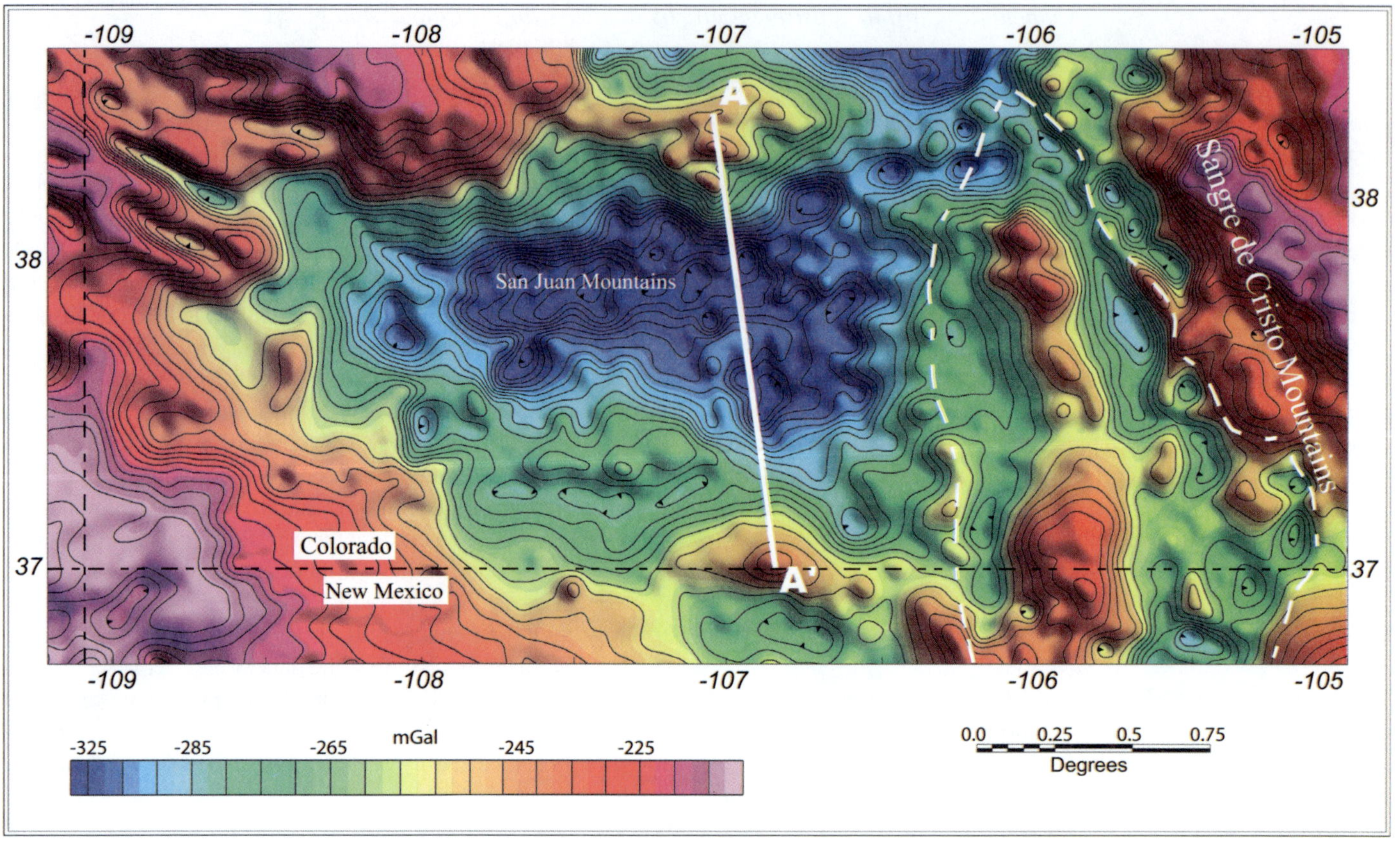

Figure 3. Simple Bouguer gravity anomaly map of the San Juan Mountains–San Luis basin–Sangre de Cristo Mountains area of southern Colorado. A–A′ is the profile modeled in Figure 6. Contour line interval—4 mGal.

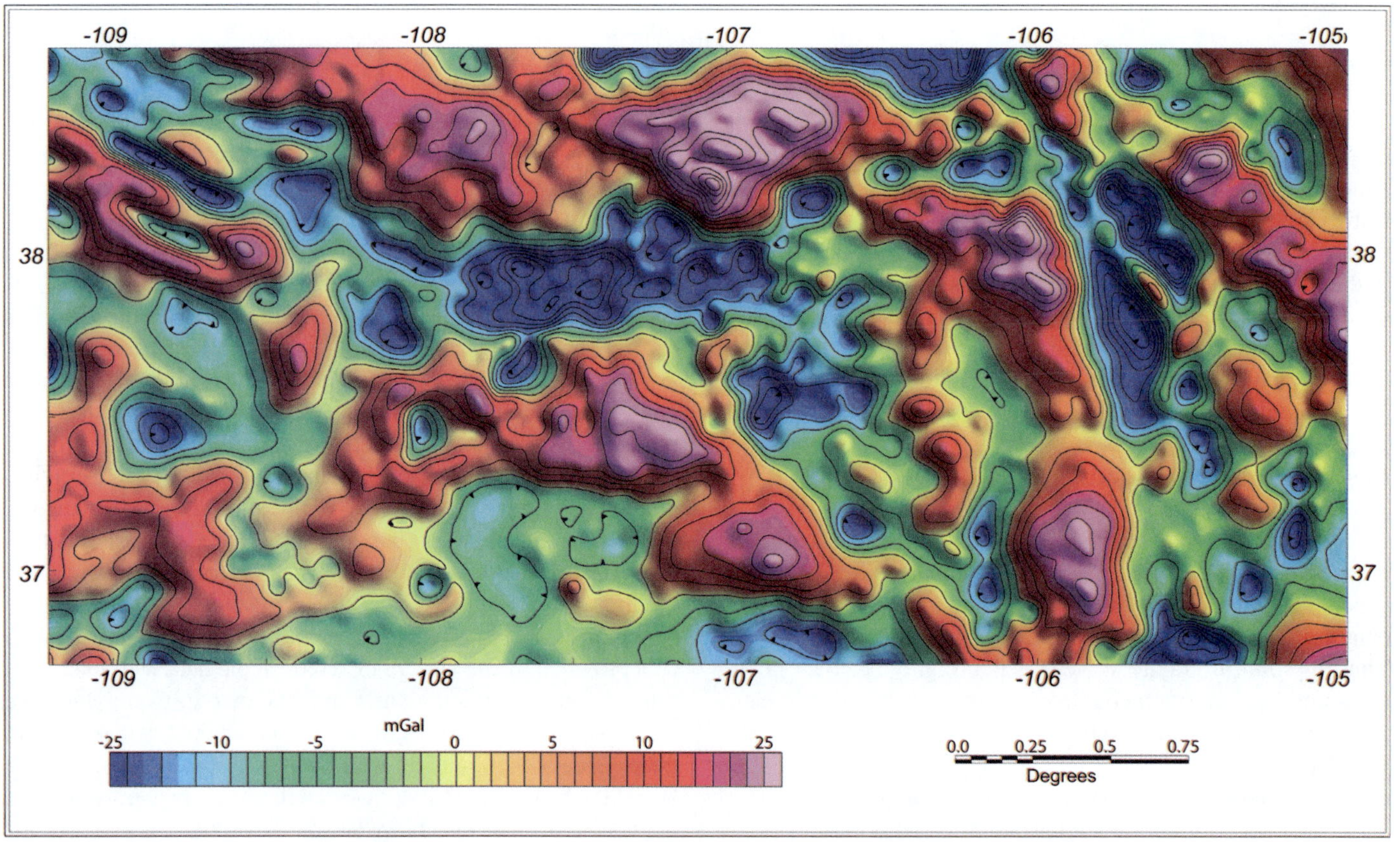

Figure 4. Filtered gravity map (10–150 km wavelengths passed) of the San Juan Mountains–San Luis basin–Sangre de Cristo Mountains area of southern Colorado. Contour line interval—4 mGal.

Construction of Maps and Filtering

The first step in the interpretation of gravity data is to construct an anomaly map, usually of the complete Bouguer anomaly. The Generic Mapping Tools (GMT) computer cartography package (Wessel and Smith, 1995; http://gmt.soest.hawaii.edu/) and the user interface (iGMT) developed by Becker and Braun (1998; http://www.seismology.harvard.edu/~becker/igmt/) is an ideal example of the type of public domain software envisioned in the Geoinformatics initiative. The map shown in Figure 3 was constructed using this package. In order to construct a contour map, the data must be gridded, and GMT includes a selection of gridding algorithms.

Once gridded, gravity maps can be enhanced by many approaches that emphasize anomalies of various wavelengths, sources, and trends (strikes). All of these approaches can be thought of as filtering in a general sense. GMT has the capability of wavelength filtering; an example is shown in Figure 4. This map is the result of a simple bandpass filter (10–150 km) and clearly shows the effects of upper crustal features better than the Bouguer anomaly map in Figure 3. We will also convert selected elements of a package of filtering and processing algorithms compiled by Phillips (1997) so that they will work in a Windows environment.

Profile Extraction

Users may want to extract gravity data along profile to model or simply compare with other geoscience data, such as seismic lines. Computer modeling in a cross-sectional sense requires extraction of a profile of gravity readings that should be tied to points of independent control such as outcrops and drill holes if possible. One should also avoid profiles with large anomalies on the ends. We have developed a profile extraction algorithm that selects points along a profile defined by two end points provided by the user. Points within a user-prescribed distance from the profile are projected onto it, and both the distance along the profile and the projected distance are included in the resulting file that is an input to the modeling algorithm. The profile of anomaly values modeled in Figure 5 was extracted using this algorithm.

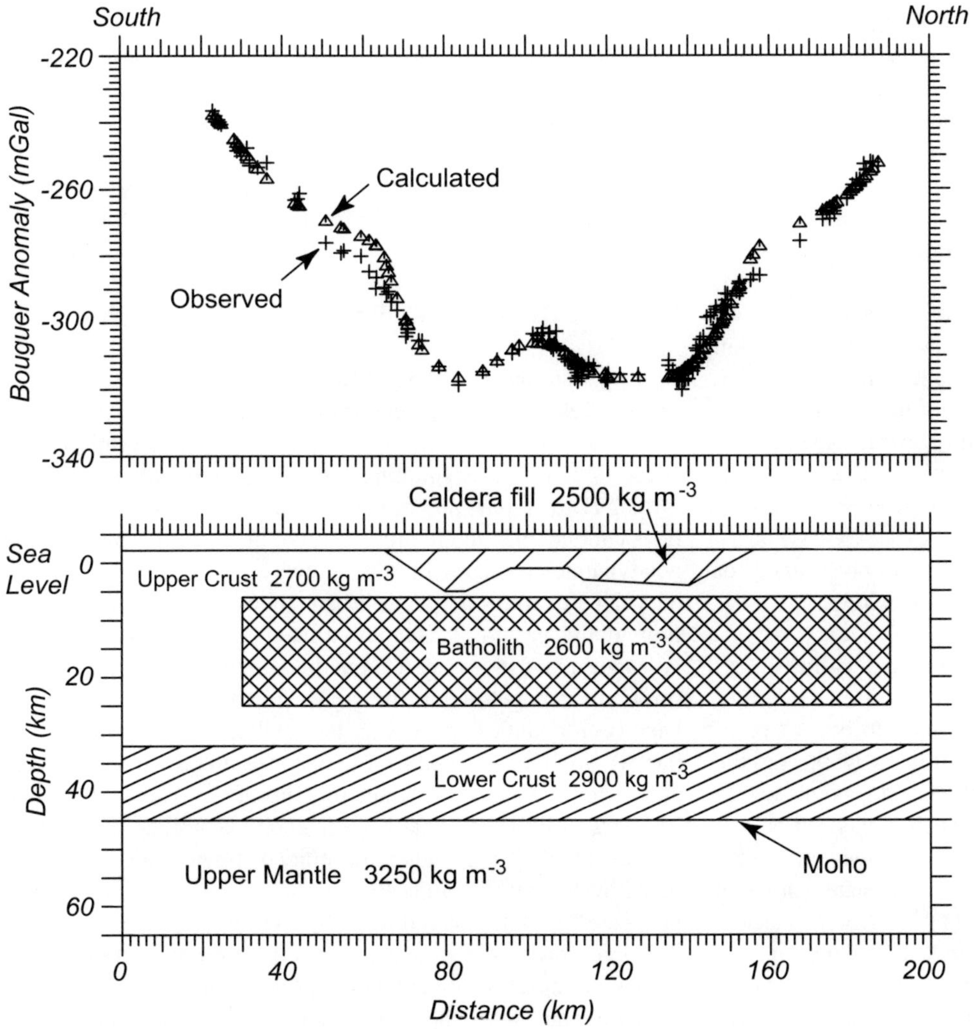

Figure 5. Gravity model of Profile A–A´, location shown on Figure 3.

Modeling

Modeling of gravity data is a very common interpretation technique. There are a number of approaches to computer modeling in two and three dimensions using both forward (trial and error) and inverse (iterations done via various mathematical approaches) methods. Initially, we have implemented an algorithm based on the approach of Cady (1980) in which 2-1/2D models can be easily constructed in a Windows environment, and we are distributing a working but evolving version of this program. The term 2-1/2D means that the polygonal bodies that are manipulated to attain a match between calculated and observed values can have finite (rather than infinite) strike lengths. This approach has the advantage of having the feel of the construction of a geologic cross section while retaining some capability to consider three-dimensional effects. A model derived using this approach is shown in Figure 5 and illustrates the importance of the caldera fill and the batholith in producing the anomalies in the San Juan volcanic field. Another version of freeware for the modeling and inversion of gravity and magnetic data is available from Rockware (http://www.rockware.com).

Detection and Removal of Duplicate and Redundant Values

Our duplicate value detection algorithm is designed to quickly find points that lie within a prescribed distance of each other. In order to accomplish this, we devised an approach wherein the data set is highly sorted by both latitude and longitude and is thus easy to search for points that lie within a user-prescribed distance from all other data points. There is no hierarchy of data when a single data set is searched for duplicates. However, when merging two or more data sets, the order in which they are entered determines the hierarchy (Torres et al., 2004). The algorithm can be used in many ways. For example, measurements submitted for inclusion in the data system will be examined to see if they are duplicates of older data in the system. It should usually be true that new data have higher precision than older data, and in this case, we will replace the older data with the newer values. However, the issue of which data to include will be decided on a case-by-case basis. Redundant values of equally high precision will be retained for use in studies involving changes in the gravity field.

Detection and Removal of Erroneous and Bad Values

It is obviously important for any database to be as free of incorrect data as possible. The experience of many workers in the field of gravity is that every database contains a few points whose calculated anomaly values are incorrect. Such points can be due to many factors and given the large number of values in the data system (~1.3 million in the conterminous United States), it is not practical to research each questionable data point that is detected. Thus, isolated values that are clearly outliers will be removed from the main database but retained for possible future use. Our initial efforts have found many examples of groups of points that appear suspicious in that their anomaly values diverge from nearby stations by a small but consistent amount. As time allows, these points will be investigated to see if some simple problem such as an incorrect base station value can be detected and corrected. However, this procedure is laborious and will be an example of a process that will be ongoing as the database evolves. Users will be encouraged to report erroneous values that they detect and to submit corrected values for any data that they edit. After confirming the calculations, the new values will be adopted, but again the original data will be archived in case they are of interest to future investigators.

The design of an algorithm to detect spurious gravity values is challenging because the goal of gravity studies is to study anomalies. Thus, it would be easy to design an algorithm that removes all of the interesting anomalies. Our approach to this problem is to window the data set into small areas that must contain at least 50 readings and to use localized statistical approaches to identify bad and questionable values (Wen et al., 2001; Beck, 2004).

Proprietary Software Tools

There is of course a considerable amount of industry standard licensed software that will do many of the operations mentioned above. For example, mapping software products include ArcGIS produced by Environmental Systems Research Institute (ESRI; http://www.esri.com), Oasis Montaj by Geosoft (http://www.geosoft.com), and ERMAPPER by Earth Resource Mapping (http://www.ermapper.com), while gravity modeling software is available from GMSYS and GMSYS-3D by Northwest Geophysical Associates (http://www.nga.com) and Rockware (http://www.rockware.com).

UTILITY OF GRAVITY DATA

Applications of gravity databases include studies of Earth's gravity field in geodesy, geophysics, and geology. At regional scales, for example, gravity data are useful in determining the shape of Earth, in accounting for the orbits of satellites, determining Earth's mass and moment of inertia, and conducting geophysical mapping and interpretation of lithospheric structure and geodynamic processes. In local studies of the upper crust, gravity data can effectively address a broad range of basic geologic questions, delineate geologic features related to natural hazards (faults, volcanoes, landslides), and aid in the search for natural resources (groundwater, petroleum, minerals, geothermal energy). There have been countless papers on the applications of gravity data, but two volumes devoted to case histories on this subject at local and regional scales were edited by Hinze (1985) and Gibson and Millegan (1998) and contain many good examples.

In regard to local studies, the utility of gravity data can be illustrated by an analysis of the Bouguer gravity anomaly map shown in Figure 3 and the filtered map shown in Figure 4. For example, the large gravity low that dominates the western portion of this map is associated with the San Juan Mountains and volcanic field (e.g., Lipman, 2000). Plouff and Pakiser (1972) suggested that this feature is due to a batholith that underlies the nested calderas that are present there, and subsequent work supports this interpretation. Our modeling of a profile across the gravity low (A–A′, Fig. 3) shows the approximate size of the batholith (Fig. 5), but it also shows that the low-density volcanic fill in the calderas is an important contribution to the gravity anomaly. The San Luis Valley lies east of the San Juan Mountains (Fig. 3), and the early gravity interpretations (e.g., Keller et al., 1984) suggested that the basin underlying this valley consists of a deep subbasin to the east, adjacent to the Sangre de Cristo Mountains, and a shallower subbasin to the west that is separated by an almost totally buried horst. Subsequent seismic reflection studies and analysis of data from drill holes have refined this interpretation and show that the eastern subbasin is mostly filled with sediments associated with the Rio Grande rift and that the western subbasin is mostly filled with older Mesozoic and Cenozoic sediments (Kluth and Schaftenaar, 1994; Brister and Gries, 1994). The southern portion of the basin has been the focus of a new gravity and magnetic study by Grauch and Keller (2004) that shows these structures extend southward into New Mexico where another subbasin is encountered near Taos. These subbasins represent major water resources for the region and may ultimately be a source for hydrocarbons. However, the lack of a large gravity high over the southern Sangre de Cristo Mountains is enigmatic, because dense Precambrian rocks are widely exposed in the mountains and because these mountains are flanked by basins filled with low-density sediments (e.g., Quezada et al., 2004). This brief discussion shows that straightforward analysis of gravity data can easily produce results of both basic and applied scientific value as well as point out relationships that require further investigation.

STRUCTURE OF THE WEB PORTAL

We refer to the Web site that users enter (http://paces.geo.utep.edu) as a Web portal because it provides access to a fully integrated data system "populated" with high-quality data as well as an expanding set of software for analyses of the data, a tutorial on gravity data, access to gravity base station information, links to other resources, and the reduction of new data that the user uploads into the system. Information about the development of standards and other items of interest are posted for downloading by interested parties. The goal of the tutorial is to facilitate the efforts of researchers and students who want to learn about the gravity method and perhaps wish to collect data in areas currently not represented adequately in the database. To that end, this tutorial contains or will contain information on topics such as: (1) basic physical principles; (2) utility and planning of gravity surveys; (3) data reduction schemes and key parameters; (4) locations and proper use of base stations; (5) construction and interpretation of maps; and (6) preparation for integration with other data.

The design and specifications for the portal were the research project for a software engineering class at the University of Texas at El Paso during the 2002–2003 academic year, and almost 40 students worked on the project in teams. The ultimate design was the topic of an MS thesis (Rasillo, 2004). The entry to this portal is illustrated in Figure 6A. Users who want to locate data in their area of interest can simply describe the area by a bounding rectangle and then download the results of the search (Figs. 6B and 6C). The points obtained by the search can be plotted on a map online, and the map can be enlarged via a zoom-in capability.

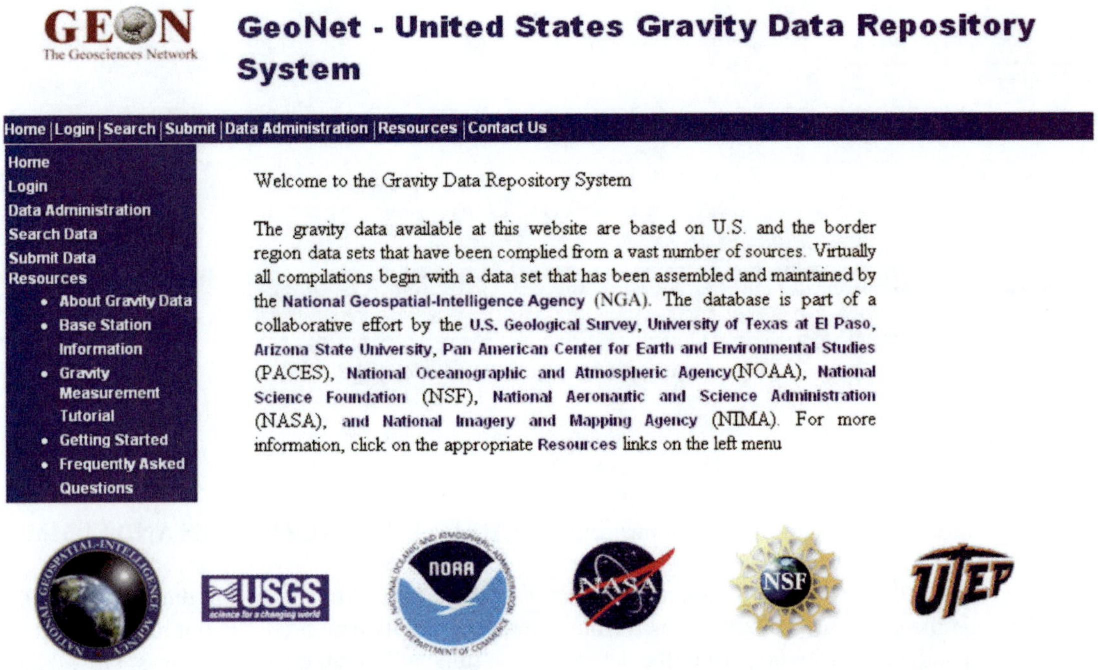

Figure 6. Screen capture images of the key pages of the Web portal. (A) Entry page (*continued on following two pages*).

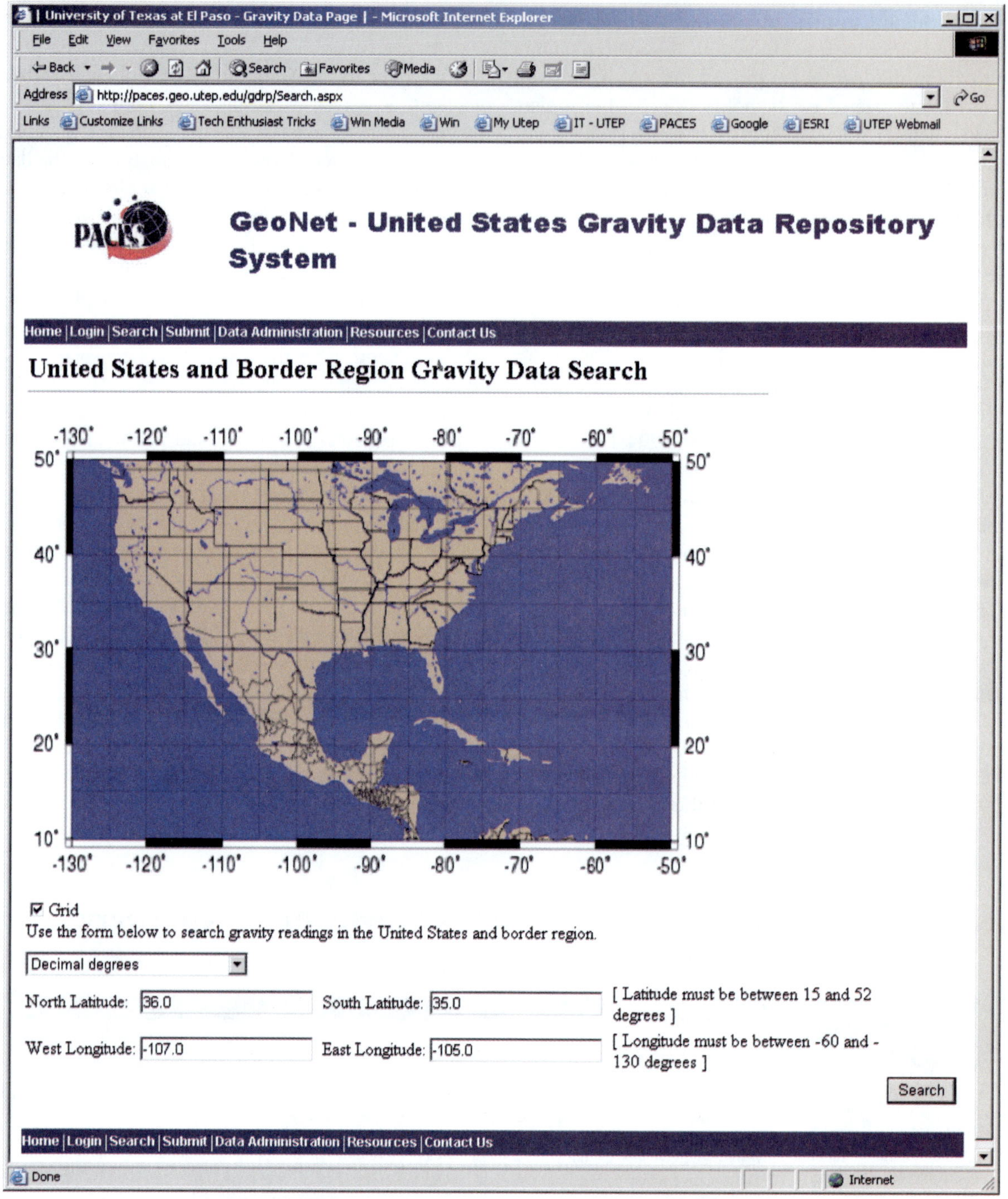

Figure 6 (*continued*). (B) Data selection page.

In addition to links to the participating organizations, our portal is registered in the GEON (www.geongrid.org) system in order to facilitate discovery of this resource by researchers. To date, our focus has been on the database, but our goal is for our set of software tools to expand via contributions from the developers and users of the data system.

SCHEDULE OF ACTIVITIES AND SUMMARY

We achieved the following during 2003–2004: (1) gathered existing gravity data presently not in the NGA unclassified database (this will be an ongoing process, but we have already produced the updated coverage that is shown in Figure 2 by search-

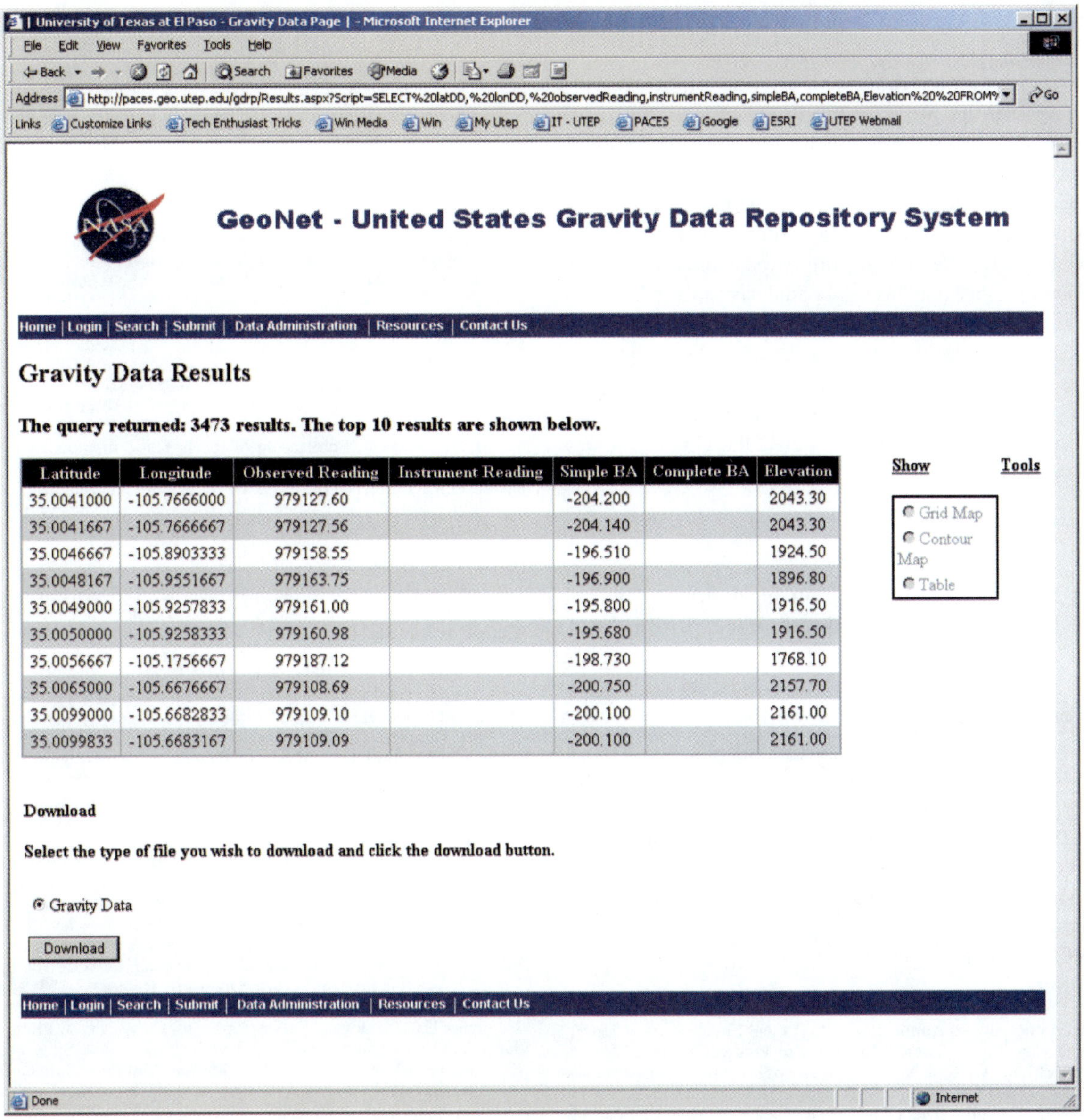

Figure 6 (*continued*). (C) Example of the results of a search.

ing the data sets maintained by the U.S. Geological Survey, the National Geodetic Survey, and the University of Texas at El Paso and merging them with the latest version of the NGA unclassified data); (2) conducted quality-control analysis including filtering out erroneous data (this has been done for the first major upgrade); (3) computed gravity anomalies and terrain corrections consistent with internationally agreed upon procedures and with full documentation; (4) constructed an efficient Web portal; (5) made the upgraded, terrain-corrected conterminous United States database available on the Web portal along with tools to search it; (6) provided easy access to current descriptions of gravity base stations in the United States, and (7) provided users the ability to upload and reduce their data online. In 2006, we plan to produce a 4 km grid of the upgraded conterminous United States database and make it available via the Web portal and release new United States Bouguer and isostatic anomaly maps. These maps will replace an existing map produced by the Society of Exploration Geophysicists (1982). We also continue to expand the list of software tools available.

The *mid-term goals* to be completed within the next three years will be to (1) develop new methods to improve data quality; (2) add bathymetry corrections; (3) extend outer terrain correction to 500 or 1000 km and compute inner terrain corrections using 10 m DEMs; (4) improve the contents and appearance of the Web portal based on input and contributions from the scientific community; (5) resolve base station network datum shifts;

and (6) complete development of a gravity data system for North America.

The *overall long-range goal* of our effort is for the open and flexible data system that we have built to grow and evolve via contributions of data, ideas, and software from the user-members of the earth science community. Community involvement will be central to the vitality and usability of both the United States and the North American data systems. The ultimate gravity data system we envision will offer greatly improved United States and North American gravity databases and analysis tools, which represent a resource fundamental to geoscience investigations.

At each state of its development, the resulting data system will be freely accessible to all on the Internet and will be linked to integrated data systems such as the GEONgrid that are being developed as part of the Geoinformatics initiative, the U.S. Geological Survey data system, and the National Geophysical Data Center.

ACKNOWLEDGMENTS

This project has benefited greatly from input from many colleagues who have provided data and technical expertise. At the University of Texas at El Paso, this ongoing effort is supported by the National Aeronautics and Space Administration via the Pan American Center for Earth and Environmental Studies (NASA NCC5-498) and the National Science Foundation via grants to support GEON (EAR-0225670) and SWGeoNet (EAR-0112960) projects. Reviews by Ross Black and V.J.S. Grauch helped us improve this paper significantly. The software engineering class at the University of Texas at El Paso helped develop the Web portal, and Joseph Miller produced the diagrams shown as Figures 3–5.

REFERENCES CITED

Beck, J., 2004, Data Processing under a Combination of Interval and Probabilistic Uncertainty and Its Application to Earth and Environmental Studies and Engineering [Ph.D. dissert.]: El Paso, University of Texas, 222 p.

Becker, T.W., and Braun, A., 1998, New program maps geoscience data sets interactively: EOS (Transactions, American Geophysical Union), v. 79, p. 505.

Blakely, R.J., 1995, Potential Theory in Gravity and Magnetic Applications: Cambridge, U.K., Cambridge University Press, 441 p.

Brister, B.S., and Gries, R.R., 1994, Tertiary stratigraphy and tectonic development of the Alamosa basin (northern San Luis Basin), Rio Grande rift, south-central Colorado, *in* Keller, G.R., and Cather, S.M., eds., Rio Grande Rift: Structure, Stratigraphy, and Tectonic Setting: Geological Society of America Special Paper 291, p. 39–58.

Cady, W.J., 1980, Calculation of gravity and magnetic anomalies of finite-length right polygonal prisms: Geophysics, v. 45, p. 1507–1512, doi: 10.1190/1.1441045.

Committee for the Gravity Anomaly Map of North America, 1987, Gravity Anomaly Map of North America: Geological Society of America, Continental-Scale Map 002, scale 1:5,000,000.

FGDC (Federal Geographic Data Committee), 1998, FGDC-STD-001–1998, Content standard for digital geospatial metadata (revised June 1998): Washington, D.C., Federal Geographic Data Committee, http://www.fgdc.gov/metadata/contstan.html (March 2005).

Gibson, R.I., and Millegan, P.S., 1998, Geologic Applications of Gravity & Magnetics: Case Histories: Tulsa, Oklahoma, Society of Exploration Geophysicists, 162 p.

Grauch, V.J.S., and Keller, G.R., 2004, Gravity and aeromagnetic expression of tectonic and volcanic elements of the southern San Luis basin, New Mexico and Colorado: Socorro, New Mexico, New Mexico Geological Society, 55th Field Conference Guidebook, p. 230–243.

Hammer, S., 1939, Terrain corrections for gravimeter stations: Geophysics, v. 4, p. 184–194, doi: 10.1190/1.1440495.

Hildenbrand, T.G., Briesacher, A., Flanagan, G., Hinze, W.J., Hittelman, A.M., Keller, G.R., Kucks, R.P., Plouff, D., Roest, W.R., Seeley, J., Smith, D.A., and Webring, M., 2002, Rationale and operational plan to upgrade the U.S. Gravity Database: U.S. Geological Survey, Open-File Report 02-463, 12 p.

Hildenbrand, T.G., Briesacher, A., Hinze, W.J., Hittelman, A.M., Keller, G.R., Kucks, R.P., Smith, D.A., and Roest, W.R., 2003, Web-based U.S. Gravity Data System Planned: Eos (Transactions, American Geophysical Union), v. 83, p. 613.

Hinze, W.J., ed., 1985, Utility of Regional Gravity & Magnetic Anomaly Maps: Tulsa, Oklahoma, Society of Exploration Geophysicists, 468 p.

Hinze, W.J., Aiken, C., Brozena, J., Coakley, B., Dater, D., Flanagan, G., Forsberg, R., Hildenbrand, T., Keller, G.R., Kellogg, J., Kucks, R., Li, X., Mainville, A., Morin, R., Pilkington, M., Plouff, D., Ravat, D., Roman, D., Urrutia-Fucugauchi, J., Véronneau, M., Webring, M., and Winester, D., 2005, New standards for reducing gravity data: The North American gravity database: Geophysics, v. 70, p. 325–332.

Jachens, R.C., and Roberts, C.W., 1981, Documentation of program, ISOCOMP, for computing isostatic residual gravity: U.S: Geological Survey Open-File Report (United States) 81-0574, 26 p.

Karki, P., Kivioja, L., and Heiskanen, W.A., 1961, Topographic isostatic reduction maps for the world to the Hayford Zones 18–1, Airy-Heiskanen System, T=30 km: Publication of the Isostatic Institute of the International Association of Geodesy, no. 35, 5 p., 20 pl.

Keller, G.R., 1988, The development of gravity and magnetic studies, emphasizing articles published in the Geological Society of America Bulletin: Geological Society of America Bulletin, v. 100, p. 469–478, doi: 10.1130/0016-7606(1988)100<0469:TDOGAM>2.3.CO;2.

Keller, G.R., Cordell, L., Davis, G.H., Peeples, W.J., and White, G., 1984, A geophysical study of the San Luis Basin: Socorro, New Mexico, New Mexico Geological Society, 35th Field Conference Guidebook, p. 51–57.

Kluth, C.F., and Schaftenaar, C.H., 1994, Depth and geometry of the Rio Grande rift in the San Luis Basin, south-central Colorado, *in* Keller, G.R., and Cather, S.M., eds., Rio Grande Rift: Structure, Stratigraphy, and Tectonic Setting: Geological Society of America Special Paper 291, p. 27–37.

Li, X., and Gotze, H.-J., 2001, Ellipsoid, geoid, gravity, geodesy, and geophysics: Geophysics, v. 66, p. 1660–1668, doi: 10.1190/1.1487109.

Lipman, P.W., 2000, Central San Juan caldera cluster; regional volcanic framework, *in* P.M. Bethke and R.L. Hay, eds., Ancient Lake Creede; Its Volcano-Tectonic Setting, History of Sedimentation, and Relation to Mineralization in the Creede Mining District: Geological Society of America Special Paper 346, p. 9–69.

Morelli, C., ed., 1974, The International Gravity Standardization Net 1971: International Association of Geodesy, Special Publication 4, 194 p.

NIMA, 2000, Department of Defense World Geodetic System 1984: St. Louis, Missouri, National Geospatial-Intelligence Agency, TR8350.2 (http://earth-info.nga.mil/GandG/publications/index.htm).

Phillips, J.D., 1997, Potential-field geophysical software for the PC, version 2.2: U.S. Geological Survey, Open-File Report 97-725, 34 p.

Plouff, D., and Pakiser, L.C., 1972, Gravity study of the San Juan Mountains, Colorado: U.S. Geological Survey Professional Paper, v. 800-B, p. B183–B190.

Quezada, O., Andronicos, C.A., and Keller, G.R., 2004, Structure of the Sangre de Cristo Mountains between Taos and Mora based on an integrated geophysical analysis: Socorro, New Mexico, New Mexico Geological Society, 55th Field Conference Guidebook, p. 257–263.

Rasillo, J., 2004, Gravity data repository and processing system: El Paso, Texas, Master's project report, Department of Computer Science, University of Texas at El Paso, 65 p.

Simpson, R.W., Jachens, R.C., Blakely, R.J., and Saltus, R.W., 1986, A new isostatic residual gravity of the conterminous U.S., with a discussion of the significance of the isostatic residual anomalies: Journal of Geophysical Research, v. 91, p. 8348–8372.

Society of Exploration Geophysicists, 1982, Bouguer gravity anomaly map of the United States (excluding Alaska and Hawaii): Society of Exploration Geophysicists, scale 1:2,500,000, 2 sheets.

Telford, W.M., Geldart, L.P., and Sheriff, R.E., 1990, Applied Geophysics: Cambridge, Cambridge University Press, 770 p.

Torres, R., Keller, G.R., Kreinovich, V., Longpre, L., and Starks, S.A., 2004, Eliminating duplicates under interval and fuzzy uncertainty: An asymptotically optimal algorithm and its geospatial applications: Reliable Computing, v. 10, p. 401–422, doi: 10.1023/B:REOM.0000032121.21617.38.

Wen, Q., Gates, A.Q., Beck, J., Kreinovich, V., and Keller, G.R., 2001, Towards automatic detection of erroneous measurement results in a gravity database: New York, New York, Proceedings IEEE Systems, Man, and Cybernetics Conference, p. 2170–2175.

Wessel, P., and Smith, W.H.F., 1995, New version of Generic Mapping Tools released: Eos (Transactions, American Geophysical Union), v. 76, p. 329.

Manuscript Accepted by the Society 18 June 2005.

Geological Society of America
Special Paper 397
2006

The Georgia Basin Digital Library for sustainable development: A digital library approach to addressing the societal relevance of geoscientific knowledge

Rob Harrap*
Queen's University GIS Laboratory, Queen's University, Kingston, Ontario K7L 3N6, Canada

Sonia Talwar
Department of Geography, University of British Columbia, 217-1984 West Mall, Vancouver, British Columbia V6T 1Z2, Canada, and *Geological Survey of Canada (Natural Resources Canada), 101-605 Robson Street, Vancouver, British Columbia V6B 5J3, Canada*

Murray Journeay
Geological Survey of Canada (Natural Resources Canada), 101-605 Robson Street, Vancouver, British Columbia V6B 5J3, Canada

Boyan Brodaric
Geological Survey of Canada (Natural Resources Canada), 234B-615 Booth Street, Ottawa, Ontario K1A 0E9, Canada

Ryan Grant
Institute for Resources, Environment and Sustainability, University of British Columbia, 2nd floor, 1924 West Mall, Vancouver, British Columbia V6T 1Z2, Canada

Joost Van Ulden
Shannon Denny
Geological Survey of Canada (Natural Resources Canada), 101-605 Robson Street, Vancouver, British Columbia V6B 5J, Canada

ABSTRACT

Digital libraries extend traditional library tools, collections, and metaphors in new and sometimes radical directions. Geoscience information, traditionally found in libraries in the form of maps, reports, books, and journal articles, can potentially gain enormous benefits from digital interfaces that allow geospatial information to be effectively searched and displayed in context. However, to many users of such a collection, the concepts within geoscience represent a barrier to use and understanding.

The Georgia Basin Digital Library (GBDL) Project examines community-based decision support for sustainable development, using digital geoscience information as a fundamental component. The context and relevance of geoscientific and other

*E-mail: harrap@geol.queensu.ca

Harrap, R., Talwar, S., Journeay, M., Brodaric, B., Grant, R., Van Ulden, J., and Denny, S., 2006, The Georgia Basin Digital Library for sustainable development: A digital library approach to addressing the societal relevance of geoscientific knowledge, *in* Sinha, A.K., ed., Geoinformatics: Data to Knowledge: Geological Society of America Special Paper 397, p. 35–51, doi: 10.1130/2006.2397(03). For permission to copy, contact editing@geosociety.org.

information in the GBDL is explained and represented through the use of semistructured representations based on architectural pattern languages, where a web of connected concepts are each contextualized using storytelling with supporting graphics. The GBDL allows multiple coexisting pattern networks, so that multiple perspectives on an idea, a situation, or a region are possible and in fact preferable.

The GBDL interface makes use of geographic information systems (GIS) components to show maps, such that the point of engagement for a user might alternatively be a story, a map, a photo, or a text heading. Linking the mapping component directly to the pattern framework allows for community-level contextualization of ideas and stories.

For geoscientists, the ideas in the GBDL are relevant both as novel approaches to sharing geoscience information with the public and as approaches that can be applied to specialized scientific discourse in the future. As the Web evolves, geoscience information systems and communication must evolve along with it; the GBDL shows one such evolutionary pathway.

Keywords: digital libraries, community GIS, Web architecture, information architecture.

INTRODUCTION

Geoscientific knowledge is highly complex. This is self-evident to geoscientists, who must deal with the complexity on a day-to-day basis and who have a high degree of training in the geosciences. The transformation in publishing that has followed the meteoric rise of the World Wide Web has had relatively little influence on how geoscientists share their knowledge to date; although many journals are available online, the structure of scientific papers and geologic reports has changed little over the last few decades. This structure is both a grounding for geoscientists, in that it is familiar, and a barrier for nonspecialists to use geoscience literature, knowledge, information, and data to any significant degree. Beyond the scale of publishing, the very nature of scientific education and practice represents a barrier to many community members with an interest in public debate; these "science-challenged" citizens are increasingly unheard as scientific knowledge becomes a prerequisite for public debate.

Web Information Architecture (cf. Wurman et al., 2000) has developed rapidly with regard to more informal means of sharing information, geoscientific or not. The original geoscience Web pages had explicit links to other resources, hand-coded and manually maintained. By the late 1990s the concept of a list of links was largely obsolete, a victim of the shear number of pages and the transitory nature of many Web addresses. The original geoscience Web pages were the equivalent of a word-processing file, with no explicit structure based on how information is used; instead, quality, style, and content varied widely. By the late 1990s, styles of pages had evolved, some by experimentation, and others based on the extensive literature on hypermedia that existed even before the Web itself (cf. Barrett, 1989). This general evolution provided necessary infrastructure for building more advanced forms of information display; this evolution continues at a rapid pace today.

The early geoscience Web pages, although furnished with hypertext navigation and so to some degree integrated into a larger body of knowledge, nevertheless mimicked print products then in existence—geological reports and scientific papers—that were already a significant barrier to nonspecialists and increasingly represented part of an information deluge even to specialists. This situation was made more complex than typical for the Web itself because of the nature of geoscience information. A key component of an information architecture for geoscience information is the map, which is a central requirement in the geosciences. Maps can range from sketches to highly detailed technical presentations using domain-specific symbology, terminology, and specialized graphic forms such as cross sections. Behavior of maps in a computer-mediated environment can range from static to highly interactive. The underlying technology that has evolved to build, store, and analyze maps in a digital environment is geographic information systems (GIS), which marry database functionality with computer graphics display functionality (cf. Bonham-Carter, 1995). To a geoscientist, the map is often the primary means to communicate geoscientific relationships, with the report essentially documenting the spatial information present on the map. This centrality must be taken into account with geoscience information architecture designs.

The main styles of Web pages that exist today, limited to those highly relevant to the expression and transmission of geoscience knowledge, include at least the following.

1. Static, topic-focused pages corresponding to a paper, memo, or article.

2. Dynamic, topic-focused pages built for a specific purpose by Web database technology.

3. Static map and content pages, using dynamic maps generated by hard-coded techniques where the maps are links to accessory information.

4. Dynamic map and content pages, using dynamic maps generated on the fly by Web GIS technology. The dynamic relationship can be links, but can also include direct manipulation of the map itself using limited GIS functionality (zooming, panning, legend expansion, and control over what content is displayed).

5. Mixed-focus pages, usually dynamic, where multiple content themes are mixed in the fashion of a newsletter or newspaper. These are often entryways for collections of single-focus pages as described above.

6. Discovery and linkage pages, always dynamic, where search technology and multiple content themes are mixed in a fashion that transcends any traditional media. These are often called portals, as they are doorways to knowledge exploration.

Translating traditional reports, maps, and journal structures onto these forms is relatively simple. Certainly all of these techniques have been applied to geoscience topics, whether on academic, government, or corporate Web sites. Yet the simplicity of the taxonomy given begs the question: can we do things with digital media that supplant what can and has been done with print media? Despite possible technical complexity, can anything be done with Web-based media that fundamentally advances how geoscience information is stored, distributed, shared, and understood?

To address this question, we need to examine who might reap the benefits of more widespread access to geoscience information and knowledge. Clearly, highly sophisticated Web-based resources for advanced geoscience use can be built and in fact have appeared or are under construction at this point (cf. GeoConnections, 2003). Some existing Web-based resources are also intended to support geoscience education, albeit in a more indirect way than proposed here (cf. DLESE, 2003). However, a different approach would be to attempt to widen the audience for geoscience information by breaking down the complexity barrier of geoscience information. This goal is arguably more difficult than building expert-level tools but is also arguably more important, especially in securing a place for geoscience knowledge in the wider understanding of decision making in an increasingly complex world. This might in turn lead to a wider appreciation for the significance of geoscience information in understanding pressing societal issues such as natural hazards, global climate change, and pathways toward a sustainable future. Geoscience information is often overlooked precisely where it is most needed because of the complex interplay of spatial, temporal, and descriptive styles of information presentation that geoscientists use; bridging this gap for interested members of the greater community will help to bring geoscience issues into the public debate.

THE GEORGIA BASIN DIGITAL LIBRARY PROJECT

The Georgia Basin Digital Library Project aims to provide a conceptual framework for specific and pragmatic attempts to explore dynamic digital libraries in the context of sustainability. A specific trial application, GB-Explorer, has been constructed and is described in detail below. The overarching goal of the Georgia Basin Digital Library (GBDL) and hence GB-Explorer is to promote an awareness and understanding of issues related to sustainability (Bruntland, 1987) and sustainable decision making at the community level.

The GBDL Project also attempts to address the question of how to best present geoscience information to community members in aid of decision making, greater understanding, and a shared sense of place and purpose. GB-Explorer thus represents a fusion of Web information architecture, GIS, computer science, community mapping, and cooperative work techniques with a goal being access and support for geoscience knowledge integrated decision making.

Digital libraries (cf. Marchionini and Fox, 1999) are a response within the library and information science community to the anthropological need for library resources and library mediation in a digital world. Existing digital libraries span the entire technical range from static document collections to entirely dynamic, multitheme presentations. Like the physical libraries that they mirror, digital libraries mix the social need for information archives with the need for directed search, information and knowledge discovery, and community focus. Just as a library patron doesn't necessarily know exactly what they are looking for when using a physical library, digital libraries support both directed search and retrieval and less directed exploration and discovery.

The GBDL reflects these needs, at least within the mandate of a sustainable decision-making focus. In particular, the project emphasizes discovery and user services of directed search and archiving. This recognizes the perceived audience of the GBDL as not being overly familiar with advanced Web technology and having little interest or capacity to understand technical material directly or to use raw data at all. The design goal of the GBDL was as a result based on building prototypes and testing with user communities for usability and effectiveness rather than developing academically interesting advanced functionality.

Finally, the GBDL does not exist in a vacuum. The scope of the project and the identification of a specific user community arise from a preexisting project at the Institute for Resources, Environment, and Sustainability, University of British Columbia. The Georgia Basin Futures Project makes use of a combination of spatial and attribute data archives, integrated process models of environmental and human systems, and modeling functionality to provide the capacity to represent how the future development of a region is linked to fundamental policy decisions made at a local and perhaps community level. This functionality has been expressed in a software application called Quest™. Quest™ operates much as a video game, reaching a very wide audience in principle, but in practice limited because the conceptual framework of what the policy trade offs, natural system constraints, and modeling consequences are for the most part lacking in the target audience. The design scope of the GBDL was thus to build a companion digital library that provided contextual support for Quest™ users who might have great desire to understand sustainable future alternatives but little background in the issues.

Since the use of Quest™ is intended to be community driven, this provides an additional requirement that the GBDL be non-prescriptive, acting as a community library where content may originate from community groups with radically different goals and perspectives.

Knowledge Requirements of the GBDL

The design of the GBDL, and in particular the exploration and discovery component of the library as constructed in GB-Explorer, are informed by an appreciation of the nature of geoscience knowledge and of the conflicting demands of human users versus computer servers. The project attempts to take fundamentally new approaches based on this perspective; the cost of this is that the GBDL represents something of an anomaly in terms of performing traditional comparative analysis of tools and techniques.

The fundamental goal of the GBDL is to provide mechanisms for users to explore concepts that are to some degree community and geographically situated in the Georgia Basin and that contribute toward a dialogue about sustainable development. The process of exploration thus needs to be a balance of directed and nonprescriptive, so things can be found and different perspectives can be explored. Furthermore, the process must explicitly recognize that the main challenge to many explorers will be a lack of framework knowledge; indeed, this is the historical problem that the GBDL seeks to address in support of the Quest™ scenario modeling experience. How can we support people learning about a region, without a preexisting understanding of the concepts themselves and without providing a formal structure that is tantamount to a structured education in subjects from a prescriptive vantage?

This amounts to two problems, which will be addressed in turn.

1. How do we support learning in the absence of existing conceptual frameworks?

2. How do we support multiple perspectives on topics?

Consequently, a series of secondary goals arise.

3. What other points of engagement are needed for the GBDL to comprise a community-oriented digital library?

4. What technical advances and challenges arise in building such a library and how can these be addressed while keeping the tool both accessible to groups and manageable to maintain?

5. How can the goals of the GBDL and Quest™ reinforce each other?

The remaining portion of this paper addresses these key problems, highlights the state of the GBDL prototype, GB-Explorer, discusses the technical framework for implementation, and draws preliminary conclusions about the key project outcomes.

The range of topics introduced above—from geographic information systems to Web information architecture—illustrates the range of complexity both in designing and, in turn, describing the construction of the GBDL. What follows is not exhaustive but is intended to illustrate some of the design goals and solutions that the project involved. If possible, it is recommended that the user examine the online result (http://www.georgiabasin.info) to see the dynamic nature of the result, which of course cannot easily be captured on paper!

Significance and Setting of the Georgia Basin Region

The Georgia Basin region, shown in Figure 1, comprises the drainage basins of the Strait of Georgia and includes the cities of Vancouver as well as many smaller communities ranging from suburban to rural. The Georgia Basin occupies nearly 46,000 km^2 of land and sea populated by nearly three million people of diverse ethnic background. The region is highly mountainous, with significant earthquake, landslide, flood, and wildfire risk, and the terrain places extreme constraints on land-use planning and places true tradeoffs on most community decisions. The population is expected to grow substantially over the coming decades, and this combined with the planning constraints provides an opportunity to examine how decision making at the community level can be influenced by notions of sustainable development and an appreciation for the relevance of earth science knowledge to problems within the human sphere.

LEARNING FRAMEWORKS FOR DIGITAL LIBRARIES

Constructing a digital library involves many layers of design; while the technical (software and hardware) layers typically receive much attention, the ultimate success or failure of a system is more profoundly influenced by the attention to the relationship between user and interface, or more properly the attention paid to human-computer interaction. In the case of the GBDL, this is doubly important given the focus on supporting learning for a wide-ranging audience.

Geoscience knowledge is much like knowledge in other domains in that it reflects a very complex interplay between abstract concepts, concrete concepts, observational data, and tacit knowledge about how these may be meaningfully tied together. The process of *becoming a geoscientist* is to a large measure a process of acquiring the concepts, the methodologies, and the tacit knowledge about the discipline; for nongeoscientists to use geoscience information is difficult since this background is lacking. While extensive supporting documentation, written at an appropriate level, may capture some of the needed context for understanding specific concepts in the realm of decision making, knowledge capture by domain experts such as geoscientists is notoriously difficult (Brodaric and Gahegan, 2002), especially for tacit knowledge.

Furthermore, even if some contextual information is present, a recursive problem of understanding typically arises. Called the scaffolding problem (Bransford et al., 2000) in the education literature, the issue arises that concepts are built on other concepts, such that without a firm place to start, it is very difficult to meaningfully interact with learning materials. The solution in education is to build structured learning in such a way that increasing complexity and conceptual depth is built in stages, as a hierarchy, with the concepts used in each successive stage having been built by earlier lessons. This can be visualized in the context of a Web environment as a staged series of pages that build toward successive levels of complexity. Seen from the perspective of a concept perhaps not understood by a user, this would be reflected as a series of suggestions of starting-point concepts for understanding, in essence a structured dictionary or encyclopedia.

In addition, there is a highly complex spatial relationship between geological and geographic concepts and space; without a deep appreciation for the spatial context, the scope of concepts in the real world is unclear. For example, knowing that a landslide can occur without an appreciation for where it is likely to occur has little use; furthermore, building an appreciation for why landslides occur allows a learner to generalize to new situations. Understanding how spatial context influences the behavior—the "why"—is a key. As a result, a concept-building tool that addresses the scaffolding problem for geoscience information needs explicit support for establishing the spatial context of concepts, information, and data.

At the most basic level, concepts exist as some form of succinct (or explanatory) description, perhaps with spatial context in some form, and linked both to other dependent concepts and to increasingly detailed, more sophisticated conceptual frameworks. The framework, in this case, is simply the linkages made explicit. Such a network of concepts is called a *semantic network* (Stefik, 1995), in that the network directly captures meaning to as great a degree as possible. Obviously, a particular semantic network exists from a specific perspective, which reflects how reality works, what is important and not, and what spatial context is relevant. This point is challenging to the normal, implicitly positivistic tone of scientific research, in that it suggests that there may be different perspectives on, say, geology that ascribes meaning in different ways and support different interpretations to real-world phenomena. This is, on the other hand, a key requirement of a community-engagement tool; the tool must support perspectives and even ideologies that are not in accord with contemporary scientific values. In fact, in an open library, many of the perspectives may be outside the experience of the creators of the library; enabling this openness is what builds trust that a system is *for the users* rather than an authoritarian instrument.

The Structure Tradeoff

Conceptual frameworks built out of semantic networks have the advantage of making relationships explicit, supporting the idea that knowledge exists as a web of connected knowledge (Fig. 2). The approach of explicit semantic network representation also goes some way toward addressing a basic issue in the design of any dynamic information system: the tradeoff between representations that are powerful for humans and ones that are tractable for computers.

As an example, a poem can have immense human meaning but it is unclear how a computer would address it except as a

Figure 1. The Georgia Basin Region of southwestern British Columbia, Canada. The area outlined in red—the drainage basin of the Straits of Georgia and Juan de Fuca—comprises the focal area of the Georgia Basin Digital Library; topics range from natural hazards to fundamental earth science, but the overarching goal is to support participatory decision making within the geographic region. Note that the Georgia Basin Digital Library Project addresses issues in the Canadian portion of the region. Image courtesy of Environment Canada.

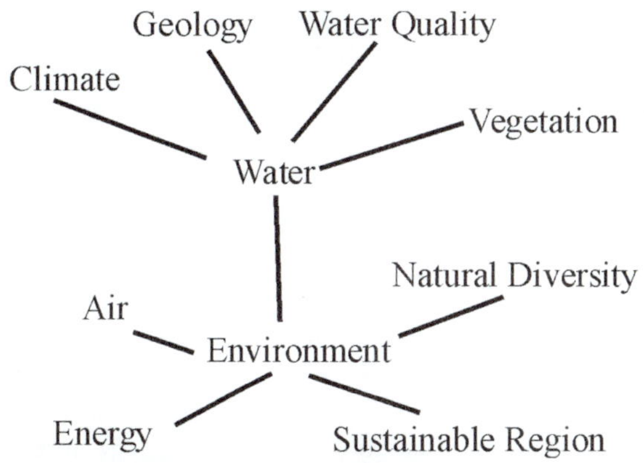

Figure 2. Fragment of a Sustainability Pattern Language. The concepts are structured to support learning increasingly detailed concepts; many arrangements of the same content might be possible, reflecting different perspectives on what is important.

simple block of text. Similarly, a concept description articulating geoscience knowledge to a GBDL user might exist as a block of text and images. This could, from the human perspective, exist simply as a large number of hand-linked Web pages, ruling out computational support and dynamic interface behavior. On the other hand, a highly structured database representation allows for flexible presentation at different levels of detail, computational comparison of different community perspectives, and the like but runs the risk of being inaccessible *because users lack not only the subject-oriented framework for understanding, but also the tool-oriented framework.*

What is needed is a format or structure for concept representation and description that links well to the idea of scaffolding-building semantic networks but has at the description level enough additional structure to support dynamic presentation, database support, and some measure of analytical possibility. This format must also be sufficiently simple and powerful that community members can use it easily yet profitably to add content to the digital library system; a format or system with a proven track record in community-based decision support is ideal.

Pattern Languages

Such a systematic format does exist, developed during the 1970s as an aid in community and individual architectural design. The original pattern language captured the essence of community-centered town, building, and room design in a series of short patterns, with a definite structure and yet expressive flexibility. The overall structure of a pattern is:

1. Title and Ranking: A short, expressive title, plus a subjective ranking on how often this particular pattern appears in a target audience. In the original pattern language, the target audience was the world. Titles are accompanied by maps, sketches, or photographs to establish context.

2. Problem Statement: A succinct statement of a real-world problem, using as little domain-specific terminology as possible.

3. Problem Discussion: A multimedia discussion, usually in a somewhat storytelling format, of what the problem context is, how different groups have addressed it, and what some solutions might then be.

4. Solution Statement: A community-sanctioned suggested solution to the stated problem, again in as simple language as possible.

5. Links To and From: Explicit hypertext links to other patterns that either embrace the current pattern or are at a greater level of detail.

Note that although the original pattern language (Alexander et al., 1977) was published long before the World Wide Web, it nevertheless contains paper-based indices that function as hypertext links forming a basic network of descriptions. In the original pattern language, approximately 250 distinct patterns that address participatory design from a scale of countries (the first pattern) to elements within a room (the last pattern) are given. Each pattern contains all of the elements above. The effect is to provide an interface to knowledge that, after experimentation of a few minutes, is remarkably fluid despite being presented on hundreds of pages of a hardbound book. The original patterns were discovered through participatory planning by a large number of graduate students who traveled the globe, living and building within a wide range of communities in order to develop a global perspective on design. Although the scope of the GBDL is more modest, the overall theme is similar.

Since their inception, pattern languages have been applied to a range of other problems, including software architecture (Buschmann et al., 1996), user interface design (Borchers, 2001), and (in parallel to this project) sustainable development (Ecotrust, 2001). They have had the largest impact in software architecture largely because the method of developing patterns brings human focus into software design, and this had been largely in the past.

From the point of view of building the GBDL, then, pattern languages match the requirement of a scaffolding-building framework that mixes structure with freeform storytelling in a way that is amenable both to human consumption and computational support. The network of linked concepts is explicitly a semantic network structured to lead from the general to the specific. The use of mixed media support such as sketches, maps, and photographs provides spatial context, familiarity, and the opportunity for a community to personalize contributions.

The Georgia Basin Digital Library includes a core exploration component based on semantic networks and pattern languages, and significant underlying support exists both for the concept-and-relation structure of data and knowledge in the system. The implemented system is described below. However, the crucial question is: how do we support multiple perspectives on topics?

COMMUNITY FRAMEWORKS FOR DIGITAL LIBRARIES

There is an enormous body of academic work on supporting cooperative decision making (cf. Beaudouin-Lafon, 1999). An entire subfield of Geographic Information Science, called Public Participation in GIS (PPGIS) is thriving, and numerous publications document attempts to develop community-centric GIS tools and databases (cf. Craig et al., 2002, Haklay and Tobon, 2003). However, given that the Georgia Basin Digital Library is primarily intended to support personal discovery and to overcome the scaffolding problem, the scope of community-framework support in the GBDL is limited to support within the context of a semantic network and pattern-language–enabled architecture. The consequences of this are discussed in more detail later in the document.

The two components described—pattern languages to capture concepts and semantic networks to capture interconcept relationships—represent a framework for knowledge construction by an individual or coherent group. The pattern-language process (Alexander, 1979) includes mechanisms for community decision making, but the final pattern language is intended to be integrated rather than multi-viewpoint. However, in a multi-dimensional environment for decision making like the Georgia Basin region, there is no hope that a single coherent point of view can arise; in fact, such an outcome is not socially desirable (cf. Honadle, 1999). The GBDL must be capable of showing multiple points of view.

The simplest approach to this is to allow the development of any number of parallel pattern language systems, some or all of which may refer to a common collection of geoscience and economic data, maps, and other resource media. However, perhaps the richest opportunity for community learning that can exist is where these languages can refer to each other for better or worse, building a higher-order network of interconnections and commentaries. Exactly this multilevel commentary and metacommentary framework was in fact part of the original specification for a knowledge system, known as the Memex, that provided the architectural blueprint for the World Wide Web (Bush, 1945; Berners-Lee, 1999)

In his seminal article, Bush (1945) pointed out that in the long-term development of any community library—he was interested in the community of scientists—the linkages between items that are built by users either directly or by navigating through the library comprise an important secondary network. Such a network transcends the conceptual level of individual resources (concepts, linkages, media) and in fact given enough development emerges as an overarching, community-wide framework for shared understanding and values. In other words, in the long-term development of a conceptual network-based digital library, it is possible to have linkages between concepts *emerge from interactions* rather than being explicitly constructed. As a conceptual network gets very large, explicit construction of relationships stops being an option: witness the disappearance of hand-constructed links pages on the World Wide Web by the late 1990s.

At a pragmatic level, such a conceptual network framework and the computational infrastructure to support it offers several possibilities.

1. A means to build semantic networks through an emergent process rather than by explicit linkages. As a corpus becomes very large—say, the entire historical geoscience library of a geological survey organization—such emergent linkages are the only possible way to maintain a framework, since no individual can hope to embrace the entire contents and contexts of the library.

2. An opportunity to build visualization or other representational tools that specifically show how community frameworks differ, in aid of building shared understanding. Understanding the degrees of difference between different communities is central to understanding how compromises and flexibility are needed in multi-stakeholder decision making.

3. An opportunity to investigate the design of reactive and adaptive interfaces where the dynamism of the interface responds to conceptual clues rather than syntactic or explicitly coded cues. As an example, many sales Web sites use adaptive interfaces to track the conceptual similarity of users and so focus marketing efforts based on user profiling. Such profiling could be done as a way to either direct learning in a digital library (focus in) or to deliberately misdirect navigation (focus out) to promote a sense of opportunistic discovery.

The design architecture of the Georgia Basin Digital Library recognizes these opportunities as supporting justification for a strong computational foundation to the digital library, discussed below.

TECHNICAL REQUIREMENTS OF THE GBDL

Given the conceptual and community-support architecture described above, a number of technical issues arise relating to Web design, integration with other Web resources, and underlying technical framework.

Client-Server Architecture of the Web

The World Wide Web is built on a client-server architecture, where relatively few servers distribute or create content to a much larger number of clients, usually known as browsers. The workload in such a system is divided between tasks on the server and tasks on the client; historically most of the load was placed on the server. When building a Web-based digital library almost any interface tool for dynamic behavior can be built to run "on the server" or "on the client." The primary concern is that the amount of information transmitted to the client should be minimized, which reflects the historical and to some degree current need to transmit across relatively slow network channels such as dial-up modems. The tradeoff here is that any interaction on the client must be transmitted back to the server and processed there, often at a significant time cost. Highly interactive interface elements—such as semantic network diagrams that move freely under user control—therefore *must* be on the client side to ensure responsiveness.

Web Interface Architecture

Web design involves significant tradeoffs between performance, universal access, and system cost. First, any system that adds significant functionality to the basic Web server and browsers runs the significant risk of only being compatible with some systems. Web standards are a moving target and often used as a means of intercompany warfare, where adherence to specific standards is used as a means to procure exclusive use. In essence, if a system will only run with one browser or one Web server, this may in the long run be a serious issue in terms of the number of users and the number of agencies interested in using the architecture.

Second, even the most benign enhancements to basic Web functionality run the risk of limiting use. One of the technical advantages of the HTML language for Web design is that it separates content and structure of a page from specific display parameters: what you see is not necessarily what they will see. The advantage of this is that some groups, such as the visually challenged, may override appearance settings to make a resource more useable. Enhanced Web interface components typically do not allow this, and such users are then left incapable of using the site.

Finally, many Web design elements, especially on the server, are very expensive. This cost is not seen by system users, but if an overall architecture is to be distributed to community groups for use, the cost of setting up a server then becomes prohibitive. The tradeoff is that many of these design elements are very useful, especially in specific roles such as the dynamic display of maps, semantic networks, and text.

Services

Just as standards exist for Web content, standards exist for the negotiation of information sharing between dynamic Web sites. For example, there are many Internet servers that provide live news feeds or news stories to other servers. The process of discovering and recovering resources from such sites depends on service protocols. A digital library site that is to be connected to the larger world of information resources needs to support such protocols both as a consumer, to go and get news feeds, and as a provider, where other digital libraries access content transparently.

Web GIS Technology

Web mapping technology is far from mature. Both the Web architecture elements and the service protocols are rapidly evolving. Approaches to putting map content on a page range from displaying static images of maps to generating images from map data on a server to providing GIS data to the client browser to be displayed and perhaps analyzed. The tradeoffs between cost, universality, and service protocols described above apply in full to this process.

In addition, analytical support is largely lacking in Web-based GIS at the current time. This means that a site that will include not only maps as context, but also opportunities to do significant analytical work with maps must link the Web server to a full-featured GIS toolkit.

The development of the GBDL took place over a number of years (principally 1999–2003). During this time, the Open GIS initiative (OpenGIS, 2005) developed a range of standards and compliant software tools that combine the services and Web GIS approach used in the GBDL. Where possible, the GBDL was made compliant with these standards. Were the system being developed again, the OpenGIS approach would be followed more closely.

Database Technology

Dynamic Web sites typically rely on database servers running side-by-side with the Web server. Different database servers support different approaches to data storage. The key to supporting the requirements described above rests in using database systems to record knowledge beyond data. This is particularly critical for geoscience in that geoscience extends beyond data collection to include interpretation of landmass through the development of new concepts, theories, and explanations about the spatio-temporal–process evolution of the landmass. Databases must therefore record not only collected data but also the interpretations that give meaning to the data, including the relationship between the data and the interpretation. To achieve the management of concepts, theories, and explanations in the database requires the adoption of novel approaches to database schema design, e.g., as in the relationship between concepts, conceptual systems, and data (see Brodaric and Hastings, 2002; Brodaric and Gahegan, 2002; this volume).

Although such designs are technically feasible, the sophistication of such a database architecture raises significant problems for community groups that may implement their own server. Furthermore, it raises potential difficulties in loading traditional data into the server database but possesses significant advantages in terms of contextualizing traditional data through more explicit representation of their interpretation.

Metadata

Metadata is data about data. Metadata provides not only a documentary trail to what resources exist on a system, where they came from, how they differ from other resources, and so on, but also supports very sophisticated searching since much of the content of metadata is in effect extended information about the usefulness of the data. Similarly, metadata can exist for conceptual material such as the patterns described earlier and are in effect part of the pattern itself in this case, albeit part that the user might not need or wish to see.

In terms of the design of a digital library, metadata is important at the level of resource finding and sharing and also at the level of tracking the evolving contents of the library. Significant standards for digital library metadata are evolving (e.g., Dublin Core, 2003) and are tied to the service protocols used in a system, to the database technology used, and to the type of data being described (e.g., FGDC, 2003).

IMPLEMENTATION OF THE GBDL DESIGN FRAMEWORK

The original design goals for the Georgia Basin Digital Library as discussed above evolved into a concept document (Harrap et al., 2000) that was used as to frame design decisions. Several iterations of the semantic-network–based interface were designed and tested, and additional modules were added to fill out the required suite of digital library tools. The technological architecture underlying the GBDL is discussed in Brodaric et al. (2003).

The current online application, known as GB-Explorer (http://www.georgiabasin.info), comprises five modules accessed via a home page, tabloid newspaper style, with dynamic content reflecting current events in the region, providing navigation to the other modules and additional information on the project and application. The modules are:

1. Ideas & Perspectives, the semantic-network and pattern-language–based exploration and learning interface.

2. Local Stories, a community perspective Web-mapping interface including online pattern entry tools.

3. News & Information, a services portal linked to online news feeds keyed to key sustainability and regional development issues.

4. Library Collections, a metadata-based data and resource search and discovery tool for advanced users.

5. Future Scenarios, a linkage to the Quest™ scenario modeling application.

Together these comprise an integrated digital library interface built to support online discovery, learning, and community-information–building and sharing. The modules are described in detail below, followed by a technical discussion of the implemented architecture.

The GB-Explorer Home Page

The tabloid-newspaper style GB-Explorer home page is a typical multitheme, multitool Web home page (cf. Nielson and Tahir, 2001). The interface is intended to both welcome new users and provide interesting contemporary links for returning users. News stories including photographs and headlines are provided. As shown in Figure 3, the main modules are accessed through a menu bar and multimedia tutorials and site help are available.

The traditional Web-homepage nature of the entry point provides consistency with other Web sites, provides a point of entry where the project and nature of the other modules can be introduced, and provides a spotlight on current issues.

Figure 3. GB-Explorer Home Page, showing links to the main modules and current sustainability issues for the region.

Ideas & Perspectives

The semantic network and pattern language architecture described in detail above, intended to address the scaffolding problem in knowledge discovery and concept exploration, is implemented as the Ideas & Perspectives module of GB-Explorer (Fig. 4). In the prototype implementation, two contrasting pattern languages are provided, one for the region of Bowen Island off the coast of Vancouver, British Columbia, and another linking an existing sustainability pattern language to regional data. This existing pattern language, developed by Ecotrust, an environmental nongovernmental agency (Ecotrust, 2001), is richly populated with stories and illustrative media. It was developed using the community pattern language development process described above and so is emblematic of the type of content the GBDL should include as it grows.

The interaction style in the Ideas & Perspectives module is based on exploration of the semantic network, which is a client-side Java-language applet (Thinkmap, 2003) that links via a database to server-side storage of pattern-language structured stories. The patterns are in turn linked to specific map displays using a server-side Web-mapping application. The result is that clicking on an item in the semantic network loads the corresponding pattern and the corresponding contextualizing map. Simple GIS interface tools allow browsing of the map, inspection of the legend, and zooming to various levels of detail. The pattern frame can contain hyperlinks to other Web resources and to documents in a digital library document archive. The linkage between the map and the semantic network of sustainability concepts reinforces a view of mapping as a means of communicating ideas within a cultural and political context (Barnes and Duncan, 1992).

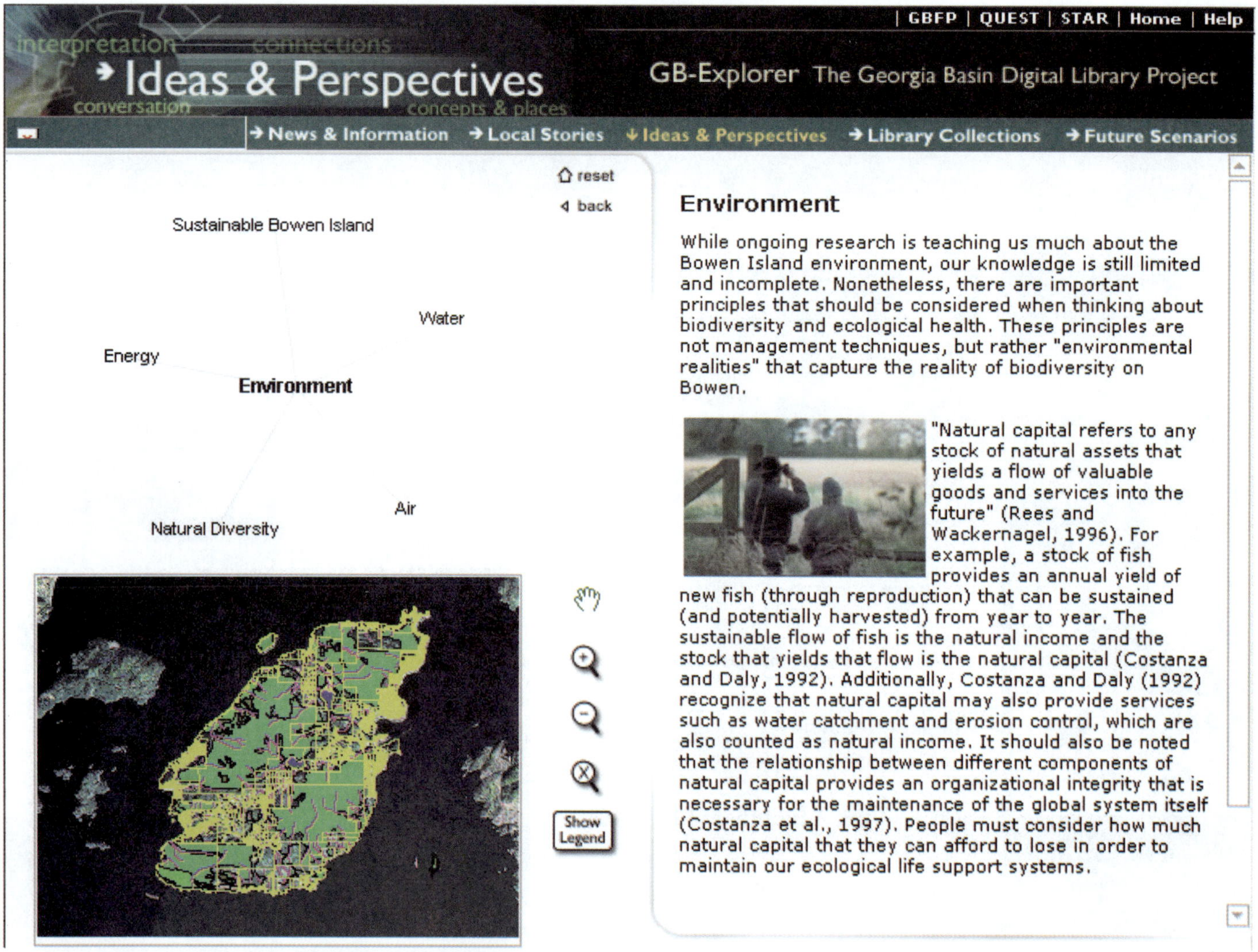

Figure 4. Ideas & Perspectives module, combining Pattern Language content with semantic network knowledge representations and Web GIS.

Since exploration paths in the Ideas & Perspectives module are user driven, the user can self-monitor what level of detail is appropriate, moving on to more complex material or to related concepts as needed. This interface style, as discussed above, is in principle capable of simultaneously showing different perspectives, though this has not yet been implemented.

Local Stories

The Local Stories interface mixes community mapping with online creation and sharing of views situated in a community (Fig. 5). It is a means of reflecting on priorities and of sharing a personal or collective connection to specific places in a community and places that are of personal or shared significance. The interface supports group-based story collections attached to specific locations in a community; the architecture is based loosely on the pattern language style described above but currently does not enforce adherence to the exact structure. The necessity of a group basis is based on the need for some level of control of content for appropriateness and relevance. By using community leaders as auditors the GBDL team avoids a prescriptive approach to control of content.

The Local Stories module is a first step toward constructing a community-centered, shared-experience mapping environment. The goal in the long term is to embrace both the structure of the environment of the Ideas & Perspectives module, which via the underlying architecture provides significant aids to discovery as discussed above, with more personal mapping technology and styles. The module is evolving as a result of collaborative work with community groups; in the long term it may become a separate community narrative module (based loosely on the metaphor of a shared community album) and a creation and editing interface for the Ideas & Perspectives module. The narrative interface is arguably the most important in building community support, but

Figure 5. Local Stories module, showing community-based data entry on bird observations on Bowen Island.

linkage to the more technical information is also crucial if that information is to become contextualized by communities.

One of the key problems of how to realize sustainable development is how to shift community awareness, thinking, and practices to allow dialogue around community development decisions such as land-use planning and growth management. The Local Stories module uses the Green Map System of icons as a way to identify spatially significant sites of interest on a geospatial base such as a digital orthophoto or vectorized base map of the region. The Green Map System is a globally connected, locally adaptable ecocultural program for community sustainability (Brawer, 1995). The system includes 125 green map icons that capture the interconnection between nature and the built environment. The icons, mapped on to a location, identify a wide-ranging assortment of features such as an ecospiritual site, a museum, a local artisan, a bike path, a community garden, bird or wildlife habitat, a fair trade shop, or a farmers' market. While there is general agreement on the meanings of icons, local mapmakers are encouraged to determine a precise definition for their purposes. The Green Map System offers a well-used set of icons that could easily be applied to categorize and represent residents' experiences.

News & Information

The News & Information module provides access to current news stories categorized according to issues related to sustainability at the Georgia Basin and on a national and global scale. The news stories are gathered from online news services using standard services. Rather than providing a full search interface, the News & Information interface uses predefined scales and keywords to dramatically simplify the interface. Since the content is retrieved on the fly when the module is accessed, the content is always up to date.

The prototype of the News & Information module accesses two online news services: (1) Tidepool (http://www.tidepool.org) and (2) Moreover (http://www.moreover.com). The Tidepool service provides a bioregional online news service that selects stories deemed to contribute to a conservation-based economy, drawing on a feed of over thirty worldwide providers. The Moreover service is more extensive, accessing over 4000 news sources. These together provide a combination of sustainability focus and global coverage. Since the categorization process happens in the GBDL interface but the content generation rests with qualified news, science, and current affairs reporters skilled in providing material for public consumption, the News & Information module provides a range of focused content of consistent quality.

The News & Information module also provides an interface for community groups to post links to online content for inclusion in the current news collection. This process is vetted through the GBDL team, though in principle it should be based on the same community-group–based process described for the Local Stories module.

Although the News & Information module is not strictly part of the concept exploration and discovery mandate of the GBDL, it provides an essential link to the wider Web of community, national, and global knowledge, with content scaled to an appropriate level for community consumption (Fig. 6). In principle an integration of current news stories into the Ideas & Perspectives interface would further highlight the broad linkages developed in the News & Information module.

Library Collections

The GBDL Project's premise is based on the integration of information to support exploration, discovery, access, and the generation of new ideas. While much of the project focused on extending relationships between Web-based mapping capabilities, semantic Web browsing, and the integration of the narrative element, a fundamental component of a knowledge resource is information and data content. With an increase in information technologies and the development of digital libraries elsewhere, the research team was more interested in supporting access to existing content than adopting a centralist approach to gathering, structuring, and organizing existing data sets. The primary purpose, then, of the Library Collections module is to provide access to existing Web-enabled information sources in a distributed networked environment (Fig. 7).

The functionality within the Library Collections module is based on compliance to metadata standards used to describe spatial data sets. National agencies such as the Federal Geographic Data Committee in the United States and Natural Resources Canada in Canada have begun the process of developing a flexible spatial data infrastructure; this involves both the creation of framework data for universal access and adherence to standards that support online search and discovery. The functionality of the Library Collections module is compatible with the Canadian Geospatial Data Infrastructure (CGDI). The underlying software tools conduct a distributed search of compliant information resources on the Web and return a combination of metadata and dynamic map layers to the interface.

Although the Library Collections module is likely to be of interest to only the most sophisticated users of the GBDL, the purpose of the library is to educate users to the level where they can both use such resources effectively and, in parallel, use the Quest™ game described above. The lack of a rich explanatory framework is understood—that is the role of the Ideas & Perspectives module. This module thus has the potential to be a door into data use and operational knowledge, and the Library Collections module provides the corresponding data access point.

Future Scenarios

As discussed above, one of the primary goals of the GBDL is to provide a learning environment about sustainable development and geoscience issues to users of the Quest™ scenario modeling tool. The intention of the Future Scenarios module is to be the

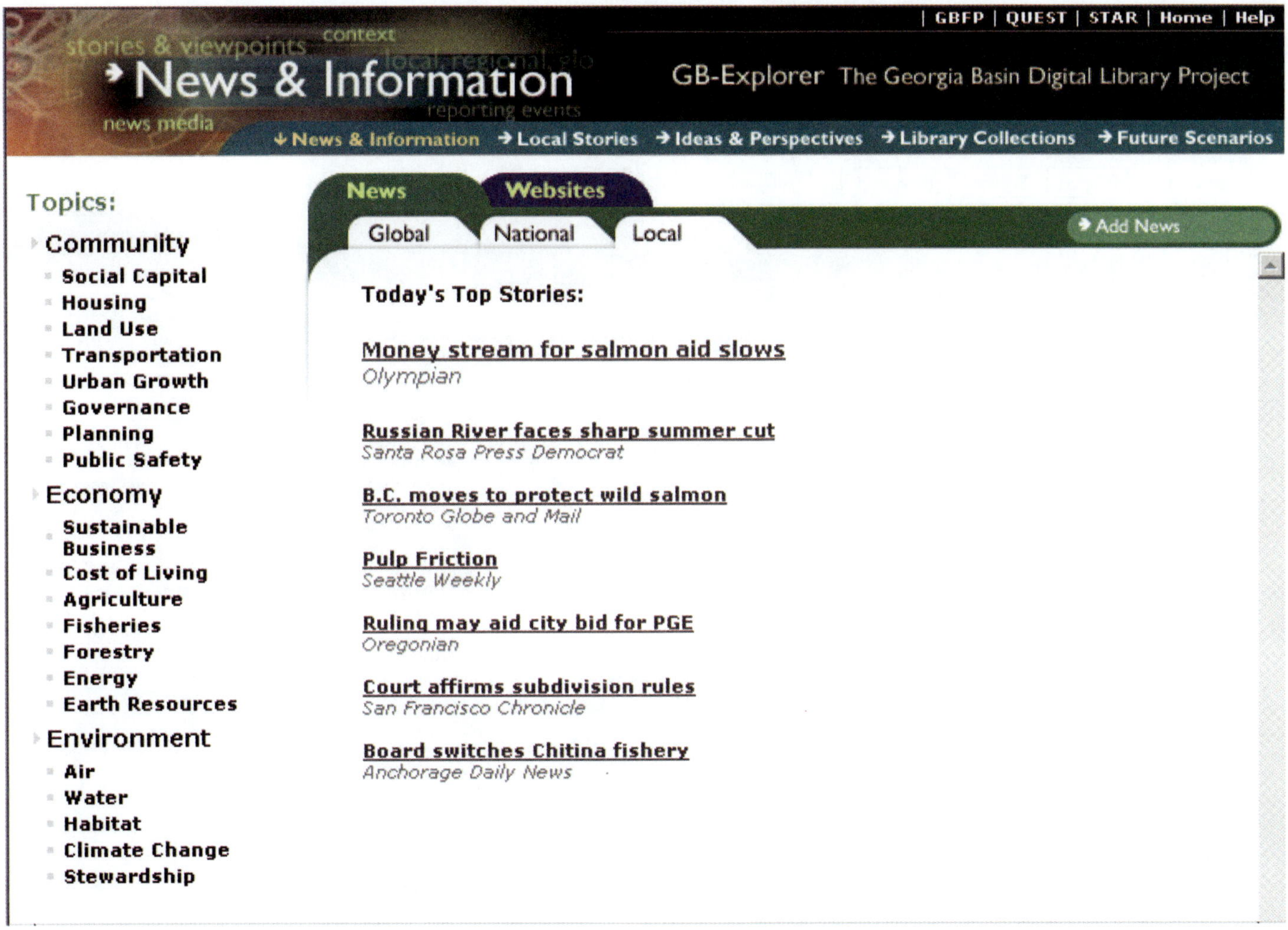

Figure 6. News & Information module, highlighting Web-services–based newsfeeds categorized by relevance to sustainability issues.

bridge between the applications, wherein Quest™ users can be brought to an appropriate spot in the GBDL or users of the GBDL can explore the consequences of ideas found in the Ideas & Perspectives module in the Quest™ game. The interface between the two applications and approaches has been defined at a preliminary level but the necessary linkage points within Quest™ have not yet been constructed. At the moment, the Future Scenarios module thus represents a tutorial on the Quest™ tool rather than an active bridge (Fig. 8).

DISCUSSION

The completed Georgia Basin Digital Library has been demonstrated and used with target communities and audiences for over a year. Initial response has been very favorable, and as a result of both development learning processes and audience response a number of issues have arisen. These include the following.

1. Does the knowledge discovery approach used in the Ideas & Perspectives module appeal to novice users, or is it in effect a "simple expert interface?"

2. Is the framework of the GBDL sufficiently general that it could be applied to a very different domain?

3. Is the framework sufficiently robust and cost effective that community groups could implement stand-alone versions of the GBDL?

4. Are there significant missing modules that would provide needed functionality?

5. Is the underlying architecture sufficiently powerful and robust to support more advanced interface modes, such as semantic networks generated on the fly and between different community pattern languages?

The Georgia Basin Digital Library Project is ongoing, and we are seeking to address each of these issues as part of research, development, and community outreach activities. In particular, ongoing work includes the examination of fundamentally

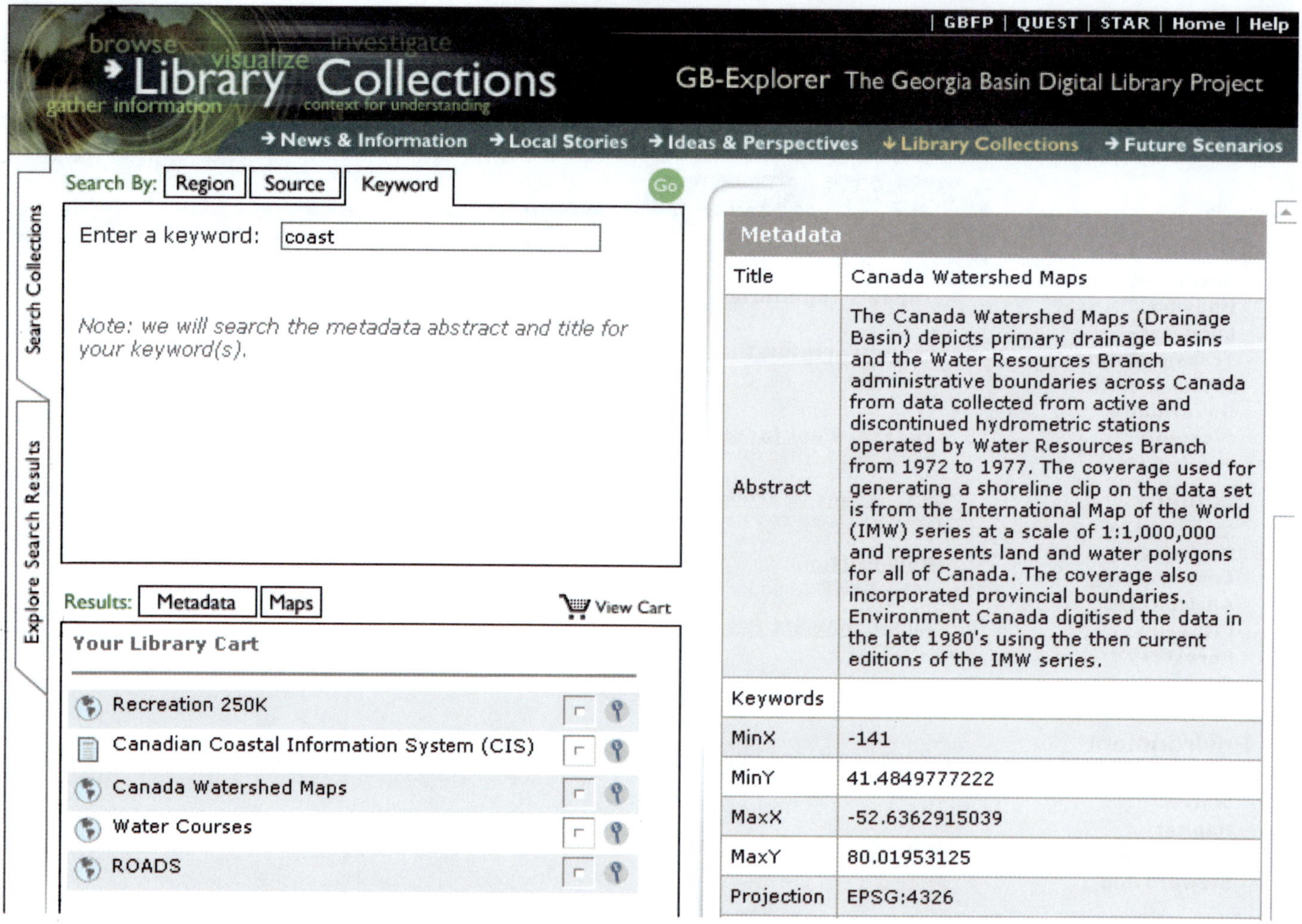

Figure 7. Library Collections module, showing metadata search for coastal information data sets available from Canadian Federal servers.

different interface modules both for experts and for novices. These include scientific visualization interfaces, three-dimensional GIS, and adaptive interface approaches.

Relevance to Geoscience

A key issue, related to the generality of the GBDL, is the relevance of the GBDL to traditional geoscience. There are several key issues here, including the following.

1. To what degree is such community-centered outreach a key requirement for the survival of academic and governmental geoscience activities?

2. To what degree is traditional publication to be replaced with dynamic, Web-based forms of communication?

3. Is the GBDL approach, perhaps with additional advanced interface elements, useful to geoscientists themselves?

There is certainly a groundswell of support for the idea that geoscience must be responsive to the needs and interests of citizenry. This is demonstrated by fundamental changes in geological survey organizations both in Canada and the United States in the late 1990s and early 2000s (Frodeman and Mitchum, 2000; Lubchenco, 1998). Information access points such as the GBDL represent a point of connection between the community and academic and governmental geoscience that addresses this general trend.

The transformation of publication in the Internet era is a subject of substantial debate in academic circles. Both the conceptual framework for the World Wide Web (Bush, 1945) and the construction of the prototype (described in Berners-Lee, 1999) were responses to the overwhelming volume of academic publications and the resulting impossibility of any scientists to keep up with a range of topics effectively. The World Wide Web has done little to address this, since it in effect provides yet another information torrent too large for consumption, but more recent efforts such as the Semantic Web initiative (Berners-Lee et al., 2001), broadly similar and coeval to the GBDL Project, offer some promise for

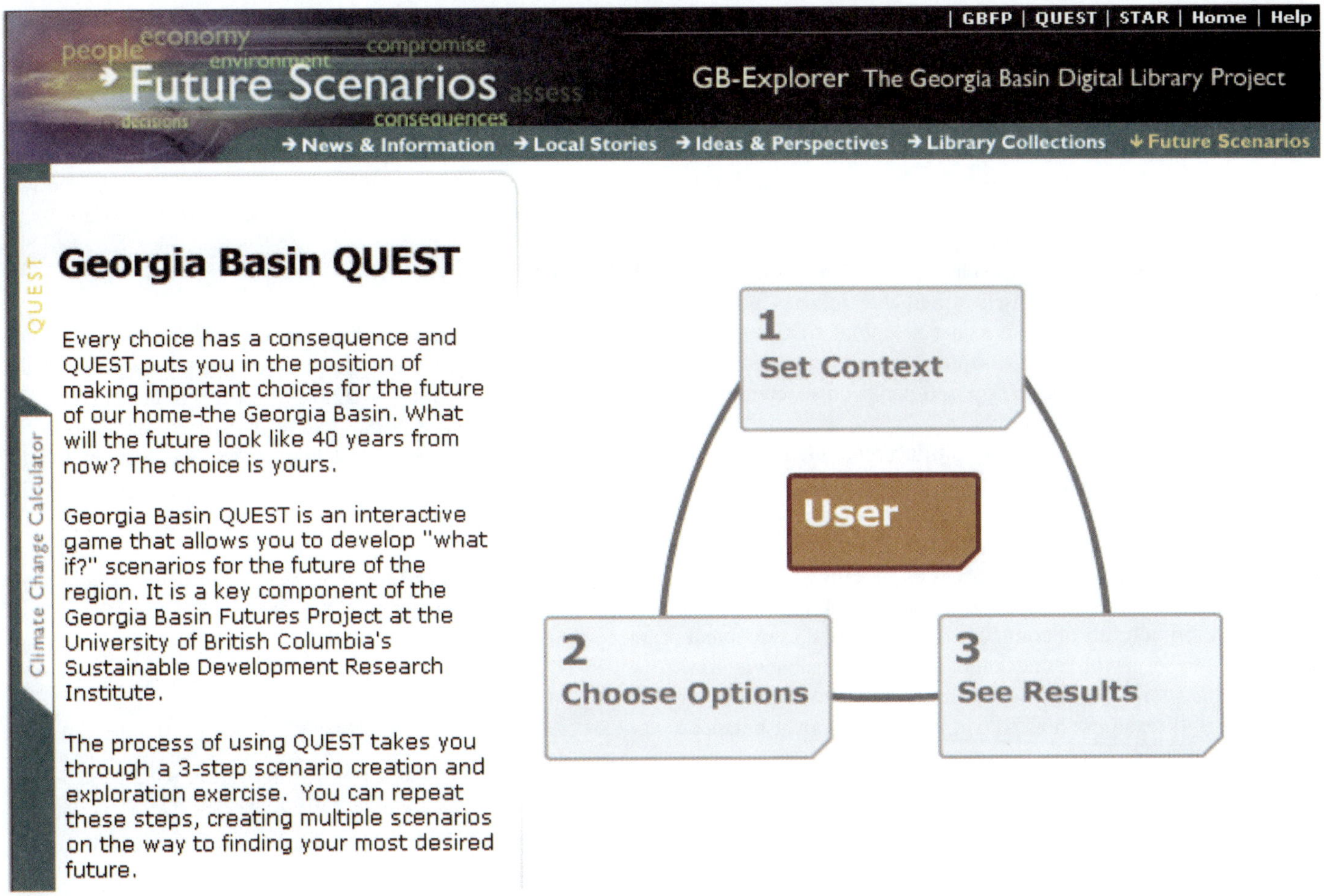

Figure 8. Future Scenarios module, a bridging point between the Georgia Basin Digital Library and the Quest™ scenario modeling game.

at least providing effective means to search the Web and thus restrict the torrent to a manageable trickle of information. These initiatives address the process of information *selection* rather than detailed *format and presentation.* The GBDL attempts to link these two domains through the idea of a semantic network browsing mechanism and so represents a genuinely new format for documents. However, the key to this format is that it does not try to be *entirely new,* unlike many Web initiatives; the GBDL makes a scaffold with stories and maps, elements that are familiar and understood even to young children. The significance is the mode of connection between elements. Whether this new mode is sufficiently engaging to be persistent, to be adopted elsewhere, and so become a regular part of the geoscience communications media remains to be seen.

Following on this, geoscientists, like any other community, are being overwhelmed by information. The additional burden of geoscience is that much of this information has a strong spatial context, making traditional Web-search mechanisms of little use in finding items or limiting searches to specific regions. The GBDL approach can address this, and ongoing work by the GBDL team working with geoscientists in South America attempts to determine if the interface is useful as a discovery, elaboration, and communication environment at this level of discourse.

Finally, as discussed above, in the long term a successful platform such as the GBDL that develops a rich, multi-stakeholder, multiperspective content pool must be based on search, categorization, and interface functionality that builds and shows linkages dynamically. The Georgia Basin Digital Library, to the limit that it is successful, will need to evolve as we learn more about representation and dynamic content; a newer, dynamic engine will be needed if the system is to grow into the future. Work on this engine is ongoing.

CONCLUSIONS

The construction of a flexible, multicomponent digital library for earth science and sustainability knowledge exploration depends on a rich understanding of both the structure of geoscience knowledge and the problems encountered by novices attempting to use such resources. The Georgia Basin Digital

Library Project is an attempt to explore this design space to provide tools relevant to both professional geoscientists and those in greater society wishing to learn about issues of sustainability and the earth sciences.

The GBDL attempts to investigate the creation of learning environments in the absence of existing conceptual frameworks and the support of multiple perspectives on topics of interest. Both of these goals are addressed through the use of explicit, semistructured knowledge forms based on architectural pattern languages. The patterns themselves and the linkages between them support the creation of a rich conceptual framework by users, and the possibility of representing pattern networks in different forms and perhaps with opposed perspectives creates the possibility of multiple perspectives.

The pattern structure itself is multilayered, and so different perspectives can be supported by the creation of: (1) different pattern languages; (2) different solution descriptions for a given set of patterns; (3) different stories for a given set of patterns; and (4) different illustrations for a given set of patterns.

The pattern framework provides a universe of discourse that supports different levels of community agreement, disagreement, and personalization within one computationally feasible environment. Furthermore, ongoing work on interface development on the GBDL will examine how to compare and contrast perspectives at all levels.

Relevance to Geoscience

The GBDL was originally conceived as a resource point for understanding issues related to sustainability and sustainable decision making at the community level. It was and is intended to interface to other approaches in the Georgia Basin Futures Project such as Quest™, as described above. However, the GBDL and GB-Explorer are highly relevant within geoscience as well, in a number of ways.

First, communicating geoscience concepts to the public and involving the public in dialogue about the Earth is of obvious relevance to the future of a strong geoscience community in government and in academia. Although this is broadly supported by the GBDL, the project is not primarily intended for this.

Second, the future of geoscience research lies within a digital world of data collection, data analysis, and communication. The GBDL involves core research on powerful new approaches to store highly contextual, highly interrelated geoscience information in a way that is compatible both with existing forms of analysis and with entirely new approaches wherein explanatory frameworks and perspectives can be treated explicitly within the system.

Third, the effective communication of geoscience information requires a framework that acknowledges that geoscience is storytelling as much as it is traditional science. Interlinked explanatory stories, often site-based, provide the framework not only for learning about geoscience as a novice but for doing geoscience as a professional. Although much of the content of GB-Explorer is aimed at novices learning issues related to sustainable development, the same interface could be used equally for complex decision support tasks in the geosciences, for the creation of virtual interfaces to scientific experiments, and for disseminating geoscience data and information at a professional level. In fact, ongoing work is applying the GBDL framework to all of these tasks at the present time, with the hope that lessons learned will feed back into a better core design in the immediate future.

ACKNOWLEDGMENTS

We gratefully acknowledge financial support for this work from the GEOIDE Network Centre of Excellence (grant RES#52). We also wish to thank the Earth Sciences Sector of Natural Resources Canada for collaborative and logistical support. In addition, we thank the Island Pacific School, Bowen Island, British Columbia for user-testing a preliminary version of the Local Stories module.

REFERENCES CITED

Alexander, C., 1979, The Timeless Way of Building: New York, Oxford University Press.

Alexander, C., Ishikawa, S., and Silverstein, M., 1977, A Pattern Language: New York, Oxford University Press.

Barnes, T., and Duncan, J., eds., 1992, Writing Worlds: Discourse, Text and Metaphor in the Representation of Landscape: New York, Routledge.

Barrett, E., ed., 1989, The Society of Text: HyperText, Hypermedia, and the Social Construction of Information: Cambridge, Massachusetts, MIT Press, 350 p.

Beaudouin-Lafon, M., ed., 1999, Computer Supported Cooperative Work: Hoboken, New Jersey, John Wiley and Sons Ltd.

Berners-Lee, T., 1999, Weaving the Web: The Original Design and Ultimate Destiny of the World Wide Web by Its Inventor: New York, HarperCollins.

Berners-Lee, T., Hendler, J., Lassila, O., 2001, The Semantic Web: Scientific American, v. 284, no. 5, p. 34–43.

Bonham-Carter, G.F., 1995, Geographic Information Systems for Geoscientists: Modeling with GIS: Oxford, UK, Pergammon Press, 398 p.

Borchers, P., 2001, A Pattern Approach to Interaction Design: Hoboken, New Jersey, J. Wiley and Sons.

Bransford, J.D., Brown, A.L., and Cocking, R.D., eds., 2000, How People Learn: Brain, Mind, Experience, and School: Report of the Commission on Behavioral and Social Sciences and Education, National Research Council: Washington, D.C., National Academy Press.

Brawer, W., 1995, The Green Map System: http://www.greenmap.org (October 2005).

Brodaric, B., and Gahegan, M., 2002, Distinguishing instances and evidence of geographical concepts for geospatial database design, *in* Egenhofer, M., and Mark. D., eds., Geographic Information Science—2nd International Conference, GIScience 2002, LNCS 2478: Berlin, Springer, p. 22–37.

Brodaric, B., and Hastings, J., 2002, An object model for geologic map information, *in* Richardson, D., and van Oosterom, P., eds., Advances in Spatial Data Handling, 10th International Symposium on Spatial Data Handling: New York, Springer, p. 55–68.

Brodaric, B., Journeay, M., Harrap, R., Talwar, S., van Ulden, J., Grant, R., and Denny, S., 2003, The Architecture of the Georgia Basin Digital Library: Using geoscientific knowledge in sustainable development: Geological Survey of Slovenia, Geologija, v. 46, p. 343–348.

Bruntland, G., ed., 1987, Our Common Future: The World Commission on Environment and Development: Oxford, Oxford University Press.

Buschmann, F., Meunier, R., Rohnert, H., Sommerlad, P., and Stal, M., 1996, Pattern-Oriented Software Architecture—A System of Patterns: Hoboken, New Jersey, John Wiley and Sons Ltd.

Bush, V., 1945, As We May Think: The Atlantic Monthly, v. 176, no. 1, p. 101–108, http://www.theatlantic.com/unbound/flashbks/computer/bushf.htm (October 2005).

Craig, W.J., Harris, T.M., and Weiner, D., 2002, Community Participation and Geographic Information Systems: New York, Taylor and Francis.
DLESE, 2003, Digital Library for Earth Science Education: http://www.dlese.org (October 2005).
Dublin Core, 2003, The Dublin Core Metadata System: http://www.dublincore.org (October 2005).
Ecotrust, 2001, Conservation Economy.Net: http://www.conservationeconomy.net (October 2005).
FGDC, 2003, Federal Geographic Data Committee resource site: http://www.fgdc.gov (October 2005).
Frodeman, R., and Mitchum, C., 2000, Beyond the social contract myth: Issues in Science and Technology, v. 16, no. 4, p. 37–41.
Geoconnections, 2003, GeoConnections Discovery Portal: http://geodiscover.cgdi.ca (October 2005).
Haklay, M., and Tobon, C., 2003, Usability Evaluation and PPGIS: Towards a user centered design approach: International Journal of Geographical Information Science, v. 17, p. 577–592, doi: 10.1080/1365881031000114107.
Harrap, R., Journeay, M., Talwar, S., and Brodaric, B., 2000, Discovering the Community of our Future: Decision support and digital library tools supporting informed community decision-making in the Georgia Basin Region: White Paper, December 2000 (http://geol.queensu.ca/harrap).
Honadle, G., 1999, How Context Matters: Linking Environmental Policy to People and Place: West Hartford, Connecticut, Kumarian Press.
Lubchenco, J., 1998, Entering the century of the environment: A new social contract for science: Science, v. 279, p. 491–497, doi: 10.1126/science.279.5350.491.
Marchionini, G., and Fox, E.A., 1999, Progress toward digital libraries: Augmentation through integration: Information Processing & Management, v. 35, p. 219–225, doi: 10.1016/S0306-4573(98)00058-2.
Nielsen, J., and Tahir, M., 2001, Homepage Usability: 50 Websites Deconstructed: Indianapolis, Indiana, New Riders Publishing.
OpenGIS, 2005, Open Geospatial Consortium Inc. Web Resource: http://www.opengeospatial.org (October 2005).
Stefik, M.J., 1995, Introduction to Knowledge Systems: San Francisco, California, Morgan Kaufmann, 870 p.
Wurman, R.S., Sume, D., and Leifer, L., 2000, Information Anxiety 2: Indianapolis, Indiana, Que Books, 308 p.

Manuscript Accepted by the Society 18 June 2005

Geological Society of America
Special Paper 397
2006

An integrated chronostratigraphic data system for the twenty-first century

Paul J. Sikora*
Energy & Geoscience Institute, University of Utah, 423 Wakara Way, Suite 300, Salt Lake City, Utah 84108, USA

James G. Ogg*
Purdue University, Department of Earth & Atmospheric Sciences, Civil Building 1397, West Lafayette, Indiana 47907-1397, USA

Anthony Gary*
Energy & Geoscience Institute, University of Utah, 423 Wakara Way, Suite 300, Salt Lake City, Utah 84108, USA

Cinzia Cervato*
Iowa State University, Department of Geological and Atmospheric Sciences, 253 Science Hall I, Ames, Iowa 50011, USA

Felix Gradstein*
University of Oslo, Museum for Geology and Paleontology, Sars' gate 1, N-0562 Oslo, Norway

Brian T. Huber*
Smithsonian Institution, National Museum of Natural History, Department of Paleobiology, MRC NHB-121, 10th and Constitution Ave., NW, Washington, D.C. 20560, USA

Charles Marshall*
Harvard University, Department of Earth and Planetary Sciences, 20 Oxford Street, Cambridge, Massachusetts 02138, USA

Jeffrey A. Stein*
BP America, 501 WestLake Park Blvd., P.O. Box 3092, Houston, Texas 77253, USA

Bruce Wardlaw*
U.S. Geological Survey, MS 926A National Center, 12201 Sunrise Valley Drive, Reston, Virginia 20192, USA

ABSTRACT

Research in stratigraphy is increasingly multidisciplinary and conducted by diverse research teams whose members can be widely separated. This developing distributed-research process, facilitated by the availability of the Internet, promises tremendous future benefits to researchers. However, its full potential is hindered by

*E-mails: psikora@egi.utah.edu; tgary@egi.utah.edu; jogg@purdue.edu; cinzia@iastate.edu; felix.gradstein@geolgi.uio.no; huber.brian@nmnh.si.edu; marshall@eps.harvard.edu; SteinStop@aol.com; bwardlaw@usgs.gov

Sikora, P.J., Ogg, J.G., Gary, A., Cervato, C., Gradstein, F., Huber, B.T., Marshall, C., Stein, J.A., and Wardlaw, B., 2006, An integrated chronostratigraphic data system for the twenty-first century, *in* Sinha, A.K., ed., Geoinformatics: Data to Knowledge: Geological Society of America Special Paper 397, p. 53–59, doi: 10.1130/2006.2397(04). For permission to copy, contact editing@geosociety.org.

the absence of a development strategy for the necessary infrastructure. At a National Science Foundation workshop convened in November 2001, thirty quantitative stratigraphers and database specialists from both academia and industry met to discuss how best to integrate their respective chronostratigraphic databases. The main goal was to develop a strategy that would allow efficient distribution and integration of existing data relevant to the study of geologic time.

Discussions concentrated on three major themes: database standards and compatibility, strategies and tools for information retrieval and analysis of all types of global and regional stratigraphic data, and future directions for database integration and centralization of currently distributed depositories. The result was a recommendation to establish an integrated chronostratigraphic database, to be called Chronos, which would facilitate greater efficiency in stratigraphic studies (http://www.chronos.org/). The Chronos system will both provide greater ease of data gathering and allow for multidisciplinary synergies, functions of fundamental importance in a variety of research, including time scale construction, paleoenvironmental analysis, paleoclimatology and paleoceanography. Beyond scientific research, Chronos will also provide educational and societal benefits by providing an accessible source of information of general interest (e.g., mass extinctions) and concern (e.g., climatic change). The National Science Foundation has currently funded a three-year program for implementing Chronos.

Keywords: stratigraphy, chronostratigraphy, database, education, research.

INTRODUCTION

Fundamental to stratigraphy is the study of rocks, their record of the past (fossils, climatic proxies, magnetic polarity, etc.), and their distribution in space and time. The object of stratigraphy is to reconstruct the history of Earth. In order to do so, stratigraphy must be referenced to time. Stratigraphy within an established time frame is called chronostratigraphy. Essential to the geosciences, chronostratigraphy is the tool whereby the findings of the other disciplines are integrated via age ties. This vital integrative function has been facilitated by the assembly of vast amounts of stratigraphic information into databases, creating important tools for understanding past climatic, evolutionary, and geological processes and their implications for future trends on our planet.

However, in order to conduct such research, chronostratigraphers have need of three basic components: data, applications (analytical tools), and knowledge (expert judgments and interpretations). Private and public publishers, universities, and other agents are active in the Web-enabled dissemination of applications and knowledge. A largely unresolved need, however, is access to raw data. Although there are many options for publishing data, in many cases data are not published at all. For a researcher to determine whether particular data are available or even exist can be a very time-consuming and error-prone process. There are no centralized depositories or Web-enabled search engines for locating and retrieving existing data for published research results. Even if needed data are located, with no widely accepted standard for storage media and formatting, additional resources are often required to retrieve the data and perform the necessary conversions. Although difficult, this data reconnaissance/acquisition process is critical for conducting any study that builds upon and incorporates the data of prior research. Considering that this task is performed continually by researchers worldwide, it represents a large but inefficiently used part of the stratigraphic community's resources. Aside from financial cost, the existing unstructured nature of this process can result in critical data being overlooked, the corrupting of data during downloading and conversion, and the redundant regeneration of data that already exist. Further complicating the search for data specific for chronostratigraphic research is the frequent need to obtain and integrate data sets from many diverse subdisciplines (e.g., biostratigraphy, radiometric dating, magnetostratigraphy, chemostratigraphy, cyclostratigraphy). Such complex integration often taxes the financial and temporal resources available for chronostratigraphic study. An obvious solution is to find, obtain, and integrate chronostratigraphic data from preexisting studies to supplement new interpretations and applications.

It is therefore critical to establish a widely recognized procedure whereby chronostratigraphic data will be maintained as a valuable resource for the geoscience community. To work toward this aim, thirty quantitative stratigraphers and database specialists, from both academia and industry, met to discuss how best to integrate their respective chronostratigraphic databases at a National Science Foundation (NSF) workshop convened at the University of Massachusetts, Amherst, in November 2001. The workshop was titled "Integration of Chronostratigraphic Databases for the 21st Century" (Ogg and Sikora, 2001).

THE AMHERST WORKSHOP

The workshop participants considered the scientific and information technology aspects of database development for three main focus areas: coordination of database standards and compatibility, designing strategies and tools for information retrieval and analysis of all types of global and regional stratigraphic data, and consideration of future directions for database integration and centralization of currently distributed depositories. An early consensus developed for wide inclusiveness as to the type of chronostratigraphic data that would be part of the developmental effort. Accepted data could be point-sourced (e.g., a radiometric date) or sequential (e.g., biostratigraphy), quantitative (e.g., normalized numerical data) or qualitative (descriptive terms such as "rare" or "common"), derived from as wide a range of facies and sample types and from as many disciplines as possible. Envisioned participating disciplines include biostratigraphy, chemostratigraphy, radiometric dating, magnetostratigraphy, cyclostratigraphy, event stratigraphy (e.g., tephra layers), sequence stratigraphy, and dendrochronology. The one overriding requirement for inclusion, however, would be access to raw data, not just the resulting interpretations.

After some discussion, another consensus developed that the most efficient and cost-effective method to coordinate standards and ensure access, integration, and preservation of valuable raw data was through construction of an integrated system that would link existing chronostratigraphic databases. The major portion of the workshop was then devoted to defining the structure, functions, user base, and administration of this system, provisionally dubbed Chronos.

THE CHRONOS SYSTEM—A DATABASE TOOL FOR THE GEOSCIENCE COMMUNITY

Chronos is envisioned as an Internet-based suite of cooperative databases that are accessed through distributed thematic nodes linked to a central hub. The nodes will be composed of existing chronostratigraphic databases for relevant disciplines, including biostratigraphy, radiometric dating, chemostratigraphy, magnetostratigraphy, and cyclostratigraphy (Fig. 1). The thematic nodes will be responsible for coordinating the standards and formats for the participating databases in their subdiscipline, as well as publicizing the system and helping the developing databases to arrange necessary financial and technical support to ensure the continued growth and development of Chronos. An important consideration for the node design process is to try to incorporate as wide a variety of data types as possible, while requiring the least amount of costly revisions to the contributory databases because of standards or formatting requirements. The central hub will act as the user interface as well as coordinating a continually expanding network of individual databases, thereby providing the user not only access to data from the thematic nodes and their contributory databases, but also the ability to integrate, manipulate, and download the data. However, data archiving will not be a central-hub function. That responsibility will remain with the contributory databases within the distributed thematic nodes.

Ultimately, Chronos will provide a geospatially and temporally referenced research network with the goal of providing a dynamic, interactive, and time-calibrated framework for earth system history. A variety of data search capabilities will be available, with all data referenced geographically, stratigraphically, and by discipline. Easy integration of this large amount of existing data will be a boon for multidisciplinary regional and global earth science applications such as time scale calibration, paleoceanography, paleogeography, and geohistory analysis. The Chronos user interface will employ Web-based visualization utilizing integrated, interactive maps and stratigraphic columns, constructed through geographic information system (GIS) and Internet map server (IMS) technologies. Systems such as GIS/IMS will reduce user interaction, as much as possible, to simple point-and-click procedures, thus increasing system friendliness and efficiency.

The distributed structure of Chronos will have major advantages. Control of the data remains local, with the preexisting contributory databases. Having no requirement for database merging increases the potential for continued growth in the number of data owners willing to link to the central hub. At the same time, such local control ensures that data archival costs also remain with the preexisting databases, thereby not adding to the administrative costs of the Chronos system. However, in special circumstances in which a database deemed as essential by the chronostratigraphic community is in danger of becoming inaccessible (e.g., due to lack of funding, or due to the death or retirement of its owner), the Chronos central hub could assume administration of such "orphan" systems.

Central-Hub Role

The principal database function of the Chronos "central hub" will be to provide oversight and a common integrative

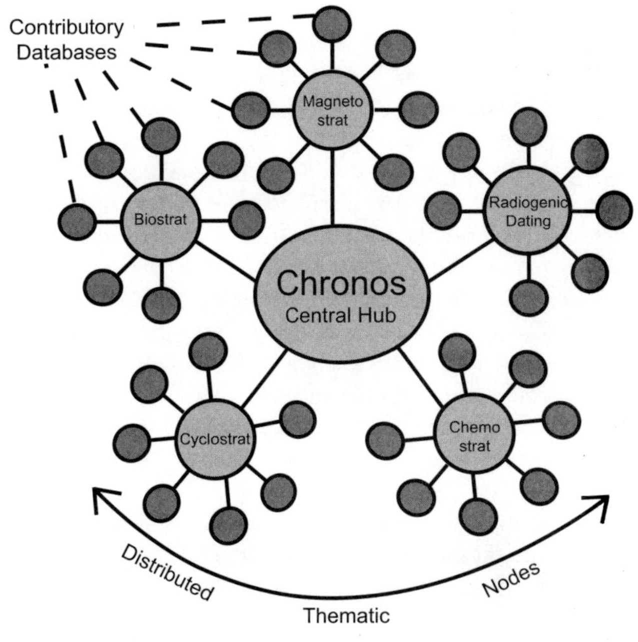

Figure 1. Diagram illustrating the basic structure of the Chronos system.

functionality for the distributed nodes. Oversight entails two major responsibilities: ensuring basic data quality standards, and maintaining and promoting data format standards and translation tables. Although Chronos will be designed to include as wide a range of chronostratigraphic data as possible, there will be by necessity a few basic data requirements for contributory databases. These include full attribution of data, ranking of data quality and error values (where possible), and a clear delineation between raw and interpreted data. The central hub will also act, through an advisory board, as a facilitator of interest groups establishing common formats for their subdiscipline nodes. The Chronos central hub will then store and update these formats as needed. An example is the development of a taxonomic synonymy and translation file for the biostratigraphic subnode. Once format questions are largely resolved, these subdiscipline interest groups will then act to continually update their distributed nodes as well as to eliminate redundant data and obsolete or superseded interpreted data.

In its role as user interface, the central hub for Chronos will consist of a frontispiece with a user-interactive tool (in the form of a chronostratigraphic chart and interactive maps), providing access to the essential elements of a consensus standard geological time scale derived from integration of data contained within the databases of the thematic nodes. This time scale will be the central achievement of the Chronos system (the current time scale can be obtained at http://www.stratigraphy.org/GTS04.pdf). In conjunction with the International Commission of Stratigraphy (ICS), Chronos will work to maintain a time scale that will evolve frequently as new information is generated, analyzed, and integrated. The time scale will require complex data integration that will provide a test bed for the Chronos system during development. Four other test bed projects have been implemented as part of the initial three-year funding of Chronos by the NSF: Cambrian life explosion, Permian-Triassic mass extinction, middle Cretaceous super-greenhouse world, and middle Miocene climate transitions. These four studies were chosen as test beds because they address long-standing scientific questions of societal relevance and interest. Each involves different types and qualities of data and will improve and refine marine and continental data correlations at the global scale. Eventually, all test beds will be user-interactive from the Chronos home page, allowing the user to drill down to access the raw data upon which the interpretations are based by linking directly to the thematic node databases that house the data. Users will also be able to integrate and download data for their own use, with different portals designed specifically for the geoscientist and for the general public.

Thematic Node Role

Thematic nodes will be created and maintained by the subdiscipline interest groups promoted by the central hub. The thematic nodes with their own contributory databases, will not be end nodes in themselves, but will also serve to link Chronos to additional data systems. The interest groups, composed of independent researchers, will act to coordinate the integration of the contributory databases, housed at their various institutions, within their thematic node. These groups will also be tasked with format determination and development of translation tables for their subdisciplines, although to be compiled and standardized at the node level, these network operation files will be stored, shared, and maintained by the central-hub administration as part of its best-practice guidelines.

Chronos Administration

The initial development of the Chronos system is being funded by a three-year grant from the NSF under the directorship of Cinzia Cervato of Iowa State University. Chronos will operate under the aegis of the International Commission on Stratigraphy, a global body of expert stratigraphers founded by the International Union of Geological Sciences for the purpose of promoting and coordinating long-term international cooperation and establishing standards in stratigraphy. The database components for Chronos will be modeled after that of successful multicomponent programs such as the Ocean Drilling Program.

Direct oversight for the system will be undertaken by an advisory board, selected for a diverse membership from the International Commission on Stratigraphy, academia, industry, and government. The advisory board will be charged with coordinating scientific community support for Chronos, developing funding strategies and industry contacts, and overseeing the distributed nodes and their interest groups. A list of members of the advisory board can be obtained at www.chronos.org/community/advisory_board.html. Advisory board members are chosen from geoscience representatives among academia, industry, government, and museums, each noted for the leading role that they have played in community involvement and in the promotion of time data collection, analysis, and preservation.

Ultimately, the daily administrative functions of the central hub will be the responsibility of a permanently funded, central office with an international, permanent core staff. Composed of a minimal number of both professional chronostratigraphers and information technology specialists, the central office staff will run the main database system (central hub and linkages to the distributed nodes), handle system maintenance, be a source of technical assistance to the distributed nodes, and provide user support. With regard to user support, the central office staff will be required to interact with a highly variable user base, from general public inquiries to aiding researchers with complex data transformations. Therefore, in order to provide high-quality user support, the Chronos staff will establish linkages to volunteer experts beyond the central office, including education professionals, outside specialists in temporary project-oriented working groups, and members of the thematic node interest groups. Funding for this central office will likely be administered through the ICS. The contributory databases composing the distributed thematic nodes will be expected to rely upon their own funding sources. However, in extraordinary cases in which a thematic node requires major development or upgrading, the central hub may issue grants for database development.

The overall goal of Chronos of designing a system that will centralize access to chronostratigraphic data is complementary to current trends in geoscience data organization. For example, the Geosciences Network (GEON, http://www.geongrid.org/) initiative, currently funded for five years (2002–2007) by the NSF, is a pilot study designed to show the feasibility of establishing a central hub for the integration of all types of geoscience data (e.g., seismologic, petrologic, hydrologic). The data will be organized in a distributed system similar in structure to Chronos. Assuming the success of both initiatives, Chronos could eventually evolve into the stratigraphic node of a larger GEON structure, helping to provide data access for the entire earth science community (Fig. 2).

ANTICIPATED BENEFITS

Once the Chronos system is established, several major advantages will accrue to stratigraphic researchers, the larger geoscience community dependent upon stratigraphic data, and the general public. Foremost of these benefits will be greater efficiency. Formulation and implementation of research projects will be much easier through access to a central database containing most of the relevant existing data sets, with resultant time and cost savings. Furthermore, centralized access to data will facilitate data integration among varied disciplines to help address studies of complex natural systems (e.g., climate change, evolution) and likely reveal previously unrecognized empirical relationships between the many different subdisciplines. These previously unrecognized synergies will create valuable new knowledge from data that already exist, simply because there will be a capability for consistent multidisciplinary data integration.

Through this greater efficiency and integrative ability, Chronos will advance knowledge and understanding in the geosciences. It will provide a focal point for data access, fostering greater scientific interaction between researchers, both internationally and across the United States. As a common data model for chronostratigraphy, coupled with advanced Web-based applications, Chronos is an original concept for the exploration of new, creative, multidisciplinary research in the geosciences.

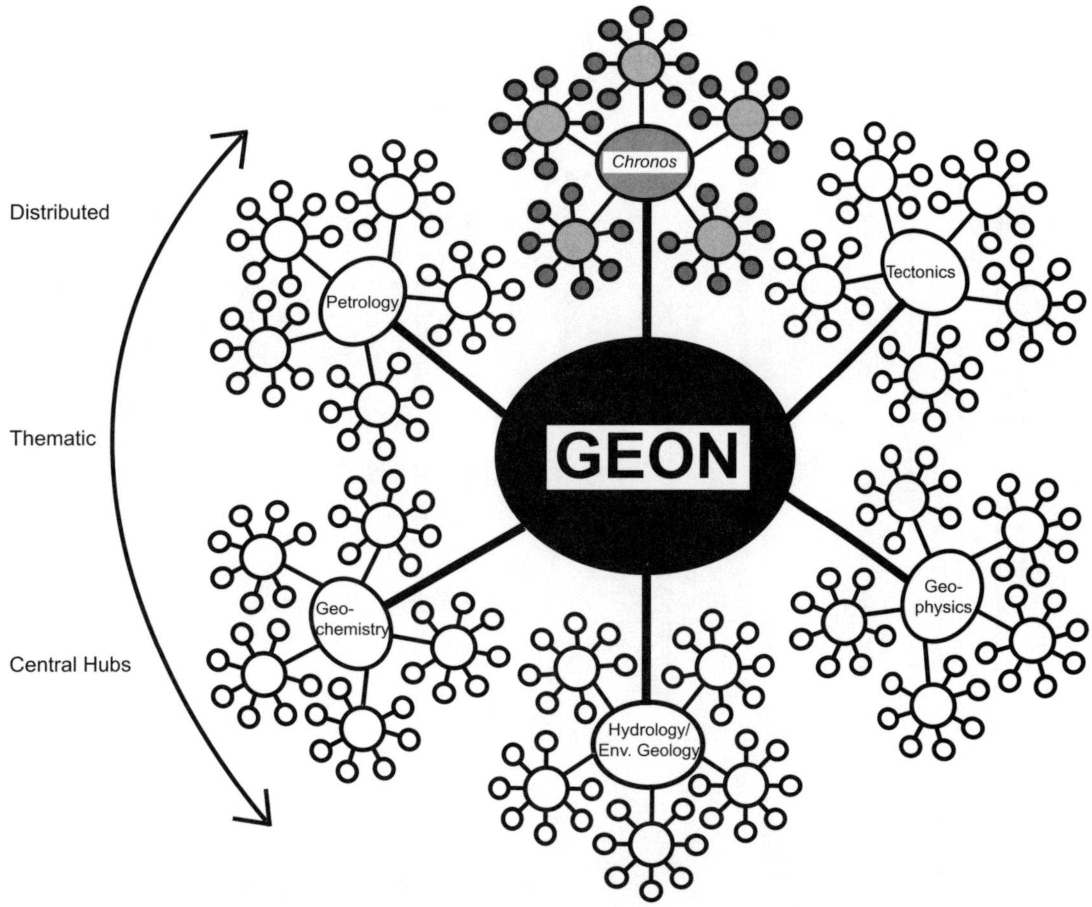

Figure 2. Possible future design for a central database for the earth sciences, with the completed Chronos system serving as the stratigraphic central hub for the Geosciences Network (GEON) initiative.

TABLE 1. OUTLINE OF PROVISIONAL PLAN FOR THE DEVELOPMENT OF THE CHRONOS SYSTEM

Task	Year 1	Year 2	Year 3	Year 4	Year 5
Existing database integration	Biostrat test integration Formatting, system analysis	Biostrat test integration Search engines, data output	Multinode development using biostrat model	Multinode workspace and visualization tools	Fully integrated systems
Multidiscipline stratigraphic projects	Permian-Triassic boundary	Permian-Triassic boundary Mid-Miocene biostrat	Mid-Miocene biostrat Middle Cretaceous anoxic events	Middle Cretaceous anoxic events Fourth project to be named	Fourth project completed
Thematic node development	Radiometric, chemostrat, magnetostrat	Radiometic, chemostrat, magnetostrat, cyclostrat	Cyclostrat, eustatic-climatic	Eustatic-climatic	Other methods
Central hub development	Staffing, infrastructure Thematic node IT support Development of formats, standards and protocols Integration with GEON	Load and maintain translation tables Archive thematic node data Alpha test biostrat node Integration with GEON	Beta test biostrat node Alpha test multiple thematic nodes Thematic node IT support Load and maintain translation tables Archive thematic node data	Biostrat node deployment Beta test multiple thematic nodes Thematic node IT support Load and maintain translation tables Archive thematic node data	Multiple thematic node deployment Thematic node IT support Load and maintain translation tables Archive thematic node data
Public outreach and access	Geological time scale frontispiece developed Pointers to databases	Biostrat node demo at ICS meeting in Florence Central hub deploys simple biostrat search engines Time scale accessible, but with limited user interaction	Standard geological time scale user interaction deployed Access to radiometric and magnetostrat goes online	All node search engines online Visualization and analytical tools available, but without tie to user workspace	Capability to access and visualize data from all nodes via users' workspace Remote data contribution to Chronos deployed Search engines fully deployed
Primary and secondary education	Pointers to useful stratigraphic teaching modules Tutorial of geological time scale Visuals installed, illustrating biostrat methods and fossils	Biostrat exercise module Microfossil visuals, including habitats Paleogeographic pointers linked to the standard geological time scale	Basic instructional modules utilizing thematic nodes go online	Basic instructional modules linking to multidiscipline projects	Advanced visualization and analysis Tool modules for integrated datasets Exercises tied to multidiscipline project results

Another major value provided by Chronos will be enhanced educational opportunities. As previously noted, simple centralization of data sets will create a potential for recognition of interrelated empirical relationships that cannot be determined within isolated databases. Such "data mining" will provide numerous opportunities for graduate theses and dissertations. In addition, through online support, Chronos will provide education in chronostratigraphic methodology and data integration on both the graduate and postgraduate levels. Chronos will also provide educational benefits to the general public by providing a readily accessible source of information on topics of interest (e.g., evolution, frequency of catastrophic events in earth history) with download capabilities. A further societal benefit will be the facilitation of research into issues of modern concern (e.g., atmospheric change, biodiversity). A provisional development plan for Chronos is presented in Table 1.

CALL FOR PARTICIPATION

In the original NSF-sponsored workshop and in subsequent meetings and communications, the Chronos initiative has made a conscientious effort to be as inclusive as possible of the entire chronostratigraphic community. Such efforts have continued under the current initial implementation stage. However, it is likely that many chronostratigraphic databases currently in existence remain unknown to current Chronos participants. The greater the participation of the stratigraphic community, the greater will be the chances of success for the Chronos initiative. It will also be of great benefit to Chronos to have the input of potential nonstratigraphic users (e.g., evolutionary biologists, tectonicists, paleoclimatic modelers) so that Chronos application developers can gain a better understanding as to the type and format of data and data integration necessary to aid their research. Therefore, the authors strongly urge anyone who would like to participate in the Chronos project, either as a database owner, scientific or information technology expert, or potential user/advisor to please contact the project administrator, Cinzia Cervato, at Iowa State University.

ACKNOWLEDGMENTS

The authors are most grateful to the chronostratigraphic researchers who attended the Amherst workshop in November 2001 (Ogg and Sikora, 2001) and provided many of the ideas upon which this project is based. We are thankful for the support of the National Science Foundation for their funding of the Amherst workshop as well as for their three-year funding of the early organizational phase of the project. We also thank Mark Leckie of the University of Massachusetts, Amherst, for hosting the workshop. The American Geological Institute has been instrumental in providing contact between Chronos and complementary initiatives by organizing and hosting two workshops on linking chronostratigraphic databases. Finally, thanks are also due to the International Commission on Stratigraphy for their advisory support and agreeing to act as the governing body for the Chronos system.

REFERENCE CITED

Ogg, J.G., and Sikora, P.J., 2001, Integration of chronostratigraphic databases for the 21st century: White paper for the NSF-sponsored workshop at the University of Massachusetts, Amherst, November 8–10, 2001, 30 p. (available as a pdf download at http://www.eas.purdue.edu/chronos [June 2005]).

Manuscript Accepted by the Society 18 June 2005

Printed in the USA

Geological Society of America
Special Paper 397
2006

A geochemical database for western North American volcanic and intrusive rocks (NAVDAT)

J. Douglas Walker
Todd D. Bowers
Ross A. Black
Department of Geology, University of Kansas, Lawrence, Kansas 66045, USA

Allen F. Glazner
Department of Geology, University of North Carolina, Chapel Hill, North Carolina 27599, USA

G. Lang Farmer
Department of Geological Sciences, University of Colorado, Boulder, Colorado 80309, USA

Richard W. Carlson
Carnegie Institution of Washington, Department of Terrestrial Magnetism, Washington, D.C. 20015, USA

ABSTRACT

The Western North American Volcanic and Intrusive Rock Database (NAVDAT—http://navdat.geo.ku.edu and http://navdat.geongrid.org) is a relational database that serves as a repository for chemical and age data for igneous rocks in this region and allows, for the first time, the voluminous, high-quality, igneous rock data generated over the past thirty years to be integrated into time-space-composition models of western North American igneous activity. As in other continental regions, the analysis of geochemical, geochronological, and geospatial data from igneous rocks has played a critical role in the understanding of the tectonic evolution of western North America. However, the seminal, regional-scale studies of the tectonic evolution of this region remain primarily those undertaken in the 1970s, prior to the explosion in the number of high-quality chemical and isotopic studies of North American igneous rocks that occurred during subsequent decades. There has been little concerted effort to integrate these new data into regional tectonic models of western North America, due in large part to the lack of a unified, high-quality database for igneous rocks in this region. As a result, the potential that these data have for providing new fundamental insights into the tectonic evolution of the continent remains largely untapped. The NAVDAT effort is designed to remedy this situation. At its core, NAVDAT is a Web-accessible petrologic database with a schema compatible with existing petrologic databases (GEOROC, PetDB), but with an emphasis on a structure that will easily allow these data to be queried and manipulated as functions of age, composition, and geographic position. The NAVDAT user interface includes a number of advanced query options including both graphical and map-plotting analysis tools

Walker, J.D., Bowers, T.D., Black, R.A., Glazner, A.F., Farmer, G. L., and Carlson, R.W., 2006, A geochemical database for western North American volcanic and intrusive rocks (NAVDAT), *in* Sinha, A.K., ed., Geoinformatics: Data to Knowledge: Geological Society of America Special Paper 397, p. 61–71, doi: 10.1130/2006.2397(05). For permission to copy, contact editing@geosociety.org.

as well as close linking between data and reference information for the source of the data. NAVDAT allows investigations of continent-scale patterns in magmatism and provides the means to address how local variations in the age, composition, and location of igneous rocks fit into a regional context. The latter ability will produce more thorough and rigorous assessments of the relationships between volcanism and lithospheric deformation, including continental mantle delamination, the controls of subducted slab orientation on space-time-composition patterns of magmatism, and the role of mantle plumes in continental evolution. NAVDAT also will provide basic information on volcanic hazards and geothermal resources in western North America. The overall goal is for NAVDAT to become a rich data tool, an important part of the national geoinformatics effort, and a functioning component of an overall national geoscience cyberinfrastructure.

Keywords: geoinformatics, igneous, petrology, tectonics, Cenozoic.

INTRODUCTION

Over the past thirty years, geochemical, geochronological, and geospatial data analysis of igneous rocks has proven invaluable in the development of an understanding of the plate-tectonic evolution of western North America. Early in this time period, efforts were made to interpret the age-composition patterns of Cenozoic volcanism in the western United States relative to the emerging understanding of plate tectonics (Lipman et al., 1971, 1972; Christiansen and Lipman, 1972; Atwater, 1973; Coney and Reynolds, 1977). The studies of Lipman and Christiansen, for example, recognized an eastward migration in the loci of western U.S. magmatism during the Late Cretaceous and Early Tertiary, which these authors connected to the changing interaction of North America with the various oceanic plates to the west. Through this work, Christiansen and Lipman showed that an understanding of volcanic loci, volumes, and composition is critical in developing an understanding of the tectonic evolution of continents.

There has been an explosion of high-quality age and geochemical studies of Phanerozoic igneous rocks since these cited works were published. A casual search of the GeoRef database, for example, yields over 700 references published since 1970 dealing with geochronologic studies of igneous rocks in western North America, and over 2000 dealing with the geochemistry of such rocks. Furthermore, over the past ten years the breadth of elemental geochemical data has increased by a factor of 3 or more because of the routine application of ICP-MS (inductively coupled plasma–mass spectrometry) analyses (Lehnert et al., 2000), and over the past 25 years the precision of age analyses has increased by a factor of 10 or more. There also is a much wider geographic distribution of these high-quality geochemical data. There have been many new studies from the Pacific Northwest (Zoback et al., 1994; Brandon and Goles, 1995; Othberg et al., 1995; Hart et al., 1997; Edwards and Russell, 2000), the Great Basin (Walker and Coleman, 1991; McKee et al., 1993; Mattox, 1994; Miller and Wrucke, 1995; Yogodzinski et al., 1996; Nealey et al., 1997; Nelson and Tingey, 1997; Dodson et al., 1998; Henry and Boden, 1998; Smith et al., 1999), the Mojave Desert (Glazner, 1990; Glazner et al., 1991; Farmer et al., 1995; Miller et al., 2000), the southern Basin and Range in both the United States (Bradshaw et al., 1993; Johnson, 1993; Hawkesworth et al., 1995; Cascadden, 1997; Hallett et al., 1997; Leighty, 1997; Lopez and Cameron, 1997; Stroud, 1997; DePaolo and Daley, 2000; McMillan et al., 2000), and northern Mexico (Lynch et al., 1993; McDowell and Mauger, 1994; Luhr et al., 1995; Suter et al., 1995; Grijalva-Noriega and Roldan-Quintana, 1998; Ferrari et al., 1999; Nagy et al., 1999; Gonzalez-Leon et al., 2000; Henry and Aranda-Gomez, 2000).

Although the new data have provided important insights into magma generation and evolution for a given volcanic center or volcanic area, there have been few efforts to apply these data to reassessing western North American age and chemical patterns in magmatism. For example, newly available Mexican igneous rock data have not been integrated into space-time models of magmatism across the entire Basin and Range. Those data compilations that have been attempted since the late 1970s involve either spatially restricted portions of western North America or specific data types (e.g., Glazner and Bartley, 1984; Axen et al., 1993; Livaccari and Perry, 1993; Zoback et al., 1994; Ferrari et al., 1999). Furthermore, few studies have generated high-quality, public-domain electronic databases, particularly in geographic information system (GIS)–accessible formats.

The authors of this paper are attempting to rectify this situation by compiling existing age, chemical, and isotopic data from latest Cretaceous to Holocene igneous rocks from the western United States, British Columbia, and northern Mexico into a Web-accessible database. The final product, the Western North American Volcanic and Intrusive Rock Database (NAVDAT), will contain the most complete geochemical and geochronological data available for western North America. Besides age and geochemical information, geospatial data will be a fundamental requirement of analyses that populate this database. The geospatial component will be critical to the analysis of patterns in the age-composition variations in igneous rocks throughout western North America. Through the NAVDAT user interface, the patterns can be explored by the database user through a variety of relational and graphical tools. Exploration of NAVDAT data should yield important insights into the role of changing plate orientations, mantle plumes, and lithospheric deformation in controlling the production, distribution, sources, and composition of magmatism in western North America. This will provide a basis for general models of continental magmatism worldwide. Inversely, exploration of the database also will be a means to investigate the origin of local volcanic centers by showing how these centers relate to regional trends in magmatism. Finally, NAVDAT will provide the compilation necessary for identifying areas for additional studies of igneous rocks and for archiving new data generated by these studies.

TECTONIC AND OTHER CONTROLS ON IGNEOUS ACTIVITY

To demonstrate the utility of the NAVDAT database, we review below some specific examples where data compilations have increased our understanding of the time-space-composition patterns of Cenozoic igneous activity in western North America in relation to tectonic and crust-mantle processes. These studies, like almost all of the work appearing in the literature in the past two decades, do not possess the large geographic breadth of the pioneering studies of Lipman and coworkers (e.g., Lipman et al., 1971, 1972; Christiansen and Lipman, 1972). However, they illustrate how far the field has advanced in 30 years and indicate the potential of more regionally inclusive studies.

Delamination of Sierran Lithosphere

Analysis of a GIS-based compilation of Cenozoic volcanic rock age, chemical, and isotopic data for eastern California shows clear space-time-composition patterns in magmatism (Glazner et al., 1999; Manley et al., 2000). This eastern California database shows that highly potassic volcanism in the southern Sierra Nevada Range was restricted to a short time interval in the Pliocene and affected only a certain portion of the southern Sierra Nevada and adjacent areas. Interestingly, the time interval in the Pliocene during which the potassic magmatism occurred has also been suggested as the time when portions of the deep continental lithosphere of the Sierra delaminated (Manley et al., 2000; Farmer et al., 2002). As a result, one possible explanation for the Sierran potassic magmatism is that it represents the product of decompression melting of K-metasomatized portions of the continental mantle lithosphere triggered by the delamination event itself (Farmer et al., 2002). Although the exact relationship between the small-volume potassic magmatism and delamination remains speculative, it seems clear that deformation of the deep continental lithosphere was involved in the generation of Sierran potassic magmatism (Farmer et al., 2002). For this reason, an evaluation of space-time patterns in potassic igneous rock elsewhere in western North America could lead to fundamental insights both into the origin of potassic magmatism and into the physical behavior of the mantle lithosphere.

Volcanism and Glaciation

The eastern California database, limited in scope though it was, revealed a possible anticorrelation between volcanism and glaciation in the past 800,000 yr. Although such a link may seem surprising, climatic influence on volcanism has been proposed in many parts of the world (e.g., Hall, 1982; Nakada and Yokose, 1992). Recent precise Ar/Ar dating in the Long Valley caldera is consistent with the hypothesis that volcanic rocks are preferentially erupted during warmer periods (Ring, 2000). The full NAVDAT database will allow us to more rigorously test this relationship.

Controls over Space-Time-Composition Variations

The regional space-time relationships of volcanism in western North America show clear systematics that were interpreted as representing the interplay between the North America plate, the Pacific plate, and other oceanic microplates to the west (e.g., Lipman et al., 1971, 1972). Using the data already contained in the NAVDAT database, the so-called "ignimbrite wave" originally proposed by Christiansen and Lipman (1972) stands out clearly in animations available on the NAVDAT Web site, segments of which are reproduced as Figure 1. Christiansen and Lipman (1972) proposed that this migrating wave of silicic volcanism reflected retreat of the subducting Farallon plate as it transitioned from flat to normal-dip subduction. Alternative explanations include peeling away of a flat slab (Humphreys, 1995) and complicated patterns of reheating of the young subducted plate (Severinghaus and Atwater, 1990; Axen et al., 1993).

Many other time-space relationships are present in the western North American volcanic record, some still without convincing explanations. Perhaps the most obvious is the northeastward migration of silicic volcanism up the Snake River Plain over the past 12 m.y. that some (Suppe et al., 1975; Pierce and Morgan, 1992), though not all (Christiansen et al., 2002), workers have attributed to movement of North America over the Yellowstone plume (Fig. 2). Complicating the plume explanation for the Snake River Plain volcanism is the corresponding northwestward propagating silicic volcanism that has formed the Oregon High Lava Plains over the past 12 m.y. (MacLeod et al., 1975). In general, Cenozoic-Mesozoic rocks in the Pacific Northwest display a number of spatial and temporally correlated features (Fig. 3), including (1) strong east-west control over the radiogenic isotopic composition of magmas, i.e., the $^{87}Sr/^{86}Sr = 0.706$ line (Armstrong et al., 1977) of rock types ranging from granite through primitive basalt, and (2) a dominance of calc-alkaline volcanism prior to the Miocene, switching to large-volume, Fe-enriched basalts during the ca. 16 Ma Columbia River–Steens flood basalt event, followed by small-volume eruptions of primitive tholeiitic

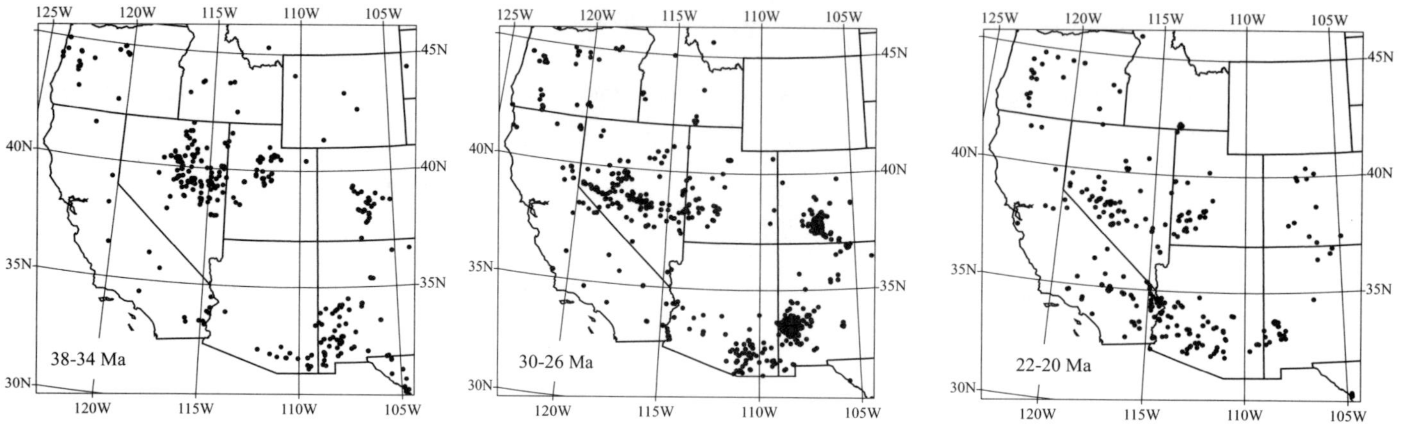

Figure 1. Time slices showing sweeps of magmatism southwest through Nevada and northwest through Arizona and southern California during the late Oligocene and early Miocene. These sweeps are visualized far more clearly by an animation (see NAVDAT Web site). Animations clearly show the Nevada sweep originating in Montana in the Early Tertiary and jumping across the Snake River Plain (where rocks of the relevant age are covered by younger lavas) to Nevada. The sweep out of Arizona clearly originates in Mexico, for which there are currently no data in the database. Preliminary data in these panels from Reynolds et al. (1986), Zartman et al. (1995), and Wilks and Chapin (1997).

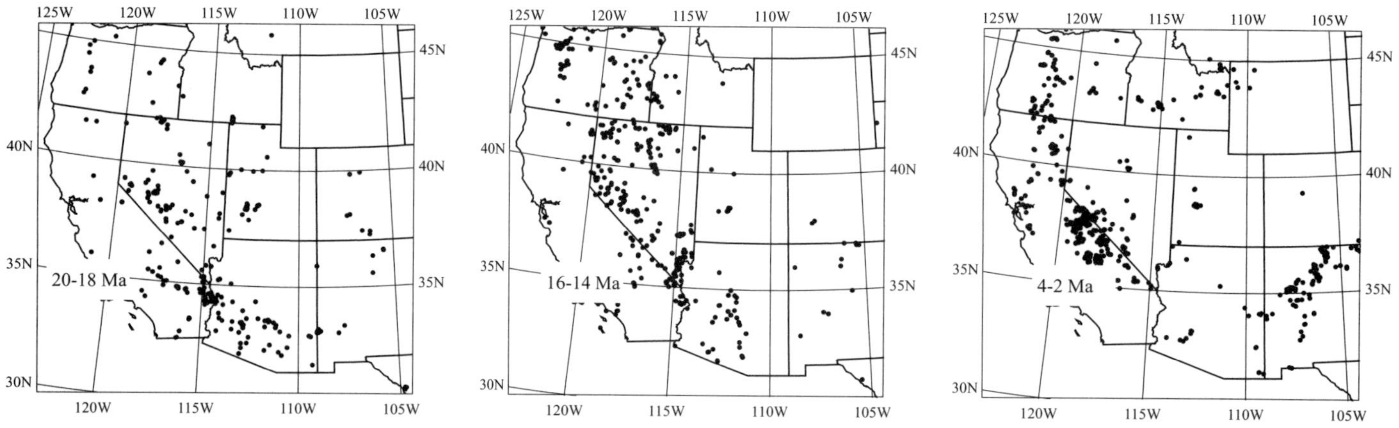

Figure 2. Time slices showing the development of middle Miocene volcanism in northern Nevada, perhaps heralding arrival of the Yellowstone plume. In the 20–18 Ma period, volcanism was nearly absent from northern Nevada. Around 16 Ma, volcanism commenced abruptly across a large area including northern Nevada, southeastern Oregon, and southwestern Idaho. By the Pliocene (4–2 Ma time slice), volcanism had largely shut off in this area, and the track of Snake River Plain volcanism leading to Yellowstone is evident. Data sources as in Figure 1. See animation at the NAVDAT Web site for clearer visualization of these trends.

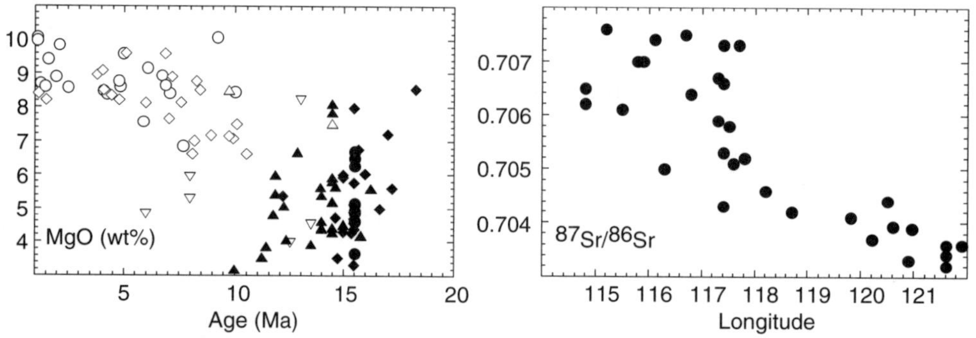

Figure 3. MgO-$^{87}Sr/^{86}Sr$-age-longitude relationships for Cenozoic basaltic rocks from eastern Oregon and westernmost Idaho. Filled symbols at the Steens flood basalts; open symbols are for widely scattered high-aluminum tholeiitic basalts. Data from Carlson and Hart (1987).

basalt accompanied by large-volume high-silica eruptive centers since 10–12 Ma.

The line of marked west-to-east increase in $^{87}Sr/^{86}Sr$ near the Washington-Oregon-Idaho state borders is widely regarded as the western boundary of the Precambrian basement of North America (Armstrong et al., 1977; Manduca et al., 1992). The explanation for the changing eruptive volume and composition throughout this area is still a matter of debate. One school relates the initiation of the flood basalt event to the arrival of the Yellowstone plume beneath this area with subsequent migrating volcanism reflecting the outflow of the plume head and its interaction with the preexisting continental lithosphere and the retreating Juan de Fuca subduction zone (Geist and Richards, 1993; Camp, 1995). The other school interprets the localization of flood basalt centers along the western edge of Precambrian North America as characteristic of a backarc rift, with the subsequent volcanism showing patterns consistent with expansion of a backarc basin caused by changing rates of convergence between North American and the oceanic plates to the west (Carlson and Hart, 1987; Humphreys et al., 2000; Christiansen et al., 2002). Better definition of the time–space–eruptive volume–composition relationships of the type that will be allowed by a fully populated NAVDAT database will help to more accurately define the changing characteristics of this volcanism and perhaps settle the debate over its ultimate cause.

Origin of Laramide Volcanism

The origin of Laramide magmatism (45–65 Ma) has remained an unresolved question, both in terms of why magmatism occurred up to 1500 km inboard of the continental margin and in terms of why this magmatism was concentrated along the NE-trending Colorado Mineral Belt (Fig. 4). One possible explanation is that hydration of the mantle or lower crust via fluids derived from subducted oceanic lithosphere during and following an episode of shallow subduction triggered Laramide magmatism. Alternatively, magmas produced over a wide area at greater depths in the lithosphere or underlying asthenosphere may have been channeled along lithospheric weaknesses along the Colorado Mineral Belt (Lipman, 1980). Lastly, deformation related to Laramide contraction may have led to lithospheric mantle delamination and thus initiated Colorado Mineral Belt magmatism: Deformation coincided with the NE grain of the Precambrian crust in Colorado.

Distinguishing between the above possibilities requires an understanding of the sources of the magmas parental to the Colorado Mineral Belt (asthenosphere versus lithosphere versus lower crust) as a function of time and position. Older compilations of age, petrographic, and geochemical data (Mutschler et al., 1987; Stein and Crock, 1990; Cunningham et al., 1994) suggest that regular spatial variations in both the age and compositions of Laramide magmatism do exist. Mutschler et al. (1987) demonstrated that alkalic magmatism (perhaps reflecting smaller degrees of melting of mantle or crustal sources) was concentrated at each end of the Laramide Colorado Mineral Belt, while calc-alkaline magmatism occurred in the center portions of the belt (Fig. 4). This may reflect the fact that the thermal flux varied considerably along the length of the belt (highest in the middle and diminishing to SW and NE) resulting in variations in the sources involved

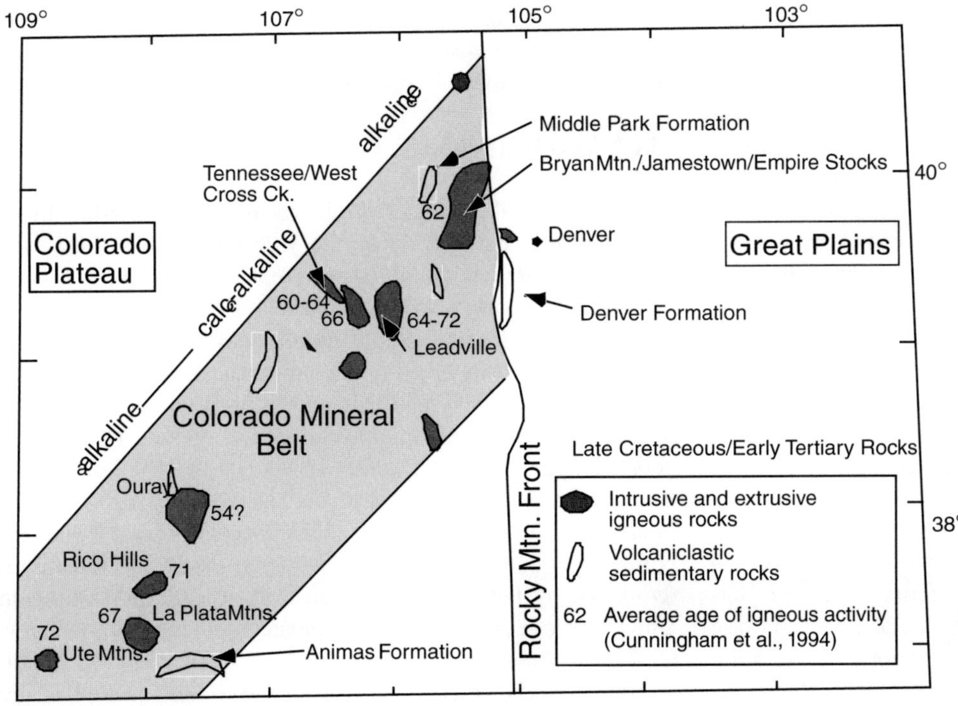

Figure 4. Spatial variations in compositions and ages of Laramide (65 Ma to 45 Ma) igneous activity in the Colorado Mineral Belt. Data from Cunningham et al. (1994), Mutschler et al. (1987) and Stein and Crock (1990).

in the magmatism as a function of position. However, existing compilations of K-Ar and fission-track age determinations from Laramide intrusive igneous rocks suggest that there is a regular, SW to NW younging in the age of the Colorado Mineral Belt magmatism (Fig. 4) (Cunningham et al., 1994). Remaining unanswered by these studies is why a regular progression in position of magmatism through time in the Colorado Mineral Belt should exist while the magma compositions vary only with position. NAVDAT will play an important role in addressing this issue by bringing to bear all the currently available data for Colorado Mineral Belt magmatism and by helping to identify critical regions for additional study.

Slab Windows

NAVDAT will allow testing of hypotheses for the effect of slab windows on magmatism. Such windows, which are thought to develop at subduction zones when a ridge meets a trench, may (e.g., Dickinson and Snyder, 1979; Hole et al., 1991; Dickinson, 1997; Atwater and Stock, 1998) or may not (Glazner et al., 2002) control the location and timing of subsequent mafic volcanism. Regional correlation of magmatism with plate-tectonic events at the continental margin should allow this question to be more fully addressed. The link between the development of the San Andreas transform fault and coastal volcanism in California (e.g., Johnson and O'Neil, 1984; Cole and Basu, 1995; Dickinson, 1997) should also be clarified.

THE NAVDAT PROJECT

The examples given above show clearly that a more comprehensive compilation of data on Cenozoic igneous rocks in western North America is needed. The pressing need for better management of solid earth science data was recognized by the Geoinformatics Working Group (http://www.geoinformatics.org/), and the need for improved geochemical databases, in particular, has been the subject of a variety of recent workshops (Integrated Solid Earth Sciences—CyberInfrastructure, Kansas, March 2003; Cyberinfrastructure for Solid Earth Geochemistry, Washington, October 2003; Linking Information Systems in Marine and Terrestrial Geosciences, Washington, June 2004 [workshop reports available at http://www.earthchem.org]).

Web-based data management in the fields of geochronology, petrology, and geochemistry began with PetDB (http://www.petdb.org/), which serves geochemical data for igneous rocks of the ocean floor, and GEOROC (http://georoc.mpch-mainz.gwdg.de/georoc/), which contains geochemical data for igneous rocks from ocean islands, convergent margins, and large igneous provinces. Before the addition of these and the NAVDAT database efforts, existing databases generally resided on individual computers, in piecemeal and incomplete form. Large data sets produced by the U.S. Geological Survey (USGS) over the past 20 years exist in electronic form (cf. the National Geochronological Database; Great Basin geoscience database), but these were produced before any data standards were implemented, are fraught with problems (e.g., locations recorded only to the nearest quadrangle corner), and are largely unexploited by the geologic community. The slow growth of electronic data management for geochemical and petrologic data largely is a consequence of the large time investment involved in producing the necessary regional database. With the sheer volume of information available, in various paper or electronic forms, it is now beyond the capability of individual investigators to compile literature data in any comprehensive fashion.

Over the past 20 years, great advances have been made in geophysical models for the evolution of western North America based on geodetic and seismic experiments, new understanding of the basic geologic development of the region, and improved understanding of interactions between the North America plate, the Pacific plate, and various microplates (e.g., Atwater and Stock, 1998; Bohannon and Parsons, 1995). These improved geophysical models provide a new context in which comprehensive geochemical and geochronological data for igneous rocks of western North America can be added to better understand the driving forces and geologic evolution of the western North America plate boundary. As earth scientists prepare for the huge geophysical initiative represented by the EarthScope National Science Foundation–MREFC project, the need for a complete and easily accessible database of the Mesozoic-Cenozoic volcanic record in western North America is critical. Both the USArray and Plate Boundary Observatory (PBO) components of EarthScope will need a geologic context within which to place their geophysical observations. NAVDAT will provide the record of western U.S. volcanism over the last 70 m.y. that will form an important element in the interpretation of EarthScope measurements. Without a compilation of the igneous rock data, it is not possible to fully assess on a regional scale the role of lithospheric deformation, mantle plumes, and subducted lithosphere in the production and eruption of continental magmas (see discussion above).

The NAVDAT Database

The NAVDAT effort was initiated in the fall of 2001. Since that time we have established the Web interface (Bowers) and data entry nodes at the University of Kansas (Walker, Black), University of North Carolina (Glazner), University of Colorado (Farmer), and the Carnegie Institution of Washington (Carlson). This initial work involved relational database design as well as development of data standards and input protocols. The database can be viewed at http://navdat.geo.ku.edu and http://navdat.geongrid.org. The user interface developed by NAVDAT serves not only the NAVDAT database, but also a variety of other compilations of geochemical and geochronologic data for western North American igneous rocks, including the PETROS data set (compiled by Felix Mutschler and staff at Eastern Washington University, see http://www.ngdc.noaa.gov/mgg/geology/petros.html) and the National Geochronological Database (DDS-14, see http://wrgis.wr.usgs.gov/open-file/of03-236/) compiled by the USGS and the New Mexico Bureau of Geology and Mineral Resources geochronology database (see http://geoinfo.nmt.edu/data/ims/geochron.html). Including these other compilations, NAVDAT currently serves data on over 80,000 samples.

NAVDAT uses a relational database approach, which was chosen for two reasons: (1) to maintain compatibility with existing geochemical databases, and (2) to take advantage of the multifold increase in performance inherent in the use of a well-designed relational database over typical flat databases (e.g., Excel spreadsheet). The Web server utilizes Cold Fusion™. This was chosen so that text files could be entered directly into the database using Structured Query Language (SQL) calls. This is advantageous in that it does not require any intermediate data management system to be involved and can accommodate data in fairly unstructured formats. In response to user queries, NAVDAT presents the selected data in a simple table-based format or as graphic plots using similar SQL calls. In addition, NAVDAT data can be returned in an XML string formatted appropriately to allow the data to be plotted on top of a variety of interactive GIS maps. The maps are created at an Internet map server run by the Kansas Geological Survey.

Another important addition to the NAVDAT approach has been the inclusion of an American Geological Institute GeoRef identifier (as well as the full citation) for each data set entered into the database. GeoRef is the standard index of the geological literature for North America, and as such contains all of the bibliographic information that is commonly available for journal articles, books, theses, and government publications related to geology. This allows a database user to locate the original reference containing the data and to perform searches based on the extended bibliographical data in GeoRef. When the transition to electronic publishing is complete, this should allow database users possible linked access to an electronic version of the paper.

The NAVDAT system will also serve other data sets appropriate to the analysis of volcanic rocks. Gravity and magnetic data for the western United States have been assembled from existing sources, such as the Decade of North American Geology compilations. A collection of existing digital implementations of 1:500,000 geologic maps and mosaicked 1:250,000, 1:500,000, and 1:1,000,000 digital elevation models is also available. The geological mapping information was downloaded via the Web from a variety of sites (see NAVDAT Web site for more documentation). The GIS data files were in a variety of ARC/INFO-compatible formats but were not consistent in their geographic referencing. The data were all converted to a standardized coordinate system and file type. The data were converted to ARC coverage format and then the coverages converted to ARC e00 export files. The coordinate systems were all transformed to decimal degrees on a Clarke 1866 Spheroid/NAD27 datum. Digital elevation models come from the USGS.

The NAVDAT user interface currently includes or will soon be adding a number of online geochemical tools and GIS routines such as:

1. A petrology toolkit. This comprises a basic geochemical classification and modeling package including Harker diagrams, total alkali-silica diagrams and bulk-silicate-earth normalized trace element profiles. For example, the user can select volcanic rocks based on any set of criteria and then perform plotting or analysis using standard petrology methods. Normative calculations and tools for analysis of magma generation and evolution (e.g., links to the MELTS program [http://melts.ess.washington.edu/]) will be added to the toolkit.

2. Time-space GIS tools. The database was initiated to examine geochemical trends with respect to time and space. Current GIS tools are lacking in this regard (e.g., it is difficult to incorporate age errors when plotting data). Of particular interest are tools for making movies that step through the database in time, displaying geochemical data on a map base (see Figs. 1 and 2). We have already developed a number of such animations, which are available on the NAVDAT Web site. We are working on adding the capability for the user to create animations online. With the vast improvement in accuracy of both age and spatial information, such tools should provide many new insights into the interplay of magmatic and tectonic processes.

3. Geochemistry-geophysics-crustal structure visualization tools. The combination and cross-analysis of geochemical and geophysical properties is potentially very powerful. We intend to construct tools that will make this analysis easier and more intuitive. This effort will attempt to visualize primary variations in time-space-composition patterns with respect to such ancillary data as present-day topography, slope, gravity and aeromagnetic measurements, earthquake data, global positioning system (GPS) and very long baseline interferometry (VLBI) motion studies, crustal thickness and seismic-wave velocity information, and remote sensing data (e.g., ASTER, Landsat).

Relation to Existing Geochemical Database Efforts

NAVDAT is a significant attempt to compile geochemical and geochronological data for western North American Cenozoic igneous rocks into a single Web-accessible database. This effort is being done in consideration of several other igneous rock geochemical databases, specifically PetDB (http://www.petdb.org/) for ocean basalts and GEOROC (http://georoc.mpch-mainz.gwdg.de/georoc/), whose focus is on volcanic rocks from ocean islands, convergent margins, and large igneous provinces worldwide. In order to ensure compatibility with these efforts, with the guidance of Kerstin Lehnert of the PetDB group, we developed a NAVDAT schema (Fig. 5) that is functionally identical to that of PetDB, but with the some important additions. Because PetDB is primarily aimed at oceanic igneous suites, some of the critical information needed for examining continental samples was not accounted for in the PetDB database. We have extended some tables of the PetDB structure by adding additional flexibility for age information, as such data are critical for elucidating continental volcanic-tectonic evolution. The largest addition is an additional table for stratigraphic descriptors. The end result is a

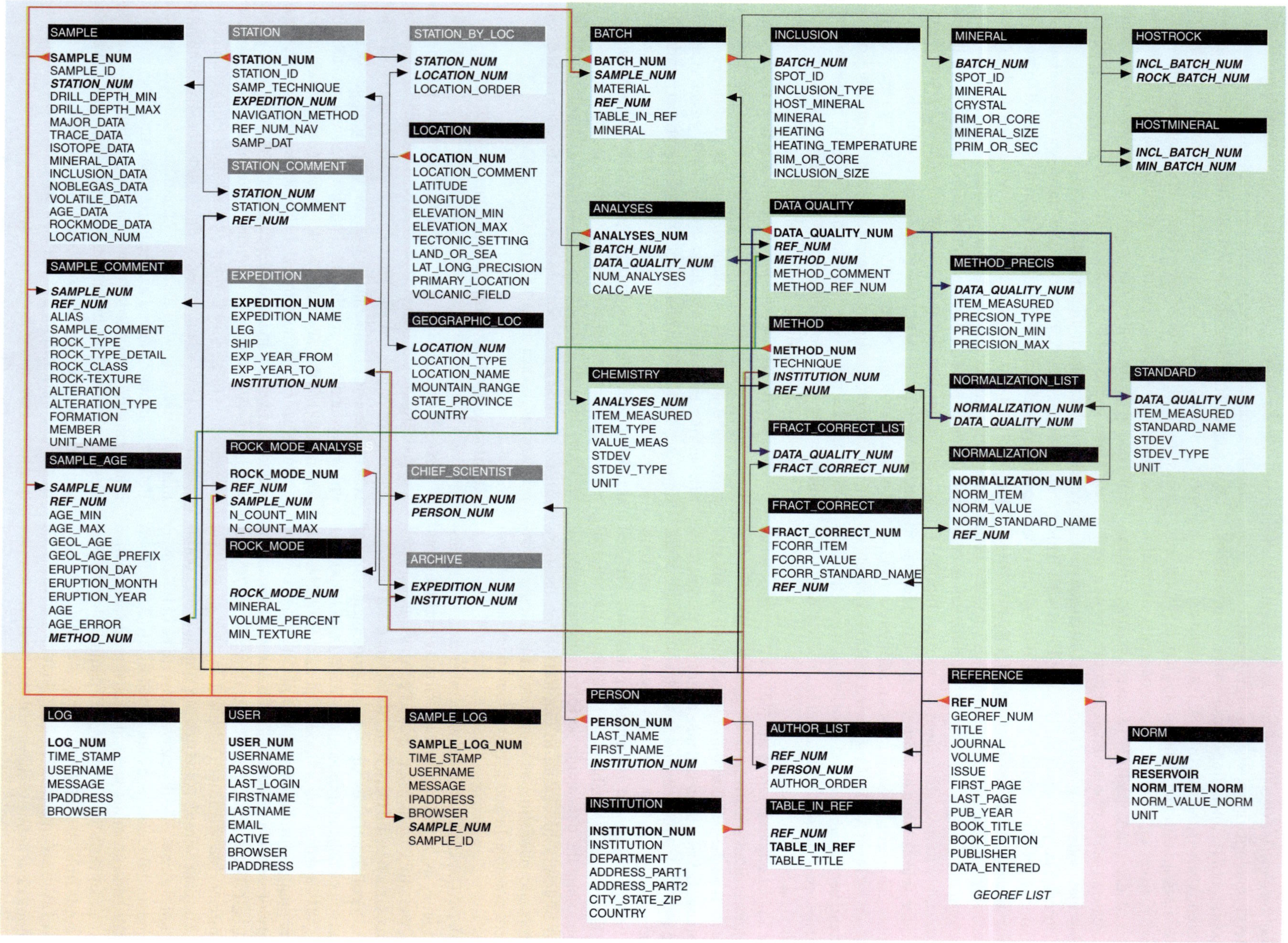

Figure 5. Schema used for the NAVDAT effort. This is modified from the PetDB schema (PetDB, 2002) with additions to make it more appropriate to on-land samples. In addition, we have added a number of log functions (lower left part of the diagram) in the anticipation that the database may be open to external contributions from the geologic community at large. Lines and arrows show the general flow of primary and secondary keys for each table in the relational database. Grayed fields in the upper left portion of the diagram are not important to the NAVDAT emphasis; these mostly apply to samples collected on cruises.

structure that is seamless between the three databases. To further this collaboration between the igneous rock geochemical databases, NAVDAT has become part of the EarthChem consortia (www.earthchem.org), whose goals are to (1) develop a common query portal to geochemical databases, (2) expand data coverage, and (3) add new functionality to the system to bolster the utility of the data sets.

CONCLUSIONS

The strong connection between igneous activity and the general geologic evolution in western North America has been illustrated by several examples pertaining to tectonics. A better delineation of these relationships can be made through a complete, and easily accessible, catalog of geochemical and geochronological data for the igneous rocks of western North America. Specific relations to explore as soon as the data compilation is reasonably complete include

1. How is the down-going plate that is subducting eastward beneath North America deformed in response to changing tectonic setting?
2. How are tectonism and volcanism related over the past 70 m.y.?
3. What are the roles of mantle plumes in volcanism/tectonism in western North America?
4. What caused the transition from continental to ocean-like mantle in the Basin and Range?
5. What factors determine the present location of potentially hazardous volcanic and earthquake activity in the western United States?
6. What is the relationship of volcanism to geothermal resources and ore deposits?

For NAVDAT to be most useful in addressing these questions, it must allow users to explore the relation of magmatism to tectonics in ways never before possible. To do this, we intend to make available the most accurate, comprehensive, standardized database management system possible and to develop the ability to interact with allied databases utilizing queries ranging from the most straightforward to the most innovative. The system will then facilitate the visualization of the query results in ways that enable the user to address fundamental questions regarding geochemistry, magmatism, tectonics, crustal structure, and geologic history. The overall goal is to make NAVDAT available as a rich data tool while helping to develop and maintain standards necessary for making NAVDAT an important part of the geoinformatics effort and a functioning component of an overall geoscience cyberinfrastructure (Allison et al., 2002).

ACKNOWLEDGMENTS

Funding for the NAVDAT effort is provided through the National Science Foundation–ITR programs and from the Geothermal Program Office of the U.S. Navy. In addition, we are deeply indebted to Kerstin Lehnert for sharing information and philosophy about the PetDB project and helping us to get started with NAVDAT. Reviews of this paper by Cin-Ty Lee and an anonymous reviewer greatly helped the final version.

REFERENCES CITED

Allison, M.L., Snyder, W.S., and Walker, J.D., 2002, Geoinformatics: A nascent revolution in the earth sciences: GSA Today, v. 12, no. 7, p. 17, doi: 10.1130/1052-5173(2002)012<0017:GANRIT>2.0.CO;2.

Armstrong, R.L., Taubeneck, W.H., and Hales, P.O., 1977, Rb/Sr and K/Ar chronometry of Mesozoic granitic rocks and their Sr isotope compositions, Oregon, Washington, and Idaho: Geological Society of America Bulletin, v. 88, p. 397–411, doi: 10.1130/0016-7606(1977)88<397:RAKGOM>2.0.CO;2.

Atwater, T., 1973, Implications of plate tectonics for the Cenozoic tectonic evolution of western North America, *in* Cox, A., ed., Plate tectonics and geomagnetic reversals: San Francisco, W.H. Freeman, p. 583–609.

Atwater, T., and Stock, J., 1998, Pacific-North America plate tectonics of the Neogene southwestern United States: An update: International Geological Review, v. 40, p. 375–402.

Axen, G.J., Taylor, W.J., and Bartley, J.M., 1993, Space-time patterns and tectonic controls of Tertiary extension and magmatism in the Great Basin of the Western United States: Geological Society of America Bulletin, v. 105, p. 56–76, doi: 10.1130/0016-7606(1993)105<0056:STPATC>2.3.CO;2.

Bohannon, R.G., and Parsons, T., 1995, Tectonic implications of post-30 Ma Pacific and North American relative plate motions: Geological Society of America Bulletin, v. 107, p. 937–959, doi: 10.1130/0016-7606(1995)107<0937:TIOPMP>2.3.CO;2.

Bradshaw, T., Hawkesworth, C.J., and Gallagher, K., 1993, Basaltic volcanism in the southern Basin and Range: No role for a mantle plume: Earth and Planetary Science Letters, v. 116, p. 45–62, doi: 10.1016/0012-821X(93)90044-A.

Brandon, A.D., and Goles, G.G., 1995, Assessing subcontinental lithospheric mantle sources for basalts: Neogene volcanism in the Pacific Northwest, USA as a test case: Contributions to Mineralogy and Petrology, v. 121, p. 364–379, doi: 10.1007/s004100050102.

Camp, V.E., 1995, Mid-Miocene propagation of the Yellowstone mantle plume head beneath the Columbia River basalt source region: Geology, v. 23, p. 435–438, doi: 10.1130/0091-7613(1995)023<0435:MMPOTY>2.3.CO;2.

Carlson, R.W., and Hart, W.K., 1987, Crustal genesis on the Oregon Plateau: Journal of Geophysical Research, v. 92, p. 6191–6206.

Cascadden, T.E., 1997, Quaternary volcanism in the Colorado Plateau-Basin and Range transition zone: Zuni-Bandera and nearby volcanic fields [Ph. D. thesis]: Albuquerque, University of New Mexico, 407 p.

Christiansen, R.L., and Lipman, P.W., 1972, Cenozoic volcanism and plate-tectonic evolution of the western United States, II: Late Cenozoic: Royal Society of London Philosophical Transactions, ser. A, v. 271, p. 249–284.

Christiansen, R.L., Foulger, G.R., and Evans, J.R., 2002, Upper-mantle origin of the Yellowstone hotspot: Geological Society of America Bulletin, v. 114, p. 1245–1256, doi: 10.1130/0016-7606(2002)114<1245:UMOOTY>2.0.CO;2.

Cole, R.B., and Basu, A.R., 1995, Nd-Sr isotopic geochemistry and tectonics of ridge subduction and middle Cenozoic volcanism in western California: Geological Society of America Bulletin, v. 107, p. 167–179, doi: 10.1130/0016-7606(1995)107<0167:NSIGAT>2.3.CO;2.

Coney, P.J., and Reynolds, S.J., 1977, Cordilleran Benioff zones: Nature, v. 270, p. 403–406.

Cunningham, C.G., Naeser, C.W., Marvin, R.F., Luedke, R.G., and Wallace, A.R., 1994, Ages of selected intrusive rocks and associated ore deposits in the Colorado Mineral Belt: U.S. Geological Survey Bulletin 2109, p. 1–31.

DePaolo, D.J., and Daley, D.E., 2000, Neodymium isotopes in basalts of the southwest Basin and Range and lithospheric thinning during continental extension: Chemical Geology, v. 169, p. 157–185, doi: 10.1016/S0009-2541(00)00261-8.

Dickinson, W.R., 1997, Tectonic implications of Cenozoic volcanism in coastal California: Geological Society of America Bulletin, v. 109, p. 936–954.

Dickinson, W.R., and Snyder, W.S., 1979, Geometry of subducted slabs related to San Andreas transform: Journal of Geology, v. 87, p. 609–627.

Dodson, A., DePaolo, D.J., and Kennedy, B.M., 1998, Helium isotopes in lithospheric mantle: Evidence from Tertiary basalts of the Western USA:

Geochimica et Cosmochimica Acta, v. 62, p. 3775–3787, doi: 10.1016/S0016-7037(98)00267-1.

Edwards, B.R., and Russell, J.K., 2000, Distribution, nature, and origin of Neogene-Quaternary magmatism in the northern Cordilleran volcanic province, Canada: Geological Society of America Bulletin, v. 112, p. 1280–1295, doi: 10.1130/0016-7606(2000)112<1280:DNAOON>2.3.CO;2.

Farmer, G.L., Glazner, A.F., Wilshire, H.G., Wooden, J.L., Pickthorn, W.J., and Katz, M., 1995, Origin of late Cenozoic basalts at the Cima volcanic field, Mojave Desert, California: Journal of Geophysical Research, v. 100, p. 8399–8415, doi: 10.1029/95JB00070.

Farmer, G.L., Glazner, A.F., and Manley, C.R., 2002, Did lithospheric delamination trigger late Cenozoic potassic volcanism in the southern Sierra Nevada, California?: Geological Society of America Bulletin, v. 114, p. 754–768, doi: 10.1130/0016-7606(2002)114<0754:DLDTLC>2.0.CO;2.

Ferrari, L., Lopez-Martinez, M., Aquirre-Díaz, G., and Carrasco-Nunez, G., 1999, Space-time patterns of Cenozoic arc volcanism in central Mexico, from the Sierra Madre Occidental to the Mexican volcanic belt: Geology, v. 27, p. 303–306, doi: 10.1130/0091-7613(1999)027<0303:STPOCA>2.3.CO;2.

Geist, D., and Richards, M., 1993, Origin of the Columbia Plateau and Snake River Plain: Deflection of the Yellowstone plume: Geology, v. 21, p. 789–792, doi: 10.1130/0091-7613(1993)021<0789:OOTCPA>2.3.CO;2.

Glazner, A.F., 1990, Recycling of continental crust in Miocene volcanic rocks from the Mojave block, southern California, *in* Anderson, J.L., ed., The nature and origin of Cordilleran magmatism: Geological Society of America Memoir 174, p. 147–168.

Glazner, A.F., and Bartley, J.M., 1984, Timing and tectonic setting of Tertiary low-angle normal faulting and associated magmatism in the southwestern United States: Tectonics, v. 3, p. 385–396.

Glazner, A.F., Farmer, G.L., Hughes, W.T., Wooden, J.L., and Pickthorn, W., 1991, Contamination of basaltic magma by mafic crust at Amboy and Pisgah craters, Mojave Desert, California: Journal of Geophysical Research, v. 96, p. 13,673–13,691.

Glazner, A.F., Manley, C.R., Marron, J.S., and Rojstaczer, S., 1999, Fire or ice: Anticorrelation of volcanism and glaciation in California over the past 800,000 years: Geophysical Research Letters, v. 26, p. 1759–1762, doi: 10.1029/1999GL900333.

Glazner, A.F., Walker, J.D., Bartley, J.M., and Fletcher, J.M., 2002, Cenozoic evolution of the Mojave block and environs, *in* Glazner, A.F., et al., eds., Geologic evolution of the Mojave Desert and southwestern Basin and Range: Geological Society of America Memoir 195, p. 19–41.

Gonzalez-Leon, C.M., McIntosh, W.C., Lozano-Santacruz, R., Valencia-Moreno, M., Amaya-Martinez, R., and Rodriquez-Castaneda, J.L., 2000, Cretaceous and Tertiary sedimentary, magmatic, and tectonic evolution of north-central Sonora (Arizpe and Bacanuchi quadrangles), Northwest Mexico: Geological Society of America Bulletin, v. 112, p. 600–610, doi: 10.1130/0016-7606(2000)112<0600:CATSMA>2.3.CO;2.

Grijalva-Noriega, F.J., and Roldan-Quintana, J., 1998, An overview of the Cenozoic tectonic and magmatic evolution of Sonora, northwestern Mexico: Revista Mexicana de Ciencias Geologicas, v. 15, p. 145–156.

Hall, K., 1982, Rapid deglaciation as an initiator of volcanic activity: An hypothesis: Earth Surface Processes and Landforms, v. 7, p. 45–51.

Hallett, R.B., Kyle, P.R., and McIntosh, W.C., 1997, Paleomagnetic and $^{40}Ar/^{39}Ar$ age constraints on the chronologic evolution of the Rio Puerco volcanic necks and Mesa Prieta, west-central New Mexico: Geological Society of America Bulletin, v. 109, p. 95–106, doi: 10.1130/0016-7606(1997)109<0095:PAAAAC>2.3.CO;2.

Hart, W.K., Carlson, R.W., and Shirey, S.B., 1997, Radiogenic Os in primitive basalts from the northwestern USA: Implications for petrogenesis: Earth and Planetary Science Letters, v. 150, p. 103–116, doi: 10.1016/S0012-821X(97)00075-7.

Hawkesworth, C.J., Turner, S., Gallagher, K., Hunter, A., Bradshaw, T., and Rogers, N., 1995, Calc-alkaline magmatism, lithospheric thinning and extension in the Basin and Range: Journal of Geophysical Research, v. 100, p. 10,271–10,286, doi: 10.1029/94JB02508.

Henry, C., and Boden, D.R., 1998, Eocene magmatism: The heat source for Carlin-type gold deposits of northern Nevada: Geology, v. 26, p. 1067–1070, doi: 10.1130/0091-7613(1998)026<1067:EMTHSF>2.3.CO;2.

Henry, C.D., and Aranda-Gomez, J.J., 2000, Plate interactions control middle-late Miocene, proto-Gulf and Basin and Range extension in the southern Basin and Range: Tectonophysics, v. 318, p. 1–26, doi: 10.1016/S0040-1951(99)00304-2.

Hole, M.J., Rogers, G., Saunders, A.D., and Storey, M., 1991, Relation between alkalic volcanism and slab-window formation: Geology, v. 19, p. 657–660, doi: 10.1130/0091-7613(1991)019<0657:RBAVAS>2.3.CO;2.

Humphreys, E.D., 1995, Post-Laramide removal of the Farallon slab, western United States: Geology, v. 23, p. 987–990, doi: 10.1130/0091-7613(1995)023<0987:PLROTF>2.3.CO;2.

Humphreys, E.D., Dueker, K.D., Schutt, D.L., and Smith, R.B., 2000, Beneath Yellowstone: Evaluating plume and nonplume models using teleseismic images of the upper mantle: GSA Today, v. 10, no. 12, p. 1–7.

Johnson, C.M., 1993, Mesozoic and Cenozoic contributions to crustal growth in the southwestern United States: Earth and Planetary Science Letters, v. 118, p. 75–89, doi: 10.1016/0012-821X(93)90160-B.

Johnson, C.M., and O'Neil, J.R., 1984, Triple junction magmatism: A geochemical study of Neogene volcanic rocks in western California: Earth and Planetary Science Letters, v. 71, p. 241–262, doi: 10.1016/0012-821X(84)90090-6.

Lehnert, K., Su, Y., Langmuir, C.H., Barbas, S., and Nohl, U., 2000, A global geochemical database structure for rocks: Geochemistry, Geophysics, Geosystems (G^3), v. 1.

Leighty, R.S., 1997, Neogene tectonism and magmatism across the Basin and Range–Colorado Plateau Boundary, Central Arizona [Ph.D. thesis]: Tempe, Arizona State University, 1019 p.

Lipman, P.W., 1980, Cenozoic volcanism in the western United States: Implications for continental tectonics: Continental Tectonics: Washington D.C., National Academy of Sciences of the United States, National Research Council, p. 161–174.

Lipman, P.W., Prostka, H.J., and Christiansen, R.L., 1971, Evolving subduction zones in the western United States, as interpreted from igneous rocks: Science, v. 174, p. 821–825.

Lipman, P.W., Prostka, H.J., and Christiansen, R.L., 1972, Cenozoic volcanism and plate-tectonic evolution of the western United States, I: Early and middle Cenozoic: Royal Society of London Philosophical Transactions, ser. A, v. 271, p. 217–248.

Livaccari, R.F., and Perry, F.V., 1993, Isotopic evidence for preservation of Cordilleran lithospheric mantle during the Sevier-Laramide orogeny, western United States: Geology, v. 21, p. 719–722, doi: 10.1130/0091-7613(1993)021<0719:IEFPOC>2.3.CO;2.

Lopez, R., and Cameron, K.L., 1997, High-Mg andesites from the Gila Bend Mountains, southwestern Arizona: Evidence for hydrous melting of lithosphere during Miocene extension: Geological Society of America Bulletin, v. 109, p. 900–914, doi: 10.1130/0016-7606(1997)109<0900:HMAFTG>2.3.CO;2.

Luhr, J.F., Pier, J.G., Aranda-Gomez, J.J., and Podosek, F.A., 1995, Crustal contamination in early basin-and-range hawaiites of the Los Encinos volcanic field, central Mexico: Contributions to Mineralogy and Petrology, v. 118, p. 321–329, doi: 10.1007/s004100050018.

Lynch, D.J., Musselman, T.E., Gutmann, J.T., and Patchett, P.J., 1993, Isotopic evidence for the origin of Cenozoic volcanic rocks in the Pinacate volcanic field, northwestern Mexico: Lithos, v. 29, p. 295–302, doi: 10.1016/0024-4937(93)90023-6.

MacLeod, N.S., Walker, G.W., and McKee, E.H., 1975, Geothermal significance of eastward increase in age of upper Cenozoic rhyolitic domes in southeastern Oregon: Proceedings of the Second United Nations Symposium on the Development and Use of Geothermal Resources, San Francisco, California, May 20–29, v. 1, p. 465–474.

Manduca, C.A., Silver, L.T., and Taylor, H.P., 1992, $^{87}Sr/^{86}Sr$ and $^{18}O/^{16}O$ isotopic systematics and geochemistry of granitoid plutons across a steeply dipping boundary between contrasting lithospheric blocks in western Idaho: Contributions to Mineralogy and Petrology, v. 109, p. 355–372, doi: 10.1007/BF00283324.

Manley, C.R., Glazner, A.F., and Farmer, G.L., 2000, Timing of volcanism in the Sierra Nevada of California: Evidence for the Pliocene delamination of the batholithic root?: Geology, v. 28, p. 811–814, doi: 10.1130/0091-7613(2000)028<0811:TOVITS>2.3.CO;2.

Mattox, S.R., 1994, Geochemistry and tectonic significance of late Cenozoic basalts in the Basin and Range-Colorado Plateau transition zone, Utah: Utah Geological Association Publications, v. 23, p. 125–138.

McDowell, F.W., and Mauger, R.L., 1994, K-Ar and U-Pb zircon chronology of Late Cretaceous and Tertiary magmatism in central Chihuahua State, Mexico, with Supplemental Data 9403: Geological Society of America Bulletin, v. 106, p. 118–132, doi: 10.1130/0016-7606(1994)106<0118:KAAUPZ>2.3.CO;2.

McKee, E.H., Best, M.G., Barr, D.L., and Tingey, D.G., 1993, Potassium-argon ages of mafic and intermediate-composition lava flows in the Great Basin of Nevada and Utah: Isochron/West, v. 60, p. 15–18.

McMillan, N.J., Dicking, A.P., and Haag, D., 2000, Evolution of magma source regions in the Rio Grande Rift, southern New Mexico: Geological Society of America Bulletin, v. 112, p. 1582–1593, doi: 10.1130/0016-7606(2000)112<1582:EOMSRI>2.0.CO;2.

Miller, J.S., Glazner, A.F., Farmer, G.L., Suayah, I.B., and Keith, L.B., 2000, Middle Tertiary magmatism across the Mojave Desert and southeastern California: A Sr, Nd, and Pb isotopic study of mantle domains and crustal structure: Geological Society of America Bulletin, v. 112, p. 1264–1279, doi: 10.1130/0016-7606(2000)112<1264:ASNAPI>2.3.CO;2.

Miller, R.J., and Wrucke, C.T., 1995, Age, chemistry, and geologic implications of Tertiary volcanic rocks in the Last Chance Range and part of the Saline Range, northern Death Valley region, California: Isochron/West, v. 62, p. 30–36.

Mutschler, F.E., Larson, E.E., and Bruce, R., 1987, Laramide and younger magmatism in Colorado: Colorado School of Mines Quarterly, v. 82, p. 1–45.

Nagy, E.A., Grove, M., and Stock, J.M., 1999, Age and stratigraphic relationships of pre- and syn-rift volcanic deposits in the northern Puertecitos volcanic province, Baja California, Mexico: Journal of Volcanology and Geothermal Research, v. 93, p. 1–30, doi: 10.1016/S0377-0273(99)00080-3.

Nakada, M., and Yokose, H., 1992, Ice age as a trigger of active Quaternary volcanism and tectonism: Tectonophysics, v. 212, p. 321–329, doi: 10.1016/0040-1951(92)90298-K.

Nealey, L.D., Budahn, J.R., Maldonado, F., and Unruh, D.M., 1997, Geochemistry and petrogenesis of Quaternary basaltic rocks from the Red Hills and western Markagunt Plateau, southwestern Utah: U.S. Geological Survey Bulletin, p. 177–198.

Nelson, S.T., and Tingey, D.G., 1997, Time-transgressive and extension-related basaltic volcanism in southwest Utah and vicinity: Geological Society of America Bulletin, v. 109, p. 1249–1265, doi: 10.1130/0016-7606(1997)109<1249:TTAERB>2.3.CO;2.

Othberg, K.L., Bonnichsen, B., Swisher, C.C., and Godchau, M.M., 1995, Geochronology and geochemistry of Pleistocene basalts of the western Snake River Plain and Smith Prairie, Idaho: Isochron/West, v. 62, p. 16–29.

PetDB, 2002, Schema for PetDB database: http://www.petdb.org/.

Pierce, K.L., and Morgan, L.A., 1992, The track of the Yellowstone hot spot: Volcanism, faulting, and uplift, *in* Link, P.K., et al., eds., Regional geology of eastern Idaho and western Wyoming: Geological Society of America Memoir 179, p. 1–53.

Reynolds, S.J., Florence, F.P., Welty, J.W., Roddy, M.S., Currier, D.A., Anderson, A.V., and Keith, S.B., 1986, Compilation of radiometric age determinations in Arizona: Arizona Bureau of Geology and Mineral Technology Bulletin 197, 258 p.

Ring, J.H., 2000, Young volcanism in western Long Valley Caldera, California [master's thesis]: Stanford, California, Stanford University, 202 p.

Severinghaus, J., and Atwater, T., 1990, Cenozoic geometry and thermal state of the subducting slabs beneath western North America, *in* Wernicke, B. P., ed., Basin and Range extensional tectonics near the latitude of Las Vegas, Nevada: Geological Society of America Memoir, v. 176, p. 1–22.

Smith, E.I., Sanchez, A., Walker, J.D., and Wang, K., 1999, Geochemistry of mafic magmas in the Hurricane volcanic field, Utah: Implications for small- and large-scale chemical variability of the lithospheric mantle: Journal of Geology, v. 107, p. 433–448, doi: 10.1086/314355.

Stein, H.J., and Crock, J.G., 1990, Late Cretaceous–Tertiary magmatism in the Colorado Mineral Belt: Rare earth element and samarium-neodymium isotopic studies: Geological Society of America Memoir, v. 174, p. 195–223.

Stroud, J.R., 1997, The geochronology of the Raton-Clayton volcanic field, with implications for volcanic history and landscape evolution [M.S. thesis]: Socorro, New Mexico, New Mexico Institute of Mining and Technology, 116 p.

Suppe, J., Powell, C., and Berry, R., 1975, Regional topography, seismicity, Quaternary volcanism, and the present-day tectonics of the western United States: American Journal of Science, v. 275A, p. 397–417.

Suter, M., Carillo-Martinez, M., Lopez-Martinez, M., and Farrar, E., 1995, The Aljibes half-graben: Active extension at the boundary between the trans-Mexican volcanic belt and the Basin and Range Province, Mexico: Geological Society of America Bulletin, v. 107, p. 627–641, doi: 10.1130/0016-7606(1995)107<0627:TAHGAE>2.3.CO;2.

Walker, J.D., and Coleman, D.S., 1991, Geochemical constraints on mode of extension in the Death Valley region, with Supplementary Data 91-28: Geology, v. 19, p. 971–974, doi: 10.1130/0091-7613(1991)019<0971:GCOMOE>2.3.CO;2.

Wilks, M., and Chapin, C.E., 1997, New Mexico geochronological database: New Mexico Bureau of Geology and Mineral Resources Digital Data Series DB1 (http://geoinfo.nmt.edu/data/ims/geochron.html).

Yogodzinski, G.M., Naumann, T.R., Smith, E.I., Bradshaw, T., and Walker, J.D., 1996, Evolution of a mafic volcanic field in the central Great Basin, south central Nevada: Journal of Geophysical Research, v. 101, p. 17,425–17,445, doi: 10.1029/96JB00816.

Zartman, R.E., Bush, C.A., and Abston, C.C., 1995, National geochronological and natural radioelement data bases: U.S. Geological Survey Digital Data Series DDS-0014 (http://pubs.usgs.gov/publications/1995-02.digital.html).

Zoback, M.L., McKee, E.H., Blakely, R.J., and Thompson, G.A., 1994, The northern Nevada rift; regional tectono-magmatic relations and middle Miocene stress direction: Geological Society of America Bulletin, v. 106, p. 371–382, doi: 10.1130/0016-7606(1994)106<0371:TNNRRT>2.3.CO;2.

Manuscript Accepted by the Society 18 June 2005

Printed in the USA

Geological Society of America
Special Paper 397
2006

Building the American Mineralogist Crystal Structure Database: A recipe for construction of a small Internet database

Hareesh Rajan
Department of Computer Science, University of Arizona, Tucson, Arizona 85721-0077, USA

Hinako Uchida
Deborah L. Bryan
Department of Geosciences, University of Arizona, Tucson, Arizona 85721-0077, USA

Ranjini Swaminathan
Department of Computer Science, University of Arizona, Tucson, Arizona 85721-0077, USA

Robert T. Downs
Michelle Hall-Wallace
Department of Geosciences, University of Arizona, Tucson, Arizona 85721-0077, USA

ABSTRACT

Crystal structure data represent one of the most important resources for developing scientific knowledge and should be archived in ways that make them easy to access and preserve. The American Mineralogist Crystal Structure Database currently contains every crystal structure published in *American Mineralogist, The Canadian Mineralogist, European Journal of Mineralogy,* and *Physics and Chemistry of Minerals*. It is maintained by the American and Canadian mineralogical societies and is freely accessible through the Internet. The database consists of the data, server-side search and retrieval software and user-side analysis software. It is managed through a partnership of PHP and MySQL programming that provide dynamic construction of Web pages and search procedures. The purpose of this paper is to describe the database and its implementation and to illustrate how to construct similar small, interactive Internet databases.

Keywords: crystal structure, database, visualization.

INTRODUCTION

The literature in many disciplines is full of small data sets created through individual studies, which over decades and even centuries have defined the core of understanding for the discipline. These data sets are fully recoverable by an experienced individual researcher, but the recovery often requires significant amounts of time and effort. The move to digital scientific journals is one step toward making current data sets more readily available, as is the creation of many new digital libraries. We have created a comprehensive database of all crystal structure data published in the journals of the Mineralogical Society of America, the Mineralogical Association of Canada, and the European societies through a time-consuming yet richly rewarding process. The database not only provides specific information about any one of 8000 different crystal structures in one location, but it also

Rajan, H., Uchida, H., Bryan, D.L., Swaminathan, R., Downs, R.T., and Hall-Wallace, M., 2006, Building the American Mineralogist Crystal Structure Database: A recipe for construction of a small Internet database, *in* Sinha, A.K., ed., Geoinformatics: Data to Knowledge: Geological Society of America Special Paper 397, p. 73–80, doi: 10.1130/2006.2397(06). For permission to copy, contact editing@geosociety.org.

provides freely available analysis software that is fully integrated with the database's data retrieval system.

The American Mineralogist Crystal Structure Database consists of (1) a collection of crystal structure data, (2) an Internet site that allows search and retrieval of data sets using a browser, (3) server-based PHP: Hypertext Preprocessor (or PHP) and FTP (file transfer protocol) software to provide direct access to the data using PHP queries, and (4) user-based software that provides for interactive analysis. The database can be found at http://www.geo.arizona.edu/AMS/amcsd.php or from the Web sites of the mineralogical societies. In this paper, we describe the unique characteristics of the data and their storage, development of the database, and the tools for integrated data retrieval and analysis. Our goal is to describe the functionality of the American Mineralogist Crystal Structure Database and to provide information on how to construct a similar database and Web interface. Downs and Hall-Wallace (2002) provide a detailed description of the purpose of the database and its user-based analysis software.

The structure of a crystal represents a minimum energy configuration adopted by a collection of atoms at some set of conditions. In principle, all the physical and chemical properties of any crystalline substance can be computed from its crystal structure. The determination of crystal structures and the deduction and understanding of these computational algorithms constitutes a major part of scientific research in physics, chemistry, biology, medicine, mineralogy, geology, and material sciences. As such, crystal structure data represent one of the most important resources for developing our scientific knowledge and thus should be archived in ways that make them easy to access and preserve. However, these data are often cumbersome to retrieve and verify from the literature, and even more difficult to analyze without a strong background in crystallography. Thus, the National Science Foundation, in cooperation with the Mineralogical Society of America, the Mineralogical Association of Canada, and the University of Arizona, has established a collection of mineralogically important crystallographic data sets that are freely accessible through the Internet. This database is an important resource to both of these mineralogical societies, and they are committed to maintain it as part of their outreach programs.

GATHERING AND PREPARING THE DATA SETS

The database contains every experimentally determined crystal structure reported in *American Mineralogist*, *The Canadian Mineralogist*, *European Journal of Mineralogy*, and *Physics and Chemistry of Minerals*. At present there are about 8000 individual data sets representing about 7000 unique mineral and chemical phases. Data from *Acta Crystallographica* are now being added. The data represent structures determined at ambient conditions as well as at temperature or pressure. Assembling the database is a multistep process that includes:

- examining each volume of the journals to identify all papers that report crystal structures;
- manually entering the reported data in the database;
- computing various crystal chemical parameters and verifying the consistency of these parameters with those that are reported;
- contacting the authors about irresolvable inconsistencies between reported and computed parameters; and
- incorporating comments when changes are made to the originally published data.

Each record in the database consists of a bibliographic reference, cell parameters, symmetry, and atomic positions, as well as displacement parameters and site occupancies when applicable. An example of retrieved data sets is provided in Figure 1.

The first part of each data set contains identifying information, bibliography, and notes, while the second part of a data set contains the crystallographic parameters. The first line of a data file contains an identifier, such as the name of the mineral or formula of the chemical species. The next lines contain the names of the authors, followed by the journal reference, title of the paper, and any additional notes. The crystallographic data begin with a listing of the cell parameters and space group. The rest of the data set is a listing of the atoms, their positional and displacement parameters, and occupancies. A header is provided that defines right-justified columns. The name of each atom identifies the occupying elements, with additional identifiers added when appropriate. For instance, "Oco" identifies a particular oxygen atom in the albite structure. Each data set is stored in a separate and uniquely named ASCII file with the extension .amc and is put in a read-only directory of an FTP server.

Quality control is a major concern, and considerable effort is expended to ensure that the data are correct. Each data set is verified for accuracy and omissions before being added to the database. We use METRIC, software developed by Bartelmehs et al. (1993) at Virginia Tech, to compute the crystal chemical parameters that are typically found in many crystal structure publications. The crystal chemical parameters include bond lengths, angles, polyhedral volumes, distortion indices, and such, as recently reviewed in Hazen et al. (2000). The crystal chemical parameters calculated by METRIC are compared with those included in the original publication to provide a check of the internal consistency of the data. If a discrepancy is found, then we try to determine its source and correct it. Many errors are typographical, while others are more complex. The authors are contacted to correct an error, if possible. The database contains only the corrected data set.

Developing a fast and efficient search engine requires the data to be well organized. Storing all the data sets within a large single file was initially considered but rejected in order to simplify the transfer of individual data sets between our server and a user's computer. Hence, the data sets are stored and maintained in individual files, with each file containing the data for a single crystal structure. Individual data files are simple to manage and edit. More importantly, in the event of unforeseen needs, the format of a single data file can be easily modified. The data sets are stored as individual ASCII files, averaging ~20 lines each. It is anticipated that the final set of data will consist of ~50,000 files.

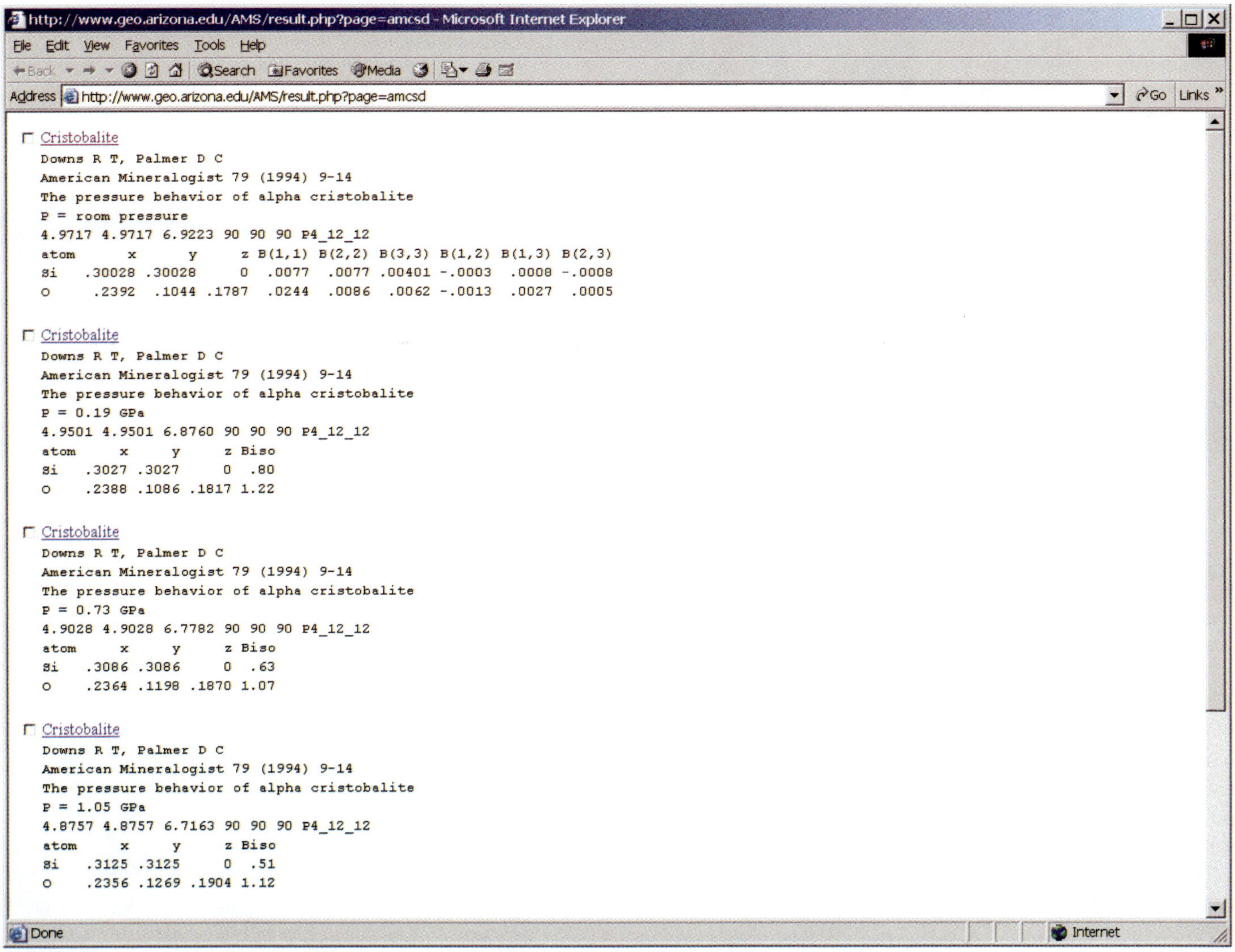

```
http://www.geo.arizona.edu/AMS/result.php?page=amcsd - Microsoft Internet Explorer
File Edit View Favorites Tools Help
Address http://www.geo.arizona.edu/AMS/result.php?page=amcsd

Cristobalite
Downs R T, Palmer D C
American Mineralogist 79 (1994) 9-14
The pressure behavior of alpha cristobalite
P = room pressure
4.9717 4.9717 6.9223 90 90 90 P4_12_12
atom      x      y      z B(1,1) B(2,2) B(3,3) B(1,2) B(1,3) B(2,3)
Si    .30028 .30028     0  .0077  .0077 .00401 -.0003  .0008 -.0008
O      .2392  .1044 .1787  .0244  .0086  .0062 -.0013  .0027  .0005

Cristobalite
Downs R T, Palmer D C
American Mineralogist 79 (1994) 9-14
The pressure behavior of alpha cristobalite
P = 0.19 GPa
4.9501 4.9501 6.8760 90 90 90 P4_12_12
atom     x     y     z Biso
Si   .3027 .3027     0  .80
O    .2388 .1086 .1817 1.22

Cristobalite
Downs R T, Palmer D C
American Mineralogist 79 (1994) 9-14
The pressure behavior of alpha cristobalite
P = 0.73 GPa
4.9028 4.9028 6.7782 90 90 90 P4_12_12
atom     x     y     z Biso
Si   .3086 .3086     0  .63
O    .2364 .1198 .1870 1.07

Cristobalite
Downs R T, Palmer D C
American Mineralogist 79 (1994) 9-14
The pressure behavior of alpha cristobalite
P = 1.05 GPa
4.8757 4.8757 6.7163 90 90 90 P4_12_12
atom     x     y     z Biso
Si   .3125 .3125     0  .51
O    .2356 .1269 .1904 1.12

Done                                        Internet
```

Figure 1. This example of a retrieved set of data for cristobalite illustrates the format of the data. Four data sets are displayed in the figure. The underlined mineral names indicate FTP hyperlinks that provide automatic launching of user-defined software. Check boxes are provided for each data set to permit the download of a single file that contains several selected data sets.

THE VISUALIZATION AND ANALYSIS SOFTWARE

Crystallographic parameters are the foundation of research in material properties, and being able to visualize and manipulate crystal structures is fundamental to understanding them. A suite of Windows®-based visualization and analysis tools, fully integrated with the database, has been developed to make the database more useful as a scientific and educational tool. Three key applications are (1) XtalDraw, an application that creates graphical representations of a crystal structure, (2) METRIC, described above, and (3) XPOW, an application that computes and displays a theoretical powder diffraction pattern of the crystal. The software currently only operates in Windows®.

When a crystal structure in the database is accessed, its image can automatically be displayed with the XtalDraw application (Bartelmehs et al., 1993; Downs and Bartelmehs, 1996; Hazen and Downs, 1996; Downs, 1998; Hazen et al., 2000). Currently, this stand-alone software contains options to draw the crystal structure with ball-and-stick, polyhedral, and thermal ellipsoidal renderings (Fig. 2A). The user can rotate the image with the arrow keys, or manually enter directions to view in direct or reciprocal space coordinates. Users can expand or shrink the image, add or delete atoms from the field of view, change the colors of atoms, their sizes, the bonds, and so forth. The software produces publication-quality bitmaps. Carefully selected default viewing parameters are one very powerful feature of XtalDraw that make it easy to use. For instance, after a data file is read, the structure is initially displayed in a default orientation established by the International Tables for Crystallography. The atoms are drawn in a set of colors suggested by Lipson and Cochran (1957)

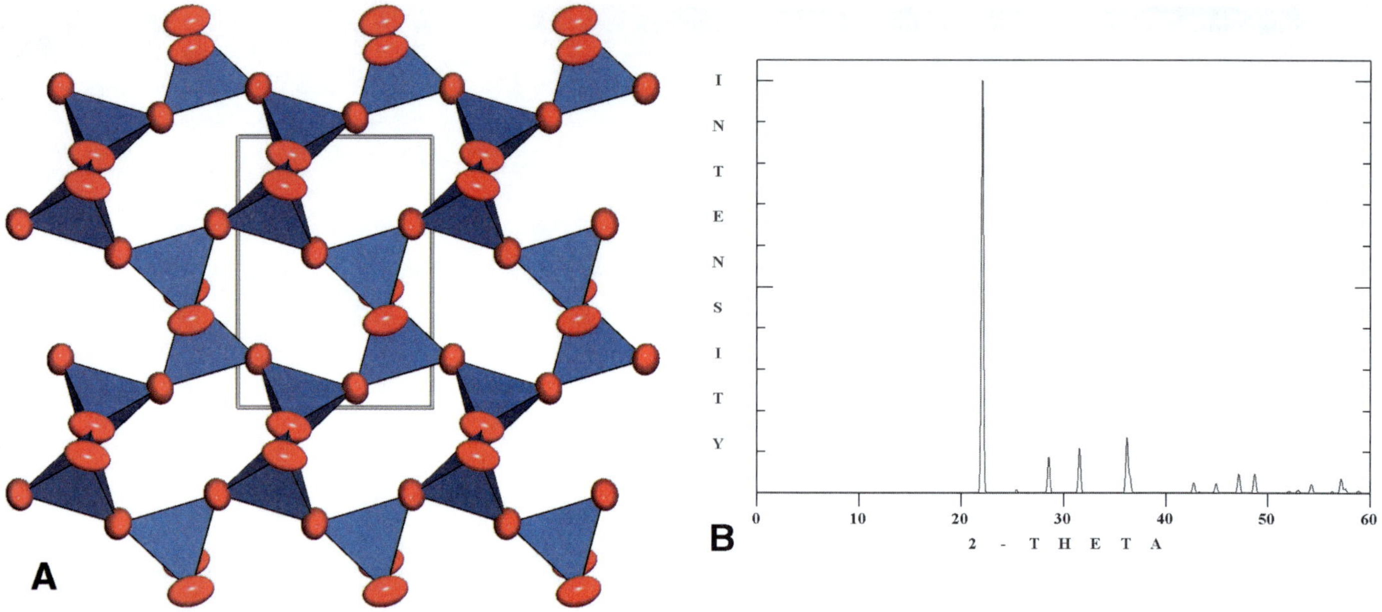

Figure 2. Two images that illustrate the output produced by the accessory software. A: A mixed polyhedral and thermal ellipsoid rendering of the crystal structure of cristobalite produced by XtalDraw. B: A simulation of an X-ray powder diffraction pattern computed for the structure of cristobalite by XPOW.

to represent elemental species, and in sizes that scale to the Shannon and Prewitt (1969) radii. Atoms can be classified as cations or anions, and the coordination sphere is automatically estimated for polyhedral representations. Users have the option to change these settings at any time and to establish their own default settings. As a feature of XtalDraw, the METRIC software can be launched to calculate many of the important crystal chemical parameters that are used by today's researchers, such as bond lengths and angles, polyhedral volumes and distortion parameters, and vibrational amplitudes and orientations (Hazen et al., 2000).

Another important property of a crystal that can be computed from its structural data is its diffraction pattern. Neutron and X-ray powder diffraction patterns, generated from a fixed source or from an energy-dispersive source, such as at a synchrotron, can be computed with the software XPOW (Downs et al., 1993). XPOW can calculate the diffraction pattern for any crystalline substance and provide an interactive display of the pattern, as it would be generated by a conventional powder diffractometer (Fig. 2B). The user can alter the radiation source wavelength, the peak widths, limits, and orientation of the display. It is also of great use to identify unknown materials by providing a complete diffraction pattern, and not just the few d-spacings listed in a search-match table.

By using the FTP hyperlink, discussed in greater detail below, the software can be automatically launched by a visit to the site and selection of a data set. The browser or operating system must be configured to automatically associate .amc files with the application. The simple structure of the data sets and database also allows any other software application capable of receiving instructions through "command line arguments" (as is typical in drag-and-drop applications) to open and manipulate the data.

DATA RETRIEVAL METHODS

A good database should be comprehensive, accurate, and easily analyzed; it must also be easily accessed, and data must be easily retrieved. Currently, access is provided through three methods: (1) a browser interface with an interactive search procedure, (2) retrieval by sending a PHP query directly to the database without use of the Web search procedure, and (3) direct FTP from an anonymous FTP site.

The interactive search procedure consists of an HTML page (Fig. 3) that allows the user to specify several criteria, including the names of the minerals or chemical species, the authors and titles of the relevant journal articles, mineral chemistry, unit cell dimensions, and space group symmetry. There is also a tool to search for arbitrary words or phrases. Furthermore, search criteria can be combined with the logical conditions *or* and *and*. The interface is constructed dynamically using PHP programming, and the search is conducted with the database management software MySQL. We use an Apache Web server running on a Linux platform, which is well suited for PHP and MySQL.

After the search criteria are defined and the search is initialized, the PHP program calls MySQL, which returns the names of appropriate .amc data files. A search may result in a single data set or multiple data sets. The PHP program collates the contents of the matching data files and displays them in a new browser window. Each of the data sets has an FTP hyperlink

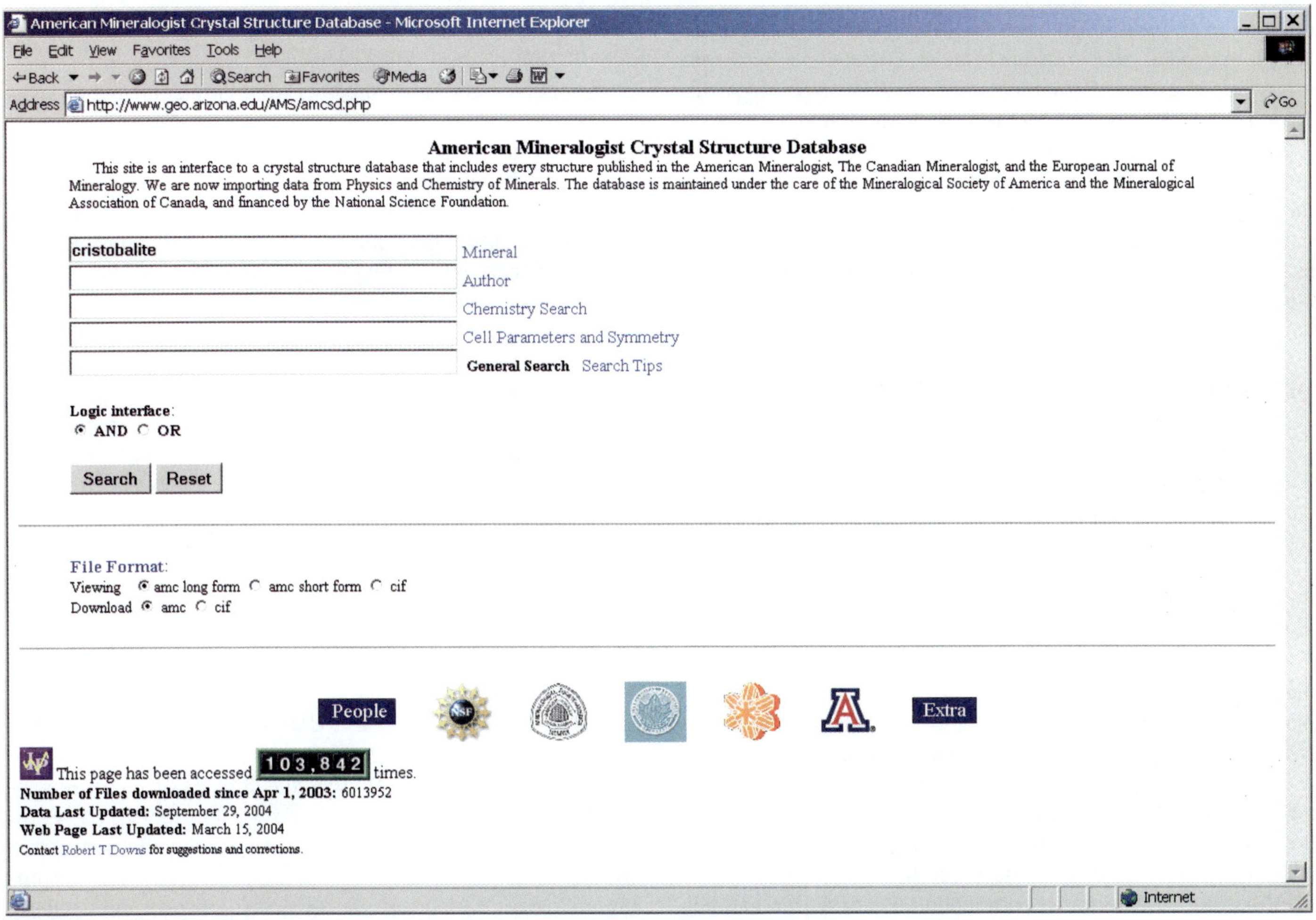

Figure 3. The American Mineralogist Crystal Structure Database Web page. This is the interface that can be used to search for crystal structure data files. The HTML code for this page is created dynamically with PHP programming. In this image, the mineral name "cristobalite" was been typed in the edit box. Clicking the *search* button will retrieve the data sets for cristobalite and display the data as shown in Figure 1.

associated with it. To retrieve the data, the user can (1) cut and paste the relevant text from the browser into another application, or (2) select the FTP hyperlink(s) and download the entire set of matching data or any subset of it. The FTP hyperlink has the form, for example,

<a href = ftp://ftp.geo.arizona.edu/pub/xtal/data/AMCfiles /01553.amc>Cristobalite</a>.

When activated, this hyperlink automatically transfers the data file from our server to the user's computer. The user is relieved from the trouble of directly dealing with the transfer. Furthermore, the FTP link provides the ability to launch an application directly on a user's computer simply by clicking the link. If the file name extension of the data, .amc, has been associated with an application on the user's local computer, then the application will be automatically launched and will open the file. For instance, XtalDraw and XPOW can be launched in this manner. This feature is modeled after the .pdf interface with Adobe Acrobat Reader®.

In addition to accessing data through the interactive search procedure, sending a PHP query directly to the Web server can initialize a search. A successful search brings up the data, as in Figure 1, without going through the search interface Web site. This method for accessing the data is currently being employed by databases such as that found at http://www.webmineral.com and is also a good way to provide access to the data through educational Web pages that need to retrieve a specific file. Access is initialized through a hyperlink that launches a PHP query, for example,

http://www.geo.arizona.edu/AMS/result.php?mineral=crist obalite.

This command produces a search for cristobalite using the *mineral* table in the database. The details for constructing searches of this type are given at the Web site.

Users can also gain direct access to the entire database, through an FTP connection to our anonymous FTP site. This enables the secure transfer of files between two computers and is often more efficient when large numbers of files are being transferred. In general, FTP software is more cumbersome to use than a browser, thus we also incorporated the FTP hyperlinks, discussed earlier, into our interactive search procedure.

Appropriate security measures are taken to prevent malicious users from damaging the data files. These include constraining the storage directory as the only directory on the server to have public access, and defining this directory and the file attributes as "read only." Users are not permitted to upload their files into our storage directory. These security issues are operating system dependent and must be addressed on an individual basis.

DESIGNING AND MANAGING THE DATABASE

We use MySQL, fast and reliable database management software, combined with the PHP scripting language to create our database structure, to populate and maintain the database, and to support data retrieval. These are two of the most widely accepted tools for Internet database applications. MySQL runs on all platforms and performs exceptionally well for small databases such as ours. Furthermore, it is open-source and hence cost-effective.

The type of data we accommodate and the requirements of the interactive search procedure guide the database design and management. The database must associate data set file names with the search criteria, which include *minerals*, *authors*, *titles*, *chemical elements*, *cell parameters*, *space group symmetry*, and a general *keyword* search. For each searchable criterion, we have created a table using the MySQL *Create* command. Each table is mainly composed of individual records containing the entity identifier in one field and a corresponding data set file name in another. For example the *minerals* criteria table contains records with a mineral name (e.g., cristobalite) in one field, and a file name containing the crystal structure data set for that mineral (e.g., 01553.amc) in another field. The *authors* criteria have a table with records that contain a field with an author's name and another field with a file name containing the associated data set. The database may contain several different data sets for each mineral or for each author, and each one of these data sets constitutes a separate record in the table.

The MySQL *Create* command constructs the table but does not put any data into it. PHP scripts read the individual data files, extract the corresponding entities, and insert them into the corresponding criteria table. The *minerals* criteria table thus contains all the minerals present in the data files along with the names of their associated files. The *chemical elements* table contains all the atoms present in each of the crystal structures along with their associated file names. Similarly, the table for *cell parameters* and *space group symmetry* contains the values for a, b, c, α, β, γ, and *space group*, along with their corresponding file names. Our database undergoes regular updating because new data sets are continually being added. The PHP scripts read all the data files and rebuild the entire MySQL database during each update.

Once the tables are created and the appropriate data inserted into them, the database is complete and ready for query purposes. Data are retrieved and viewed in a variety of ways, using the MySQL *Select* commands. For instance, a simple query

Select filename **from** minerals **where** mineralname = "Cristobalite"

will retrieve all the file names associated with the mineral "cristobalite" from the *minerals* table. Other queries are performed using the refinements of the *Select* commands, including *Order By* (sorting routines) and *Distinct* (removes duplicate entities) clauses.

The method of searching for cell parameters is a little different than for the other fields and requires different handling in the database design. The cell parameters are stored as real numbers rather than as text strings. The table for the cell parameters has seven different fields: one for file name and six others for the six different cell parameters. A user searching for cell parameters usually would not know their exact values, and instead could only provide an estimate. The search criteria are established through a Web page as shown in Figure 4. MySQL supports searches for real numbers that lie within a given range or tolerance. This option is used for searching for cell parameters and is invoked with a command such as

Select filename **from** cell_parameters **where** $a >= 4.97$ **and** $a <= 4.99$.

MANAGING THE WEB INTERFACE THROUGH PHP

The user interface to the searchable fields includes interactive HTML elements such as text boxes, drop-down list boxes, radio buttons, push buttons, etc., created with the server-side scripting language PHP. More importantly, PHP securely supports the MySQL database. It is the link between the database and the Web page.

The Web search interface is dynamically created with PHP using the *Select* commands to access the criterion tables in the MySQL database. The results of the queries populate the search criteria fields in the HTML form. The user is presented a detailed listing in a drop-down list box of all the *minerals*, *authors*, *titles*, and *space groups* available in the database. The *chemical elements* search provides a periodic table from which the user can select elements to constrain a search (Fig. 5) and is described in detail below. The *cell parameter* and *space group* search is done through another Web page illustrated in Figure 4 and described above.

Each of the different search tools is used to create search requests that are processed using scripts written in PHP, which provides numerous built-in functions to access the MySQL database. Any interaction with the database requires PHP to establish a connection to the MySQL server using the *mysql_connect ()* function. Once connected, our database is selected using the *mysql_select_db ()* function. All the queries in the script are subsequently directed to the database. Database queries are made using the *mysql_query ()* function, which creates SQL statements to be executed by the MySQL server. The results of the query are retrieved using the *mysql_fetch_row ()* function.

The Web server, on receiving a search request from the user, processes the PHP scripts to connect, select, query, and fetch the

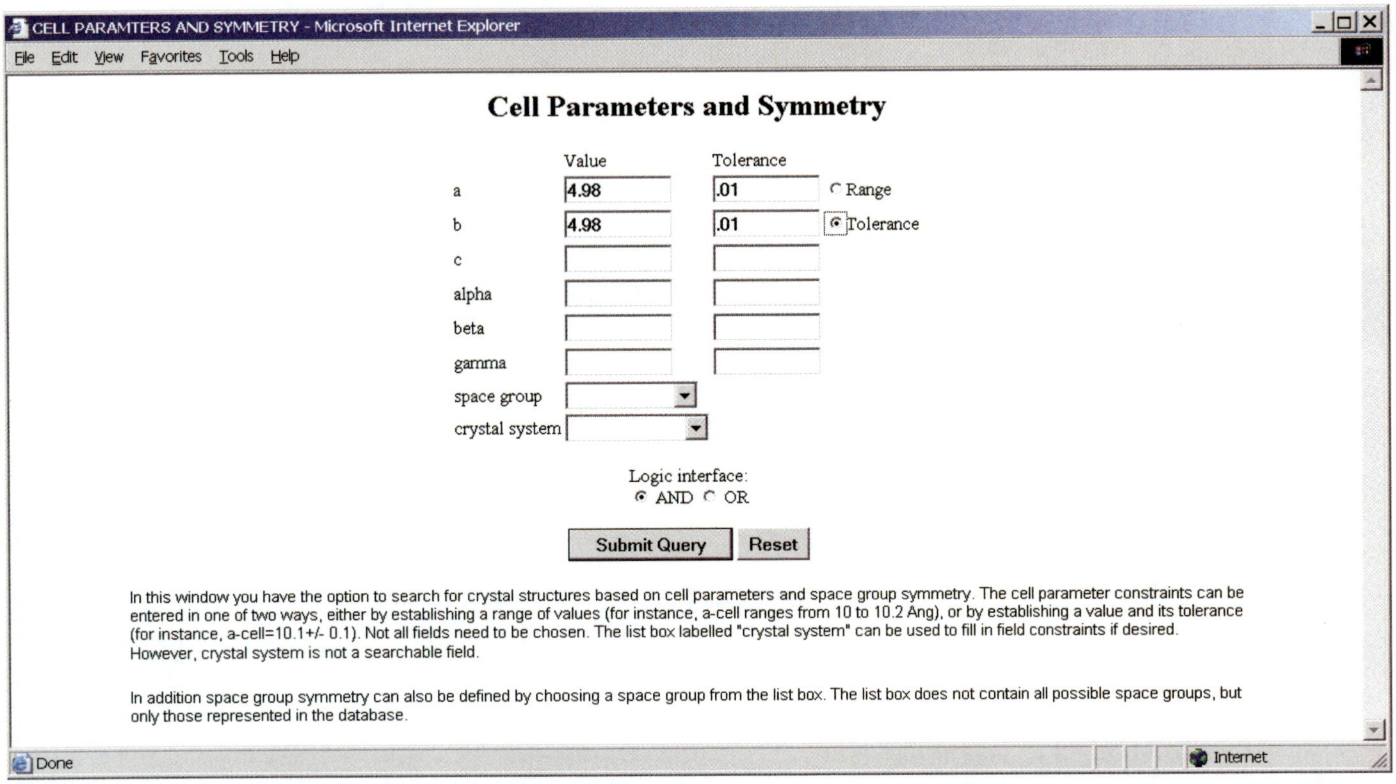

Figure 4. The Web page that is used to define cell parameters and space group symmetry. A search for cell parameters is handled in a different way than a search for the other fields because cell parameters are real numbers that normally can only be estimated by a user. Thus, the user defines a search for cell parameters by providing a search interval, or by providing a median value with a tolerance.

data. The MySQL database responds by executing the queries and sending the names of the files containing the selected items in a PHP script. PHP stores the results in one or more variables and uses the *print()* function to display the data files in HTML format, as output to the user. The *Limit* clause of the MySQL *Select* command is used to define the number of data files displayed on a single page. We designed a single page to display up to 100 data sets at a time. Individual check boxes associated with the output of each data set allow the user to choose a subset of the retrieved data. The user can select all or a subset of the data by clicking the check boxes. Then, clicking the *process the data* button launches the FTP command to transfer the selected data sets (Fig. 1).

Even though handling the search for chemistry can be done in the same way as a search for *mineral name*, *author*, or *title*, the complexity of a chemical search requires a different interface. This is because a good chemical search routine must permit the selection of many different elements at a time, as well as the exclusion of many elements. Therefore, we constructed an interactive and user-friendly periodic table interface (Fig. 5) for the *chemical elements* search, again managed by PHP. Users make their selection of elements directly from the periodic table by clicking the elements. This is mainly designed with the help of CSS (cascading style sheets), which offer great flexibility in controlling the overall presentation of the HTML document and also significantly reduce the amount of coding required for formatting. Each chemical element has an associated button that records one of three conditions for the element: (1) Find data that include this element, (2) find data that do not include this element, and (3) ignore the presence of this element in the data set. For example, all silicate minerals can be obtained by clicking only Si and O. All other elements are, by default, in state (3) and thus can be present in a data set. To obtain only the SiO_2 silica polymorphs, all chemical elements other than Si and O must be put into the exclude state (2). Instead of clicking each and every element to put them into this state, an *exclude others* button is provided that defines all other elements to be in the exclude state.

Lastly, a general *keyword* search is conducted using the individual data set files rather than the database tables. The general *keyword* search is done using the familiar *grep* command of Unix, which searches through the directory of individual data files for the desired words or phrases and returns the name of the files containing it. The *grep* command can be invoked from PHP using suitable system functions. The general search requires the user to input either words or phrases in the text box, with the words being case-insensitive. Since the search request for the *grep* command comes from the user input, care is taken to prevent the user from tricking the system

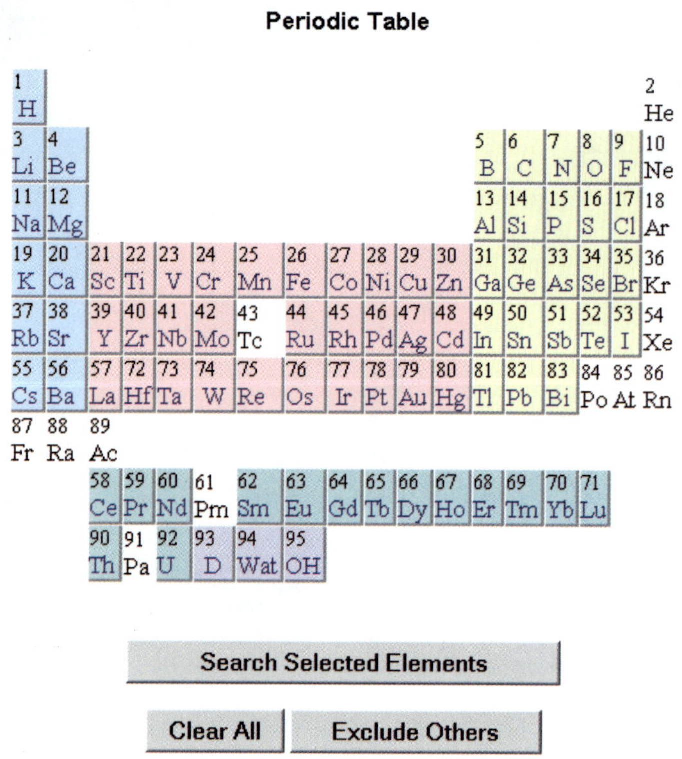

Figure 5. The Web page that is used to select chemical elements for the database search. It is created dynamically with PHP programming and cascading style sheets. Each element is in a table cell that behaves like a button and can be clicked into one of three different selection states.

into executing arbitrary commands. The *grep* utility is fast for searching, easy to implement, and fits our application's requirements. Other means for searching through a collection of files include search engines such as Web Glimpse or Google.

REFERENCES CITED

Bartelmehs, K.L., Gibbs, G.V., Boisen, M.B., Jr., and Downs, R.T., 1993, Interactive computer software used in teaching and research in mineralogy at Virginia Tech: Geological Society of America Abstracts with Programs, v. 25, no. 6, p. A-347.

Downs, R.T., 1998, Computer graphics simulation of compression mechanisms in crystals: IUCR-HP98, Argonne, Illinois, Abstracts, v. 21.

Downs, R.T., and Bartelmehs, K.L., 1996, Computer visualization of temperature and pressure effects on crystal structures: Eos (Transactions, American Geophysical Union), Spring Meeting Supplement, v. 77, no. 17, p. S261.

Downs, R.T., and Hall-Wallace, H., 2002, The American Mineralogist Crystal Structure Database: American Mineralogist, v. 87, p. 247–250.

Downs, R.T., Bartelmehs, K.L., Gibbs, G.V., and Boisen, M.B., Jr., 1993, Interactive software for calculating and displaying X-ray or neutron powder diffractometer patterns of crystalline materials: American Mineralogist, v. 78, p. 1104–1107.

Hazen, R.M., and Downs, R.T., 1996, Systematic crystal chemistry of high-pressure silicates: An interactive graphics demonstration: International Union of Crystallography XVII Congress and General Assembly, Seattle Washington, C-543.

Hazen, R.M., Downs, R.T., and Prewitt, C.T., 2000, Principles of comparative crystal chemistry, *in* Hazen, R.M., and Downs, R.T., eds., Comparative crystal chemistry: Washington, D.C., Mineralogical Society of America, Reviews in Mineralogy and Geochemistry, v. 41, p. 1–33.

Lipson, H., and Cochran, W., 1957, The determination of crystal structures: London, G. Bell and Sons, 345 p.

Shannon, R.D., and Prewitt, C.T., 1969, Effective ionic radii in oxides and fluorides: Acta Crystallographica, v. B25, p. 925–945.

Manuscript Accepted by the Society 18 June 2005

Printed in the USA

Geological Society of America
Special Paper 397
2006

Geoscience concept models

Stephen M. Richard*
Arizona Geological Survey, 416 W. Congress, #100, Tucson, Arizona 85701, USA

ABSTRACT

Computer information systems that will archive, query, retrieve, and display geologic information tailored to specific requirements are the foundation for a leap in scientific development similar to that fostered by the invention of the geologic map. In order to achieve this potential, some standardization of the conceptual model for basic elements of geoscience is required to provide a consistent framework for developing interoperable systems. Models for three concepts that are central to geologic science, Earth material, geologic units, and geologic structure, are proposed as a starting point for this framework. An Earth material is a substance defined by chemical constituents, in concert with crystal structure, physical properties, or properties related to the nature and arrangement of constituent particles. Earth material is a mass noun, not countable. A geologic unit is a part of Earth located and distinguished from other parts of Earth based on geologic properties. Geologic units are countable. A geologic structure is a configuration of Earth material within Earth, and may or may not be countable. The existence of a geologic structure requires the existence of some Earth material substrate. A top-level vocabulary defines subclasses of these concepts, and description schemas specify relationships and attributes used to characterize defined classes and instances that extend the top-level subclasses. The geologic models in this paper are consistent with the NADM-C1 model of the North American Geologic-Map Data Model Steering Committee with minor modifications, and addition of detail on a number of important points. A distinct knowledge-representation framework provides a foundation for the geoscience domain models. In particular, modeling of a geologic map as a knowledge representation device is separated from the problem of geoscience knowledge representation in general. Models are presented as Unified Modeling Language static class diagrams.

Keywords: knowledge representation, data model, ontology, database, geologic time, geologic unit, earth material, geologic structures.

INTRODUCTION

This paper describes models developed as a means to design and implement an information system for archiving, accessing, and analyzing geoscience data. In the course of developing these models, it has become apparent that geology as a science is in need of a greater degree of formalization in order to take advantage of evolving computer-aided knowledge representation and analysis systems. The complexity of Earth has precluded much formalization thus far in the development of geoscience. Rapid evolution in the development of knowledge representation systems, along with the maturing of global geoscience theory is

*E-mail: steve.richard@azgs.az.gov

Richard, S.M., 2006, Geoscience concept models, *in* Sinha, A.K., ed., Geoinformatics: Data to Knowledge: Geological Society of America Special Paper 397, p. 81–107, doi: 10.1130/2006.2397(07). For permission to copy, contact editing@geosociety.org.

creating a situation in which such formalization is becoming possible. The exercise of defining exactly what the various basic geologic concepts mean and schematically representing the relationships between them has certainly sharpened my thinking with respect to the work I do as a geologic mapper. The models presented here are meant to stimulate thinking about the basis of our science and to contribute to the development of the next generation of geologic map tools.

Geologic maps have been the primary means of transmitting and archiving geologic data since the presentation of the first such map by William Smith in 1815. Invention of the geologic map provided a knowledge tool with the flexibility necessary to accommodate the many ways of seeing and describing the rocks that form Earth. This new way of synthesizing and analyzing geologic information led the rapid development of the science of geology. Modern geographic information systems provide a new, more flexible knowledge tool that permits storage of great quantities of information and allows users to customize information display according to their particular interest and evolving understanding. The ability to explore data in ways not possible with paper maps, fostering new approaches to archiving, representing, analyzing, and synthesizing geoscience knowledge, brings the promise of new developments in the science of geology.

Two major factors impede achievement of this potential for geoscience information systems. First, the proliferation and rapid evolution of software and hardware has led to a wide variety of different data formats. Second, the absence of any standard model for constructing geoscience databases has resulted in incompatibilities between the many databases that currently exist or are being developed. These problems are not unique to the geoscience community, and the information science community has been working diligently to solve these data interoperability problems (Open GIS Consortium Technical Committee, 1998). A fundamental requirement is the development of implementation-independent models of knowledge domains, which have come to be known as ontologies (Parent et al., 1997; Guarino, 1998; Fonseca and Egenhofer, 1999).

This paper presents a conceptual model for some fundamental aspects of geology that are central to the understanding and use of geologic map data. These include lithology, and geologic units, surfaces, and structures. The models are, with few exceptions, static in that process is not modeled. Then the vocabulary of geoscience phenomena to be modeled is presented, followed by a discussion of models for the description of these phenomena. Appendix A provides a glossary of terms that may be unfamiliar to geoscientists. Throughout the paper, names of model elements are written in TitleCase. If the name includes more than one word, spaces between the words are left out, for example EarthMaterial. This is to emphasize that the term has a specifically defined meaning in the context of the model. The models presented here were originally developed using the Object-Role Modeling (ORM) notation (metamodel) (Halpin, 1995, 2001), but are presented here using the Unified Modeling Language (UML) notation because it is familiar to a wider audience. For details on the syntax see Appendix B and Object Management Group (2001, 2003). Appendix C introduces the interested reader to some of the philosophical issues associated with conceptual modeling, and provides more background for understanding of the models in the body of the paper.

This work has been done in support of the development of a geologic map database system for the Arizona Geological Survey (Richard and Thieme, 1997; Richard, 1998; Richard, 1999; Richard and Orr, 2001). The ideas presented here have been evolving in conjunction with the evolution (Johnson et al., 1999; Brodaric and Hastings, 2002; Data Model Design Team, 2002) of a proposed standard North America Geologic map data conceptual model that will be referred to as NADM-C1 (North American Geologic-Map Data Model Steering Committee, 2004a). This paper was submitted during the early phases of the development of the NADM-C1 model. Final revision of this paper has been completed while developing an implementation of the NADM-C1 model in the context of a U.S. Geological Survey–Association of American State Geologists project to prototype a National Geologic Map Database (Richard, 2003; Richard et al., 2004). The geologic models in this paper are consistent with NADM-C1 with minor modifications, and add detail to the NADM-C1 models at a number of important points. The knowledge-representation framework for the models is developed in greater depth. In particular, modeling of a geologic map as a knowledge representation device is separated from the problem of geoscience knowledge representation in general.

GEOSCIENCE KNOWLEDGE REPRESENTATION

The purpose of a computer database system is to store information in a systematic manner such that it may be retrieved according to the requirements of future users. As such, the database system is a sort of knowledge representation. The electronic bits in the hardware memory constitute a collection of symbols that record geologic knowledge. The database serves the same function as traditional geologic maps, diagrams, and text reports.

Figure 1 presents the various agents and abstractions involved in this knowledge representation. Keeping the various contexts in mind is useful in understanding the development of the conceptual models that follow. Firstly the "real" world ("World" box in Fig. 1) includes human observers, and the realm of natural phenomena. The perception of a Phenomenon by an Observer is an Observation. Based on one or more observations, the observer makes generalizations, or categories, of phenomena, represented in their consciousness as Concepts (see Brodaric and Gahegan, this volume). Different observers may interpret the same Phenomenon in different ways to generate different Observations and Concepts for the same Phenomenon in the world. Collections of Concepts, including those for objects and relationships, are organized into an internal model of reality, modeled as a Theory. Using some suitable system of notation (metamodel) the Observer may

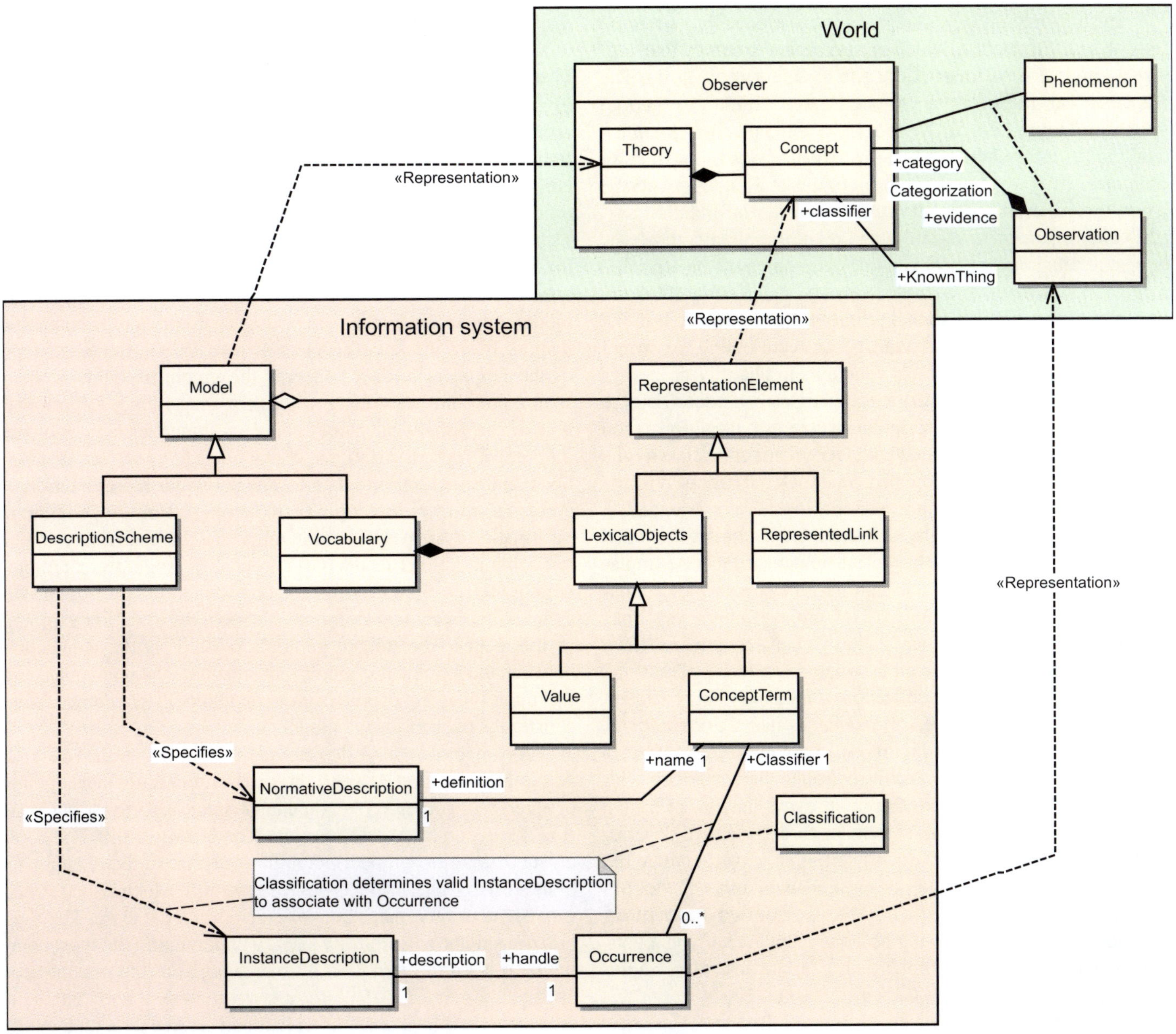

Figure 1. Geoscience knowledge representation. This model asserts that human observers in the "world" interact with phenomena there to make observations. Based on these observations, the observer formulates mental concepts that categorize the observations, and organizes these into theories about the world that are the basis of human knowledge. The purpose of an information system is to represent the observer's internal knowledge in a medium outside the observer, such that it may be transmitted to other observers and persisted beyond the lifetime of the original observer. The information system uses a collection of LexicalObjects (material symbols) and some mechanism for encoding relationships between these to represent the observer's concepts. Theories are represented as a model, which has a structural aspect and a vocabulary that enumerates things described by the model. This and other models in this paper use Unified Modeling Language notation (Object Management Group, 2001, 2003).

then construct a representation of a Theory as a data Model (see Appendix B, MetaModels). This has two aspects, one of which is structural, including kinds of objects, their properties, relationships between objects, and rules that determine valid states, which is modeled by DescriptionSchema. The other is a collection of symbols (e.g., words, images, graphic elements), each representing some Concept, that constitutes an enumeration of things thought to exist in the domain or possible values for properties. This collection is represented in the information system as a Vocabulary. The Vocabulary also defines relationships between the represented concepts, in particular a kind of hierarchy where appropriate.

The information system represents knowledge through RepresentationElements, which are persistent symbols that represent an observer's ideas (Concept in Observer, Fig. 1). The RepresentationElements may be LexicalObjects or RepresentationLinks. LexicalObjects have some physical manifestation in the information system—whether as words on paper, or character strings or numbers in a computer. RepresentationLinks are materialized in the structure of the information system—the organization of tables and relationships in a database, or spatial relationships between lines and polygons on a printed map. LexicalObjects include Values and ConceptTerms. Value instances are conceptual primitives (as represented in the information system)—a LexicalObject representing a concept that has no attributes, and only functions to quantify properties of other LexicalObjects. ConceptTerms are LexicalObjects that have an associated description structure in the information system, and may be associated with Occurrences to classify the kind of Observation represented by the Occurrence. A ConceptTerm has properties and relationships that are arranged in a data structure specified by a DescriptionSchema included in the information system. In typical relational database implementations, these schemas are built into the structure of a database through the field definitions for tables, foreign key links between tables, and uniqueness and cardinality constraints. Each ConceptTerm instance has an associated NormativeDescription that ideally specifies a collection of necessary and sufficient attributes (property-value pairs) unique to that ConceptTerm. Occurrences are symbols that represent an observer's observations. Classification relationships correlate Occurrences with ConceptTerms. In practice, ConceptTerms may be primitive concepts in the information system, with a normative description that consists only of human intelligible text to convey the meaning of the concept to a person. For many other ConceptTerms, the attributes specified by the NormativeDescription may not be unique, comprising necessary, but not sufficient conditions. This reflects the complex nature of human categorization (Lakoff, 1987; Brodaric and Gahegan, 2001).

Each Occurrence has an associated InstanceDescription. In practice, an Occurrence must be classified in order to determine the InstanceDescription used to describe the Occurrence. The InstanceDescription for an Occurrence must include at least the properties and relationships in the NormativeDescription for the ConceptTerm linked to the Occurrence through a Classification relationship. These dependencies are not explicitly modeled in Figure 1. Typically, the schema for an InstanceDescription will include incidental properties and relationships not present in the associated NormativeDescription because they are not defining.

The Geologic Map

The most important mechanism for representing geoscience knowledge is the geologic map. A geologic map is a graphic knowledge representation that records some understanding about the spatial distribution of rocks and the relationships between them in a particular area. The medium of representation is a visual display with a collection of colored polygons, lines and glyphs. Geologic maps are a vital form of geologic knowledge representation because the map efficiently represents spatial relationships that are fundamental to interpreting subsurface geometry and determining geologic history. A database structure can efficiently represent the interpreted relationships based on the map geometry, but is not well suited to fully record the details of the spatial relationships and the knowledge implicit in the spatial patterns. Careful distinction must be made between the geologic model represented by the map, and representation of the geologic map itself as a graphical artifact (Richard, 2000). Representation of the graphical artifact concerns the graphical symbols, their size, color, and positioning within the map frame. Representation of the geological model concerns things such as geologic map units, rock descriptions, contact and fault surface descriptions, and structural orientation. The computer representation of graphical artifacts is a topic of wide interest in the geographic information system (GIS) community, and there are many graphics, computer-aided design (CAD), and GIS programs that implement a variety of models for this representation (e.g., Open GIS Consortium Technical Committee, 1998); these are not explored in this paper. The unique geologic concern is the connection between the graphical symbols that constitute the map image and the underlying conceptualization of geologic phenomena that constitutes the geoscience model.

Any geologic map depicts some physical surface (Fig. 2, ActualMapHorizon, lower right) on or within Earth that contains or intersects geologic phenomena of interest to form outcrops (Fig. 2, section in "World" box; Bain and Giles, 1997). The geologist (Observer in model) identifies outcrop of the phenomena of interest to make an OutcropObservation, which is then stored (represented) on a map sheet or entered in a database. The map horizon may be any arbitrary surface within Earth, but most commonly it is Earth's surface. Other map horizons are possible, for example, a vertical surface for a cross section, a horizontal surface for a mine level map, or an inclined surface for a fault cutoff map. A single phenomenon (a particular fault or contact) may be observed on multiple map horizons, for example, different levels in a mine, in different places in a mine pit as mining progresses and the shape of the pit changes, or in different places due to active tectonic modification of Earth. The MapHorizon concept may be generalized to include drill holes as a linear map horizon, and volumes as the map horizon for three-dimensional (3-D) models. In the following discussion, MapHorizon is assumed to be Earth's surface in order to better correspond with what is typically thought of as a geologic map. The exact traces of polygon boundaries and other lines shown on the map, in concert with knowledge of the map horizon geometry (typically conveyed by topographic contours or shaded relief on a base map) provide information about the geometry of rock bodies, their bounding surfaces, and structures that may bound or transect the rock bodies. These traces represent the outcrop of the geologic features

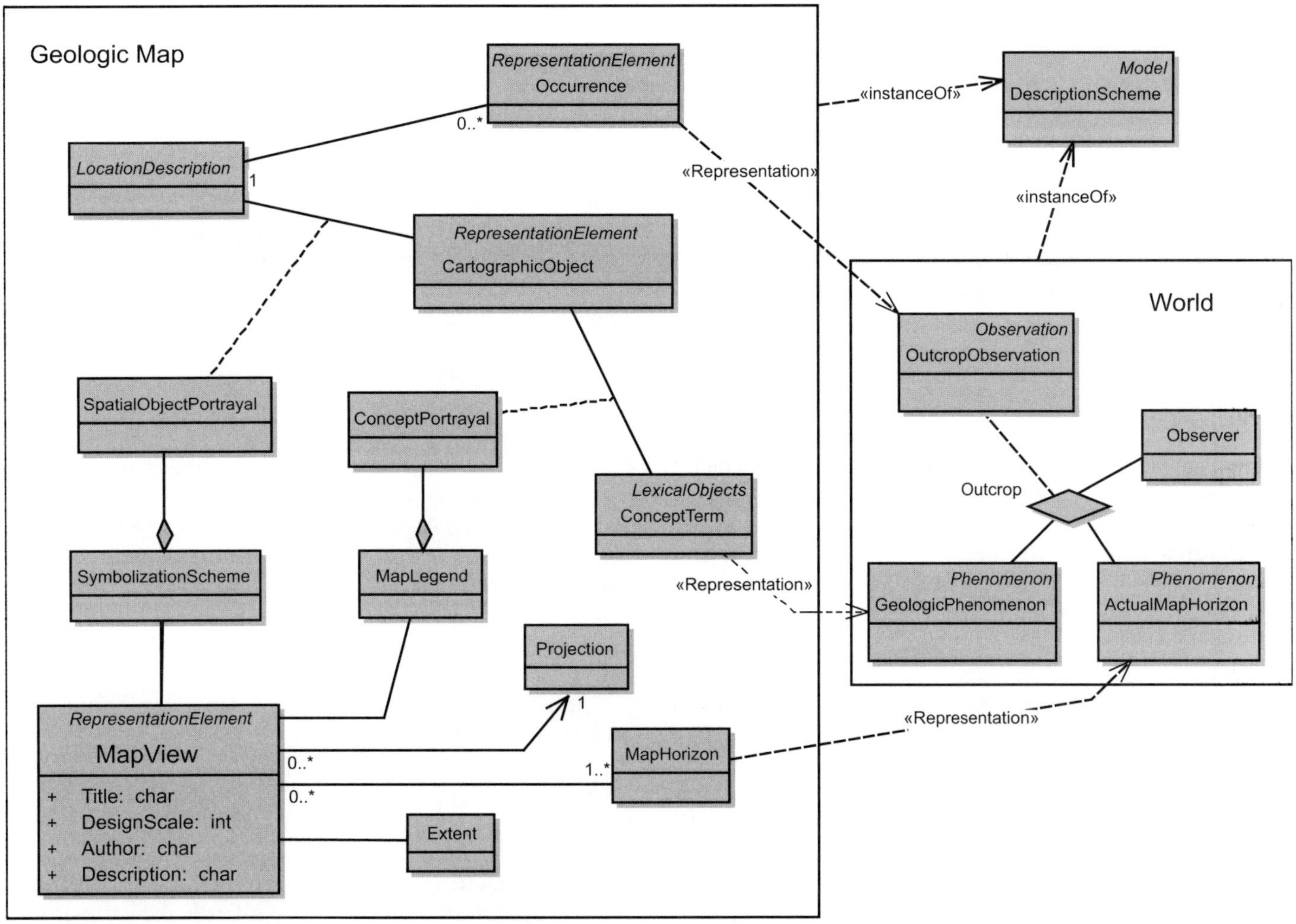

Figure 2. Geologic Map conceptual model. A geologic map is a kind of geoscience knowledge representation. Symbols on the map represent outcrop observations made by the geologist (Observer) in the field. The symbols are given meaning by association with ConceptTerms through a MapLegend. Many geologically important spatial relationships are represented by the arrangement of the symbols in the map view. Vocabulary of class names and properties used in this schema is not explicitly modeled. Class names in italics in upper right corner of class boxes in this and subsequent figures are generalizations of the class represented by the box.

of interest (GeologicPhenomena)—that is, the intersection of rock bodies, their boundaries, and faults, with the map horizon as interpreted by some Observer (OutcropObservation in Fig. 2). It is this outcrop that the geologist recognizes, classifies, describes, and locates in the field and depicts on the map as both a record and a tool for developing understanding of the geology. The outcrop pattern is the basis of geologic interpretation.

The map visualization consists of polygons and lines symbolized using color or pattern, and point observations symbolized using graphical symbols that may be text or selected from a collection of accepted graphical elements (glyphs). Each symbol conveys some combination of geologic concept, location confidence, identification confidence, and orientation. Each different color and pattern combination used to symbolize polygons on the map is associated with a geologic unit. Geologic units fall on a continuum between units defined based on properties of a rock volume and units defined based on properties of a surface. Polygons depicting the distribution of a volume unit on the map represent the intersection of its 3-D extent with the map horizon. A surface unit represents identifiable, locatable parts of the map horizon itself. The colored polygons on a geologic map either represent the intersection of rock volumes with Earth's surface, or represent regions of Earth's surface that have some particular combination of properties. Lines on a geologic map represent boundaries of geologic units in the map horizon surface or the intersection of some surface with the map horizon. To a geologist interested in the processes and characteristics of Earth's surface, the lines on the map represent boundaries of closed regions in that surface (e.g., soils, geomorphic features, depositional environments). Fault and rock-body contact traces (lines) on a geologic map represent the intersection of two surfaces (fault or rock body boundary and Earth surface), projected onto a two-dimensional (2-D) plane—the map surface. Because of the variety of features that may be symbolized, the significance of the symbols on the map must be made explicit in an explanation that accompanies the map graphic.

A geologist interested in the rock bodies beneath Earth's surface uses the 3-D geometry implied by the lines on the map, along with measurements of surface orientations noted on the map, to understand the 3-D model of Earth depicted by the map. This 3-D analysis requires geologic reasoning to determine the relationships between rock bodies. For example, consideration of rock type (e.g., granite) and rock age for adjacent rocks allows the geologist to identify a contact as a conformable depositional contact, non-conformity, angular unconformity, or intrusive contact. Accompanying text sometimes describes the nature of contacts that do not conform to a standard classification. The geometry of faults and contacts is either encoded on the map using a symbol (dip-direction arrow and dip magnitude), or it must be extracted by analyzing the relationship between the trace of the fault or contact and the geometry of the map horizon. Use of the law of Vs is the typical application of this logic. The geometry of faults and contacts can be used to determine underlying and overlying relationships between rock bodies. Once the geometrical relationships of depositional and intrusive contacts are understood, the offset of intersection points of contacts at faults on the two sides of the fault for a series of contacts can be used to constrain possible magnitude and sense of displacement on a fault. The validity of a geologic map is based on the internal consistency of the geologic relationships implied by the geometry of the lines and polygons on the map horizon (Ady, 1993).

The classification of geologic structures (e.g., bedding, foliation, contacts) shown on the map is indicated by a set of symbols representing standard feature types, commonly with little or no explanation of the criteria used to assign features to a particular class. In areas of complex structure, much of the structural orientation data collected in the field is not shown on the map because of space limitations. Polygons on the map are colored or labeled to indicate the geologic unit that is found at (or below!) Earth's surface at points within the polygon. The system used to define the geologic units depicted on the map is described in accompanying text. The map units are often associated with geologic units that are only described in a cursory fashion with reference to technical literature that provides information that is more complete. The printed geologic map thus contains a great deal of information that is not explicitly stated, and probably contains only a subset of the data and observations made by the geologist in the field. All of this information can be recorded in a usable form in a computerized map database.

Every geologic map is subjective to some degree and represents the knowledge and interpretations of the geologist who made the map. The map composer makes decisions about what features to display on the map based on their interests and judgment of what is geologically important. The location of geologic phenomena (contacts, faults, fold hinge surface traces, and structure symbols) in the map view is determined by the actual location of their outcrop on or in Earth. The graphical symbols that represent these on a map represent the geologic nature of the feature depicted and may additionally encode information about location confidence, classification confidence, or orientation. Other symbols are placed on the map to provide additional information for understanding the map. The location of these symbols on the map does not represent the location of a particular feature on or in Earth. Such symbols include map unit labels, symbols showing the nature of a fold represented by a hinge-surface trace and symbols that indicate the separation across a fault trace. These are referred to here as cartographic map elements (or annotation). Although their location on the map is not determined by actual outcrop, the positioning of these map elements effects the information communicated by the map. The actual map composition—colors, cartography, line weight, distribution of labels and structure symbols—is a semiotic device designed for human communication. It is thus essential that the graphical aspects of individual map design can be encoded in a geoscience database.

Geologic Map Representation

The model of a geologic map as a representation of geoscience knowledge presented in Figure 2 comprises a "World" box on the right, and a "Geologic Map" box on the left, with representation links between them, analogous to the representation links in Figure 1. Bear in mind that this model pertains to a single geologic map as a visualization of a geologic data set, which is a collection of Occurrences, ConceptTerms, and Classification links between them (Richard, 2000). As such, the geologic map is a particular kind of knowledge representation within the scope of geoscience representation modeled in Figure 1. The essential element of the map as a visualization is a collection of CartographicObjects (symbols) that depict Occurrences representing OutcropObservations. The CartographicObjects (e.g., color, line, glyph) have an explicitly declared correspondence with geologic phenomena (represented by ConceptTerms) identified by the geologist (Observer, in World). The CartographicObjects are spatially arranged in the map visualization to correspond to their spatial distribution on the actual map horizon, represented by MapHorizon in the model.

The MapView object represents a single geologic data set visualization (Fig. 2, lower left). Every MapView has an associated Title, Author, Description, Projection, DesignScale, Extent, and MapHorizon. Title, author, and description are self-explanatory. The Projection specifies how locations on (or in) Earth are mathematically mapped to coordinates in the visualization display space, which for a standard 2-D map is a plane. The DesignScale is the viewing scale for which the cartographic design is optimized; this is specified because computer systems allow zooming of the display to any scale. Extent specifies the part of the map horizon that is depicted. MapHorizon specifies the surface (ActualMapHorizon in the

World) upon which outcrops of geologic phenomenon are displayed in the MapView. Extent and MapHorizon are modeled as classes because they may be specified in different ways. For instance, MapHorizon may be an image of a topographic map, a digital elevation model, or simply a term (name) identifying the actual map horizon (e.g., Earth surface).

The position of each CartographicObject in the MapView is determined by an associated LocationDescription through the SpatialObjectPortrayal relationship class (Fig. 2, center left). In the simple case of a map that may be displayed in a plane, the LocationDescription is a 2-D map spatial object—one of the points, lines, or polygons of standard GIS systems. In general, Occurrences may have one or more LocationDescriptions (as part of their InstanceDescription; see Fig. 1) depending of the scale of depiction, the dimensionality (2-D or 3-D), and the projection. In a particular MapView, each Occurrence has exactly one LocationDescription, determined based on the MapHorizon being depicted, the DesignScale, and the Projection. Annotation on the map consists of symbols that do not represent OutcropObservations and thus are not associated with Occurrences. Annotation symbols (unit labels, antiform symbol along fold axial surface trace, separation indicator along a fault) are modeled as CartographicObjects with a LocationDescription not associated with an Occurrence (thus the 0..n cardinality on the association). The association of the LocationDescription with a symbol (CartographicObject) is represented by the SpatialObjectPortrayal link. CartographicObjects are associated with LocationDescription, not Occurrence, because symbols have scale-dependent properties, and in this model, Occurrence is considered scale-independent. The SymbolizationScheme collection is a collection of SpatialObjectPortrayal links that specifies exactly what is displayed in the MapView.

ConceptTerm represents a category (Concept in Fig. 1) of GeologicPhenomenon that has been observed within the extent of the MapView. Every CartographicObject used in the MapView must be associated with a ConceptTerm through a ConceptPortrayal link that is part of the MapLegend for the MapView, so that every symbol on the map has an explicitly stated meaning. Because Occurrences are associated with ConceptTerms through Classification links (Fig. 1, lower right), there are two paths from Occurrence to CartographicObject: (1) via LocationDescription through the SpatialObjectPortrayal link; and (2) via ConceptTerm through the Classification and ConceptPortrayal links. The SpatialObjectPortrayal link takes precedence over the ConceptPortrayal link, allowing unique symbolization of individual Occurrences independent of their classification. This is necessary because symbols do not map into unique ConceptTerms; they also convey information about orientation (horizontal, inclined, and vertical bedding use different symbols to represent the concept "bedding") or location uncertainty (dashed versus solid lines). Thus, a ConceptTerm may be associated with more than one CartographicObject in the same MapView. This modeling construct follows from the view that geologic phenomena are not defined by location or existential uncertainty, and in most cases, not by orientation.

In summary, automated representation of a geologic map as a visual artifact that can convey meaning requires three fundamental elements. There must be a mechanism to specify a collection of location descriptions used to position symbols on the map, a mechanism to associate symbols with the location descriptions, and a mechanism to associate the symbols with meaning for the viewer. This model uses two collections of links to achieve this representation, one that selects location descriptions by associating them with symbols, and a second that associates symbols with ConceptTerms to assign meaning. A variety of mechanisms may be used to implement these associations (see Richard and Orr, 2001).

CONCEPTUAL MODELS OF GEOSCIENCE FEATURES

This section presents a model of some fundamental aspects of the earth science domain. As indicated in Figure 1, two kinds of model form the foundation of an information system as a knowledge representation: a DescriptionSchema and a Vocabulary (Giles et al., 1997; Richard et al., 2003). A DescriptionSchema specifies the relationships and constraints between LexicalObjects identified in the Vocabulary; it is analogous to a grammar. The Vocabulary serves two major purposes. First, it enumerates the Concepts in the model (through their LexicalObject representations) and their hierarchical ("is a" or "kind of") relationships. Second, it defines ranges of values for properties associated with LexicalObjects. Theories of Earth evolve over time, and are to some extent dependent on the individual experience of an observer (von Engelhardt and Zimmerman, 1988; Brodaric and Gahegan, 2001; Brodaric and Hastings, 2002). Thus, there may be multiple GeologicVocabularies (Vocabularies of terms for geologic Concepts). In this paper, only the top-level part of a GeologicVocabulary is developed that represents a categorization of major geologic phenomena in the domain of the model. The Observer context for these Concepts is meant to be the entire geologic community, but is of course most strongly influenced by the author's perspective. The top level ConceptTerms (each representing a geologic Concept) in the GeologicVocabulary are EarthMaterial, GeologicUnit, and GeologicStructure (Fig. 3).

The domain of this model is the solid Earth—a composite object with a shape defined by solid matter (this discussion is limited to a static model). Earth is composite because it consists of many definable parts. Its shape is a spheroid defined by the solid matter of the lithosphere. The objects in the model are considered to represent phenomena and relationships that exist or occur (in the present or past) in or on the solid Earth. The model is based on an underlying philosophical assumption that there is a world external to human observers, and that different observers may see the same phenomenon and interpret it differently (make different observations in terms of Fig. 1). The next section discusses the geologic vocabulary that is the basis for the model, presented as

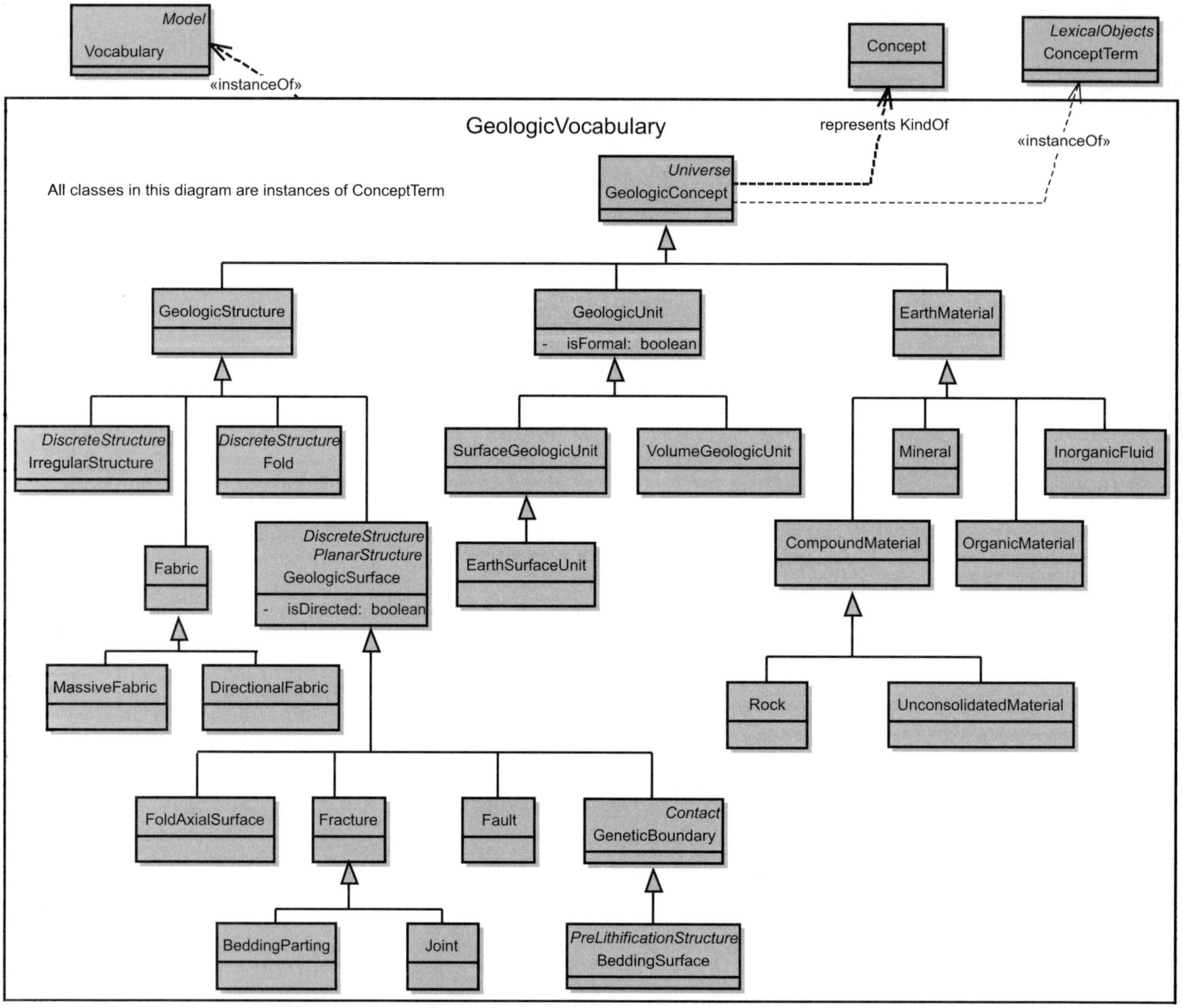

Figure 3. The Geologic Vocabulary that is the foundation for conceptual model is an instance of Vocabulary (Fig. 1). The hierarchy depicts asserted subclass relationships, that is, subclasses based on a priori distinctions that cannot be determined by properties included in the model. Rock and UnconsolidatedMaterial are exceptions to this rule (the subclass distinction is based on the Consolidation property), but is geologically significant enough that it is included at this top level. EarthMaterial description is modeled in Figures 4, 5, and 6. GeologicUnit description is modeled in Figures 7 and 8, and GeologicStructure description is modeled in Figures 9 and 10.

a taxonomy of high-level geologic phenomena. In following sections, description schema for Earth materials, geologic unit, and geologic structure are presented.

Geologic Vocabulary

The GeologicVocabulary underlying the models presented includes three top-level ConceptTerms: EarthMaterial, GeologicUnit, and GeologicStructure. GeologicVocabulary is an instance of Vocabulary (Fig. 2), and each class in GeologicVocabulary is an instance of ConceptTerm (Fig. 2). EarthMaterial is a substance (see Appendix A, Glossary) defined by chemical constituents, in concert with crystal structure, physical properties, or properties related to the nature and arrangement of constituent particles. EarthMaterials are the substance of the solid Earth (e.g., rocks, minerals, organic material, glass), defined based on intrinsic properties independent of their disposition within Earth. In a top-level ontology, they are independent, physical continuants (Sowa, 2000) with an identity condition, but no unity condition (Guarino and Welty, 2000a). EarthMaterial is a mass noun. GeologicUnits are identifiable parts of Earth. They have defining characteristics that allow for the identification and location of parts of Earth that are part of the GeologicUnit. In a top level ontology, they are independent, physical, continuants (Sowa, 2000) with a unity and identity constraint (Guarino and Welty, 2000a). GeologicUnits are composed of EarthMaterials. GeologicStructures are configurations of EarthMaterial within Earth. The existence of a GeologicStructure requires the existence of some EarthMaterial substrate. GeologicStructure instances are not composed of EarthMaterial, but have a material substrate; the same structure may be present in many different kinds of material substrate. GeologicStructures are dependent, physical continuants (Sowa, 2000) and may have a unity condition (non-penetrative), or may lack a unity condition (penetrative) (Guarino and Welty, 2000a). The intention is that the high-level vocabulary presented in Figure 2 includes only asserted classes—that is, classes whose membership can not be determined based on properties included in the subsequent description schema. Class names in italics in the upper left-hand corner of the class boxes in Figure 3 (and other schema throughout this document) identify generalization (kind of) links that are not explicitly shown in the schema.

Earth Material Vocabulary

EarthMaterial has four subclasses: Mineral, InorganicFluid, OrganicMaterial, and CompoundMaterial. This expanded vocabulary of EarthMaterial subclasses follows the NADM-C1 model except that Void is not included based on the idea that open space (pore space) is better modeled as a constituent type composed of water or whatever InorganicFluid or OrganicMaterial is in the open space. Kinds of Mineral, InorganicFluid, or OrganicMaterial are defined by chemical composition, physical properties, and crystal structure (Fig. 3, right side). Different minerals may have the same chemical composition with different physical properties (e.g., andalusite, sillimanite, and kyanite). Examples of physical properties that define minerals include refractive index, hardness, and density. InorganicFluid represents the general class of non-reduced carbon-based liquid or gaseous substances that may change shape to conform to a containing structure. Water is the most common subtype of InorganicFluid, but carbon dioxide and other inorganic gasses that are found in pore space in EarthMaterials are also included. Glass, considered as a kind of very high viscosity inorganic fluid, is included in this class. OrganicMaterial includes reduced carbon based liquids and solids that are the direct product of living things; this class includes plant and animal body parts that are composed of organic carbon compounds, as well as derivative organic carbon materials such as petroleum, lignite, and coal. OrganicMaterial does not include materials produced by or part of living things that have inorganic equivalents, such as phosphate-based bone material (apatite), or calcareous shell (aragonite, calcite); these are considered Minerals. In this model, Mineral, InorganicFluid, and OrganicMaterial are not analyzed in further detail.

CompoundMaterial represents any substance that is an aggregation of constituent parts that are composed of EarthMaterial and includes rocks and non-consolidated deposits that are the substrate for most phenomena of geologic interest. The constituent parts may be composed of Mineral, InorganicFluid, OrganicMaterial, or other CompoundMaterial (as in a conglomerate).

CompoundMaterial is subclassed into Rock and UnconsolidatedMaterial, based on the degree of consolidation (this is a defined subclass, but considered central enough to geologic thinking to include in the high level vocabulary). Rock is a consolidated aggregation of particles in which the constituent particles are bound together sufficiently that the aggregated collection may be considered a distinct, solid material. The identity of a rock is based on the size, shape, composition, and arrangement of its constituent particles. A simple rock consists of an aggregate of distinct mineral particles (granite). Rock that is more complex may consist of an aggregation of rock fragments (conglomerate, breccia) or of distinct lithologic components (migmatite, gneiss), each of which may have its own description. UnconsolidatedMaterial consists of particles that are not bound together with sufficient strength to consider the aggregation a distinct solid material. Thus, the definition of an UnconsolidatedMaterial may be based solely on the size, shape, and composition of the constituent particles. An UnconsolidatedMaterial may form an outcrop (e.g., a trench in wet beach sand) that displays GeologicStructure, but the structure is not inherent in the material. Cross bedding may be present in beach sand, but the sand can be shoveled into a pile, losing the cross bedding, but it is still sand. Clearly there is a continuum from UnconsolidatedMaterial to Rock (e.g., gravel in an active stream to conglomerate, or laterite to underlying granite).

Geologic Unit Vocabulary

A GeologicUnit is a part of the solid Earth that is identified by its geologic characteristics, has definable, locatable boundaries, and is persistent in time. Three-dimensional units are contiguous volumes of rock that have some identifying set of characteristics. Two-dimensional units are contiguous regions on a geologic surface. Ideally, every geologic unit would have a necessary and sufficient definition that specifies how the unit is recognized and what characteristics can be assumed for materials classified to this unit. In practical terms, geologic units are defined with reference to some particular set of outcrops. For a formal geologic unit, this is the type section. For informal units, the "defining" outcrops may or may not be identified, but exist in the mind of the geologist mapping the unit. Classification is based on a subjective assessment of similarity to the defining outcrops. In the simplest case, an unknown outcrop is assigned to the geologic unit that has defining outcrops deemed most similar to observed characteristics in the new outcrop. There may be no necessary or sufficient conditions for inclusion in a geologic unit. Qualifiers in the unit definition description may indicate characteristics that are more or less likely to be observed. These qualifiers, along with the geologist's experience, provide a basis for subjectively weighting the importance of observations. A further complication is the influence of location. An outcrop that is quite different from the defining outcrops, but is contiguous with other rocks assigned to a particular geologic unit may be included with that unit based on spatial continuity (Brodaric and Gahegan, 2001).

GeologicUnits may be either SurfaceGeologicUnits or VolumeGeologicUnits (Fig. 3, center) depending on whether the unit subjectively represents a surface within Earth or a volume of material. SurfaceGeologicUnits are commonly defined by the nature of material in a layer that is so thin relative to the horizontal extent of the unit that it is treated as a surface. A SurfaceGeologicUnit is defined on a GeologicSurface, which may be a boundary between rock volumes, or a discontinuity within Earth that has some set of identifying characteristics (e.g., a fault). Such units may represent the distribution of fault rocks on a fault surface, or mineralization along a vein. EarthSurfaceUnits are defined based on the characteristics of the boundary layer between solid earth and atmosphere or hydrosphere. The boundary layer has some thickness, defined as the vertical extent of solid Earth modified by direct interaction with the ocean or atmosphere. EarthSurfaceUnits may relate to surface morphology, surface age (as opposed to deposit age), depositional environment, or soil development.

VolumeGeologicUnits represent bodies of rock that have some identifying set of characteristics. The International Stratigraphic Guide (Salvador, 1994) and the North American Stratigraphic Code (North American Commission on Stratigraphic Nomenclature, 1983) define criteria for different kinds of VolumeGeologicUnits.

Geologic Structure Vocabulary

GeologicStructure represents kinds of configurations of matter in Earth. A useful way to understand this concept is to consider what would be a "structureless EarthMaterial." To be structureless, a material would have to be homogeneous, isotropic, and lack any sort of internal discontinuities. Thus, GeologicStructure in a purely descriptive sense is based on describable inhomogeneity, pattern or discontinuity in a rock mass. A classification of GeologicStructure is based on the abstract geometric form of the inhomogeneity. The identity of a GeologicStructure is independent of the material that is the substrate for the structure. There are usually strong dependencies between the nature of the EarthMaterial substrate and the kinds of GeologicStructure that may be present. The treatment of GeologicStructure here is an attempt to develop a purely descriptive approach that is not based (at least at the top levels) on the genetic interpretation of the structure.

The hierarchy for GeologicStructure in Figure 3 (left side) is based on distinctions that are asserted, not based on properties included in the model. Other hierarchies for GeologicStructure may be constructed based on properties included in the model, such as Pervasiveness, GeometricAspect, and Genesis (see "Geologic Structure Description," below). These other hierarchies may be derived from associated hierarchical property values, and do not need to be explicitly modeled. The top level subclasses of GeologicStructures may be separated into two broad groups that distinguish discrete from continuous GeologicStructures. Because distinction between these groups is defined by the Pervasiveness property (Fig. 9) the discrete structure class is not explicitly shown in Figure 3. Discrete GeologicStructures have a unity condition (Guarino and Welty, 2001). For each instance of such a structure, criteria may be specified that conceptually identify the "whole" structure. Large-scale examples include depositional contacts between particular geologic units, particular faults, and large-scale folds. Small-scale examples include channels, mud volcanoes, veins, and microfaults. Discrete GeologicStructures include GeologicSurfaces, IrregularStructures, and Folds. Continuous GeologicStructures are grouped in the Fabric class. Fabric is GeologicStructure defined by relationships between the parts of a volume of material in a pattern repeated such that it may be considered everywhere present throughout the volume at some scale. It makes sense to speak of a whole DiscreteStructure instance (e.g., the whole Keystone Thrust fault), but it does not make sense to speak of a whole Fabric instance (e.g., the whole schistosity).

GeologicSurface represents the concept of a phenomenon that has some spatial extension, but is treated as having no thickness at the scale of interest, and is identified based on geologic characteristics. Contact represents a surface that separates two GeologicUnits; Contact is not modeled as a separate class because it is logically a subset of the union of GeneticBoundary and Fault, defined by topologic relationships in a particular area. GeneticBoundary is a Contact that is not a Fault, thus is a record of the genesis of the GeologicUnits it separates. BeddingSurface is a Contact between beds (which are small scale geologic units). Fault represents a geologic surface across which rock bodies have been displaced relative to each other at the scale of interest. Some Faults are Contacts, but this is not necessary to the definition of Fault. Fracture represents a GeologicSurface across which material continuity has been lost—it is defined by a crack or zone of incohesion. Faults are commonly coincident with Fractures, but the existence of a Fracture is not necessary to define a Fault. Subclasses of Fracture in this model are BeddingParting, which represents a fracture formed along a BeddingSurface or parallel to BeddingFabric (see Fig. 9), and Joint, which is a Fracture across which there is no displacement (and is not a BeddingParting). There is a continuum between Fault and Joint with the boundary between represented concepts determined by the scale of investigation. A surface with several cm of displacement would be considered a Joint for most standard geologic mapping purposes, but for a detailed structure study might be considered a Fault. FoldAxialSurface is the locus of points that occupy the hinge of a single fold structure; the surface itself does not necessarily have a material manifestation, but is locatable. Marker beds, dikes and veins are commonly thin enough to be thought of as surfaces, but these are more accurately modeled as GeologicUnits.

Folds are bends in preexisting GeologicStructure, and are thus second-order (or higher-order) structures. The existence of a Fold implies the existence of some GeologicStructure that has been bent to define the Fold.

IrregularStructure represents a variety of other discrete structures that are not described by plane or line geometry. These are generally associated with rocks of a particular genetic origin, and include schleiren and miarolitic cavities in igneous rocks, and channels, flute casts, and mud cracks in sedimentary rocks. Because of the wide variety of such structures, they are generally classified and described by a terminology based on comparison to a well-described group of examples. Systematic classification of such structures is not attempted here.

Fabric denotes a describable pattern that is present throughout a rock body when considered at some scale. The elementary parts of the rock that define a Fabric instance (fabric elements) are distributed throughout the rock volume at the scale of observation (Passchier and Trouw, 1996). The fabric elements are repeated at distances so small, compared with the scale of the whole, that they can be considered to pervade the whole uniformly and be present at every point (Turner and Weiss, 1963, p. 21–24; Hobbs et al., 1976, p. 73; Bates and Jackson, 1987; Passchier and Trouw, 1996). Fabric is defined by the relationships between fabric elements in an average sense throughout the rock. Schistosity in micaceous rocks is typically penetrative at a hand sample scale. Folds may be penetrative at the hand-sample scale, kilometer scale, or regional scale (e.g., the Valley and Ridge in eastern North America, or Zagros, Iran).

Fabric is subclassed into DirectionalFabric and MassiveFabric, based on whether the Fabric defines surfaces or lines with an orientation that may be characterized, or whether the fabric is essentially directionless. Other hierarchies for Fabric may be constructed based on properties included in the model.

One or more of several independent characteristics may define a Fabric instance. These include the shape of particles in the rock, the distribution of particles in the rock, the distribution of other structures, or the distribution of discontinuities in the rock. Interpretation of the process responsible for the origin of non-spherical particles and their arrangement in a pattern to define a Fabric instance introduces a genetic aspect to fabric classification. Because a Fabric instance is defined in a volume of material considered at some scale, the definition of a fabric always carries at least an implicit scale connotation. Fabric may have an associated asymmetry that can be used to define a positive and negative sense. Examples include facing direction in bedding, sense of shear associated with a lineation in a mylonitic foliation, and the fold asymmetry associated with axial planar cleavage.

Particle-shape fabrics may be due to particle deformation, to preferred geometry of particle growth during crystallization, or to mechanical orientation and distribution of particles resulting from rock-forming process such as sedimentation. Examples of fabric dominated by particle shape patterns include clast imbrication in a conglomerate, mica schistosity, and mylonitic foliation. Patterns defined by the distribution of particles in a rock are probably the most common fabrics. Examples include layering in metamorphic rocks and many kinds of bedding in sedimentary rocks. The elements of the fabric are collections of particles that are distinguished based on some criteria. The most typical distinctions of the fabric elements are based on mineralogical or textural variations. Patterns defined by the distribution of structures in a rock represent second-order structures. A common example is continuous crenulation cleavage, which requires the existence of some preexisting fabric that is folded. Fabric may be defined by discontinuities in a rock body. Common examples are pressure-solution cleavage in metamorphic rocks, bedding defined by bedding partings in sedimentary rocks, and joint sets in any rock body. Because there are commonly changes in particle texture or mineralogy across discontinuities in rock bodies, there is a continuum between fabric defined by discontinuities and those defined by distribution of particles.

Description Schema

The schemas presented in the following sections model the semantic relationships between ConceptTerms in the GeologicVocabulary, and are instances of DescriptionScheme (Fig. 1) that specify how ConceptTerms may be defined, and how Occurrences classified with a ConceptTerm may be described. The properties that are introduced in these schemas constitute an additional Vocabulary of properties, and these properties would be quantified in description instances by terms from a more extensive Vocabulary of property value terms (broadly construed to include numbers). In the schema, association links for properties have stick arrowheads at the property end of the link. Many of these links have 0..* cardinalities indicating that multiple property values may be associated with the described class. A more sophisticated knowledge representation will add a frequency qualifier to these property links denoting the likelihood that the property will have that value. This might be done using terminology, with terms such as *always*, *common*, *rare*, *never*, or it might be done using a numerical "truth degree" to implement a fuzzy logic representation. This is an area of ongoing development (see Appendix D). Some of the property links have black diamonds on the association end for the described class. These indicate that the lifecycle of the property instance is linked to the class, meaning that if the class is deprecated, the property value is deprecated as well. This rule holds for most measured value specifications, because they are explicitly unique to the class instance they describe.

Earth Material Description

Figure 4 presents the basic properties that may be used to characterize any EarthMaterial. These include GeologicAge, Genesis, ChemicalDescription, PhysicalProperty, and Color. The GeologicAge property specifies the time at which an EarthMaterial is inferred to have originated with its assigned identity.

GeologicAges are determined using a wide variety of techniques, including stratigraphic ranges of fossils, superposition, cross-cutting relationships, and isotopic age dating, resulting in a wide variety of possible specifications for the GeologicAge property. These are not modeled here, but the reader is referred to GeologicAge in the NADM-C1 model (North American Geologic-Map Data Model Steering Committee, 2004a).

The Genesis property specifies the history of an EarthMaterial. The NADM-C1 model (North American Geologic-Map Data Model Steering Committee, 2004a) represents genesis as an ordered aggregation of geologic events, each with an associated environment, age, and process(s). Because the most commonly used subdivision of rock types—igneous, sedimentary, or metamorphic—is fundamentally linked to interpretation of genetic origin (e.g., Williams et al., 1954), it is also useful to specify genesis using a collection of terms that imply some simple genetic history. The model for description of Genesis is discussed in more detail with GeologicUnit (below).

The ChemicalDescription property represents any kind of chemical characterization of an EarthMaterial. Its cardinality on EarthMaterial is 0..* because different kinds of chemical analyses may be associated with the EarthMaterial as a description of the material. Note that individual analyses are associated with the particular sample of material that was analyzed, and each sample would be assigned to an EarthMaterial class (the concept of sample is outside the scope of schema presented in this paper). If multiple samples associated with the same EarthMaterial are analyzed using the same method, an aggregate description based on those analyses might be associated directly with the EarthMaterial instance. Various possible methods of specifying a ChemicalDescription include major chemical analyses, trace element analyses, and composition-related terms (e.g., peraluminous, alkalic). In this model, ChemicalDescription is broadly construed to include isotopic characterizations as well. Modeling the various possible chemical analyses is outside the scope of this model.

The PhysicalProperty class models any of the possible quantitative physical properties that may be assigned to an EarthMaterial. Examples of physical properties that might be assigned to an EarthMaterial include density, magnetic susceptibility, sound velocity, and refractive index. Specification of PhysicalProperty values is modeled as an association class between PhysicalProperty and EarthMaterial because this moves determination of the allowable physical properties out of the model and into a Vocabulary of PhysicalProperty ConceptTerm instances. Each PropertyQuantifier association link between

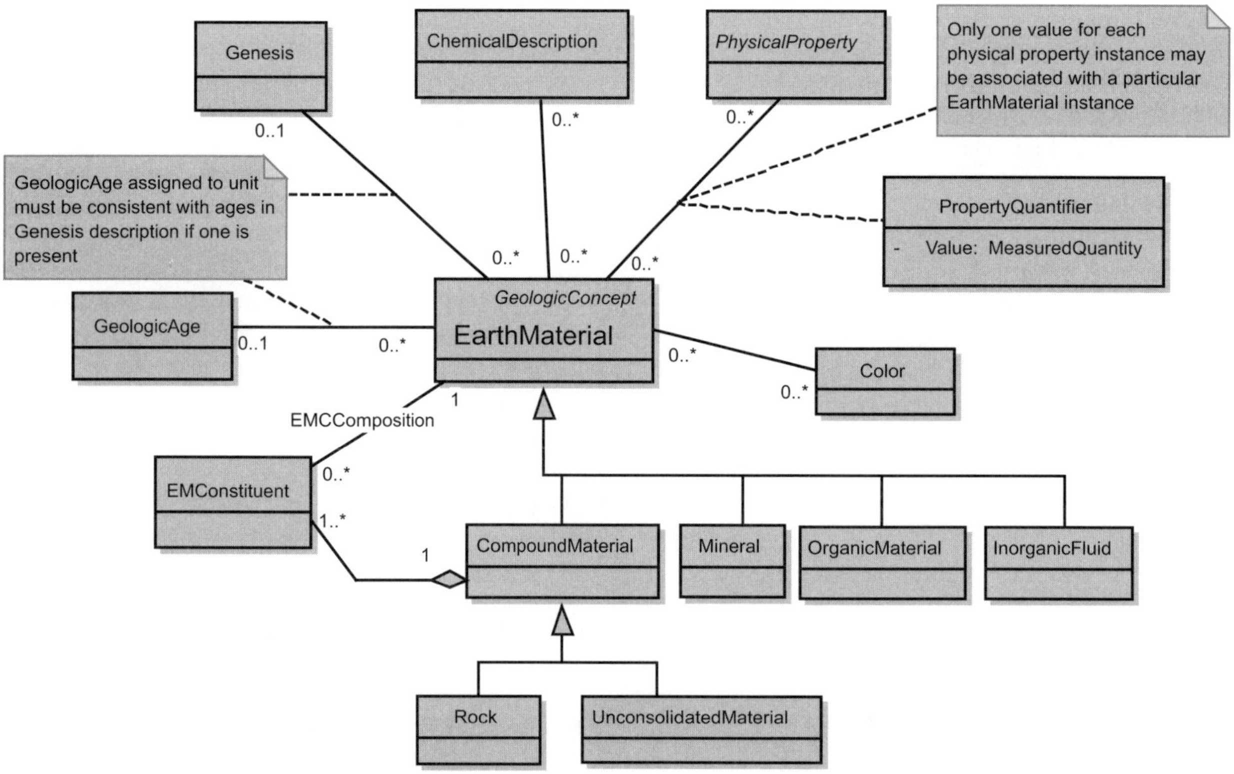

Figure 4. EarthMaterial is the substance of Earth, defined by chemical constituents in concert with crystal structure, physical properties, or properties related to the nature and arrangement of constituent particles. Properties allowed for all EarthMaterial instances include GeologicAge, Genesis, ChemicalDescription, PhysicalProperty and Color. CompoundMaterials are EarthMaterials that are aggregations of parts (EMConstituent) composed of EarthMaterial.

a PhysicalProperty instance and an EarthMaterial instance has a MeasuredQuantity attribute that specifies the value for that property on that EarthMaterial. MeasuredQuantity is not modeled in detail here, but is quite complex because of the variety of possible specifications. Numeric values may represent a single determination, a range, or an average; they may be scalar, vector, or array. This construct allows as many PhysicalProperties to be specified as necessary. As with ChemicalDescription, the intent of the PropertyQuantifier link is to characterize the EarthMaterial. Multiple measurements of the same PhysicalProperty instance may be determined from one or more samples associated with a particular EarthMaterial instance, but these are summarized by a single PropertyQuantifier link.

Although Color is a physical property, it is rarely or never specified using quantitative values and is thus not modeled using PhysicalProperty. The specification of color may take many forms and must account for natural color variation in EarthMaterials and for color variations under different observation conditions (e.g., wet, dry, fresh, weathered, natural vs. ultraviolet light). Again, the details of this specification are not modeled here.

The only properties modeled for Mineral, OrganicMaterial, and InorganicFluid are those inherited from the description of EarthMaterial.

The defining aspect of CompoundMaterial is composition as an aggregation of parts composed of EarthMaterial (lower part of Fig. 4). Because CompoundMaterial is itself a kind of EarthMaterial, CompoundMaterial instances may be constituents in other CompoundMaterial instances. This allows representation of the inclusion of different rock types as clasts in a conglomerate, or the parts of a composite rock such as a migmatite. The description of CompoundMaterial is modeled in detail in Figure 5. All CompoundMaterial descriptions must specify several basic properties—ParticleDiscernibility, RepresentativeSize,

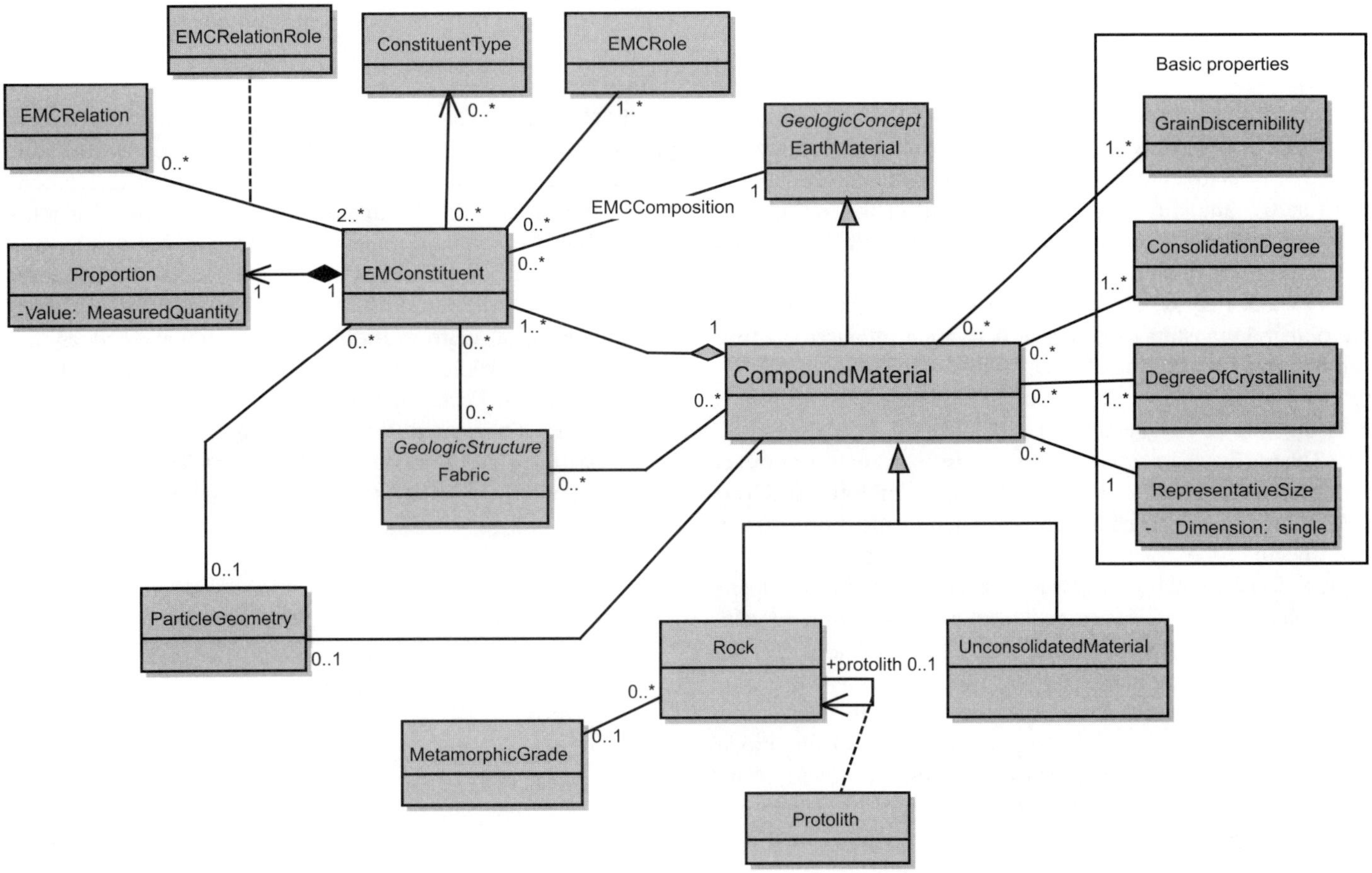

Figure 5. CompoundMaterial Description. CompoundMaterial includes Rock and UnconsolidatedMaterial, representing a group of EarthMaterial of central interest in the study of geology. All CompoundMaterials have basic properties of GrainDiscernibility (aphanitc, phaneritic), ConsolidationDegree (consolidated, non-consolidated), DegreeOf Crytallinity (crystalline, granular) and RepresentativeSize (dimension of unit sphere that characterizes material). These properties along with Genesis, constrain the domain of many other properties in the description of CompoundMaterial. ConstituentType specifies the packaging of the material in constituent parts (e.g., clast, fossil, open space, crystal), and EMCRole specifies relationship of each constituent part to the whole aggregate (e.g., framework, phenocryst, matrix, cement). See text for discussion of other properties. ParticleGeometry is modeled in greater detail in Figure 6.

ConsolidationDegree, and DegreeOfCrystallinity. ParticleDiscernibility specifies the degree to which individual particles in the aggregate may be discerned visually—with the unaided eye or hand-held magnifier. Following Schmid et al. (2002), particle sizes <0.1 mm are taken to be undiscernible. An aphanitic material is one in which all constituent particles are undiscernible. This distinction is important because the properties that can be described for an aphanitic rock are relatively restricted; the material is homogeneous for our purposes. Rocks that have discernible constituents (phaneritic) are described based on the nature of the discernible parts that make the whole, and the relationship between the parts.

RepresentativeSize specifies the linear dimension of a representative volume of the material described (Richard, 2001). A simple way to think about this would be the length of the side of a cube that is big enough to be considered representative of the rock being described. ConsolidationDegree specifies the degree to which the aggregation of particles that compose the material are bonded together to form a distinct solid. This property is the basis for distinguishing the Rock and UnconsolidatedMaterial subclasses of CompoundMaterial. DegreeOfCrystallinity specifies the degree to which particles in the rock are intergrown by crystallization. Igneous rocks that are crystallized from melt have the highest degree of crystallinity, while clastic sedimentary rocks that are only weakly cemented would have the lowest degree of crystallinity. This property allows distinction of crystalline versus granular rocks proposed by Struik et al. (2002). These four basic properties provide a variety of constraints on the more detailed aspects of a lithology description.

A Rock may be associated with another Rock through the protolith relationship to represent derivation of one rock from another. In addition, Rock has an optional MetamorphicGrade property. If no metamorphic grade is specified, a Rock instance is assumed to be not metamorphosed.

The EMConstituent association class represents occurrence of an EarthMaterial as part of a CompoundMaterial. Every EMConstituent instance has an associated Proportion instance that specifies the fraction of the whole CompoundMaterial aggregation formed by the constituent. The Proportion property has a MeasuredQuantity attribute that specifies a numeric value for the proportion. Because specification of the abundance of different constituents in many rock descriptions takes a wide variety of often very qualitative forms, the quantifier in this case needs to be even more flexible. Actual implementation of this model will require accounting for quantity terms such as *abundant* and *sparse*, relative quantity specifications (in order of abundance), incomplete specification (the rock consists of plagioclase, microcline, quartz, and 3% biotite), and quantity ranges.

Because the constituents represent units of EarthMaterial in the CompoundMaterial aggregation, the part units must have some definition, and relationships may be described between the parts. The ConstituentType property specifies the nature of the units of each constituent. Examples include clasts, crystals, interstitial fill, fossil, and pore space. An EMConstituent has one or more associated EMCRoles. The role property specifies how the constituent relates to the material aggregation as a whole (Table 1). Different EMConstituent instances composed of the same EarthMaterial may be present, playing distinct roles in the rock volume. For example, EMConstituent instances with EMCRole "groundmass" and "phenocryst," both composed of feldspar, may be present within a single igneous rock. Distinction of ConstituentType from EMCRole is subtle, and not always clear. An operational test is that if particles of the constituent are removed from the context of the rock, their ConstituentType may still be determined, but the EMCRole is lost.

Relationships between constituents in a CompoundMaterial are represented using the EMCRelation class. This is modeled as a metalevel relationship. Different kinds of constituent relationships are instantiated as EMCRelation instances. In an implementation, the different kinds of relationships would constitute a Vocabulary. A binary relationship between two constituents is represented by two links (EMCRelation-EMConstituent association instances) to the same EMCRelation instance, one from the EMConstituent instance in each role. Roles in the relationship are specified by EMCRelationRole. Table 2 gives examples of some possible relationships between constituents in a CompoundMaterial.

ParticleGeometry represents the specification of a particle geometry description, which includes particle size, particle sorting, and particle shape. Fabric describes geometrical relationships between the particles that make up the material. ParticleGeometry is commonly referred to as texture, but this term is laden with alternate definitions in the geologic and material science domains, and is therefore avoided. The ParticleGeometry description of a CompoundMaterial aggregation does not change if the particles are disaggregated and rearranged, but the nature of the Fabric would be completely changed by the same process. Fabric and ParticleGeometry are linked both to EMConstituent and to CompoundMaterial. This provides the ability to apply descriptions to the entire aggregation as a whole

TABLE 1. EXAMPLES OF EMCRole

Matrix
Cement
Framework
Phenocryst
Porphyroblast

TABLE 2. EXAMPLES OF EMRelation ASSOCIATIONS

Replaces
Forms a rim on
Crystallized before
Interstitial to
Enclosed in

(CompoundMaterial link) or to a particular constituent in the aggregation (EMConstituent link). For instance, a sand might be described as "fine-grained" referring to the entire aggregation of particles. A particle geometry description for a particular constituent might specify that the plagioclase phenocrysts are 5–7 mm in diameter, or the schistosity (Fabric) in a quartzofeldspathic metamorphic rock might be specifically associated with the biotite particles in the rock.

ParticleGeometry description is modeled in more detail in Figure 6. Particle geometry is specified through a combination of particle size description and particle shape description. Both aspects of geometry might have quantitative or qualitative (terminological) property values. Quantitative particle size descriptions might specify a minimum and maximum particle diameter, or an average particle size diameter (QuantitativeParticleSize). The most detailed particle size analysis would be a sieve analysis with weight fractions of the aggregate assigned to several numerically bounded size ranges (GrainSizeAnalysis). Terms used to describe particle size address some combination of mean/modal/median particle size (QualitativeParticleSize) and sorting (ParticleSorting). Simple particle size terms such as "fine-grained" are useful for unimodal distributions that are moderately to well sorted. Bimodal particle size distributions (a porphyritic rock for instance) would have to be described using a conjunction of such terms (very fine-grained and coarse-grained). In very poorly sorted aggregates, use of terms from regions on ternary diagrams with mud, sand, and gravel corners (e.g., Folk, 1980) interpreted strictly as grain size terms might be most useful. Terms might be used to specify the minimum and maximum particle size, corresponding to something like 5% and 95% on a cumulative size frequency plot, for which the size distribution is not known. Finally terms that pertain strictly to sorting may be used (ParticleSorting) with size distributions that are unimodal and not too skewed.

Particle shape may be specified by one or more numerical parameters related to average ratios of long intermediate and short diameters of particles, or the curvature of particle profiles. More commonly, qualitative terms are used by comparison with standard particle shape pictures. The two principal dimensions of particle shape description pertain to the form of the particle and its roundness. ParticleForm specifies the shape in a range from prolate (strongly elongate) to spherical (sphere-like) to oblate (flattened, tabular). ClastRoundness specifies the surface roughness or angularity of the profile of particles in clastic (granular) rocks. The analogous property in igneous or metamorphic (crystalline) rocks is ParticleCrystalShape, which ranges from anhedral to euhedral (in igneous rock). The "xor" constraint on the association of these properties with

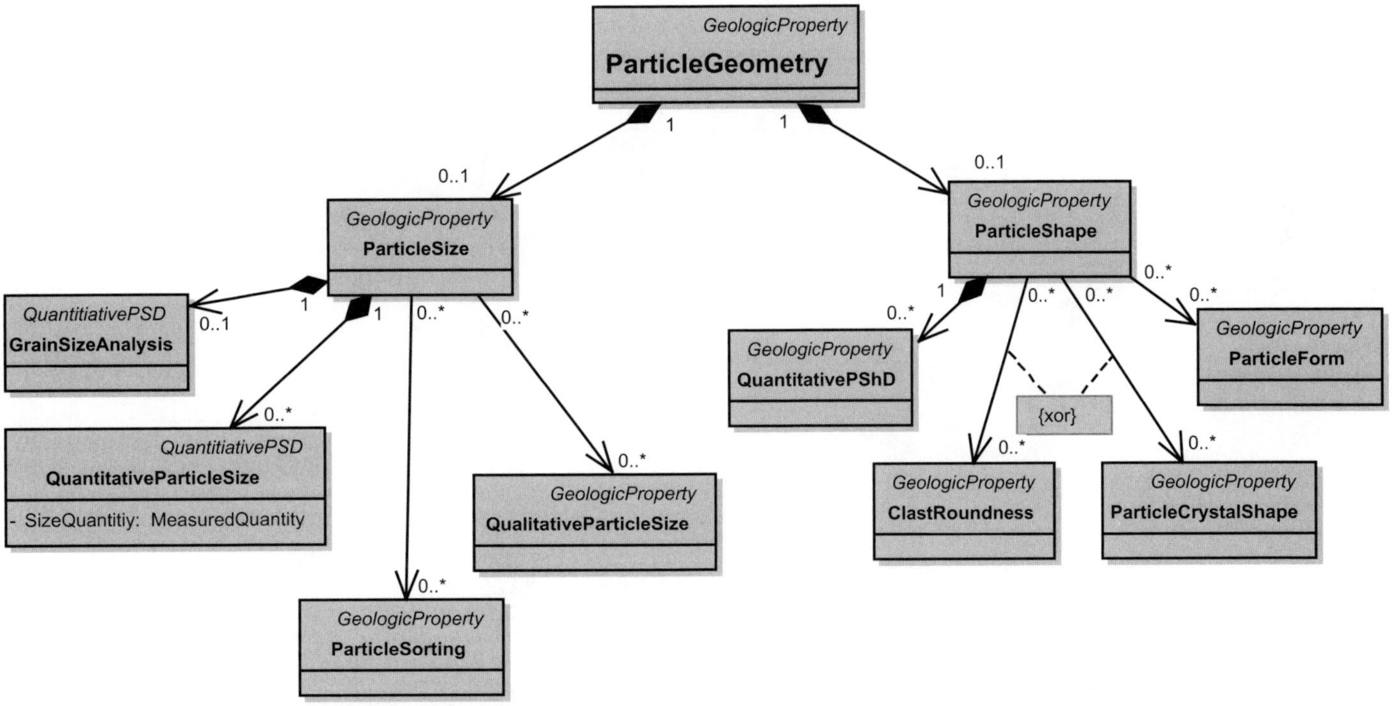

Figure 6. ParticleGeometry description. ParticleGeometry description is a collection of related properties that characterize the particle size and particle shape distribution in an aggregation of particles. ParticleGeometry properties are not dependent on the arrangement of the particles relative to each other in the aggregation. Black diamonds at the end of association links indicate that the life cycle of the class at the other end of the association is linked to the class at the end with the black diamond. Thus, if a ParticleSize instance is deprecated, any associated GrainSizeAnalysis is deprecated as well.

ParticleShape indicates that a ParticleShape instance may have values for either ClastRoundness or ParticleCrystalShape, but not for both.

Geologic Unit Description

GeologicUnit represents a part of Earth identified based on geologic criteria. All GeologicUnits have an associated Extent object that represents the distribution of the unit in Earth (Fig. 7, top center). If knowledge of Earth were complete, this extent would be a LocationDescription (Fig. 2) representing the entire 3-D distribution of the unit. In practice, it may be as simple as a bounding box for known distribution of outcrops of the unit or the convex hull of known outcrop locations. The following discussion moves clockwise through the various description elements associated with GeologicUnit in Figure 7.

A GeologicUnit has a Rank property (Fig. 7, top center) that specifies its position in a hierarchy of geologic units ranging from most specific to most inclusive. The lowest rank units identify particular parts of rock bodies that are either very thin in the case of stratified units, or of limited spatial extent for non-stratified units. Units of higher rank include larger and larger bodies of rock. In lithostratigraphic nomenclature, rank is classified by terms such as bed, member, formation, group, supergroup (North American Commission on Stratigraphic Nomenclature, 1983), but there is no formalized scheme for determining the absolute extent or dimension of a geologic unit. Rank is always defined relative to the recognized units in a particular area.

Genesis represents a description of the geologic history of the unit. Genesis is also associated with EarthMaterial (Fig. 4) and with GeologicStructure (Fig. 10) to allow description of the geologic history for those classes. A Genesis instance is an ordered collection GeologicEvents each of which may have an associated GeologicAge, one or more GeologicEnvironments, and at least one GeologicProcess. At least one associated GeologicProcess instance is required because a process is considered inherent in any event of interest in a geologic history. The action of a process on an Earth material, geologic unit or geologic structure is essential to preserve evidence the event in the geologic record. GeologicEvent instances may be shared by several Genesis instances, because the same event may leave a record in multiple materials, geologic units or structures. Multiple Genesis instances may be associated with

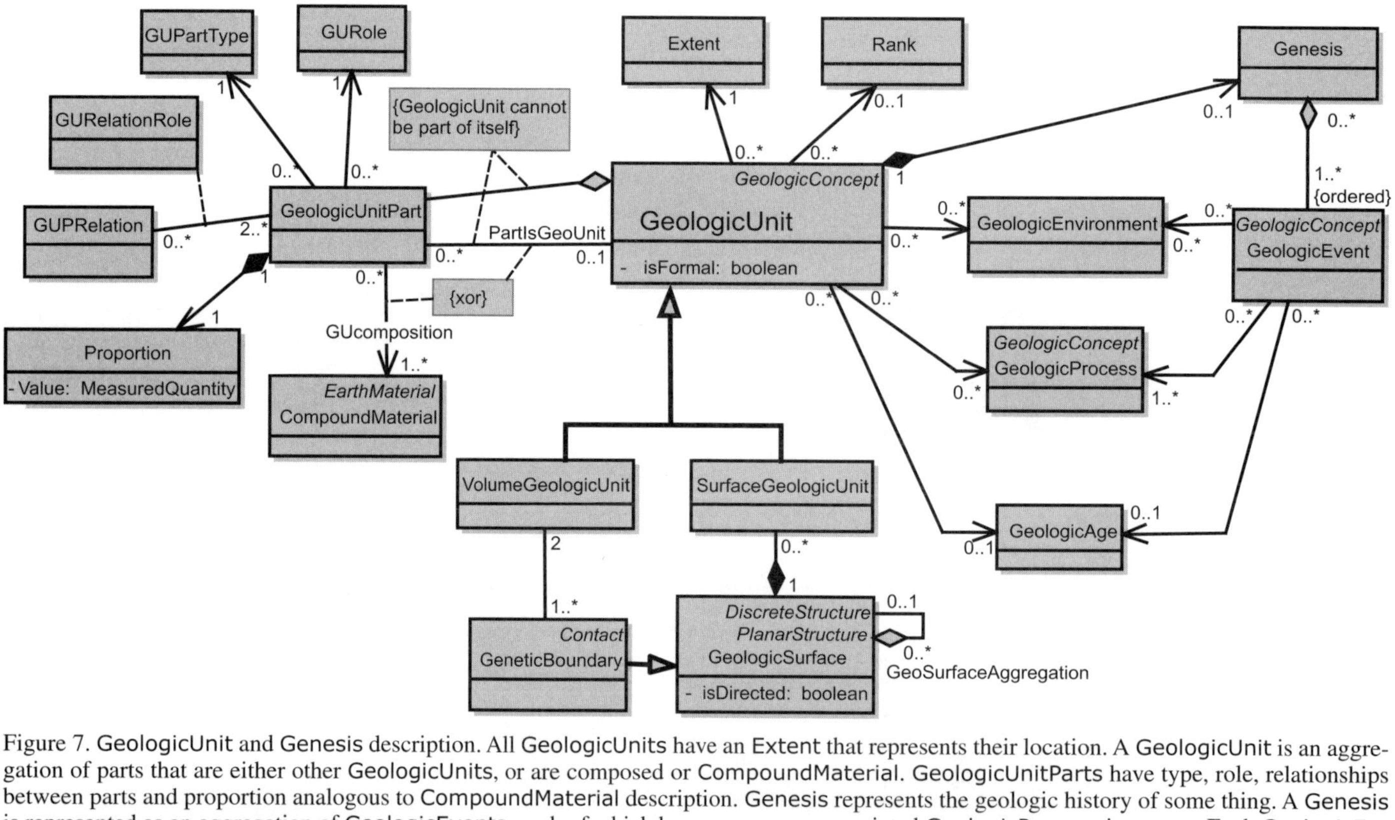

Figure 7. GeologicUnit and Genesis description. All GeologicUnits have an Extent that represents their location. A GeologicUnit is an aggregation of parts that are either other GeologicUnits, or are composed or CompoundMaterial. GeologicUnitParts have type, role, relationships between parts and proportion analogous to CompoundMaterial description. Genesis represents the geologic history of some thing. A Genesis is represented as an aggregation of GeologicEvents, each of which has one or more associated GeologicProcess instances. Each GeologicEvent may also have a GeologicAge, and be associated with one or more GeologicEnvironment instances. If a GeologicUnit does not have an associated Genesis description, GeologicEnvironment and GeologicProcess instances may be linked directly to the GeologicUnit. VolumeGeologicUnit represents some three-dimensional body of rock, and must have at least one GeneticBoundary that separated that material from adjacent material when the unit formed. SurfaceGeologicUnit represents some part of a GeologicSurface, and must be associated with an owning (substrate) GeologicSurface.

a GeologicUnit instance if there are varying interpretations of the sequencing of events in its history.

GeologicEnvironment and GeologicProcess are properties of both GeologicEvent and GeologicUnit. Because the model allows these properties to be both directly and indirectly (through GeologicEvent and Genesis) associated with GeologicUnit, there is a logical constraint that the GeologicEnvironment(s) and GeologicProcess(s) of GeologicEvents in the Genesis of a GeologicUnit must be consistent with those assigned directly to that unit. For many GeologicUnits, the geologic history is not known in sufficient detail to construct a Genesis description, but some of the processes and environments relating to the origin of the unit may be known. GeologicEnvironment specifies the physical setting in which a GeologicEvent occurred, or the physical setting in which a GeologicUnit originated. Environments may be defined based on physical conditions, or combinations of climate, physiography, tectonics, active processes, and relationship to bodies of water, hillslopes or other geographic features. The cardinality of GeologicEnvironment is zero to many in both cases because different environment descriptions may specify different, independent aspects of environment, a GeologicUnit may have originated in multiple environments, or multiple environment interpretations may be possible. GeologicProcess specifies the processes that were active during a GeologicEvent or during the genesis of a GeologicUnit. In this context, process is a naturally occurring, connected sequence of changes that produce some outcome; processes are generally directed in time, with some input and output thing or things. A process is identified by the changes it effects. Environment and process are interrelated because some processes are uniquely associated with particular environments, whereas other processes may operate in many environments. Both GeologicEnvironment and GeologicProcess are complex, hierarchical concepts that require further analysis to treat consistently in various kinds of geologic description, and to integrate with other conceptual models in the ecology, geography, and physics communities.

GeologicAge is considered one of the more significant properties of a GeologicUnit, meriting a more direct association. A GeologicAge instance directly linked to a GeologicUnit specifies the time at or during which the GeologicUnit originated. The definition of "origin" varies for different kinds of GeologicUnits. Origin normally spans a time interval, but some volcanic units may be considered to have a geologically instantaneous origin. These are not modeled here; the reader is referred to GeologicAge in the NADM-C1 model (North American Geologic-Map Data Model Steering Committee, 2004a). Because the model allows GeologicAge to be both directly and indirectly (through GeologicEvent and Genesis) associated with GeologicUnit, there is a logical constraint that the GeologicAge of GeologicEvents in the Genesis of a GeologicUnit must be consistent with the GeologicAge assigned to that unit.

Geologic units may represent surfaces (SurfaceGeologicUnit, bottom, Fig. 7) or volumes of material (VolumeGeologicUnit). A SurfaceGeologicUnit must have a link to a GeologicSurface that is the substrate for the unit (Earth's surface is considered a GeologicSurface in this context). The lifecycle of a SurfaceGeologicUnit is linked to the lifecycle of its associated substrate. Multiple SurfaceGeologicUnits may be defined in a single GeologicSurface. A VolumeGeologicUnit must have a link to one or more GeneticBoundary surfaces that form the boundaries of the unit. GeneticBoundary represents a discrete geologic surface defined by the contact between geologic units formed during the genesis of at least one of the units. This concept is commonly referred to as a contact by field geologists, but the term *contact* is defined in a broader sense in the NADM-C1 model (North American Geologic-Map Data Model Steering Committee, 2004a) to mean any surface that separates geologic units, independent of its nature. The identity of a GeneticBoundary is defined by the geologic units that are juxtaposed and by inferred (or actual) spatial continuity of the surface. A GeneticBoundary is always associated with exactly 2 GeologicUnits. If the boundary is a directed surface (e.g., younging is known across the boundary, or some other criteria are defined to give the surface a top and bottom), the association between GeneticBoundary and the associated bounded units is ordered.

A GeologicUnit is an aggregation (1..*) of GeologicUnitParts, each of which is either composed of a CompoundMaterial or is a different GeologicUnit instance (Fig. 7, left side). Each GeologicUnitPart has a GURole property that specifies the relationship of the part to the whole geologic unit, and a GUPartType property that specifies what kind of part. Example roles include interbedded-part, facies, and dominant constituent. Example part types include lithosome, subunit, and material constituent. This representation of the parts of a GeologicUnit is analogous to the description schema for CompoundMaterial. Each GeologicUnitPart has a Proportion property that specifies the abundance of that part in the GeologicUnit. GUPRelation provides a mechanism to describe relationships between parts within a GeologicUnit, for example internal cross-cutting, gradation, or superposition. Because such relationships are within a GeologicUnit, in general they will not be recorded in the geologic map pattern. The structure and semantics of GUPRelation is exactly analogous to EMCRelation in Figure 5.

Using this recursive description schema, unit descriptions of arbitrary complexity can be built, starting at the outcrop scale and working up to regional scale. Individual outcrops can be described as units, specifying a collection of parts composed of CompoundMaterial, the geometry of each part (specified by GUPartType and Morphology), and relationships between the parts. The geologic unit parts might be individual beds in a stratigraphic section, the Morphology property (Fig. 8) could specify their thickness, and the GUPRelation relationship links could be used to indicate the stacking sequence of the beds. A formation-scale description might aggregate member-scale units. This description schema can be applied to stratigraphic sections, formations, groups, supergroups, terranes, etc.

Figure 8 is a more detailed model of the properties of geologic units. Starting in the center of the schema, GeologicUnit has an isFormal attribute. If this value is true, the geologic unit is formally defined and must have a FormalUnitDescription. This description specifies a published definition through a Citation instance, and a type locality through an Extent instance (Fig. 8, lower right) that locates the outcrops that define the unit. A GeologicUnit instance may have one or more thickness determinations associated with it (GUThickness). Each GUThickness instance associated with a particular GeologicUnit has a different Extent, which specifies either the location over which the thickness determination is valid, or the location at which the thickness was measured. Thickness is generally only associated with units that have tabular geometry, but may also be used to describe the vertical thickness of an irregular rock body at a point location.

A GeologicUnit instance may be associated with one or more GeologicStructure instances, to indicate geologic structures that are present in the geologic unit. This relationship would be used to indicate the presence of mud cracks, large-scale cross beds, erosional channels in sedimentary rock, or miarolitic cavities or schleiren in igneous rocks, to name a few examples. Geologic units that are bedded may have BeddingFabric description instances. Bedding is a very important geologic structure in sedimentary rocks, and the structure of the BeddingFabric description is modeled in Figure 10. Multiple BeddingFabric instances may be associated with a GeologicUnit to specify different kinds of bedding features observed in the unit.

OutcropCharacter specifies the typical character of rock outcrops considered on a scale of tens of meters, generally having to do with the shape of the rock outcrop. Some terms used to populate this property include blocky, ledge-forming, rib-forming, and massive. WeatheringCharacter specifies the nature of weathered outcrop of a GeologicUnit. It is a property of GeologicUnit, not EarthMaterial because weathering is specific to the location of the unit on Earth. Example values include cavernous weathering, friable, saprolitic, and resistant. SoilDevelopment specifies the nature of soils developed on outcrop of the GeologicUnit, and the degree (intensity) of soil-profile development for a given soil, usually relative to a local soil development sequence. GUSurfaceDescription specifies the surface nature of geologic unit, especially late Tertiary and Quaternary units. This property represents a complex description that includes specification of surface dissection, surface armoring (pavement development), surface varnish (rock varnish development), clast weathering degree, and clast weathering style, among other things.

MetamorphicGrade specifies the metamorphic grade of rocks in the associated GeologicUnit. Multiple MetamorphicGrade instances may be associated with a GeologicUnit instance because metamorphic grade may be variable over the mapped extent of the unit. Morphology specifies the geometry of a GeologicUnit. In the case of an EarthSurfaceUnit (Fig. 3), the Morphology property describes surface morphology or landform associated with the GeologicUnit. In the case of a

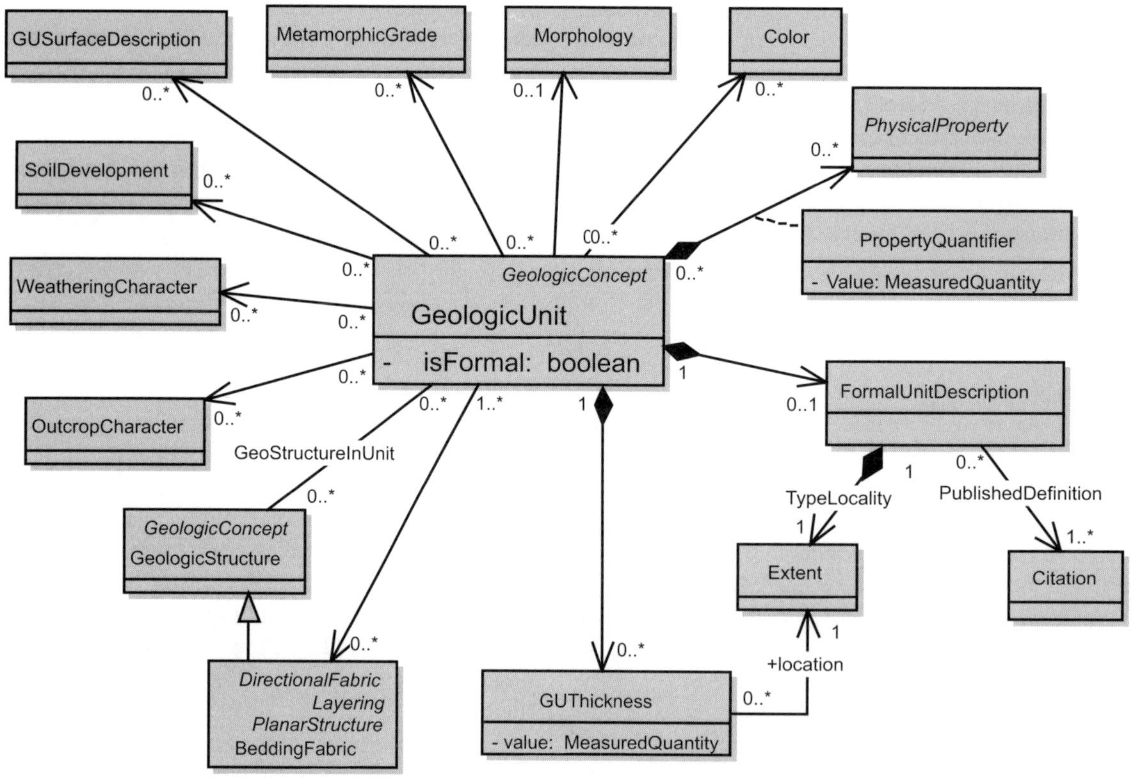

Figure 8. GeologicUnit description detail. Numerous different attributes may be assigned to GeologicUnit instances.

VolumeGeologicUnit, Morphology specifies the 3-D form of the volume occupied by the GeologicUnit. Example Morphology terms include laccolith, stock, layer, and lense. A GeologicUnit may only have one Morphology property value. Color and PhysicalProperty are treated in the same way as the Color and PhysicalProperty descriptions for EarthMaterial. Color specifies the general color of outcrops of the GeologicUnit. PropertyQuantifier links serve to specify representative values for any PhysicalProperty instances that are of interest and for which data are available.

Geologic Structure Description

All GeologicStructure instances must have Genesis, Pervasiveness, and GeometricAspect properties specified (Fig. 9, left side). Genesis specifies the geologic history of a GeologicStructure. The schema for a Genesis description is modeled in Figure 7 (left side). As with EarthMaterial, it is useful to have a Vocabulary of ConceptTerms (Fig. 1) that represent some often used, very general geologic structure genesis concepts such as *igneous structure*, *sedimentary structure*, and *tectonic structure*. The multiple cardinality of Genesis on GeologicStructure allows multiple interpretations of a structure to be represented. Pervasiveness specifies the degree to which a GeologicStructure is present throughout the rock volume that contains the structure. This property would be quantified by ConceptTerms such as *penetrative/continuous* and *discrete* and serves to distinguish Fabric from DiscreteStructure. GeometricAspect specifies whether a GeologicStructure conforms to a planar or linear abstraction, or to neither, quantified by ConceptTerms *planar*, *linear*, and *irregular*. The value of this property plays a role in determining the kind of GeometricDescription instances that are valid to associate with a GeologicStructure. GeologicAge specifies the time of origin of a GeologicStructure (see GeologicAge in North American Geologic-Map Data Model Steering Committee, 2004a).

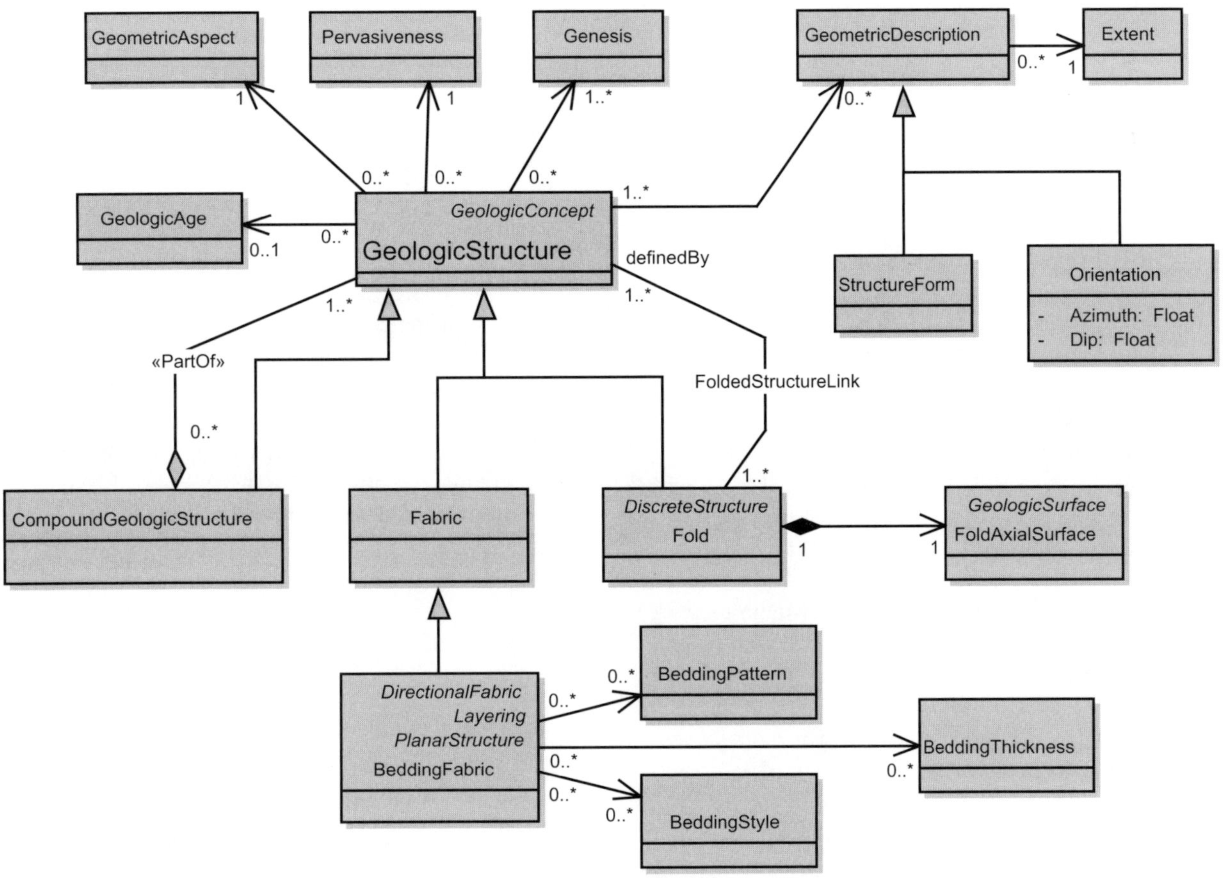

Figure 9. GeologicStructure description. All GeologicStructure instances have GeometricAspect, Pervasiveness, and Genesis attributes. GeometricAspect distinguishes planar, linear, and irregular structures. Pervasiveness distinguishes discrete (countable) structures from fabric (continuous, not countable structure). Genesis distinguishes tectonic from non-tectonic structures. These basic properties determine many of the other aspects of structure description. Structures may have a GeologicAge. Some GeologicStructures are aggregations of GeologicStructures. BeddingFabric represents bedding in stratified rocks. A Fold is always associated with some pre-existing structure that is folded. Folds always have a FoldAxialSurface. Any GeologicStructure may have a GeometricDescription that specifies the geometry (StructureForm) and Orientation of the structure. The Extent of a GeometricDescription locates the orientation description (e.g., bedding orientation at a station) or the geometric form description.

A CompoundGeologicStructure is a GeologicStructure represented as an aggregation of other GeologicStructures. A simple example is a Fabric that consists of a planar element (foliation) and a linear element (e.g., stretching lineation). The NADM-C1 model (North American Geologic-Map Data Model Steering Committee, 2004a) includes a class "GSDefiningElement" that includes ConceptTerm instances for a number of "atomic" structural elements such as oriented particle, material boundary, and layer. These could serve as the basis for modeling most GeologicStructures as an aggregation of parts, but this aspect of modeling geologic structure requires further development (see discussion of Fold, below).

Bedding in sedimentary rocks presents a special problem because the term can mean one of several things. Bedding defined by lamination or other structure that is penetrative at the scale of interest is represented by BeddingFabric in this model. This term represents the concept of bedding defined by the North American Geologic-map Data Model Steering Committee, (2004b). The individual parting surfaces in a rock body that define what are typically thought of as individual beds are BeddingParting surfaces, which are kinds of Fractures. These are non-penetrative at a hand-sample scale, but may define a BeddingFabric considered at a 10–100 m scale. A BeddingSurface is a GeologicSurface that visibly separates individual beds, which are small-scale GeologicUnits.

BeddingFabric has properties BeddingPattern, BeddingStyle, and BeddingThickness (Fig. 9, center bottom). BeddingPattern specifies patterns in a BeddingFabric. Terms used to quantify this property include bed downlap, bed onlap, and thinning upward. BeddingStyle specifies the general character of the geometry of packages of beds, and might be quantified by terms such as amalgamated, channelate, lenticular, or cross-bedded. BeddingThickness specifies the thickness of individual beds in an average sense, analogous to the specification of ParticleSize (Fig. 6) in a CompoundMaterial. As with ParticleSize, this might take the form of a quantitative specification of average thickness, minimum and maximum thickness, or an analysis of the frequency distribution of beds of varying thickness. More commonly, BeddingThickness is quantified by a ConceptTerm instance such as thin-bedded or medium-bedded, which maps into some quantitative definition (NormativeDescription associated with ConceptTerm, Fig. 1)

The description of Fold is complex and not developed in detail in this model, except to specify that every fold has a unique FoldAxialSurface (Fig. 9, lower right), and that a fold must have an associated GeologicStructure that is folded. A detailed model for the description of folds would need to represent that Folds have a hinge line, which is the locus of points of maximum curvature in a folded surface, and limbs, which are the relatively more planar (less curved) parts of a folded structure located between the hinges. The StructureForm description (see next paragraph) of a fold would specify properties such as interlimb-angle, wavelength, amplitude, similar versus concentric geometry, and other aspects of fold description.

One or more GeometricDescription instances may be associated with a GeologicStructure (Fig. 9, top right). Two kinds of GeometricDescription are modeled. StructureForm specifies the geometry of a structure and may include both quantitative (dimensional or angular) and qualitative descriptions of the relationship between the parts. Full development of this aspect of the model awaits future work. The other modeled aspect of GeometricDescription is Orientation, which specifies the geometry of the structure in space relative to Earth. Each GeometricDescription instance is associated with an Extent instance that specifies the part of Earth over which the description is valid. Orientation instances are normally associated with a point at which the orientation was measured but may be associated with a line, surface, or volume to indicate consistency over an extended space.

Figure 10 models some additional aspects of the description of GeologicSurface and Fault. Any GeologicSurface has a direction property specified by a binary attribute isDirected, which is true if the surface has an intrinsic "top" defined by younging or other criteria. A GeologicSurface may have one or more associated Thickness instances, each with an Extent property that specifies the location within which the determination is valid. A Thickness instance associated with GeneticBoundary specifies the thickness of the boundary layer between rock that clearly belongs on one side and rock that clearly belongs on the other side of the surface. This thickness may be in the mm range for sharp sedimentary contacts, or be tens of meters for some gradational contacts. A Thickness instance associated with Fault specifies the thickness of the fault zone. The Thickness property determines an inherent uncertainty in the accuracy to which a surface may be located in outcrop. A Fault may be associated with one or more CompoundMaterial instances that describe fault rocks found within the fault zone. GeologicalSurfaces may be aggregated to form composite surfaces (geoSurfaceAggregation link on GeologicSurface).

Faults represent discontinuities in Earth across which there has been discernible displacement. If the isDirected property of a Fault instance is true, the Fault has an associated Displacement description instance that specifies either the fault slip or separation. Each Displacement instance has an associated Extent that specifies the location at which the determination is made, and may have an associated GeologicAge instance that specifies when the displacement occurred. Multiple Displacement instance associated with a fault may be used to represent displacement variations along a fault, or multiple displacement events. The age range for different periods of movement must be non-overlapping.

Fault slip is specified by a SlipVector instance that represents the direction and magnitude of finite displacement across the fault. The SlipVector value is a vector quantity or vector field, depending on whether slip is defined at one point or many points on the fault surface. If the Displacement is specified by a SlipVector, there must be one or more DisplacedStructureLink association instances to DiscreteStructure instances used to

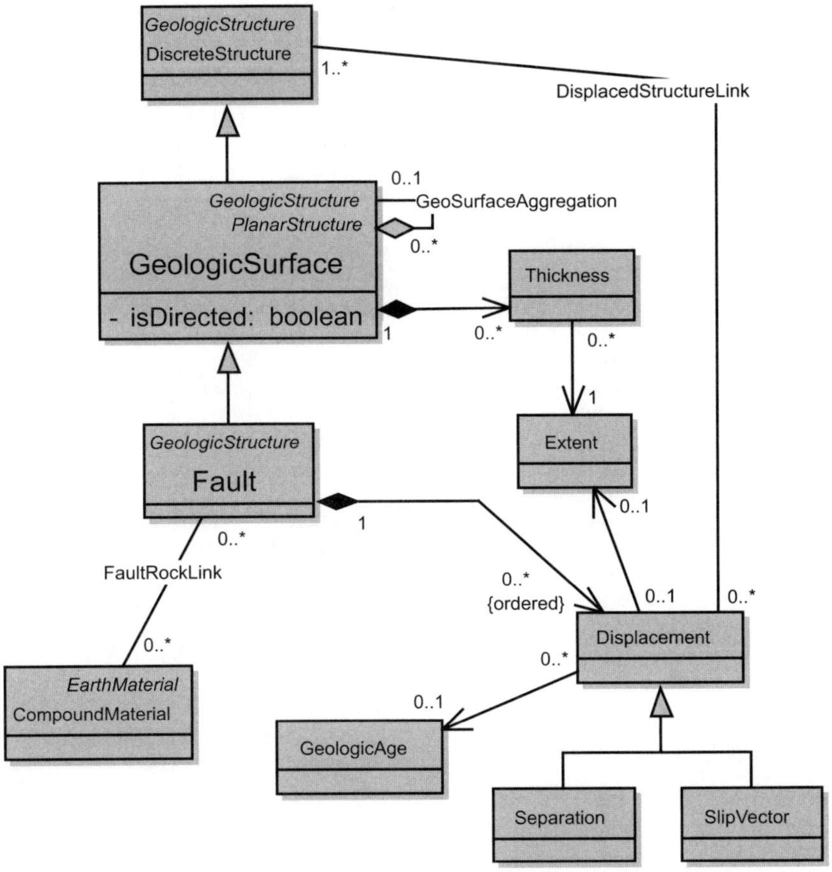

Figure 10. GeologicSurface and Fault description detail. A GeologicSurface may have an associated Thickness, and the thickness determination is located by an Extent instance. A Fault may have one or more associated Displacement descriptions, determined by either Separation or SlipVector, and each displacement has an associated GeologicAge specifying when the displacement occurred, and an Extent specifying where the displacement was determined. Fault rocks, which are kinds of CompoundMaterial, may be associated with a Fault.

determine the slip. Fault separation is specified by a Separation instance, which may consist only of a sense of separation (left, right, normal, reverse), or may include a measured quantity. If a separation quantity is reported, there must be a DisplacedStructureLink association instance to a DiscreteStructure that constrains the separation.

GeologicAge is associated with Fault and GeneticBoundary by inheritance from GeologicStructure. The GeologicAge instance associated directly with a Fault specifies the entire time interval during which the fault was active. For an active Fault the lower bound of this range is 0. The GeologicAge associated with a GeneticBoundary represents the time interval that separates rocks on either side of the boundary. This GeologicAge instance could be a derived from the ages of rock bodies adjacent to the GeneticBoundary.

CONCLUSIONS

The framework for geoscience knowledge representation presented in this paper provides a logical structure to guide the design and implementation of computer information systems for geologic data. The aspects of the science modeled here are particularly germane to information commonly associated with geologic maps. Many aspects of geoscience knowledge are not included in the model, particularly geologic process and details of structure description. The U.S. Geological Survey National Geologic Map Database project is currently implementing a database system based on this model (Richard et al., 2004). The model also provides a foundation for designing markup languages using XML or RDF for interchange of geoscience information. These and other implementations of the model presented here will test their usefulness and guide further development. Formal modeling of geoscience concepts is still in its infancy, and as more geologists become familiar with the methods, and as techniques are improved for representing fuzziness, uncertainty, and multiple interpretations, there will no doubt be a continuing evolution in the models.

ACKNOWLEDGMENTS

Thanks to Gary Raines for introduction to the intricacies of geoscience data modeling. My thinking over the years has been shaped by instructive and stimulating conversations with members of the North American Geologic Map Steering Committee Data Model Design team, including Eric Boisvert, Jordan Hastings, James McDonald, Peter N. Schweitzer, Ronald R. Wahl, Gerald A Weisenfluh, and especially Boyan Brodaric and Bruce R. Johnson. Discussions about science language with Jon Matti have stimulated much progress with the model. Interaction with geologists at the Arizona Geological Survey, including Jon Spencer, Charles Ferguson, Erin Moore, Tim Orr, and Jennifer Thieme, has helped keep my feet on the ground and provided a practical sounding board for many of the ideas here. Thanks to my family for their patience and support. Special thanks go to State Geologist Larry D. Fellows and David R. Soller for intellectual and institutional support of this activity.

APPENDIX A. GLOSSARY

Some definitions are presented to clarify the terminology used here.

Association: A concept that defines what a thing has to do with one or more other things. An association has no independent existence (Angus and Dziulka, 1998). Associations correspond to logical predicates (Halpin, 1995).

Attribute: An inherent characteristic (Merriam-Webster Dictionary, 2002). A value pair that includes a property and a value for that property.

Category: A kind of thing defined based on specific occurrences and used to classify new observations.

Characteristic: A property or relationship that may be observed and determines the classification of some phenomenon.

Class: A set of things sharing common attributes defined by a set of criteria (a type) (Angus and Dziulka, 1998).

Collection: A contingent aggregation of things, interpreted as an individual (Allgayer and Franconi, 1994).

Concept: An abstract or generic idea generalized from particular instances (Merriam-Webster Dictionary, 1999). The notion of a mental phenomena that human beings use in their internal representation of the world. Webster's dictionary (1996) uses the terms *idea* and *object of thought* to convey the meaning of concept. Concepts exist in the mind of human observers.

Continuant: A thing whose identity is preserved over some time interval; it has only spatial parts and no temporal parts. At any time t when a continuant x exists, all of x exists at the same time t. The identity conditions for a continuant are independent of time (Sowa, 2000).

Dependent thing: A thing whose existence depends on some other thing (Sowa, 2000).

Description: Specification of a phenomenon or concept by means of a collection of quantified attributes. A description identifies a thing by its properties.

Domain: The collection of things that may exist in a particular model.

Entity: A thing that has independent, separate, or self-contained existence (Merriam-Webster Dictionary, 1999). An instance of a type that has its own identity and can be distinguished from another instance that meets the definition of the same type. An entity has an identity property.

Feature: A physical thing that is represented in an information system.

Glyph: Graphical element associated with a point location; a simple icon that may be associated with some geologic concept.

Identity condition: A logical test that allows distinguishing a specific instance of a certain class from all other instances of that class (see Guarino and Welty, 2001). See also "unity condition."

Independent thing: A thing that is independent of any other thing. Formally, if x is an independent thing, it is not necessary that there exists an entity y such that x has y or y has x (Sowa, 2000).

Individual: A single thing in a domain (Allgayer and Franconi, 1994).

Instance: A thing that is representative of a type (American Heritage Dictionary, 1982). An instance does not have an identity property (compare to entity); two instances representative of the same type can not be distinguished (Angus and Dziulka, 1998).

Model: An explicit, symbolic representation of the concepts and relationships that constitute an observer's understanding of some knowledge domain

Occurrence: An individually identified data object in an information system that represents an observation of a phenomenon by an observer. Occurrences are information system objects that have a representation relationship to the "real world." Occurrences are subjective in that they represent a phenomenon as observed by some individual person or group.

Particle: A constituent part of an aggregate considered as a discrete object.

Phenomenon: An observable fact or event; an object or aspect known through the senses rather than by thought or intuition (Merriam-Webster, 2002).

Physical thing: A thing that is located in some place and is present or occurs at some time (Sowa, 2000).

Primitive concept: A concept that has no internal structure or cannot be defined in terms of necessary and sufficient properties (MacRandal, 1988).

Property: A phenomenon that is inherent in the nature of some other phenomenon and may be used to characterize it by specifying a value.

Prototype: A typical thing generalized from a particular set of entities, characterized by a subset of the attributes of the real-world entity(ies), and/or idealizations of those attributes. Attribute values are assigned, not measured, and are normally specified as a possible range of values (Angus and Dziulka, 1998).

Relationship: A semantic connection between two or more things.

Representation: The relationship between a symbol (in language, text, or computer bits) and a human concept.

Role: A role is a thing considered in relation to some other thing (Sowa, 2000). The thing that receives or is affected by the action of an association. A concept that indicates in what capacity some thing is involved with some other thing.

Substance: Thing that has mass and occupies space in which every part is similar to every other in every relevant (defining) respect. More precisely, something is a substance when it has only arbitrary pieces as parts—any parts have identifying properties identical to those of the whole (based on http://dictionary.reference.com; http://ontology.teknowledge.com/; http://virtual.cvut.cz/kifb/en/concepts/_substance.html).

Taxonomy: An IsA hierarchy in which the subclasses of each class are covering (i.e., every member of a class is a member of exactly one subclass of that class).

Thing: An entity, situation, association, or event that may be perceived, known, or imagined. Equivalent to *entity* of Sowa (2000).

Unity condition: A logical test that distinguishes "the parts of an instance from the rest of the world by means of a unifying relation that binds them together (not involving anything else). For example, asking 'Is that my dog?' would be a problem of identity, whereas asking 'Is the collar part of my dog?' would be a problem of unity." (Guarino and Welty, 2000a, p. 98)

APPENDIX B. METAMODELS

Figure B1 presents the metalevel model for models presented in this paper. A `Model` (Fig. 1) is a collection of `Elements`, which are the graphical symbols that are used to construct the model schema. The models presented in this paper use Unified Modeling Language (UML) notation based on UML 2.0 (Object Management Group, 2003). Figure B1 presents a simplified version of the UML 2.0 superstructure model that includes the `Elements` used in this paper. Because the models presented in this paper are static, no model elements relating to behavior or operations have been included. Because the model is purely conceptual, model elements relating to packages have not been

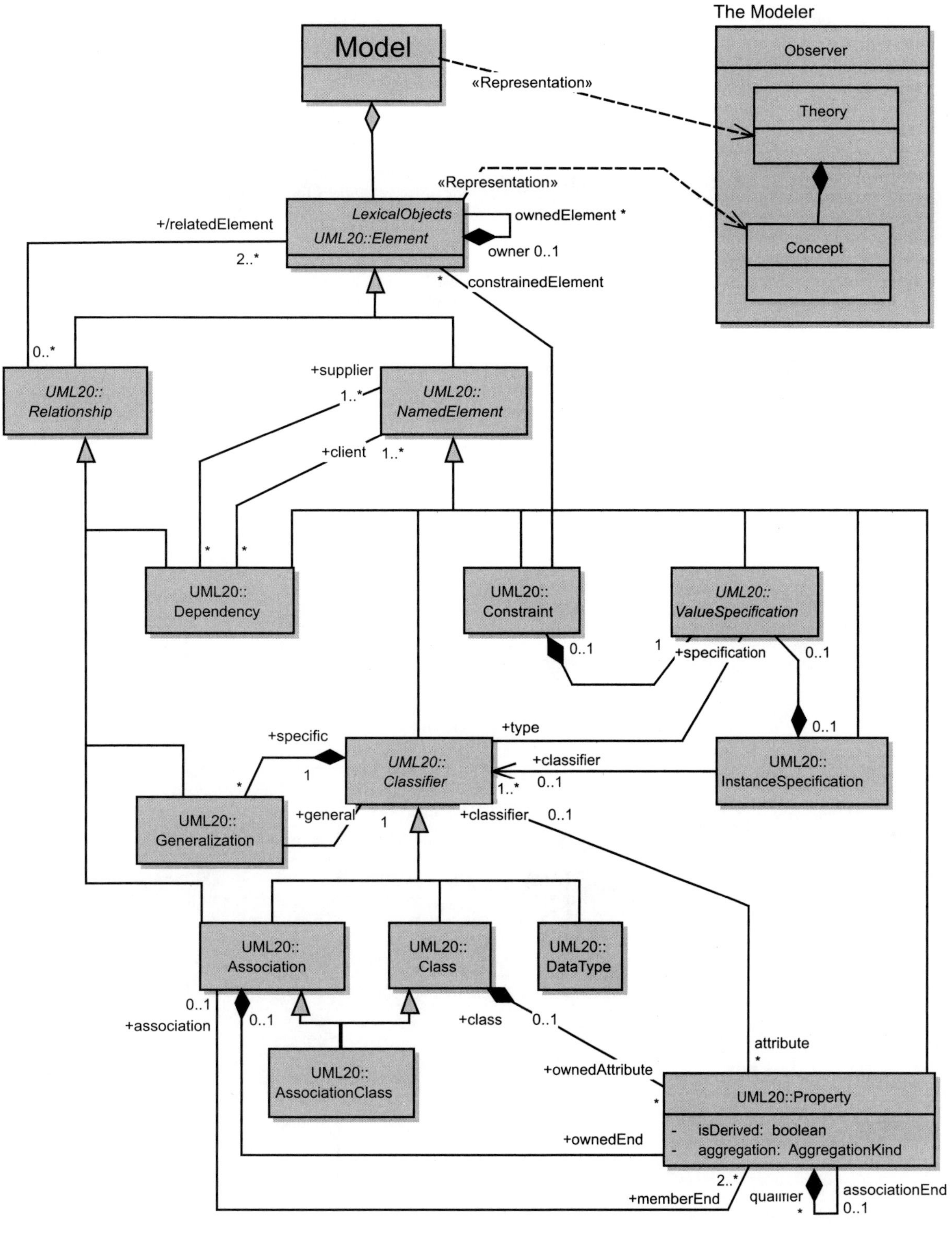

Figure B1. Model of Models (metamodel). The models presented in this paper use Unified Modeling Language (UML) notation based on UML 2.0 (Object Management Group, 2003). This figure shows key classes from the UML metamodel. Classes instantiated in models presented in this paper include Dependency, Constraint, ValueSpecification, Generalization, Association, Class, AssociationClass, and Property. Multiplicity is also used and is part of the UML metamodel, but has been omitted from this schema to simplify it.

included. Finally, to simplify the schema, elements related to Comment and Multiplicity elements are not shown. Abstract metaclasses in the superstructure model (such as Type, TypedElement, Feature, StructuralFeature) that are not of immediate interest or not required have also been suppressed. Figure B2 shows the relationship of a model instance to the metamodel elements in Figure B1. The reader is referred to the UML specification document for more in depth discussion of the metamodel.

APPENDIX C. CONCEPTUAL MODELING AND ONTOLOGY

Formalization of knowledge for representation in a computerized information system requires a mode of thinking that is unfamiliar to many geoscientists. This section presents some of the basic ideas fundamental to the development of abstract data models. Geologists may wish to skim this section on first reading, and return to it for more in depth study after reviewing the models in subsequent sections. In order to launch this endeavor, definitions of the central terms of the project—concept and model—are in order. The word *concept* is used here to represent the notion of a mental phenomenon that human beings use in their internal representation of the world. Webster's dictionary (1996) uses the terms *idea* and *object of thought* to convey the meaning of concept. A model is an explicit, symbolic representation of concepts and relationships between concepts that constitute an observer's knowledge of some domain. Any model is an abstraction. The level of abstraction is determined by the nature of the things represented by the symbols in the model. Models that are more abstract have a larger scope, and the represented things are very general. Less abstract (more concrete) models have more restricted scope, and the represented things are more specific. The most concrete model consists entirely of symbols representing particular things (ultimate instances, singular occurrences) in the domain of discourse. For example, a concrete model might represent that the human being (S.M. Richard) typed on the keyboard (serial# HP2011100023) to compose the text you are reading. A more abstract model might represent typing as a relationship between a human being and a keyboard to compose text. A further abstraction might represent a relationship between an agent, a tool, and a result. A yet more abstract model might represent a relationship as a connection between three things.

In the discussion in this paper, the term *conceptual model* refers to an explicit, symbolic representation of a knowledge domain that includes a collection of the kinds of things supposed to exist in the domain, a collection of relationships between those things, and a

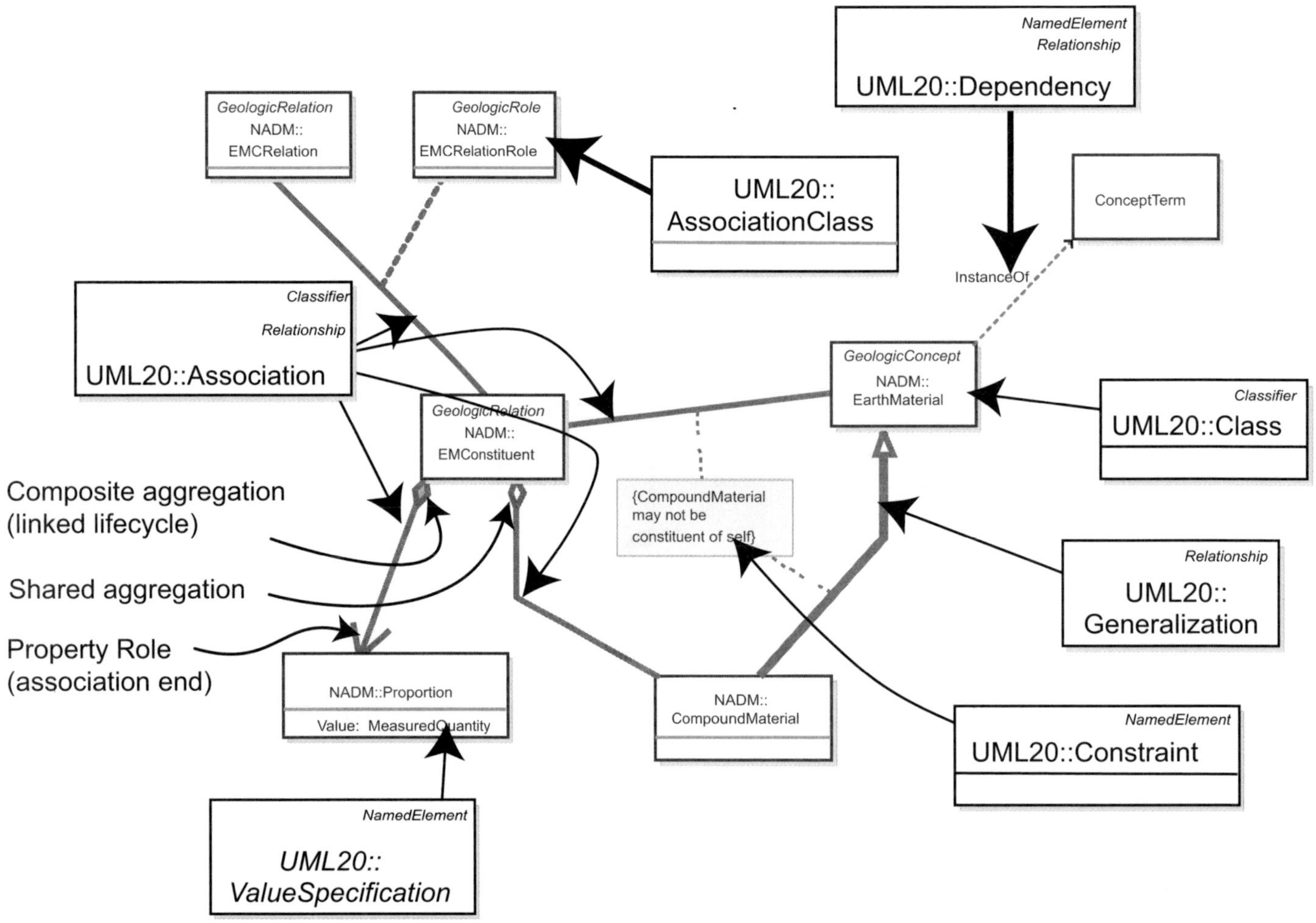

Figure B2. Relationship of Unified Modeling Language (UML) metamodel elements to a model in this paper. The model in light gray is extracted from Figure 4. Arrows link UML metaclasses with model elements that instantiate those classes.

collection of logical rules that determine possible configurations of things in the domain. Very abstract models that deal with defining the fundamental aspects of existence have traditionally been considered the realm of ontology in philosophy. In recent years, the development of more concrete domain models has become an important aspect of information management and exchange (Sowa, 2000). Such a domain model is commonly referred to as an ontology, which in this context is a conceptual model along with a controlled vocabulary that enumerates instances of things recognized in the conceptual model.

The symbolic elements used to construct a model constitute a sign system. In order to be used for communication, this sign system must have an underlying structure that determines how the system is used to represent the concepts and relationships. A metamodel is a model of a sign system used to construct other models. Language is the most commonly used sign system for representation of everyday experience, and the rules of grammar and syntax constitute a metamodel that determines how domain models (sentences) may be constructed using the language sign system. The conceptualization corresponding to a sentence in the mind of the speaker is the conceptual model represented by the sentence.

The domain ontology is based on top-level classes defined by Sowa (2000), and the ontology of properties developed by Guarino and Welty (2000). An upper-level ontology provides a philosophical foundation. Fundamentally, an upper-level ontology attempts to define the kinds of things that exist, starting from a concept that represents everything/anything or the logical concept *top* (T). *Thing* is the term used here to represent this concept.

The definition of a top-level ontology hinges on the selection of the choice of fundamental distinctions or properties used to separate different kinds of things (Gangemi et al., 2001). Sowa (2000) provides a recent review of the evolution of systems for top-level ontology since the time of the Greek philosophers, and proposed a top-level concept lattice based on three top-level distinctions: (1) independent versus relative versus mediating (firstness, secondness, thirdness of Peirce, 1891); (2) physical versus abstract; and (3) occurrent versus continuant. Guarino and Welty (2000b) identified four fundamental notions—identity, essence, unity, and dependence—that they applied to the classification of things at the top level. The dependence property appears to correspond to the distinction between independent, relative, and mediating in Sowa's upper ontology. Independent things have the property "not dependent," while relative and mediating things are "dependent." Another important kind of distinction is that between individual versus collection and instance versus class. It is not the purpose of this paper to develop an upper-level ontology. Rather, the fundamental distinctions outlined above are used as the basis for developing the domain ontology for geoscience. Readers interested in more complete top-level ontologies should see Sowa (2000), the Suggested Upper Merged Ontology (SUMO) (http://ontology.teknowledge.com/rsigma/arch.html), the EPISTLE core model (West, 1996), or DOLCE (Gangemi et al., 2002).The definitions of the fundamental properties that have proved useful in analyzing geoscience concepts are summarized in Table C1 to familiarize the reader with terminology used in the discussion.

APPENDIX D. QUALIFIED RELATIONSHIPS

For observational data, the association between attributes (property-value pairs) and instances they characterize is typically more complex than the simple assertion "has" implied by the standard UML association. Representation of the attribute-instance relationship as an association class with a variety of qualifying properties provides a mechanism to capture alternate interpretations, negation, fuzzy attribution, degree of development, or other aspects of the attribution. Figure D1 models `QualifiedAssociation` as a metalevel subclass of UML `Association`, with the properties `Basis`, `Source`, `Confidence`, `Intensity`, and `Frequency`. `Basis` specifies the evidence for asserting the association. Depending on the context, this might describe a laboratory measurement method, a field technique, or a remote analysis technique, to name a few possibilities. `Source` specifies the "author" of the attribute assertion. This could be used to differentiate alternate interpretations of geologic features. `Confidence` specifies the confidence of the attribute source in asserting the relationship. `Intensity` specifies the intensity or degree to which a property is developed. Properties such as fabric may be present, but developed to different degrees. For example, schistosity may be weakly developed in a rock in which only 10–20% of tabular mineral grains are aligned (quartzofeldspathic gneiss with 20% mica in schistosity), but strongly developed in a rock in which all mineral grains are tabular and aligned in schistosity. `Frequency`

Table C1. Fundamental ontological properties.

Distinction (alternate names)	Discussion
Independent, dependent, relative [Sowa, 2000] (Firstness, secondness, thirdness) [Peirce, 1891]	Independent entities have inherent qualities that do not require the existence of some other entity. A 'rock' is perhaps the best example. Dependent entities have an existence based on a conceptual link between two entities that have an existence independent of the connection between them. Relative entities exist in the irreducible connection between three entities.
Physical, Abstract [Sowa, 2000] (extended, non-extended; concrete, abstract [Guarino and Welty, 2000]	This distinction separates entities that consist of matter or energy from entities that are 'pure information structures' in the terms of Sowa [2000]. Physical entities have location in space-time (are extended), and may causally affect or be affected by other physical entities. Abstract entities may be encoded or represented by some physical entity without changing the mass, energy, or location of the physical entity. Abstract entities can not directly affect physical entities.
Continuant, Occurrent [Sowa, 2000] (Rigid, non-Rigid [Guarino and Welty, 2000])	This distinction has to do with persistence in time. A continuant has stable attributes or characteristics that allow it to be recognized as the same individual. An occurrent is a thing in a state of flux such that it is only identified by occupation of some region of spacetime.
Unity	Having Unity implies that a thing has some property that defines a whole instance of that thing [Gangemi et al., 2001].
Identity	Having Identity implies that a thing has some property unique to a specific individual thing [Guarino and Welty, 2001].

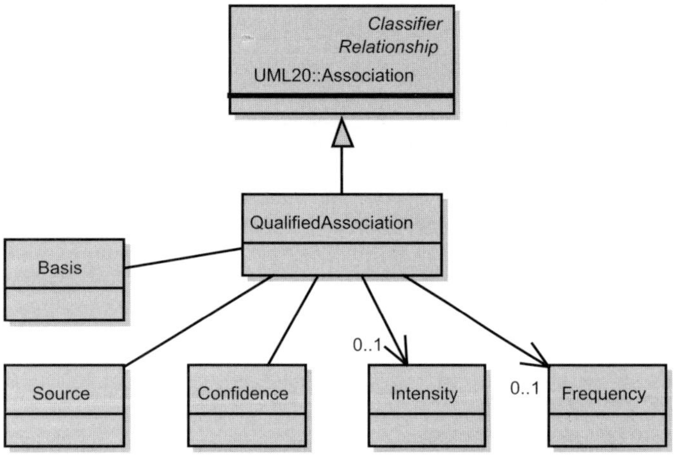

Figure D1. QualifiedRelationship. Associations between properties and attributes that have multiple values may be thought of as AssociationClasses that have properties documenting the source and scientific basis for attribution, as well as qualifiers for classification confidence, property intensity, and frequency or likelihood of occurrence of the value specified. The Frequency qualifier allows expression of negation, e.g., "primary plagioclase is never present in a lamproite."

specifies the likelihood that the property will have the particular value asserted by the relationship. This qualifier could be used to express negation—the concept that on this class, this property will never have some particular value. For instance, the definition of the rock lamproite specifies that the presence of primary plagioclase, melilite, or nepheline (among others) precludes classification as lamproite (Mitchell and Bergman,1991; Woolley et al., 1996). Representation of this assertion in a database requires the ability to express negation. The specification of Frequency might use terminology, with terms such as *always*, *common*, *rare*, or *never*, or might be done using a numerical "truth degree" to implement a fuzzy logic representation. Implementation of these properties on attribute relationships can be done by using an instance-property-value correlation table (frame) with fields (slots) for the qualifier values to represent all QualifiedAssociation instances.

REFERENCES CITED

Ady, B.E., 1993, Towards a theory of spatio-chronologic relations for geoscience: Sudbury, Ontario Geological Survey Open-File Report 5854, 76 p.

Allgayer, J., and Franconi, E., 1994, Collective entities and relations in concept languages, *in* Lakemeyer, G., and Nebel, B., eds., Foundation of Knowledge Representation and Reasoning: Berlin, Springer-Verlag, p. 13–29.

American Heritage Dictionary, 1982, Second College Edition: Boston, Houghton Mifflin Co., 1568 p.

Angus, C., and Dziulka, P., 1998, EPISTLE Framework v. 2.0, issue 1.22, Dziulka, P., ed., Keyworth Institute, Department of Mechanical Engineering, University of Leeds, Woodhouse Lane, Leeds LS2 9JT.

Bain, K.A., and Giles, J.R.A., 1997, A standard model for storage of geological map data: Computers & Geosciences, v. 23, no. 6, p. 613–620, doi: 10.1016/S0098-3004(97)00043-5.

Bates, R.L., and Jackson, J.J., eds., 1987, Glossary of Geology, Third Edition: Alexandria, Virginia, American Geologic Institute, 788 p.

Brodaric, B., and Gahegan, M., 2001, Learning geoscience categories in situ: Implications for geographic knowledge representation, *in* Aref, W.G., ed., ACM-GIS 2001, Proceedings of the Ninth ACM International Symposium on Advances in Geographic Information Systems, Atlanta, Georgia, Nov. 9–10, 2001: Association for Computing Machinery, ISBN 1-58113-443-6, p. 130–135.

Brodaric, B., and Hastings, J., 2002, An Object Model for Geologic Map Information, *in* Richardson, D.E., and van Oosterom, P., eds., Advances in Spatial Data Handling: Proceedings of the 10th International Symposium on Spatial Data Handling: New York, Springer-Verlag, ISBN 3540438025, p. 55–68.

Data Model Design Team, North American Geologic-Map Data Model Steering Committee (DMDT), 2002, Progress report: North American Geologic Map Data Model Design Team, *in* Soller, D.R., Digital Mapping Techniques '02—Workshop Proceedings: U. S. Geological Survey Open-File Report 02-370, p. 41–46.

Engelhardt, W. von, and Zimmermann, J., 1988, Theory of Earth Science: Cambridge, Cambridge University Press, 381 p.

Folk, R.L., 1980, Petrology of sedimentary rocks: Austin, Texas, Hemphill's Bookstore, 170 p.

Fonseca, F.T., and Egenhofer, M.J., 1999, Ontology-driven geographic information systems, *in* Medeiros, C.B., ed., 7th ACM Symposium on Advances in Geographic Information Systems: New York, New York, ACM Press, p. 14–19, http://www.spatial.maine.edu/~max/ACMGIS99.pdf (accessed June 2004).

Gangemi, A., Guarino, N., Masolo, C., Oltramari, A., and Schneider, L., 2002, Sweetening Ontologies with DOLCE, *in* Proceedings of the 13th International Conference on Knowledge Engineering and Knowledge Management, Ontologies and the Semantic Web, Lecture Notes in Computer Science: London, UK, Springer-Verlag, p. 166–181.

Gangemi, A., Guarino, N., Masolo, C., and Oltramari, A., 2001, Understanding top-level ontological distinctions, *in* Perez et al., eds., Proceedings of IJCAI2001 Workshop on Ontologies and Information Sharing: Seattle, USA, AAAI Press, p. 26–33.

Giles, J.R.A., Lowe, D.J., and Bain, K.A., 1997, Geological Dictionaries—Critical elements of every geological database: Computers & Geosciences, v. 23, no. 6, p. 621–626, doi: 10.1016/S0098-3004(97)00044-7.

Guarino, N., 1998, Formal ontology in information systems, *in* Guarino, N., ed., Formal Ontology in Information Systems: Amsterdam, Netherlands, IOS Press, p. 3–15, http:// www.loa-cnr.it/Papers/FOIS98.pdf).

Guarino, N., and Welty, C., 2001, Identity and subsumption: Padova, Italy, LADSEB-CNR Internal Report 01, 15 p., http://www.loa-cnr.it/Papers/Identity&Subsumption.pdf.

Guarino, N., and Welty, C., 2000a, A formal ontology of properties, *in* Dieng, R., and Corby, O., eds., Proceedings of the 12th International Conference on Knowledge Engineering and Knowledge Management: France, Springer-Verlag, p. 97–112 (http://www.loa-cnr.it/Papers/EKAW-2000.pdf).

Guarino, N., and Welty, C., 2000b, Ontological analysis of taxonomic relationships, *in* Laender, A., and Storey, V., eds., Proceedings of ER-2000, The 19th International Conference on Conceptual Modeling: Springer-Verlag Lecture Notes in Computer Science, v. 1920, p. 210–224.

Halpin, T.A., 1995, Conceptual Schema and Relational Database Design, Second Edition: Prentice Hall Australia, 547 p.

Halpin, T.A., 2001, Information Modeling and Relational Databases: From Conceptual Analysis to Logical Design: Morgan Kaufmann Publishers, 800 p.

Hobbs, B.E., Means, W.D., and Williams, P.F., 1976, An Outline of Structural Geology: New York, John Wiley and Sons, 571 p.

Johnson, B.R., Brodaric, B., Raines, G.L., Hastings, J.T., and Wahl, R., 1999, Digital Geologic Maps Data Model, v. 4.3: AASG/USGS Data Model Working Group Report, http://www.nadm-geo.org/dmdt/Model43a.pdf.

Lakoff, G., 1987, Women, Fire and Dangerous Things: Chicago, University of Chicago Press, 614 p.

MacRandal, D., 1988, Semantic Networks, *in* Ringland, G.A., and Duce, D.A., eds., Approaches to Knowledge Representation: An Introduction: New York, John Wiley and Sons, Inc, p. 45–79.

Merriam-Webster Dictionary, 2002, http://www.m-w.com/dictionary.htm.

Mitchell, R.H., and Bergman, S.C., 1991, Petrology of Lamproites: New York, Plenum Press, 447 p.

North American Commission on Stratigraphic Nomenclature, 1983, North American Stratigraphic Code: Bulletin of the American Association of Petroleum Geologists, v. 67, no. 5, p. 848–875.

North American Geologic-Map Data Model Steering Committee, 2004a, NADM Conceptual Model 1.0, A Conceptual Model For Geologic Map Information: U.S. Geological Survey Open-File Report 04-1334, 61 p., (http://pubs.usgs.gov/of/2004/1334).

North American Geologic-Map Data Model Steering Committee, 2004b, Sedimentary materials: science language for their classification, description, and interpretation in digital geologic-map databases, Version 1.0: Appendix to Digital Mapping Techniques '04, Workshop Proceedings: U.S. Geological Survey Open-File Report 2004-1451, 245 p. (http://www.nadm-geo.org/sltt/products.html).

Object Management Group, 2001, OMG Unified Modeling Language Specification, v. 1.4: Object Management Group, Inc, available at http://www.omg.org/technology/documents/formal/uml.htm (downloaded September 2002).

Object Management Group, 2003, UML 2.0 Superstructure Specification, OMG Adapted Specification 03-08-02: Object Management Group, Inc., 640 p. (PDF file).

Open GIS Consortium Technical Committee, 1998, The OpenGIS Guide, Third Edition, Buehler, K., and McKee, L. eds.: Open GIS Consortium, Wayland, Massachusetts, USA.

Parent, C., Spaccapietra, S., and Zimanyi, E., 1997, Conceptual modeling for federated GIS over the Web, *in* Proceedings of the International Symposium on Information Systems and Technology for Network Society, p. 173–182, http://lbdwww.epfl.ch/e/publications/articles.pdf/fukuoka.pdf (downloaded September 2002)

Passchier, C.W., and Trouw, R.A.J., 1996, Microtectonics: New York, Springer, 289 p.

Peirce, C.S., 1891, Review of Principles of Psychology by William James: New York, N.Y., Nation (1865), v. 53, p. 32.

Richard, S.M., 1998, Digital Geologic Database Model (Web Page): Arizona Geological Survey, http://www.azgs.state.az.us/GeoData_model.pdf, accessed June 2004.

Richard, S.M., 1999, Geologic concept modeling, with examples for lithology and some other basic geoscience features, *in* Soller, D.R., ed., Digital Mapping Techniques 1999, Workshop Proceedings: U. S. Geological Survey Open-File Report 99-386, p. 59–75.

Richard, S.M., 2000, Proposal for authorship and citation guidelines for geologic data sets and map images in the era of digital publication, *in* Soller, D.R., ed., Digital Mapping Techniques 2000, Proceedings, p. 159–168.

Richard, S.M., 2001, Rock classification for geologic map database applications, *in* Soller, D.R., ed., Digital Mapping Techniques '01—Workshop Proceedings, U.S. Geological Survey Open-file Report 01-223, p. 135–138.

Richard, S.M., 2003, Geologic map database implementation in the ESRI™ geodatabase environment, *in* Soller, D.R., ed., Digital Mapping Techniques '03—Workshop Proceedings, U.S. Geological Survey Open-file Report 2003-471 (http://pubs.usgs.gov/of/of03-471/pdf/richard1.pdf).

Richard, S.M., and Orr, T.R., 2001, Data structure for the Arizona Geological Survey geologic information system; basic geologic map data, *in* Soller, D.R., ed., Digital Mapping Techniques 2001, Workshop Proceedings, U.S. Geological Survey Open-File Report 01-223, p. 167–188, http://pubs.usgs.gov/of/of01-223/richard2.html (accessed May 2004).

Richard, S.M., and Thieme, J.P., 1997, Data structure for Arizona Geological Survey digital geologic maps: Tucson, Arizona Geological Survey Open-File Report 97-5, 15 p.

Richard, S.M., Matti, J., and Soller, D.R., 2003, Geoscience terminology development for the National Geologic Map Database, *in* Soller, D.R., ed., Digital Mapping Techniques '03—Workshop Proceedings, U.S. Geological Survey Open-file Report 2003-471.

Richard, S.M., Craig, J., and Soller, D.R., 2004, Implementing NADM-C1 for the National Geologic Map Database, *in* Soller, D.R., ed., Digital Mapping Techniques '04, Workshop Proceedings: U.S. Geological Survey Open-File Report 2004-1451, p. 111–144 (http://pubs.usgs.gov/of/2004/1451/pdf/richard.pdf).

Salvador, A., ed., 1994, International Stratigraphic Guide, Second Edition: Trondheim, Norway, International Union of Geological Sciences, and Boulder, Colorado, Geological Society of America, 214 p.

Schmid, R., Fettes, D., Harte, B., Davis, E., Desmons, J., and Siivola, J., 2002, Towards a unified nomenclature in metamorphic petrology: 1. How to name a metamorphic rock, A proposal on behalf of the IUGS Subcommission on the Systematics of Metamorphic Rocks, Web version of 31.07.2002: 13 p., http://www.bgs.ac.uk/SCMR.

Sowa, J.F., 2000, Knowledge Representation, Logical, Philosophical, and Computational Foundations: Pacific Grove, California, Brooks/Cole, 594 p.

Struik, L.C., Quat, M.B., Davenport, P.H., and Okulitch, A.V., 2002, Multi-hierarchical rock classification for use with thematic computer-based query systems: a preliminary scheme: Geological Survey of Canada, Current Research 2002-D10, 9 p.

Turner, F.J., and Weiss, L.E., 1963, Structural Analysis of Metamorphic Tectonites: New York, McGraw-Hill, 545 p.

Webster's encyclopedic unabridged dictionary of the English Language, 1996, Random House Value Publishing Inc., 2230 p.

West, M., 1996, Developing high quality data models, Issue 2.1, EPISTLE: Shell International Petroleum Co., Ltd., Report no. ISCL4-33, http://www.matthew-west.org.uk/documents/prince03.pdf.

Williams, H., Turner, F.J., and Gilbert, C.M., 1954, Petrography an Introduction to the Study of Rocks in Thin Sections: San Francisco, W.H. Freeman and Company, 406 p.

Woolley, A.R., Bergman, S.C., Edgar, A.D., LeBas, M.J., Mitchell, R.H., Rock, N.M.S., and ScottSmith, B.H., 1996, Classification of lamprophyres, lamproites, kimberlites, and the kalsilitic, melilitic, and leuctitic rocks: The Canadian Mineralologist, v. 34, p. 175–186.

Manuscript Accepted by the Society 18 June 2005

Printed in the USA

Geological Society of America
Special Paper 397
2006

*Managing scientific data: From data integration to scientific workflows**

Bertram Ludäscher†
Department of Computer Science, University of California, One Shields Avenue, Davis, California 95616

Kai Lin†
Shawn Bowers†
Efrat Jaeger-Frank†
San Diego Supercomputer Center, University of California, San Diego

Boyan Brodaric†
Natural Resources of Canada

Chaitan Baru†
San Diego Supercomputer Center, University of California, San Diego

ABSTRACT

Scientists are confronted with significant data management problems due to the large volume and high complexity of scientific data. In particular, the latter makes data integration a difficult technical challenge. In this paper, we describe our work on semantic mediation and scientific workflows and discuss how these technologies address integration challenges in scientific data management. We first give an overview of the main data integration problems that arise from heterogeneity in the syntax, structure, and semantics of data. Starting from a traditional mediator approach, we show how semantic extensions can facilitate data integration in complex, multiple-world scenarios, where data sources cover different but related scientific domains. Such scenarios are not amenable to conventional schema integration approaches. The core idea of semantic mediation is to augment database mediators and query evaluation algorithms with appropriate knowledge representation techniques to exploit information from shared ontologies. Semantic mediation relies on semantic data registration, which associates existing data with semantic information from an ontology.

The KEPLER scientific workflow system addresses the problem of synthesizing, from existing tools and applications, reusable workflow components and analytical pipelines to automate scientific analyses. After presenting core features and example workflows in KEPLER, we present a framework for adding semantic information to scientific workflows. The resulting system is aware of semantically plausible connections

*Work supported by National Science Foundation/Information Technology Research for National Priorities (NSF/ITR) 0225673 (GEON), NSF/ITR 0225676 (SEEK), NIH/NCRR 1R24 RR019701-01 Biomedical Informatics Research Network (BIRN-CC), and Department of Energy SciDAC DE-FC02-01ER25486 (SDM).
†E-mails: ludaesch@ucdavis.edu; lin@sdsc.edu; sbowers@ucdavis.edu; efrat@sdsc.edu; brodaric@nrcan.gc.ca; baru@sdsc.edu

Ludäscher, B., Lin, K., Bowers, S., Jaeger-Frank, E., Brodaric, B., and Baru, C., 2006, Managing scientific data: From data integration to scientific workflows, *in* Sinha, A.K., ed., Geoinformatics: Data to Knowledge: Geological Society of America Special Paper 397, p. 109–129, doi: 10.1130/2006.2397(08). For permission to copy, contact editing@geosociety.org.

between workflow components as well as between data sources and workflow components. This information can be used by the scientist during workflow design, and by the workflow engineer, for creating data transformation steps between semantically compatible but structurally incompatible analytical steps.

Keywords: scientific data management, data integration, ontologies, scientific workflows.

1. INTRODUCTION

Information technology is revolutionizing the way many sciences are conducted, as witnessed by new techniques, results, and discoveries in multidisciplinary and information-driven fields such as bioinformatics, ecoinformatics, and geoinformatics. The opportunities provided by these new information-driven and often data-intensive sciences also bring with them equally large challenges for scientific data management. For example, to answer a specific scientific question or to test a certain hypothesis, a scientist today not only needs profound domain knowledge, but also may require access to data and information provided by others via community databases or analytical tools developed by community members.

A problem for the scientist is how to easily make use of the increasing number of databases, analytical tools, and computational services that are available. Besides making these items generally accessible to scientists, leveraging these resources requires techniques for *data integration* and *system interoperability*. Traditionally, research by the database community in this area has focused on problems of heterogeneous systems, data models, and schemas [She98]. However, the integration scenarios considered differ significantly from those encountered in scientific data integration today. In particular, the former usually involve "one-world" scenarios, where the goal is to provide an integration schema or integrated view over multiple sources having a single, conceptual domain. Online comparison shopping for cheapest books is an example of a one-world scenario: The different database schemas or Web services to be integrated all deal with the same book attributes (e.g., title, authors, publisher, and price).

Compare this situation to the scientific information integration scenario depicted in Figure 1. A scientist (here, an igneous petrologist) is interested in the distribution of a certain rock type (say *A-type plutons*) within a specific region. He also wants to know the three-dimensional geometry of those plutons and understand their relation to the host rock structures. Through databases and analytical tools, our scientist can gather valuable information toward answering his scientific question. For example, geologic maps of the region, geophysical databases with gravity contours, foliation maps, and geochemical databases all provide pieces of information that need to be brought together in an appropriate form to be useful to the scientist (see Figure 1).

We call this integration example a *complex multiple-worlds scenario* [LGM03]. In particular, it is not possible or meaningful to integrate the schemas of the data sources into a single common schema because the various data sources contain disjoint information. However, there are often latent links and connections between these data sources that are made by scientists. Through these *implicit* knowledge structures, the various pieces of information can be "glued" together to help answer scientific questions.

Making explicit these knowledge structures—i.e., the "glue"—is therefore a prerequisite for connecting the underlying data. In scientific domains, ontologies can be seen as the formal representation of such knowledge, to be used for data and knowledge integration purposes. The problem in the geoscience domain, however, is that most of this knowledge is either implicit or represented in narrative form in textbooks, and thus not readily available for use in mediation systems as required.

As a simple example, consider a geologic map and a geochemical database (see the bottom left of Figure 1). We can link these sources thematically by establishing associations (i) between the formations in the geologic map and their constituent rock types, as indicated in a map legend and associated reports, and likewise (ii) between rock types and their mineral compositions, using, e.g., the spatial overlap of formations and geochemical samples. By making these associations explicit, and recording them in domain ontologies, we can establish interesting linkages in complex multiple-world scenarios [LGM01, LGM03].

In this paper, we focus on two important aspects of scientific data management: *data integration* and *scientific workflows*. For the former, we present a framework that extends the traditional mediator-based approach to data integration with ontologies to yield a *semantic mediation* framework for scientific data integration. In this framework, ontologies are used to bridge the gap between "raw" data stored in databases and the knowledge level at which domain scientists work, trying to answer the given scientific questions (see the left side of Figure 1).

Data integration and knowledge-based extensions such as semantic mediation deal with modeling and querying database systems as opposed to the interoperation of analytical tools, or the assembly of data sources and computational services into larger scientific workflows. As we will illustrate below, scientific workflows, however, can also benefit from knowledge-based extensions. In fact, an important goal of building cyberinfrastructure for scientific knowledge discovery is to devise integrated problem-solving environments that bring to the scientist's desktop the combined power of remote data sources, services, and computational resources from the Grid [FK99, Fos03, BFH03].

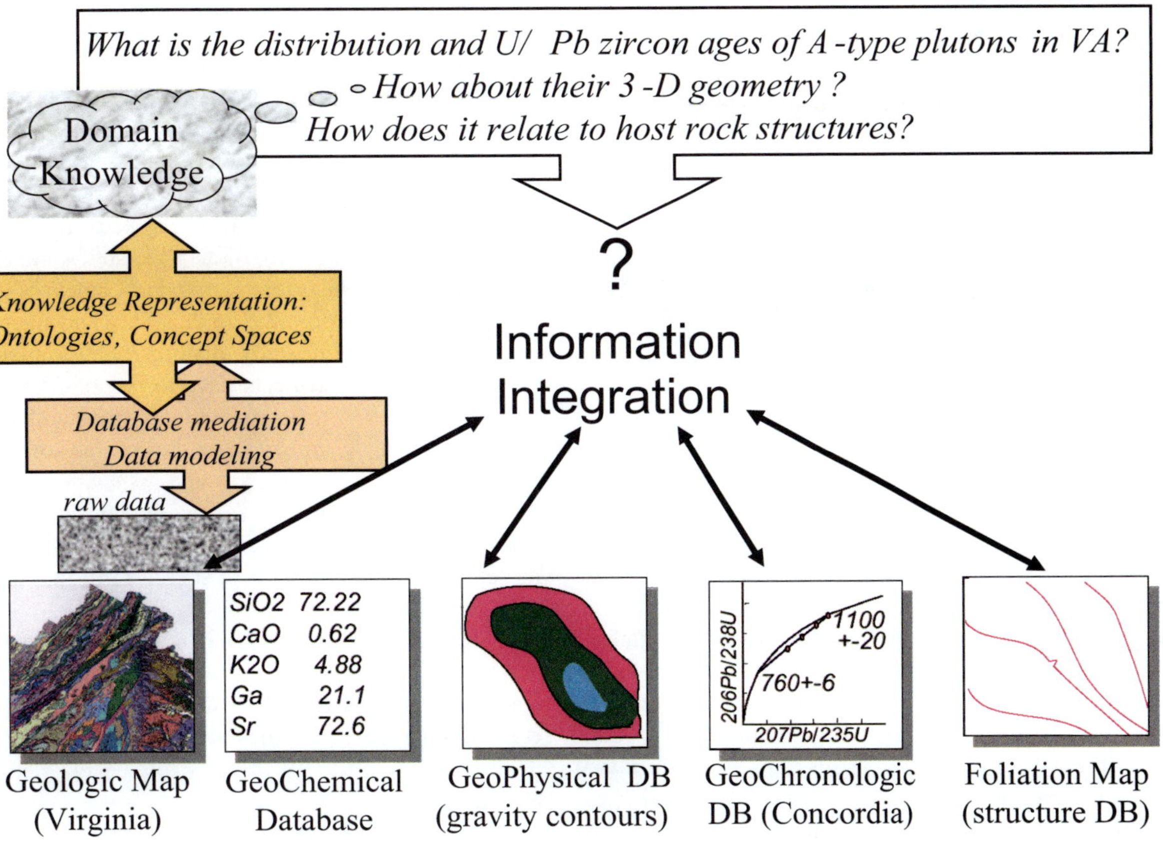

Figure 1. Complex "multiple-worlds" integration.

The rest of this paper is organized as follows. In Section 2, we provide a high-level overview of information integration and interoperability challenges in scientific data management. We then present two integration examples in Section 3. The first example illustrates the use of ontologies in scientific data integration with geologic maps (Section 3.1). Additional knowledge represented in a geologic-age or rock-type ontology is used to create different conceptual views of geologic maps and allows the user to query the information in novel ways. Section 3.2 illustrates scientific process integration and combines data-access, data-analysis, and visualization steps into a scientific workflow.

In Section 4, we give an overview of data integration approaches and discuss technical issues surrounding semantic mediation. Section 5 gives an introduction to scientific workflows and a brief overview of the KEPLER system. In Section 5.3, we show that semantic extensions not only benefit data integration, but can also provide new opportunities and challenges for scientific workflows. We conclude with a brief summary of our work and findings in Section 6.

2. INTEGRATION CHALLENGES

In this section, we provide a high-level overview of the data-integration and system-interoperability challenges that often await a scientist who wants to employ IT infrastructure for scientific knowledge discovery. Many of these challenges arise from heterogeneities that occur across systems, data sources, and services that make up a scientific data management infrastructure.

Data heterogeneity has traditionally been divided into *syntax*, *structure*, and *semantic* differences [She98, VSSV02]. Scientific data analysis and information integration scenarios such as the one depicted in Figure 1 often involve additional levels, which include low-level *system* integration issues as well as higher-level *synthesis* issues. We briefly discuss these various levels of heterogeneities and interoperability challenges below (see Figure 2).

System Aspects

By system aspects we mean differences in low-level issues relating to, e.g., network protocols[1] (e.g., HTTP, FTP, GridFTP, SOAP), platforms (operating systems), remote execution methods (e.g., Web services, RMI, CORBA), and authorization and authentication mechanisms (e.g., Kerberos, GSI). Many efforts for cyberinfrastructure aim at facilitating system interoperability by providing a common Grid middleware infrastructure [FK99, Fos03, BFH03].

[1]These are organized into a layered stack of communication protocols themselves, e.g., the TCP/IP stack and the ISO/OSI layered architecture.

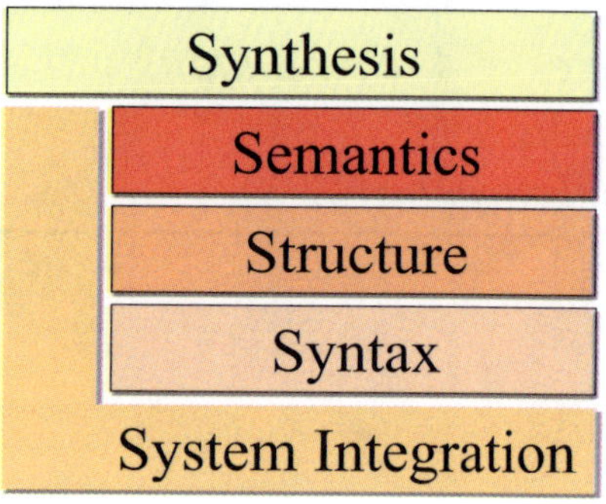

Figure 2. Interoperability challenges.

System entities requiring management include certificates, file handles, and resource identifiers.

Syntax

Data that is not stored in databases, or that is exchanged between applications, can come in different representations (e.g., raster or vector) and file formats (e.g., netCDF, HDF, and ESRI shapefile)[2]. The use of XML as a uniform exchange format can help resolve syntactic differences, but many specialized formats prevail for practical reasons (e.g., because of tool support and efficiency).

Syntactic entities requiring management include individual files in various formats and format conversion tools.

Structure

Structure differences can arise when similar data is represented using different schemas. The problem of *schema integration* is a heavily studied area in databases [She98]. There are several difficulties to overcome, e.g., how to derive an integration schema, how to find and define the right mappings between source schemas and the integration schema, how to efficiently evaluate queries against virtual schemas (i.e., which are not materialized), and how to deal with incomplete and conflicting information. In Section 4.1, we give a brief overview of the mediator approach, which addresses some of these issues.

Structure entities requiring management include database schemas and queries (e.g., in SQL or XQuery).

Semantics

Storing scientific data in a database system provides solutions to a number of technical challenges (e.g., multiuser access, transaction control) and simplifies others (e.g., query execution time can be improved, and certain structural heterogeneities can be resolved by defining queries that map between schemas, called *views*). However, languages for defining database schemas are not expressive enough to adequately capture most semantic aspects of data. For example, information about the kinds of objects being stored, and how those objects relate to each other or to general concepts of the domain, cannot be easily or adequately expressed. Some conceptual-level information can be captured during formal database design (e.g., via an ER conceptual model [Che76]). But this information is rarely connected explicitly to the database, and thus it cannot be used to query for data and is not otherwise usable by a database system. Traditional metadata can provide a limited form of data semantics and can help a scientist understand the origin, scope, and context of a data set.

However, to assess the applicability and integrate different data sets for a certain study, many possible semantic heterogeneities among the data sets have to be considered: What were the measurement or experimentation parameters? What protocols were used? What is known about the accuracy of data? And most fundamentally: What concepts and associations are encoded by the data? The usability of data can be improved significantly by making explicit what is know about a data set (i.e., by describing its semantics), and doing it in such a form that automated processing is facilitated.

Semantic entities requiring management include concepts and relationships from one or more ontologies (e.g., using a standard language such as OWL [DOS04, Usc98, OWL03] or through controlled vocabularies) and data annotations to these ontologies.

Synthesis

By synthesis we mean the problem of putting together databases, including semantic extensions, queries and transformations, and other computational services into a scientific workflow. The problem of synthesis of such workflows encompasses all previous challenges. For example, if a scientist wants to put together two processing steps A and B into the following simple analysis pipeline

$$\xrightarrow{x} \boxed{\mathsf{A}} \xrightarrow{d} \boxed{\mathsf{B}} \xrightarrow{y}$$

many questions arise: In what format does A expect its input x? Does the output d of A directly "fit" the format of B, or is a data transformation necessary? In addition to these *syntactic* and *structural* heterogeneities, *system* and *semantic* issues exist as well. For example, what mechanism should be used to invoke processes, and how should data be shipped from A to B? Is it meaningful and valid to connect A and B in this way?

The main challenges for synthesis include process composition and the modeling, design, and execution of reusable process components and scientific workflows.

3. INTEGRATION EXAMPLES

In this section, we provide two different examples, illustrating new capabilities of data integration using semantic extensions (Section 3.1), and of process integration via scientific workflows (Section 3.2). The underlying technologies are discussed in Section 4 and Section 5, respectively.

[2]There are also different database representations, or *data models*, but most scientific applications today use relational database systems.

3.1. Geologic-Map Data Integration

Geologic maps present information about the history and character of rock bodies, their intersection with Earth's surface, and their formative processes. Geologic maps are an important data product for many geoscientific investigations. In the following, we describe the Ontology-Enabled Map Integrator (OMI) system prototype [LL03, LLB+03, LLBB03] that was built as one of the first scientific data integration systems in the GEON project [GEO]. The goal of the system is to integrate information from a number of geologic maps from different state geological surveys and to provide a uniform interface to query the integrated information. Queries are "conceptual," i.e., they are expressed using terms from a shared ontology. For example, it is possible to query the integrated data using concepts such as *geologic age* and *rock type* as opposed to the terms used to describe these attributes in the underlying data. For the OMI prototype, data from nine different states were integrated. The state geologic maps were initially given as ESRI shapefiles [ESR98] (thus, there were no syntactic heterogeneities). However, each shapefile came with a different relational schema, i.e., the data had structural heterogeneity. A shapefile can be understood as a relational table having one column that holds the spatial object (e.g., a polygon), and a number of data columns holding attribute values that describe the spatial object. For example, the relational schema of the Arizona shapefile is

```
Arizona(AREA, PERIMETER, AZ_1000, AZ_1000 ID,
    GEO, PERIOD, ABBREV, DESCR,
    D_SYMBOL, P_SYMBOL)
```

Here, `AREA` is the spatial column, while `PERIOD` and `ABBREV` are data columns with information on the geologic age and formation of the represented region, respectively. The maps from the other states are each structured slightly differently. For example, the Idaho shapefile has 38 columns and contains detailed lithology information, describing rocks by their color, mineralogic composition, and grain size.

Figure 3 shows the association between the columns of the different source schemas and those of the integration schema. Note that the `Formation` attribute (column) in the integrated schema (in the center of Fig. 3) is derived from different attributes of the source schemas, i.e., `Arizona.ABBREV`, `Colorado.NAME`, `Utah.TYPE`, and so on. The Idaho and Montana West provide additional detailed information on lithology (e.g., `andesitic sandstone`). The latter can be used to derive further information on the rock type associated with a region, such as mineral and chemical *composition*, *fabric*, and *texture*. This assumes that a corresponding ontology of rock types is given. Using such an ontology, we can infer that regions of the map marked as andesitic sandstone are also regions with sedimentary rock (because sandstone is a sedimentary rock). Other logic-based rules can be used to check that a modal mineral composition is within a certain range (e.g., a range might be given as $Q/(Q+A+P) < 20\%$ or $F/(F+A+P) < 10\%$, $P/(A+P) > 90\%$ and $M < 35$ [Joh02]).

Concept-Based ("Semantic") Queries

The results of a simple conceptual-level query, asking for all regions with geologic age Paleozoic, are shown in Figure 4. Recall that all nine maps have geologic age information; nevertheless, few results are found when doing a simple lookup for

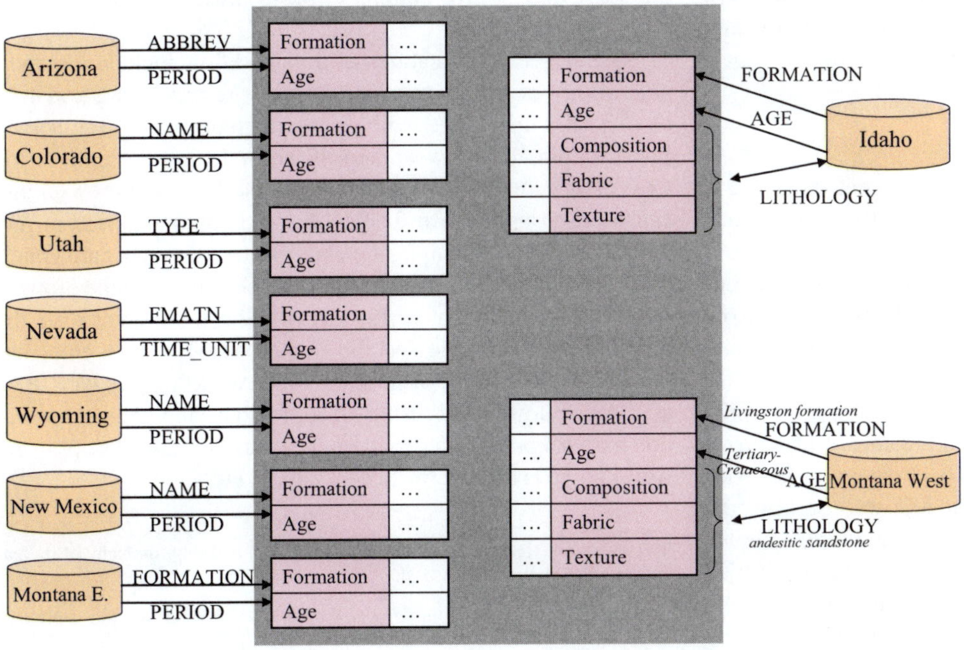

Figure 3. Schema integration in the OMI prototype. Some elements of the local schemas (outside) are mapped to the integration schema (inside).

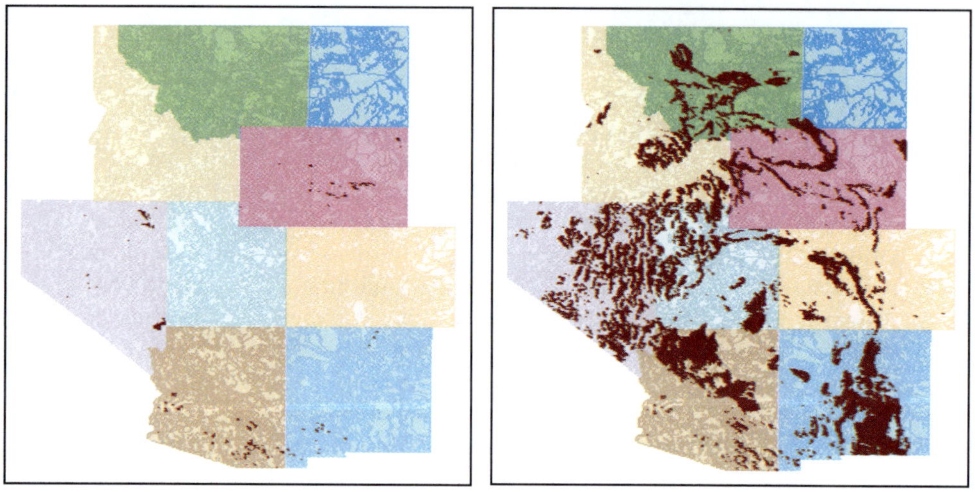

Figure 4. Results of a query for regions with "Paleozoic" age: without ontology (left), and with ontology (right) [LL03].

Paleozoic in the Age column (see the left side of Fig. 4). This occurs because by only looking for Paleozoic, we have not taken into account the domain knowledge that other geologic ages such as Cambrian and Devonian also fall within the Paleozoic. By using a corresponding geologic age ontology, the system can rewrite the original user query into one that looks for Paleozoic and all its "sub-ages." The result of this ontology-enabled query is shown on the right in Figure 4—where a more complete set of regions is returned.

An important prerequisite for such knowledge-based extensions of data integration systems is *semantic data registration* [BLL04a]. In a nutshell, data objects (here, the polygons making up the spatial regions) must be associated with concepts from a previously registered ontology. In the system, we use several such ontologies, for geologic age (derived from [HAC+89]) and for rock type classification (derived from [SQDO02] and [GSR+99]). In Section 4.3, we discuss further details of data registration and revisit the geologic-map integration example.

3.2. Mineral Classification Workflow

The previous integration example was data-centric and made use of domain knowledge by using an ontology to answer a conceptual-level query. The second integration example is process-centric and illustrates the use of a scientific workflow system for automating an otherwise manual data analysis procedure, or alternatively, for reengineering an existing data analysis tool in a more generic and extensible environment.

The upper left window in Figure 5 shows the top-level workflow, where data points are selected from a database of mineral compositions of igneous rock samples. These data, together with a set of classification diagrams, are fed into a CLASSIFIER subworkflow (bottom left). The manual process of classifying samples involves determining the position of the sample values in a series of diagrams such as the ones shown on the right in Figure 5. If the location of a sample point in a nonterminal diagram of order n has been determined (e.g., diorite gabbro anorthosite, bottom right), the corresponding diagram of order $n + 1$ is consulted, and the point is located therein. This process is iterated until the terminal level of diagrams is reached. The result is shown in the upper right of Figure 5, where the classification result is anorthosite.

This traditionally manual process has been automated in specialized commercial tools. Here, we show how the open-source KEPLER workflow system [KEP, LAB+04] can be used to implement this workflow in a more open and generic way (Fig. 5). The workflow is shown in graphical form using the VERGIL user interface [BLL+04b][3]. Note that in VERGIL, workflows can be annotated with user comments. Workflows can be arbitrarily nested, and subworkflows (e.g., shown in the bottom left of the figure) become visible by "looking inside" a composite actor[4]. The box labeled CLASSIFIER is a composite actor. VERGIL also features simple VCR-like control buttons to *play*, *pause*, *resume*, and *stop* workflow execution.

KEPLER-specific features of this workflow include a searchable library of actors and data sources (`Actor` and `Data` tabs close to the upper left) with numerous reusable KEPLER actors. For example, the Browser actor (used in the bottom right of the CLASSIFIER subworkflow) launches the user's default browser and can be used as a powerful generic input/output device in any workflow. In this example, the classification diagrams are generated on the client side as interactive SVG displays in the browser (windows on the right in Fig. 5). Moving the pointer over the diagram highlights the specific region. The BROWSER actor has proven to be very useful in many other workflows as well, e.g., as a device to display results of a previous step and as a selection tool that passes user-requested data to subsequent steps using well-known HTML forms, check boxes, etc.

[3]KEPLER is an extension of the PTOLEMY II system and inherits many of its features, including the VERGIL GUI.

[4]Following PTOLEMY II terminology, a workflow component, whether atomic or composite, is called an *actor* in KEPLER.

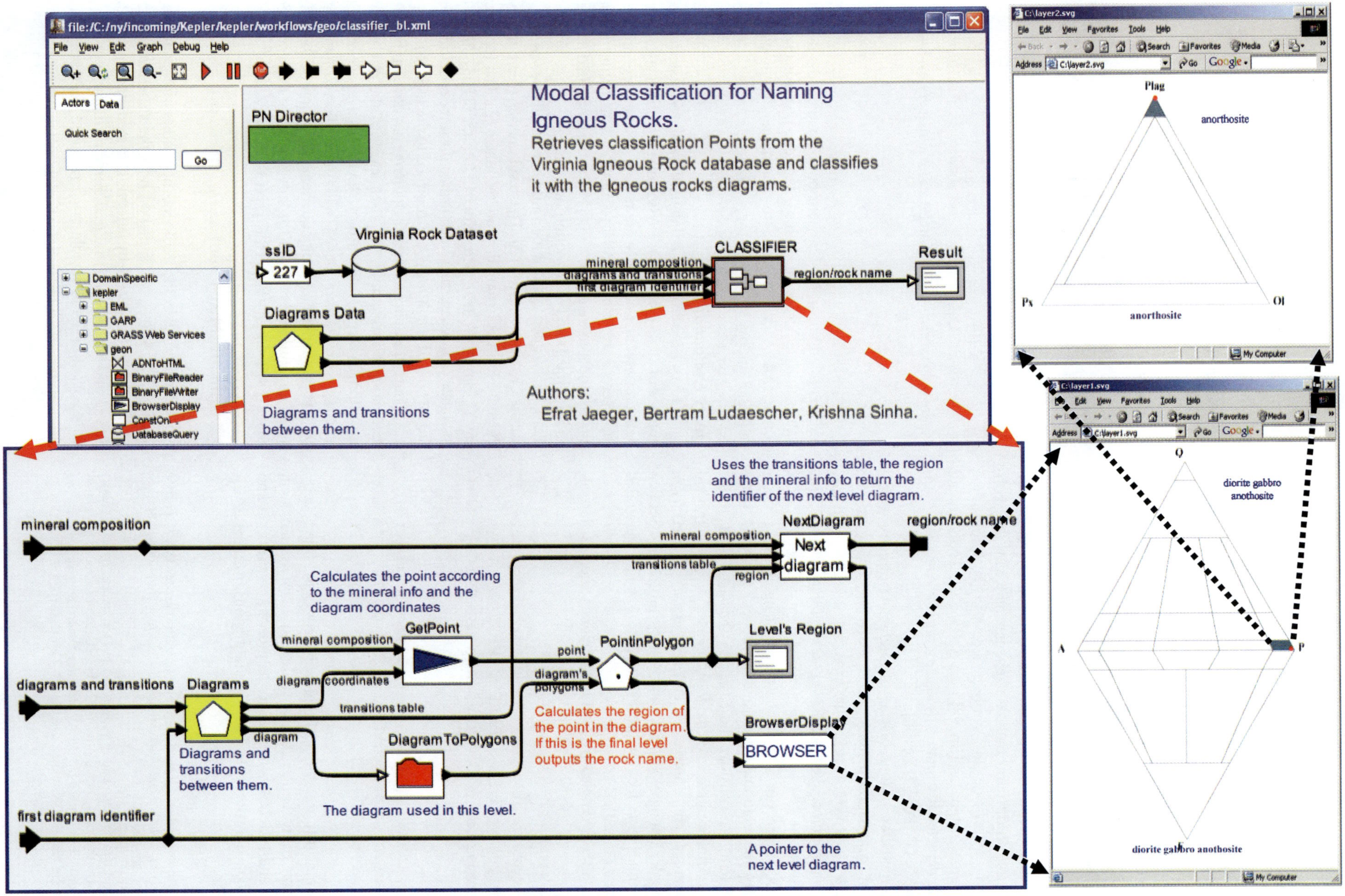

Figure 5. Mineral classification workflow (left) and generated interactive result displays (right).

4. DATA INTEGRATION

In this section, we first introduce the mediator approach to data integration and then show how it can be extended to a semantic mediation approach by incorporating ontologies as a knowledge representation formalism. We also present formal foundations of semantic data registration, which is an important prerequisite for semantic mediation.

4.1. Mediator Approach

Data integration traditionally deals with structural heterogeneities due to different database schemas and with the problem of how to provide uniform user access to the information from different databases [She98] (see Section 2). The standard approach, depicted in Figure 6, is to use a *database mediator* system [Wie92, GMPQ+97, Lev00].

In a mediator system, instead of interacting directly with a number of local data sources $S_1, \ldots, S_k$, each one having its own database schema V_i (also called the *exported view* of S_i), the user or client application queries an *integrated global view G*. This integrated view is given by an *integrated view definition* (IVD), i.e., a query expression in a database query language (e.g., SQL or Datalog for relational databases, or XQuery for XML databases)[5]. Here, we consider the case that the *integrated global view G* is defined in terms of the *local views* $V_1, \ldots, V_k$ exported by the sources. This approach is called *global-as-view* (GAV)[6]. For example, our geologic-map integration prototype (Section 3.1) integrates nine local source views into a single integrated global view (Fig. 3).

The latter is defined by query expressions such as the following:

```
G("Arizona", AZ.Aid, Formation, Age,…)←
    Arizona(Aid,…,ABBREV,…PERIOD,…),
    Formation = ABBREV, Age = PERIOD.

G("Nevada", NV.Aid, Formation, Age,…)←
    Nevada(Aid,…,FMATN,…TIME_UNIT,…),
    Formation = FMATN, Age = TIME_UNIT.
```

The first rule states that the global view G is "filled" with information from the Arizona source by mapping the local ABBREV and PERIOD attributes to the global Formation and Age attributes, respectively. Here, spatial regions from the AREA

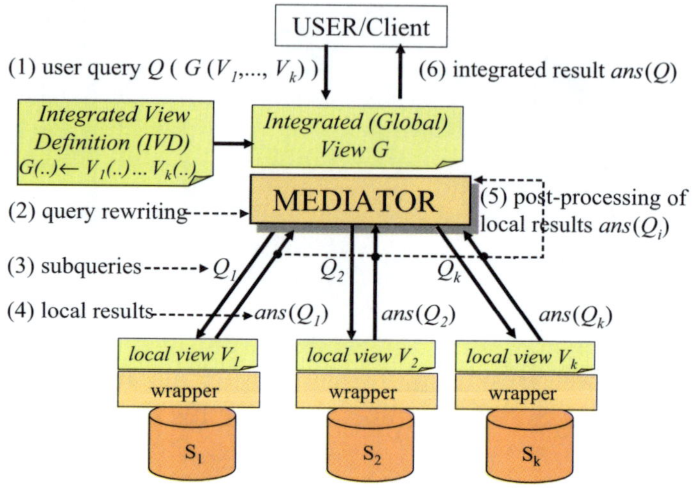

Figure 6. Database mediator architecture and query evaluation phases (1)–(6).

are identified via an Aid key attribute. To make the Aid attribute values unique across all sources, a unique prefix is used for each source (AZ.Aid, NV.Aid, etc.) to uniquely rename any potentially conflicting values.

Query Evaluation

Query processing in a database mediator system involves a number of steps (see Fig. 6). Assuming a global view *G* has been defined (which is normally the task of a data integration expert), the user or an application programmer can define a query *Q* against the integrated view *G* (1). The database mediator takes *Q* and the integrated view definitions $G(\ldots) \leftarrow \ldots V_i \ldots$ and rewrites them into a query plan with a number of subqueries $Q_1, \ldots, Q_k$ for the different sources (2). These subqueries Q_i are then sent to the local sources, where they are evaluated (3). The local answers ans(Q_i) are then sent (see [Koc01, Len02, DT03]) back to the mediator (4), where they are further postprocessed (5). Finally, the integrated result *ans(Q)* is returned to the user (6).

There are many technical challenges in developing database mediators. For example, the complexity of the query-rewriting algorithm in step (2) depends on the expressiveness of the query languages for the user query *Q* and for the allowed source queries Q_i. The problem of rewriting queries against sources with limited query capabilities is solved (or solvable) only for restricted languages; see, e.g., [VP00, Hal01, NL04b] for details.

4.2. Semantic Mediation

Consider again the geologic-map integration example from Section 3.1. Above, we sketched how the structural integration problem can be solved by adopting a mediator approach (see Figs. 3 and 6). However, the traditional mediator approach alone does not allow us to handle concept-based queries adequately. For example, in Figure 4, the conventional query for regions with

[5]In complex multiple-world scenarios, coming up with a suitable IVD is a major problem, as one usually needs domain knowledge from "glue ontologies" to link between the sources.

[6]Sometimes it can be beneficial to provide the IVD in a *local-as-view* (LAV) manner, where the local source information is defined in terms of the global schema [Hal01]. Even mixed GAV/LAV and more general approaches exist, but are beyond the scope of this paper; see [Koc01, Len02, DT03].

Paleozoic rocks yields too few results. In contrast, after "ontology-enabling" the system with a geologic-age ontology, the user query can be rewritten such that it takes into account the domain knowledge from the ontology (see the right side of Fig. 4).

The crux of *semantic mediation* is to augment the mediator approach with an explicit representation of domain knowledge in the form of one or more *ontologies* (also called *domain maps* in [LGM01, LGM03]). An ontology is often viewed as an explicit specification of a conceptualization [Gru93]. In particular, ontologies are used to capture some shared domain knowledge such as the main *concepts* of a domain, and important *relationships* between them. For example, the geologic-age ontology used in our OMI prototype can be viewed as a set of concepts (one for each geologic age) that are organized as a hierarchy, i.e., a tree in which child concepts (e.g., Devonian, Cambrian) are considered to be a subset of (i.e., a restricted set of ages of) the age described by the parent concept (e.g., Paleozoic). An "ontology-enabled" mediator can use the information from the geologic-age ontology to retrieve not only data that directly match the search concept Paleozoic, but also all data that match any subconcept of Paleozoic, such as Devonian or Cambrian. Formally, subconcepts are written using the $\sqsubseteq$ symbol, e.g., Devonian $\sqsubseteq$ Paleozoic and Cambrian $\sqsubseteq$ Paleozoic in our example.

Concept-Based Querying Revisited

A small fragment of a more complex concept hierarchy, taken from [SQDO02], is depicted in Figure 7. This "Canadian system" defines several (preliminary) *taxonomies* (classification hierarchies) of rock types, i.e., for *composition* (Figure 7 shows a small fragment dealing with Silicate rock types), *texture* (e.g., Crystalline versus Granular), *fabric* (e.g., Planar types such as Gneiss versus Nonplanar ones such as Hornfels), and *genesis* (e.g., Igneous versus Sedimentary rock). In Figure 7, subconcepts with respect to compositions are displayed as children to the right of the parent concept. For example, a Virginite is a special Listwanite; moreover, we learn that Listwanite rocks belong to the Ultramafic kind of Silicate rocks. Using this rock-type ontology, a number of new semantic queries can be formulated and executed with the prototype. Here, a semantic query means a query that is formulated in terms of the concepts in the ontologies. Specifically, the lithology information provided by two of the nine state geologic maps (Idaho and Montana West) can be linked to concepts in the *composition*, *fabric*, and *texture* hierarchies, as sketched on the right in Figure 8. On the left of Figure 8, the result of a semantic query for Sedimentary rocks is displayed. Similarly, queries for composition (e.g., Silicate), fabric (e.g., Planar), and texture (e.g., Crystalline), or any combination[7] thereof, can be executed.

To enable semantic mediation, the standard architecture in Figure 6 has to be extended to include some form of expert knowledge for linking between otherwise hard-to-relate sources. An important prerequisite is *semantic data registration*, i.e., the association of data objects in the sources with concepts defined in a previously registered ontology. Before going into the technical details of data registration, we first consider informally some ontology variants and alternatives for knowledge representation, ranging from "napkin drawings" to formal description-logic ontologies expressed in OWL [OWL03].

Ontology Variants and Alternatives

One of the reasons to use ontologies in scientific data integration is to capture some shared understanding of concepts of interest to a scientific community. These concepts can then be used as a semantic reference system for annotating data, making it easier to find data sets of interest and facilitating their reuse.

Alternatively, conventional *metadata* consist of attribute-value pairs, holding information about the data set being annotated (e.g., the creator, creation date, owner). A *metadata standard* prescribes the set of attributes that must (or can) be used in the metadata description. Metadata standards are often defined as XML schemas [XML01]. A sophisticated example is the Ecological Metadata Language (EML) [NCE], a community standard developed by ecologists that addresses several of the heterogeneity challenges discussed above. An EML description of a data set can provide information on how to parse the data file (syntax), what schema to use for queries against it (structure), and even indicate some semantic information, e.g., on unit types and measurements.

Controlled vocabularies are often part of a metadata standard and are used to constrain the values of particular attributes to come from a fixed, agreed-upon set. Thus, instead of allowing, e.g., arbitrary rock names in a geologic map shapefile, it is preferable to only use those from a controlled vocabulary. In this way, searches can be guided to only use these terms. Since the terms and definitions are (ideally) agreed upon by the community, they can also become the starting point of a more formal ontology.

Some controlled vocabularies provide relationships between the allowed terms. Similar to the relationships "narrower term"/ "broader term" in a thesaurus, the concepts may be organized into a hierarchy or *taxonomy* (for the purpose of classification), where a child concept C is linked to a parent D, if C and D stand in a specific relationship, such as C *is-a* D or C *part-of* D. For example, the geologic age ontology can be seen as a taxonomy with the child-parent relation *is-temporally-part-of*, the chemical composition ontology (Fig. 7) is a taxonomy with the child-parent relation *has-composition-like*.

Finally, (formal) *ontologies* not only fix a set of concept names, but also define properties of concepts and their relationships. *Description logics* [BCM+03] are decidable fragments of first-order predicate logic and are commonly used for specifying formal ontologies. Whereas taxonomies often result from *explicit* statements of *is-a* relationships, the concept hierarchy

[7]Domain experts will probably only ask for meaningful combinations. Data-mining techniques can be applied to this rock classification ontology to extract only those composition, fabric, and texture combinations that are consistent with the ontology. These are of course only an approximation of the actually existing combinations.

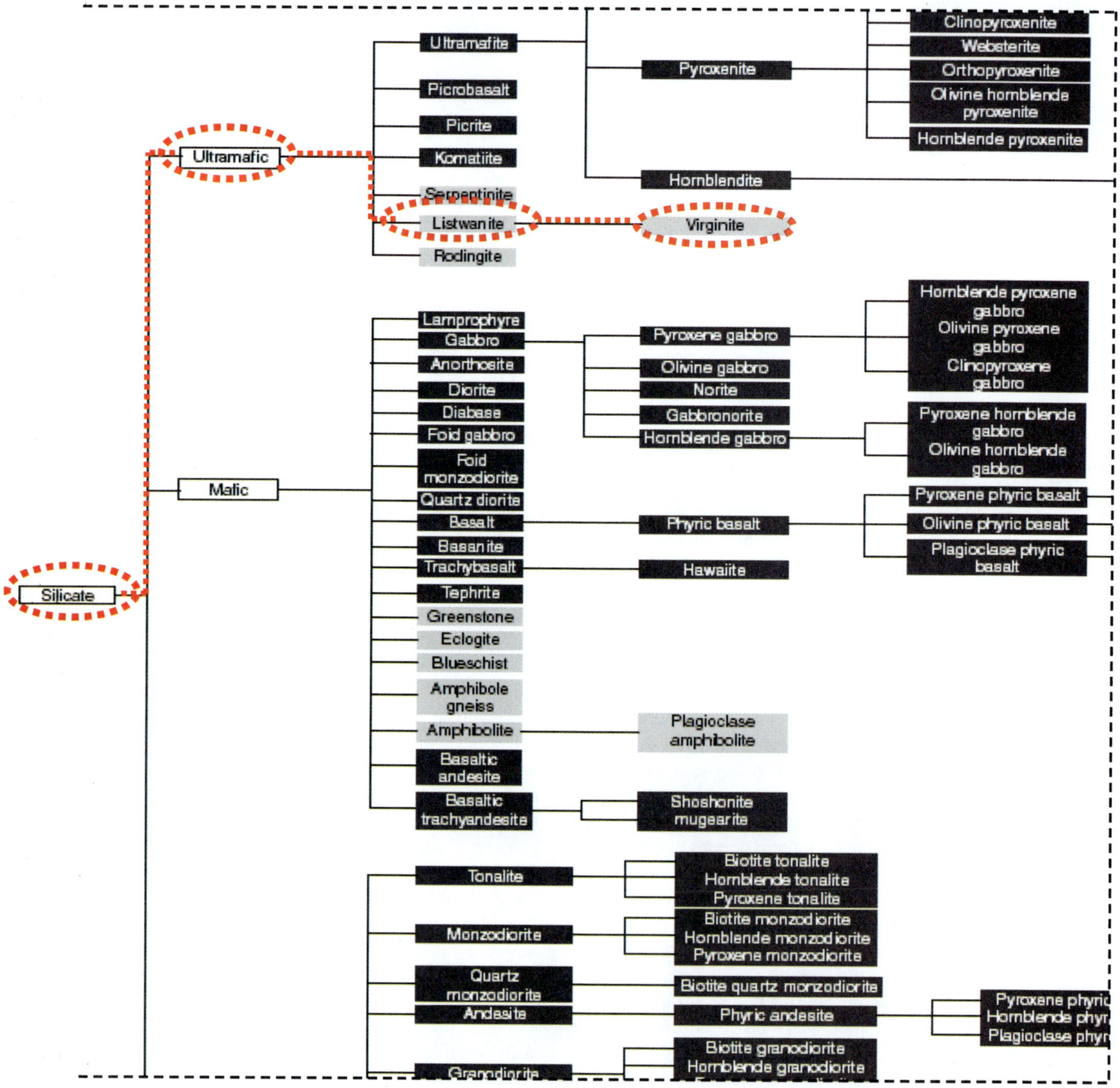

Figure 7. Rock type classification hierarchy (fragment) based on *composition* [SQDO02].

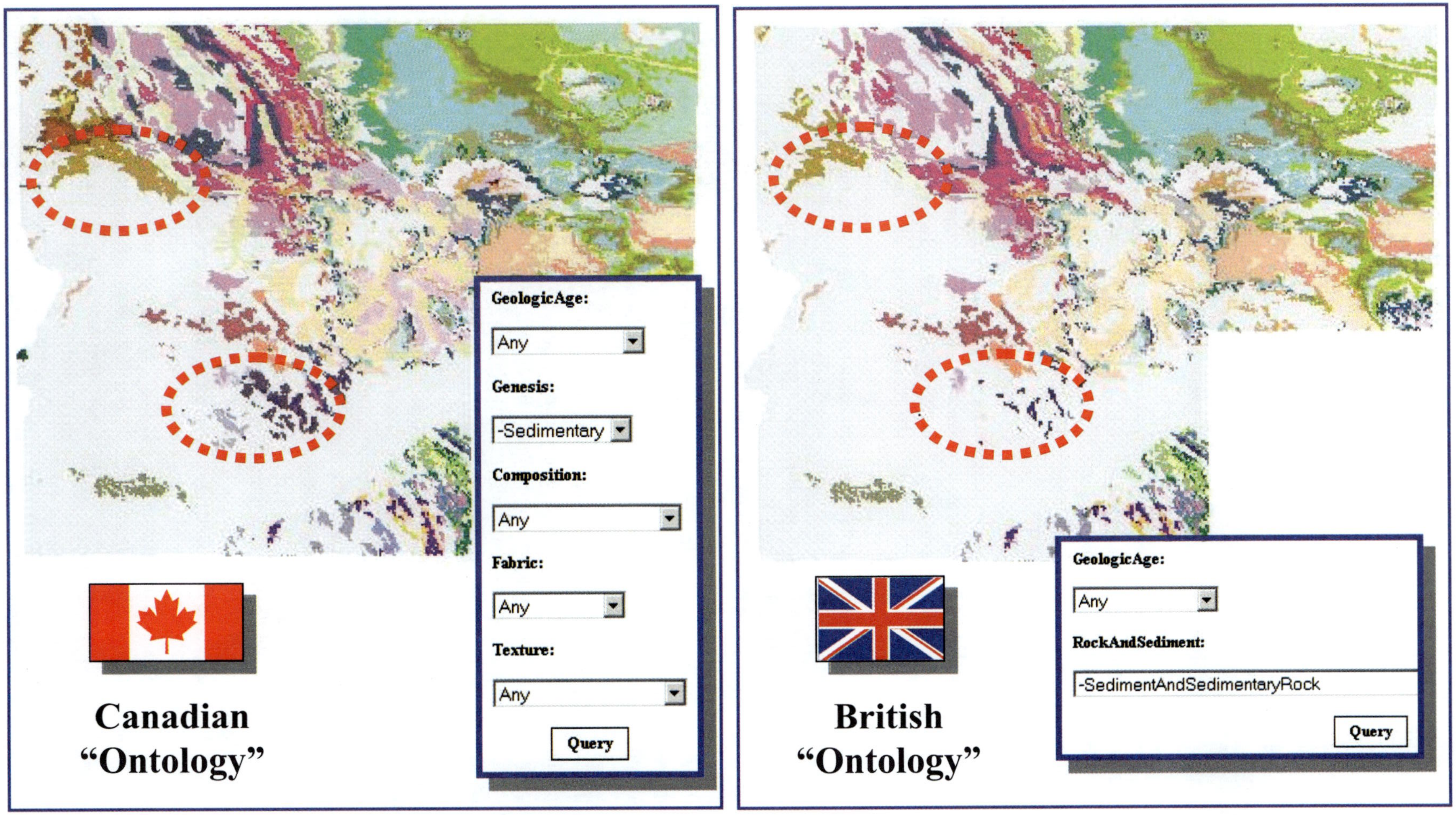

Figure 8. Different ontology-enabled views on the same geologic maps. The Canadian system [SQDO02] supports queries along several hierarchies (*genesis*, *composition*, *fabric*, *texture*); the British system [GSR+99] provides a single hierarchy (a separate *geologic age* ontology is used in both views). Via an ontology *articulation*, data registered to one ontology can be retrieved through the conceptual view of the other ontology.

in a description-logic ontology is *implicit* in the axiomatic concept definitions. In description logics, one defines concepts via their properties and relationships; the concept hierarchy is then a consequence of those definitions and can be made explicit by a description-logic reasoning system. A description-logic ontology consists of logic formulas (axioms) that constrain the interpretation of concepts by interrelating them, e.g., via binary relationships called *roles*. For example, the description-logic axiom

$$\text{Virginite} \sqsubseteq \text{Rock} \sqcap \text{Listwanite} \sqcap \exists\, \text{foundAt.Baie\_Verte}$$

states that (i) instances of Virginite are instances of Rock, (ii) they are also instances of Listwanite, and (iii) they are found at some place in Baie_Verte. It is not stated that the converse is true, i.e., that every Listwanite rock found in Baie Verte must be a Virginite (this could be stated by using "≡" instead of "⊑"). Here, the lowercase symbol "foundAt" denotes a role (standing for a binary relationship, in this case between rock types and locations); all other uppercase symbols denote concepts, each one denoting a set of concept instances, e.g., all Listwanites.

Description logics and reasoning systems for checking concept subsumption, consistency, etc., have been studied extensively over many years [BCM+03]. However, until recently there was no widely accepted syntax for description-logic ontologies. With the increased interest in using description-logic ontologies, e.g., for Semantic Web applications [BLHL01], the need for a standard Web ontology language has led to the W3C OWL standard [OWL03][8]. OWL, being an XML standard, also supports *namespaces*, a URI-based reference system. In OWL, namespaces are used to help express interontology *articulations*, i.e., formulas that relate concepts and relationships from different OWL ontologies.

Before defining a formal ontology for use in scientific data integration systems, a community-based effort may go through several intermediate steps, from informal to more formal representations. As a rule of thumb, the more formal the knowledge representation, the easier it is for a system to make good use of it, but also the harder it is usually to develop such a formal representation. A common starting point is an informal concept diagram or "napkin drawing" initially created by members of a community to give an overview of important items or concepts in the domain. Figure 9 shows such an informal concept diagram that resulted from a workshop with aqueous geochemistry experts. The diagram relates specific raw and derived data products, models, and scientific problems to one another. While such a diagram is useful as a communication means between domain experts, a data integration system cannot directly make use of it. One possible elaboration is the definition of metadata standards for data exchange between different community tools and applications, addressing syntactic and structural issues in data integration. Another possible subsequent step is the definition of one or more concept hierarchies (taxonomies) such as in the Canadian system (Fig. 7), thus enabling simple semantic mediation. A slightly more general form, *labeled, directed graphs*, use nodes for concepts and edge labels to denote arbitrary relationships between concepts. This model can be used by a data integration system to find concepts and relationships of interest via generalized path expressions (e.g., see [LHL+98]).

Finally, logic-based formalisms such as OWL ontologies can be used not only to declare concept names and their relationships, but to intentionally define new concepts relative to existing ones, and to let a reasoning system establish the logical consistency of the system and the concept hierarchy.

[8]OWL can be used not only for description logics (OWL-DL) but also for a simpler fragment (OWL-lite) or a more expressive version (OWL-full). In the OMI prototype, we have used simple concept hierarchies expressible in OWL-lite.

4.3. Semantic Data Registration

In this section, we take a closer look at the technical details of semantic mediation. In particular, we present an approach called *semantic data registration* that can facilitate data integration in such a system. Figure 10 shows an overview of the architecture of our proposed framework. The main components include services for reasoning with ontologies, database mediation, registration, and data access. We assume that a *federation registry* (not shown) stores core registry information, including the database schemas of underlying data sources, service descriptions, ontologies and ontology articulations, external semantic constraints (e.g., unit conversion rules and other scientific formulas), and registration mapping rules. We formalize the components of our framework and then discuss resource registration in more detail in the next section.

First-Order Logic

We use first-order logic as a standard, underlying formalism. The *syntax* of first-order logic is defined as usual, i.e., we consider signatures with predicate symbols Σ_P and function symbols Σ_F. By $\Sigma_{P.n}$ ($\Sigma_{F.n}$) we denote the subsets of *n*-ary predicate (function) symbols; $\Sigma_C = \Sigma_{F.0}$ are constants. *Semantics:* A first-order structure $\mathcal{I}$ interprets predicate and function symbols as relations and functions, respectively; constants are interpreted as domain elements. Given $\mathcal{I}$ and a set of formulas Φ over Σ, we say that $\mathcal{I}$ is a *model* of Φ, denoted $\mathcal{I} \models \Phi$, if $\mathcal{I} \models \varphi$ for all $\varphi \in \Phi$, i.e., all formulas in Φ are satisfied by $\mathcal{I}$. We can implement constraint checking by evaluating the query $\{\bar{x} \mid \mathcal{I} \models \varphi(\bar{x})\}$.

Ontologies (Revisited)

Given the above preliminaries, we can now consider an ontology as a certain set of first-order formulas. An ontology O is a set of logic axioms Φ_O over a signature $\Sigma = \mathbf{C} \cup \mathbf{R} \cup \mathbf{I}$ comprising unary predicates $\mathbf{C} \subseteq \Sigma_{P.1}$ (*concepts*), binary predicates $\mathbf{R} \subseteq \Sigma_{P.2}$ (*roles, properties*), and constants $\mathbf{I} \subseteq \Sigma_{F.0}$ (*individuals*). Φ_O is usually from a decidable first-order fragment, most notably *description logics* [BCM+03]. A structure $\mathcal{I}$ is called a *model* of an ontology Φ_O, if $\mathcal{I} \models \Phi_O$.

We can view controlled vocabularies and metadata specifications as limited, special cases of ontologies. A *controlled*

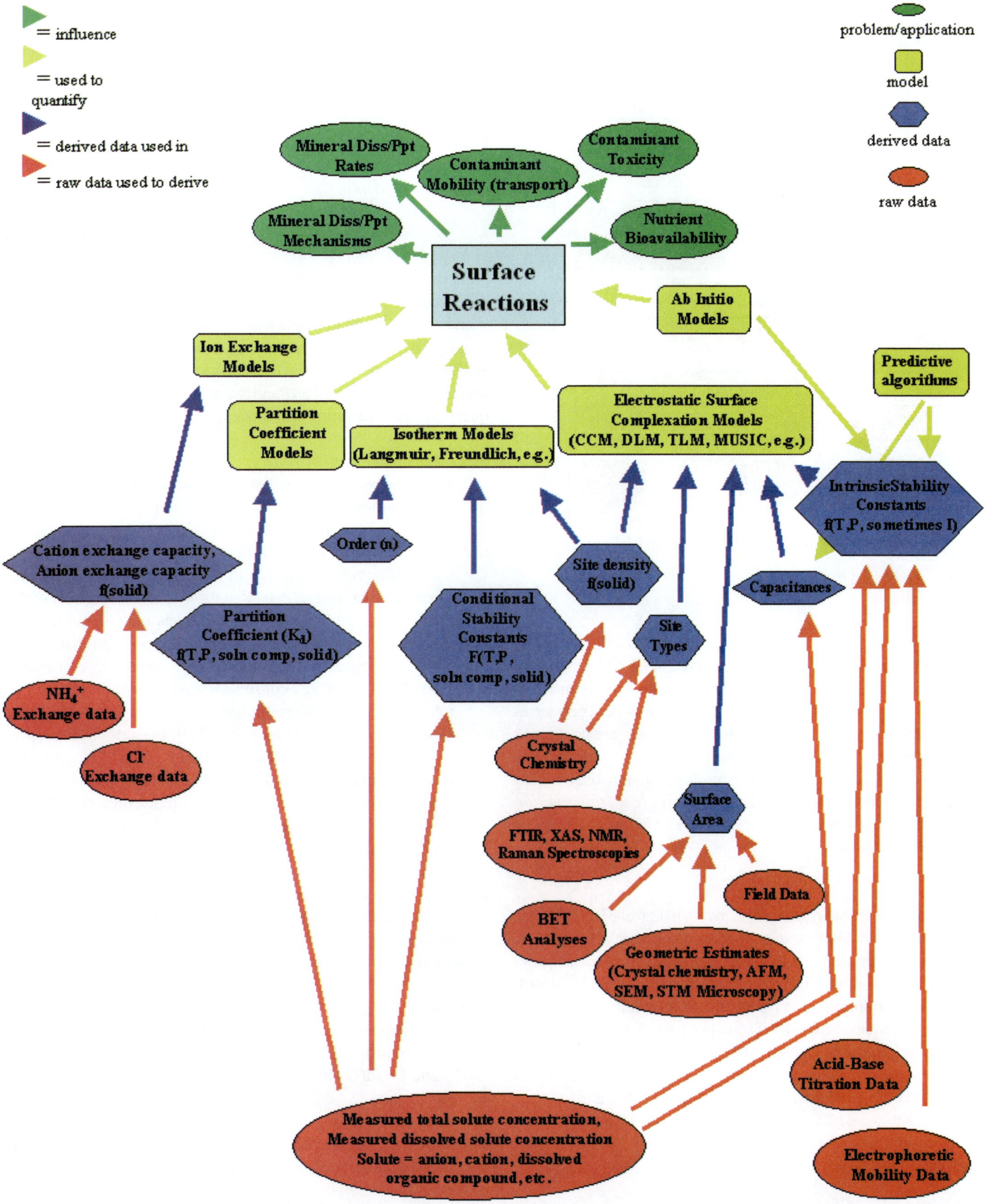

Figure 9. Informal concept diagram ("napkin drawing"), relating raw data (red ovals), derived data (blue hexagons), models (yellow rectangles), and scientific problems (green ovals) in aqueous geochemistry.

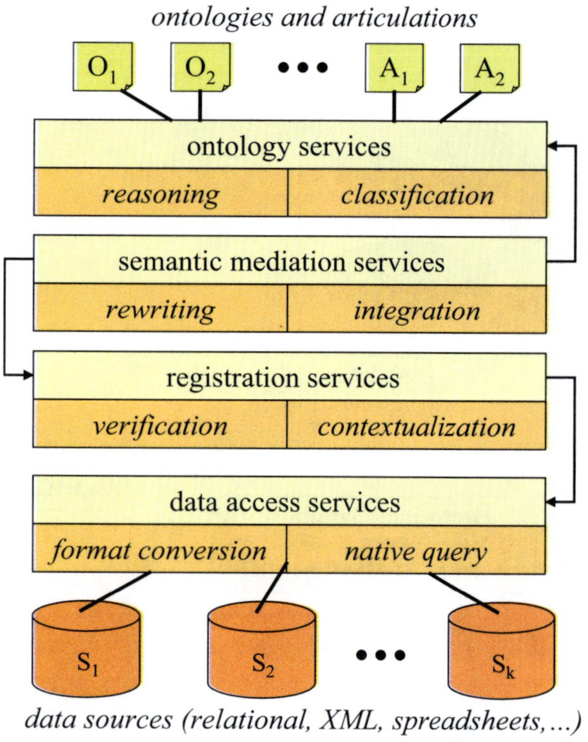

Figure 10. Overview of the registration architecture.

vocabulary can be viewed, e.g., as a fixed set $\mathbf{I} \subseteq \Sigma_{F.0}$ of *individuals* (constants); a set of named concepts **C**; or even a full ontology signature Σ (if it contains relationships between terms of the controlled vocabulary). In either case, there are no axioms and hence no defined logical semantics. A *metadata* specification can be seen as an instance of an ontology having only binary predicates **R** denoting the metadata properties (e.g., *title*, *author*, *date*; see Dublin Core). Again, the absence of axioms means that no logical semantics are defined.

Namespaces

In the federation registry, we avoid name clashes between vocabularies from different scientific resources (data sets, services, etc.) by assuming each resource has a globally unique identifier i (e.g., implemented as a URI). We then rename symbols accordingly: Every symbol in Σ_i is prefixed with its resource identifier i to obtain a unique vocabulary $\Sigma_i' := \{i.s \mid s \in \Sigma_i\}$, allowing new resources to join the federation without introducing identifier conflicts. For example, in the view definitions in Section 4.1, we disambiguated object identifiers by using a state prefix as a resource identifier: `AZ.Aid`, `NV.Aid`, etc. A resource identifier is also commonly referred to as a *namespace*. Below, by $\mathsf{id}(s)$ we denote the globally unique resource *identifier* of a symbol s.

Registering Ontologies and Articulations

An ontology O is registered by storing its logic axioms Φ_O and its signature Σ_O in the federation registry[9]. An *articulation ontology* A links two ontologies O_i and O_j and is given as a set of axioms Φ_A over $\Sigma_A = \Sigma_{Oi} \cup \Sigma_{Oj}$, thereby logically specifying interontology correspondences. For example, $i.\mathrm{C} \equiv j.(\mathtt{D} \sqcap \exists \mathtt{R}.\mathtt{E})$ is an *articulation axiom* $\varphi \in \Phi_A$ and states that the concept `C` in O_i is equivalent—in terms of O_j—to those `D` having at least one `R`-related `E`. This is expressed equivalently as follows (using first-order logic syntax):

$$\forall x : i.\mathtt{C}(x) \leftrightarrow j.\mathtt{D}(x) \wedge \exists y : j.\mathtt{R}(x,y) \wedge j.\mathtt{E}(y) \qquad (\varphi)$$

Note that φ is an intentional definition. We have not said how we can access instance objects (implicitly referred to via variables x and y), i.e., how to populate `C`, `D`, etc., as classes of objects. Finally, expressing interontology articulations as ontologies achieves closure within the framework: There is no need to manage a new type of artifact, and we can reuse the given storage, querying, and reasoning techniques.

Structural Data Registration

When registering a database, schema-level information and query capabilities should be included to facilitate queries by the end user or a mediator system. Specifically, the database registration information contains the following:

- The database *schema* Σ_D. In the case of a relational database D, we have $\Sigma_D = \{V_1, \ldots, V_n\}$, where each V_i is the schema of an exported relational view (see Figure 6).
- A set of local *integrity constraints* Φ_D. We can distinguish different types of constraints, e.g., structural constraints (such as foreign key constraints) and semantic constraints.
- A *query capability* specification Π_D. For example, Π_D may be a set of access patterns [NL04a], prescribing the input/output constraints for each exported relation. More generally, Π_D may be given as a set of view definitions (possibly with access patterns) supported by the source D. If D provides full SQL capabilities, a reserved word can be used: $\Pi_D = \{\mathtt{SQL}\}$.

To register the structural definition of the data source, a data access handler or wrapper (shown in both Figures 6 and 10) must also be provided. The wrapper provides basic services for executing underlying queries and converting data to a common format for use by the registration and mediation service.

Semantic Data Registration

A semantic data registration registers the association between data objects in a database D and a target ontology O. Let $k = \mathsf{id}(D_k)$ and $j = \mathsf{id}(O_j)$ be the unique resource identifiers of D_k and O_j, respectively.

The semantic data registration of D_k to O_j is given by a set of constraints Ψ_{kj}, where each $\psi \in \Psi_{kj}$ is a constraint formula over $\Sigma_D \cup \Sigma_O$. For example, the semantic data registration formula $\psi =$

$$\forall x \forall y : j.\mathtt{D}(x) \wedge j.\mathtt{R}(x,y) \leftarrow \exists z : k.\mathtt{P}(x,y,z) \wedge k.\mathtt{Q}(y,z)$$

[9] We use OWL as the concrete syntax for ontologies.

is a constraint saying that ontology O's concept D and its role R can be "populated" using certain tuples P(x, y, z) from D. When the semantic data registration constraint ψ and the above articulation φ are combined into $\psi \wedge \varphi$, we see that data objects from D_k can be linked to concepts such as i.C(x) in the ontology O_i, despite the fact that D_k was registered to O_j and not to O_i. The reason is that an indirect link exists to O_i via the articulation axiom φ[10]:

$$D_k \overset{\psi}{\rightsquigarrow} O_j \overset{\varphi}{\longleftrightarrow} O_i$$

For a concrete example, assume the database D_k represents a geologic map, and the ontologies O_j and O_i represent the Canadian rock classification system [SQDO02] and the British one [GSR+99], respectively. We can register D_k to the Canadian system (encoded in OWL) using mapping rules corresponding to formulas such as ψ above. This allows a semantic mediation system such as OMI to provide concept-based query capabilities: On the left in Figure 8, the resulting query interface with fields for composition, texture, and fabric is shown.

An important application of articulation axioms such as φ relating O_j and O_i is that they can be used to query and view the data from D_k through the conceptual view provided by O_i. In our geologic map example, we can query the geologic maps using the British rock classification view (a single, large hierarchy; see Figure 8, right), despite the fact that the geologic map database was originally registered to the Canadian system. Figure 8 shows the query interfaces and results of querying the same geologic map databases, using different ontology views. Note that the highlighted differences in the results shown in the figure could have different causes. For example, they might result from asking slightly different questions or from a different relative completeness of one ontology over the other or from the use of only partial mapping information in the ontology articulation[11].

Data Sets as Partial Models

This section further defines the steps involved in semantic data registration. In particular, we clarify the result of semantically registering a data set as a partial model and motivate the need for additional, data-object identification steps (see Fig. 11).

Registering Partial Models

A data set D that is registered to an ontology O contributes to the extent of the federation relative to O. Registered data sets are not materialized according to the ontology; instead the registration mappings are used to access the underlying sources when needed, similar to the way queries against an integrated view are evaluated on demand (see Fig. 6). A data set D can be registered consistently to an ontology O if it can be interpreted as a partial

[10]The actual links via φ are given by the valuations that make the formula true.

[11]In our OMI prototype, we only used a rough approximation of the data registration and articulation mappings. The mappings were based on (partial) syntactic matches and did not include a systematic review by a domain scientist.

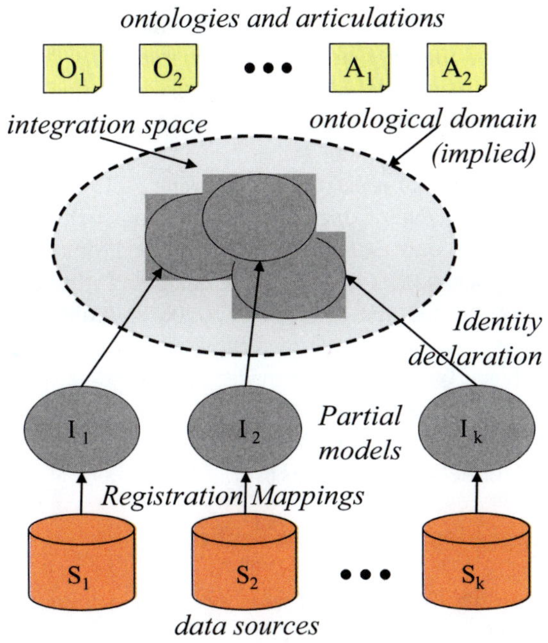

Figure 11. Result of semantic data registration: an unknown or hidden part IOD.

model $\mathcal{I}$ of O, denoted $\mathcal{I} \models_p \Psi_O$, which implies $\mathcal{I} \cup \mathcal{I}' \models \Psi_O$ for some unknown $\mathcal{I}'$.

A partial model differs from a true model of the ontology in that some required information may be missing. We denote the interpretation induced by applying a semantic data registration Ψ_D of database D to an ontology O as $\mathcal{I}_D$. If the latter is a partial model $\mathcal{I}_D \models_p \Psi_O$, then the model $\mathcal{I}_D \cup \mathcal{I}'_D \models \Psi_O$ contains an unknown or hidden part $\mathcal{I}'_D$. As more sources are registered, more of $\mathcal{I}'_D$ may become known.

When an interpretation induced by a semantic data registration is not a partial model of an ontology, we say that the interpretation is *inconsistent*. An inconsistent interpretation often violates a datatype, cardinality, or disjoint constraint in the ontology. When possible, we wish to automatically verify that a semantic data registration is consistent, e.g., by ensuring that the dataset is satisfies the axioms of the ontology, or can be extended to do so.

Identification Declaration

Semantic data registration allows a data set to be interpreted as a partial model of an ontology but does not necessarily provide enough information to identify the same domain elements (individuals) of the integration space across multiple data sets, which is essential for data integration. The ability to identify equivalent data objects across different data sets is needed in practice because each data set may only provide a portion of the information concerning a particular object. As shown in Figure 11, we consider an additional registration step called *identification declaration* that allows data providers to state how data objects should be identified across data sources.

Such an object identity can be defined in a number of ways. First, a semantic data registration can be augmented with *mapping tables*, which map individual data items to recognized individuals in **I**, i.e., individuals that are established instances within an ontology and come from an authoritative registry. For example, given an ontology of rock types, a mapping table can associate rock names in a geologic data set with the unique rock types from the ontology. Second, external rules may be used for determining identity, similar to keys in a relational database[12] . For example, we may have a rule that ISBN or DOI codes uniquely identify publications; thus, registering to such a code uniquely identifies the data object. Finally, a data provider may give data-object correspondences between registered data sets. Thus, a data object is explicitly given as equivalent to another data object (although the specific identifiers of the objects may not be authoritative).

Data Integration via Semantic Registration

We identify four classes of semantic data registration expressiveness (in terms of data integration) as follows:

- *Concept-as-keyword registration*. We can consider metadata annotations using keywords from a controlled vocabulary as (a weak form of) registration mappings. For example, we can assign a concept such as geologic age to the data set as a whole[13]. Such a mapping states that the data set contains data objects, and those data objects refer to individual geological ages. However, we cannot consider or obtain each such separate (geologic-age) data object in the data set. Clearly, such a registration cannot be used for integration; however, it can be used for data set discovery: We do not have access to the individual objects, so the best we can do is find the data set that contains such objects.
- *Local data-object identification*. Local data-object identification is the typical result of a registration mapping, where local identifiers are used to identify logical data items within a data set. In this case, the identities of the individuals are local to the source and thus cannot be used to combine data objects from multiple sources.
- *Global data-object identification*. The result of globally identifying data objects is that it becomes possible for a mediator to recognize identical individuals in multiple data sets. The result of global data-object identification is the ability to perform object fusion [PAGM96] at the mediator.
- *Property identification*. If within a given data set we relate two globally identified data objects with an ontological relation in **R**, it becomes possible to join information across data sets (assuming at least one data object occurs in at least one other relation in another source). This situation represents a stronger form of integration compared to simple object fusion and is required for wide-scale data integration.

[12]Some description logics also support keys [LAHS03].

[13]For example, by registering the data set's resource identifier with the concept, or registering all rows of the data set with the concept.

5. SCIENTIFIC WORKFLOWS

In this section, we return to the final integration challenge, synthesis (see Section 2), and illustrate how scientific workflows can be used to create new tools and applications from existing ones.

Scientific workflows are typically used as "data analysis pipelines" or for comparing observed and predicted data and can include a wide range of components, e.g., for querying databases, for data transformation and data-mining steps, and for execution of simulation codes on high-performance computers. Ideally, a scientist should be able to (1) plug in almost any scientific data resource and computational service into a scientific workflow, (2) inspect and visualize data on-the-fly as they are computed, (3) make parameter changes when necessary and rerun only the affected "downstream" components, and (4) capture sufficient metadata in the final products. For each run of a scientific workflow, when considered as a computational experiment, the metadata produced should be comprehensive enough to help explain the results of the run and make the results reproducible by the scientist and others. Thus, a scientific workflow system becomes a scientific problem-solving environment, tuned to an increasingly distributed and service-oriented Grid infrastructure.

However, before this vision can become reality, a number of technical problems have to be solved. For example, current Grid software is still too complex to use for the average scientist, and fast-changing versions and evolving standards require that these details be hidden from the user by the scientific workflow system. Web services provide a simple basis for loosely coupled, distributed systems, but core Web service standards such as WSDL [WSD03] only provide simple solutions to simple problems[14], while harder problems such as Web service orchestration remain the subject of emerging or future standards. As part of an open-source activity, members from various application-oriented research projects [GEO, SEE, SDM, BIR, ROA] are developing the KEPLER scientific workflow system [KEP, LAB+04], which aims at developing generic solutions to the process and application integration challenges of scientific workflows.

5.1. Scientific Workflows in Kepler

In Section 3.2, we have already presented a KEPLER workflow for mineral classification (Fig. 5). KEPLER extends the underlying PTOLEMY II system [BCD+02] by introducing a number of new components called actors, e.g., for querying databases, for data transformations via XSLT, and for executing local applications from a command line, from Web services via their WSDL interface, or from remote jobs on the Grid. In KEPLER, the user designs a scientific workflow by selecting appropriate actors from the actor library (or by dynamically "harvesting" new ones via the KEPLER Web service harvester) and putting them on the design canvas, after which they can be "wired" to form the desired larger

[14]For example, WSDL mainly provides an XML notation for function signatures, i.e., the types of inputs and outputs of Web services.

workflow. For workflow components that are not yet implemented (i.e., neither as a native actor nor as a Web service or commandline tool), a special *design actor* can be used. Like regular actors, the design actor has *input ports* and *output ports* that provide the communication interface to other actors. The number, names, and data types of the ports of the design actor can be easily changed to reflect the intended use of the actor. When designing a workflow, the user connects actors via their ports to create the desired overall dataflow[15]. A unique feature of PTOLEMY II and thus of KEPLER is that the overall execution and component interaction semantics of a workflow is not buried inside the components themselves, but rather factored out into a separate component called a director. For example, the PN (process network) director used in the workflow in Figure 5 (green box) models a workflow as a process network [KM77, LP95] of concurrent processes that communicate through unidirectional channels. The SDF (synchronous dataflow) director in Figure 12 is a special case of the PN director that can be used when all actors statically declare the number of tokens they consume and produce per invocation (called an actor "firing")[16]. The SDF director uses this information to statically analyze the workflow, e.g., to detect deadlocks in the workflow, or to determine the required buffer size between connected actors.

5.2. Gravity Modeling Workflow

Figure 12 shows a gravity modeling *design workflow*. Unlike the mineral classification workflow discussed in Section 3.2, this workflow involves components that are not yet implemented. This feature allows the user of the KEPLER system to seamlessly go from a conceptual design workflow to an executable version by replacing design components with implemented ones as they become available. Another benefit of this feature is that executable subworkflows (i.e., ones that do not include design actors) can already be unit-tested and debugged early on while other parts of the workflow are still in their design phase.

The workflow depicted in Figure 12 takes a spatial envelope token (given via latitude and longitude parameters in the lower left corner of the main window) and feeds it into two independent Web services that extract gridded Moho and sedimentary depths for the enveloped region, respectively, and feed them to a GRAVITYMODELER actor. This synthetic gravity model is then compared to the observed model, and the result is fed into a RESIDUALMAP Web service actor for creating the result map. The latter is shown to the user via a standard browser. Note that a subworkflow is associated with the OBSERVEDGRAVITYMODEL actor, i.e., the latter is a composite actor. In Figure 12, the subworkflow of this composite actor is shown in the lower window. It involves two components, one to translate between coordinate system (implemented) and one to access values from the UTEP gravity database (designed). While this data access is not yet implemented, the signature of the design actor reveals that it needs to take an envelope token (in degrees) and produce from the database a gravity grid of observed data.

5.3. Semantic Workflow Extensions

Important aspects when developing scientific workflows are the structural and semantic types of actors and services they represent. Here, by structural type we mean the conventional data type associated with the input and output ports of an actor. KEPLER inherits from the PTOLEMY II system a very flexible structural type system in which simple and complex types are organized into an extensible, polymorphic type structure. For example, the AddSubtract actor in Figure 12, which takes the synthetic data output from the GRAVITYMODELER and compares it to the OBSERVEDGRAVITYMODEL output, is polymorphic and can operate on integers, floats, vectors, and matrices. In the case of the gravity workflow, after connecting ADDSUBTRACT with the corresponding upstream and downstream actors, the system can determine that a matrix operation is needed. If actor ports are connected that have incompatible types, the system will report a type error.

While a structural type check can guarantee that actor connections have compatible data types, they provide no means to check whether the connections are even potentially meaningful. For example, a connection between an actor outputting a velocity vector and one that takes as input a force vector may be structurally valid, but not semantically. For physical units, a unit type system has been added to PTOLEMY II to handle such situations [BLL+04b].

The idea behind *semantic types* in KEPLER is to further extend the type system to allow the user to associate a formal concept expression with a port. Thus, semantic types in scientific workflows are related to the idea of semantic data registration (Section 4.3), which associates a data object having a structural type with a concept from an ontology. Our semantic type system[17] will detect, e.g., that a port of type `geologic-age` cannot be connected to one of type `texture`, although both ports might have the same structural type string. Similarly, `vector(force)` and `vector(velocity)` can be detected as semantically incompatible, although their structural type `vector(float)` coincides.

There are several applications of a semantic type system in scientific workflows. As was the case for semantic mediation and data registration, some domain knowledge above the structural level of database schemas and conventional data types can be captured in this way. Thus, semantic type information can be used for concept-based discovery of actors and services from actor libraries and service repositories, similar to the concept-based data queries in Sections 3.1 and 4.2. Based on the semantic type signature of an

[15]Control-flow elements such as branching and loops are also supported; see [BLL+04b, LAB+04].

[16]A token represents the unit of data flowing through a channel between actors. The workflow designer can choose to use, e.g., a single data value, a row from a database table, an XML element, or a whole table or file as a single token.

[17]The semantic type system is currently under development and being added to KEPLER. More details can be found in [BL04].

actor, the system can also search for compatible upstream actors that can generate the desired data or for downstream actors that consume the produced data. In this way, a workflow system with semantic types can support workflow design by only considering connections that are valid with respect to their semantic type.

Ontology-Driven Data Transformations

We now briefly illustrate a further application of semantic types, i.e., how they can be used to aid the construction of structural transformations in scientific workflows. Indeed, a common situation in scientific workflows is that independently developed services are structurally incompatible (e.g., they use differently structured complex types, in particular different XML schemas), although they might be semantically compatible. For example, we might know that the ports P_s (source) and P_t (target) of two actors produce and consume, respectively, a matrix of gravity values. Thus, these ports should be connectible in terms of their semantic type. However, the actors may invoke independently designed Web services that have different XML representations of gravity matrices, making the desired connection structurally invalid. The conventional approach to solve this structural heterogeneity is to manually create and insert a data transformation

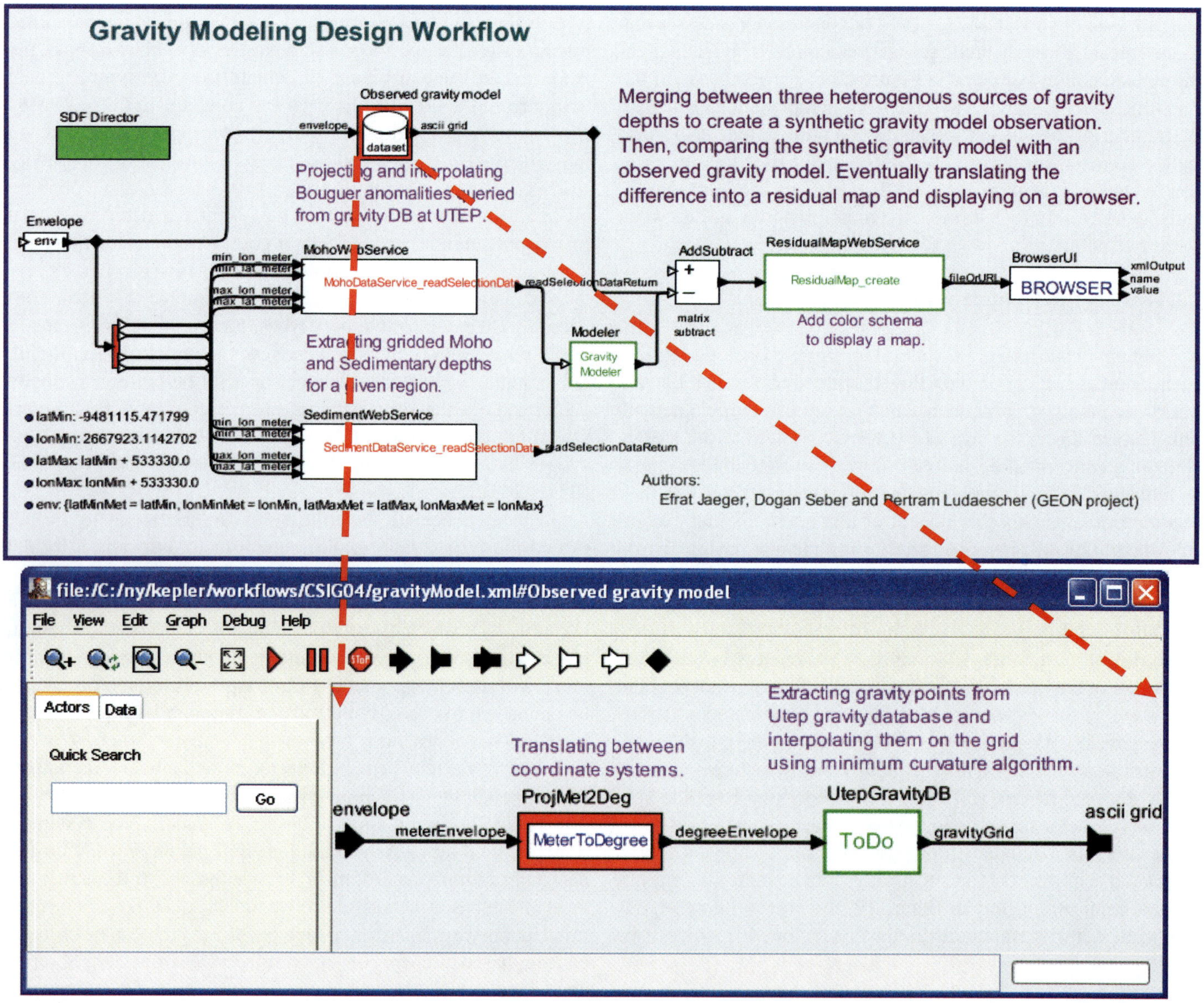

Figure 12. Gravity modeling design workflow. The main workflow (top) combines three different gravity sources into a single model: An observed model (expanded below) is compared to a synthetic model, which itself results from an integrated gravity model combining a Moho and a sediment Web service (top window, center). Outlined olive boxes are *design actors*. Unlike other actors, these are not (yet) implemented and ultimately need to be replaced by executable versions before the complete workflow can execute.

step that maps data from the XML schema used by P_s to data that conform to the schema of P_t.

The idea of ontology-driven data transformations [BL04] is to employ the information provided by the semantic type of a port to guide the generation of a data transformation from the structural type of the source port to that of the target port. Note that the semantic type alone, when considered in isolation from the structural type, cannot be used for this purpose. However, via a semantic data registration mapping (see Section 4.3), an association between structural type and semantic type is created that can be exploited for the generation of the desired transformation. Figure 13 gives an overview of the approach: The source port P_s and target port P_t are assumed to have incompatible structural types (denoted $P_s \not\preceq P_t$) but compatible semantic types (denoted $P_s \sqsubseteq P_t$). In other words, the semantic type of P_s is the same as or a subtype of the one for P_t, but this is not the case for the structural types. The goal is thus to generate a data transformation δ that maps the structural type of P_s to one that is compatible with the structural type of P_t. We call a semantically valid connection structurally feasible if there exists such a data transformation δ with $\delta\,(P_s) \not\preceq P_t$. The core idea is that the information from the semantic registration mappings induces correspondences at the structural level that can help in the generation of the desired data transformation $\delta\,(P_s)$.

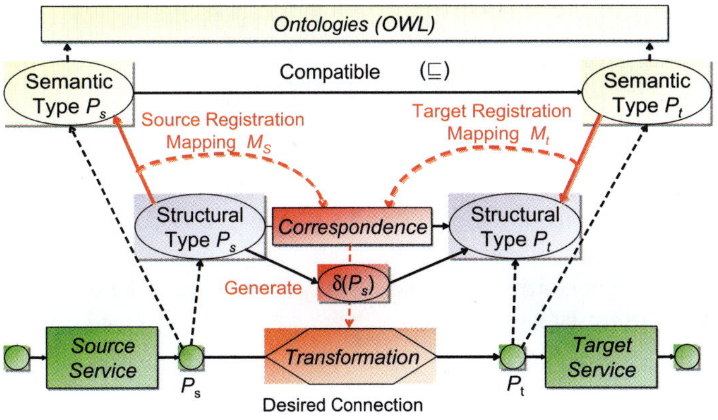

Figure 13. Ontology-driven generation of data transformations [BL04].

Correspondence Mappings

More precisely, let M_s be the source registration mapping that links between the structural type and the semantic type of P_s; similarly, let M_t denote the target registration mapping (see Fig. 13). M_s and M_t can be seen as constraint formulas Ψ_s and Ψ_t as described in Section 4.3, relating data schemas to expressions of the ontology to which the data are registered. M_s can be given, e.g., as a set of rules of the form $q_s \rightsquigarrow E_s$, where q_s is a path expression or more general query that "marks" parts of interest in the data structure of P_s, and E_s is a concept expression over the ontology O to which P_s is semantically registered. The *correspondence mapping* $M_{st} := M_s \bowtie_O M_t$ is now given as the *semantic join* of M_s and M_t with respect to the ontology O: Given a rule $q_s \rightsquigarrow E_s \in M_s$ and a rule $q_t \rightsquigarrow E_t \in M_t$, the rule $q_s \rightsquigarrow q_t$ is in the correspondence mapping M_{st} if and only if the semantic join $E_s \sqsubseteq_O E_t$ holds, i.e., if the concept expression E_s yields a semantic subtype of E_t in the ontology O.

For example, assume a scientist wants to connect various ports of two actors that deal with geologic map information. Assume that the port P_s of the source actor produces XML tokens with the structure

```
<rinfo>
    <age>...</ccomp>
    <text>...</text>
    <fab>...</fab>
</rinfo>
```

and that the port P_t of the target actor consumes XML tokens that are structured as

```
<properties>
    <lithology>...</lithology>
    <geoage>...</geoage>
    <fab>...</fab>
</properties>
```

Clearly $P_s \not\preceq P_t$, i.e., the structural types of P_s and P_t are incompatible. Now consider the source registration mapping M_s =

/rinfo/age ⇝ O.geologic_age
/rinfo/ccomp ⇝ O.lithology.composition
/rinfo/text ⇝ O.lithology.texture
/rinfo/fab ⇝ O.lithology.fabric

and the target registration mapping M_t =

/properties/lithology ⇝ O.lithology
/properties/geoage ⇝ O.geologic_age

where O is the ontology to which the structures are registered. Then the correspondence mapping M_{st} contains the rule

/rinfo/age ⇝ /properties/geoage

indicating how to transform the geologic age subelement of P_s to the one in P_t. Now assume further that lithology.composition $\sqsubseteq_O$ lithology, i.e., that in the ontology, composition is considered a special kind of lithology information[18]. Then M_{st} also includes the correspondence rule

/rinfo/ccomp ⇝ /properties/lithology

In this simple example, the correspondence mapping M_{st} may not provide detailed enough information to determine a complete transformation from P_s to P_t[19]. In particular, we do not know in this example whether the various elements involved in the mappings are complex (i.e., contain nested elements). If so, further information would be required to automatically generate the transformation.

However, even if M_{st} does not provide enough information to automatically generate the data transformation $\delta: P_s \rightarrow P_t$ (see Figure 13), the obtained correspondences are still valuable. For example, a workflow engineer who needs to develop customized data transformation actors for a scientific workflow can use these correspondences as a "semantic" starting point to define the needed transformation. Also, correspondences can be exploited by various database schema-matching tools [RB01], and used to infer additional structural mappings.

6. CONCLUSIONS

To answer specific scientific questions, a scientist often repeatedly performs similar tasks and data management steps. For example, scientists select appropriate analysis methods to address the given question, search relevant data or create new data, determine whether existing data can be used for the desired analyses, preprocess and integrate data as needed, and so on. There are a number of significant data and tool integration challenges that need to be overcome to enable more of the increasingly data-driven "e-science," and to allow the scientist to spend less time on labor-intensive, error-prone manual data management.

In this paper, we have given an overview of the different kinds of data integration and interoperability challenges in scientific data management. After reviewing the traditional mediator approach to data integration, we have presented a knowledge-based extension called *semantic mediation* that facilitates (1) the linking of hard-to-relate data sources by "going through" shared ontologies, and (2) new types of concept-based queries against the data. Semantic mediation requires a "deeper modeling" of data, which can be achieved by semantically registering existing data sources to one or more ontologies. We have presented some of the technical details of semantic data registration and illustrated semantic mediation using examples from the GEON project [GEO].

While semantic mediation addresses the problem of data integration, it does not provide a mechanism to integrate other applications and tools into data analysis "pipelines." For this kind of process integration problem, we have proposed the use of scientific workflow systems such as Kepler, to provide an open and extensible problem-solving environment for designing and executing workflows. Kepler is built on top of the Ptolemy II system and inherits from it many useful features, including an actor-oriented design methodology to create more reusable workflow components, an extensible type system, and an intuitive graphical user interface. Kepler-specific extensions include database querying and data transformation actors, and actors for executing Web services, command-line tools, and remote jobs on the Grid [LAB+04]. A large remaining problem is the generation of data transformations that are often necessary to connect two independently designed actors or Web services. To this end, we have proposed an ontology-driven data transformation approach that exploits semantic data registration information to generate correspondence mappings, which in turn aid the generation of the desired data transformations.

[18]For a detailed description of the $\sqsubseteq_O$ relation, see [BL04].

[19]Note that if `age`, `ccomp`, `lithology`, and `geoage` are "simple" elements, i.e., only contain PCDATA, the transformation can be generated from the correspondences alone.

REFERENCES CITED

[BCD+02] Bhattacharyya, S.S., Cheong, E., Davis, J., II, Goel, M., Hylands, C., Kienhuis, B., Lee, E.A., Liu, J., Liu, X., Muliadi, L., Neuendorffer, S., Reekie, J., Smyth, N., Tsay, J., Vogel, B., Williams, W., Xiong, Y., and Zheng, H., 2002, Heterogeneous Concurrent Modeling and Design in Java. Technical Report Memorandum UCB/ERL M02/23, EECS, University of California, Berkeley.

[BCM+03] Baader, F., Calvanese, D., McGuinness, D., Nardi, D., and Patel-Schneider, P.F., editors, 2003, The Description Logic Handbook: Theory, Implementation, and Applications: Cambridge University Press.

[BFH03] Berman, F., Fox, G., and Hey, A., editors, 2003, Grid Computing: Making the Global Infrastructure a Reality: John Wiley & Sons.

[BIR] Biomedical Informatics Research Network Coordinating Center (BIRN-CC), University of California, San Diego. http://nbirn.net/.

[BL04] Bowers, S., and Ludäscher, B., 2004, An ontology-driven framework for data transformation in scientific workflows, *in* Proceedings of the 1st International Workshop on Data Integration in the Life Sciences (DILS), volume 2994 of LNCS, p. 1–16.

[BLHL01] Berners-Lee, T., Hendler, J., and Lassila, O., 2001, The Semantic Web: Scientific American, May 2001.

[BLL04a] Bowers, S., Lin, K., and Ludäscher, B., 2004, On integrating scientific resources through semantic registration, *in* Proceedings of the 16th International Conference on Scientific and Statistical Databases (SSDBM).

[BLL+04b] Brooks, C., Lee, E.A., Liu, X., Neuendorffer, S., Zhao, Y., and Zheng, H., 2004, Heterogeneous Concurrent Modeling and Design in Java (Volumes 1–3), Technical report, Department of EECS, University of California, Berkeley, Technical Memoranda UCB/ERL M04/27, M04/16, M04/17.

[Che76] Chen, P., 1976, The Entity-Relationship Model: Towards a Unified View of Data: ACM Transactions on Database Systems, v. 1, no. 1, p. 9–36, doi: 10.1145/320434.320440.

[DOS04] Daconta, M.C., Obrst, L.J., and Smith, K.T., 2004, Chapter 8: Understanding Ontologies, *in* The Semantic Web: A guide to the future of XML, Web Services and Knowledge Management: Wiley, p. 181–238.

[DT03] Deutsch, A., and Tannen, V., 2003, MARS: A System for Publishing XML from Mixed and Redundant Storage, *in* International Conference on Very Large Data Bases (VLDB).

[ESR98] ESRI, 1998, ESRI Shapefile Technical Description, http://www.esri.com/library/whitepapers/pdfs/shapefile.pdf.

[FK99] Foster, I., and Kesselman, C., editors, 1999, The Grid: Blueprint for a New Computing Infrastructure: Morgan Kaufmann.

[Fos03] Foster I., 2003, The Grid: Computing without Bounds: Scientific American, April 2003.

[GEO] National Science Foundation/Information Technology Research for National Priorities: GEON: A Research Project to Create Cyberinfrastructure for the Geosciences, www.geongrid.org, 2005.

[GMPQ+97] Garcia-Molina, H., Papakonstantinou, Y., Quass, D., Rajaraman, A., Sagiv, Y., Ullman, J., Vassalos, V., and Widom, J., 1997, The TSIMMIS Approach to Mediation: Data Models and Languages: Journal of Intelligent Information Systems, v. 8, no. 2.

[Gru93] Gruber, T.R., 1993, A Translation Approach to Portable Ontology Specifications: Knowledge Acquisition, v. 5, p. 199–220, doi: 10.1006/knac.1993.1008.

[GSR+99] Gillespie, M.R., Styles, M., Robertson, S., Hallsworth, C.R., Knox, R.W.O., Evans, C.D.R., Irving, A.A.M., Merritt, J.W., Morigi, A.N., and Northmore, K.J., 1999, BGS Rock Classification Scheme, Volumes 1–4, technical report, http://www.bgs.ac.uk/ bgsrcs/home.html.

[HAC+89] Harland, W.B., Armstrong, R., Cox, A., Lorraine, C., Smith, A., and Smith, D., 1989, A Geologic Time Scale: Cambridge University Press.

[Hal01] Halevy, A., 2001, Answering Queries Using Views: A Survey: VLDB Journal, v. 10, no. 4, p. 270–294, doi: 10.1007/s007780100054.

[Joh02] Johnson, B.R., 2002, Geologic Map Unit Classification, ver. 6.1, Draft: U.S. Geological Survey.

[KEP] KEPLER: A System for ScientificWorkflows, http://kepler-project.org.

[KM77] Kahn, G., and MacQueen, D.B., 1977, Coroutines and Networks of Parallel Processes, *in* Gilchrist, B., ed., Proceedings of the IFIP Congress 77, p. 993–998.

[Koc01] Koch, C., 2001, Data Integration against Multiple Evolving Autonomous Schemata [Ph.D. thesis]: Wien, Technische Universit, p. 1–152.

[LAB+04] Ludäscher, B., Altintas, I., Berkley, C., Higgins, D., Jaeger, E., Jones, M., Lee, E.A., Tao, J., and Zhao, Y., 2006, Concurrency and Computation: Practice and Experience (in press).

[LAHS03] Lutz, C., Areces, C., Horrocks, I., and Sattler, U., 2003, Keys, Nominals, and Concrete Domains, *in* Proceedings of the 18th International Joint Conference on Artificial Intelligence IJCAI.

[Len02] Lenzerini, M., 2002, Data Integration: A Theoretical Perspective: Tutorial at PODS.

[Lev00] Levy, A., 2000, Logic-Based Techniques in Data Integration, *in* Minker, J., ed., Logic Based Artificial Intelligence: Kluwer, p. 575–595.

[LGM01] Ludäscher, B., Gupta, A., and Martone, M.E., 2001, Model-Based Mediation with Domain Maps, *in* 17th International Conference on Data Engineering (ICDE), Heidelberg, Germany, IEEE Computer Society.

[LGM03] Ludäscher, B., Gupta, A., and Martone, M.E., 2003, A Model-Based Mediator System for Scientific Data Management, *in* Lacroix, Z., and Critchlow, T., eds., Bioinformatics: Managing Scientific Data: Morgan Kaufmann.

[LHL+98] Ludäscher, B., Himmeröder, R., Lausen, G., May, W., and Schlepphorst, C., 1998, Managing Semistructured Data with FLORID: A Deductive Object-Oriented Perspective: Information Systems, v. 23, no. 8, p. 589–613, doi: 10.1016/S0306-4379(98)00030-1.

[LL03] Lin, K., and Ludäscher, B., 2003, A System for Semantic Integration of Geologic Maps via Ontologies, *in* Semantic Web Technologies for Searching and Retrieving Scientific Data (SCISW), Sanibel Island, Florida.

[LLB+03] Lin, K., Ludäscher, B., Brodaric, B., Seber, D., Baru, C., and Sinha, A.K., 2003, Semantic Mediation Services in Geologic Data Integration: A Case Study from the GEON Grid: Geological Society of America Abstracts with Programs, v. 35, no. 6, p. 365.

[LLBB03] Ludäscher, B., Lin, K., Brodaric, B., and Baru, C., 2003, GEON: Toward a Cyberinfrastructure for the Geosciences—A Prototype for Geological Map Interoperability via Domain Ontologies, *in* Workshop on Digital Mapping Techniques: AASG and U.S. Geological Survey, June 2003.

[LP95] Lee, E.A., and Parks, T., 1995, Dataflow Process Networks: Proceedings of the IEEE, May 1995, v. 83, no. 5, p. 773–799, http://citeseer.nj.nec.com/455847.html.

[NCE] NCEAS, Ecological Metadata Language (EML), http://knb.ecoinformatics.org/ software/eml/.

[NL04a] Nash, A., and Ludäscher, B., 2004, Processing First-Order Queries with Limited Access Patterns, *in* ACM Symposium on Principles of Database Systems (PODS), Paris, France, June 2004.

[NL04b] Nash, A., and Ludäscher, B., 2004, Processing Unions of Conjunctive Queries with Negation under Limited Access Patterns, *in* 9th International Conference on Extending Database Technology (EDBT), LNCS 2992, Heraklion, Crete, Greece, p. 422–440.

[OWL03] OWL Web Ontology Language Reference, W3C Proposed Recommendation, December 2003. www.w3.org/TR/owl-ref/.

[PAGM96] Papakonstantinou, Y., Abiteboul, S., and Garcia-Molina, H., 1996, Object Fusion in Mediator Systems, *in* Proceedings of the 22nd International Conference on Very Large Data Bases (VLDB), p. 413–424.

[RB01] Rahm, E., and Bernstein, P.A., 2001, A survey of approaches to automatic schema matching: VLDB Journal, v. 10, no. 4, p. 334–350, doi: 10.1007/s007780100057.

[ROA] ROADNet: Real-time Observatories, Applications and Data management Network. roadnet.ucsd.edu.

[SDM] Scientific Data Management Center (SDM). http://sdm.lbl.gov/sdmcenter/, see also http://www.npaci.edu/online/v5.17/ scidac.html.

[SEE] NSF/ITR: Enabling the Science Environment for Ecological Knowledge (SEEK). seek.ecoinformatics.org, 2005.

[She98] Sheth, A., 1998, Changing Focus on Interoperability in Information Systems: From System, Syntax, Structure to Semantics, *in* Goodchild, M., et al., eds., Interoperating Geographic Information Systems: Kluwer, p. 5–30.

[SQDO02] Struik, L., Quat, M., Davenport, P., and Okulitch, A., 2002, A Preliminary Scheme for Multihierarchical Rock Classification for Use with Thematic Computer-based Query Systems. Technical Report 2002- D10, Geological Survey of Canada.

[Usc98] Uschold, M., 1998, Knowledge level modeling: Concepts and terminology: The Knowledge Engineering Review, v. 13, no. 1, p. 5–29, doi: 10.1017/S0269888998001040.

[VP00] Vassalos, V., and Papakonstantinou, Y., 2000, Expressive Capabilities Description Languages and Query Rewriting Algorithms: Journal of Logic Programming, v. 43, no. 1, p. 75–122, doi: 10.1016/S0743-1066(99)00026-6.

[VSSV02] Visser, U., Stuckenschmidt, H., Schuster, G., and Vogele, T., 2002, Ontologies for Geographic Information Processing: Information Processing, Computers and Geosciences, v. 28, no. 1, p. 103–117.

[Wie92] Wiederhold, G., 2003, Mediators in the Architecture of Future Information Systems. IEEE Computer, 25(3):38–49, 1992.

[WSD03] Web Services Description Language (WSDL) Version 1.2. http://www.w3.org/TR/wsdl12.

[XML01] XML Schema, W3C Recommendation. www.w3.org/TR/xmlschema-{1,2}, May 2001.

MANUSCRIPT ACCEPTED BY THE SOCIETY 18 JUNE 2005

Printed in the USA

Geological Society of America
Special Paper 397
2006

Active tectonics, tectonic geomorphology, and fault system dynamics: How geoinformatics can help

J Ramón Arrowsmith
Department of Geological Sciences, Arizona State University, Tempe, Arizona 85287-1404, USA

ABSTRACT

Active tectonics, tectonic geomorphology, and fault system dynamics investigate deformation of continental crust over late Quaternary to annual time scales. These interdisciplinary endeavors require joint analysis of diverse data sets and model results. Data gathering and analysis driven by EarthScope and other large solid earth science initiatives motivates a discussion of how to apply geoinformatics tools to enhance these efforts. I use three information integration scenarios for earthquake geology site context, basic digital elevation model (DEM) analysis, and tectonic geomorphology (all located in southern California) to show the kinds of analyses, integrations, and visualizations that are commonly employed. Earthquake geology site context along the San Jacinto fault ranges in scale from the San Andreas transform boundary to high-resolution topography and balloon aerial photography. Manipulations of a DEM for the Carrizo Plain illustrate the basic record of deformation and surface processes recorded in the topography. Tectonic geomorphologic investigations at Wheeler Ridge include visualization of imagery and geology with topography, raster calculations to produce a map of local relief, and the integration of fault model calculations for surface uplift. Necessary data sets for these and other scenarios include seismicity; crustal motion; stress; topography; remotely sensed imagery; and maps of Quaternary faults, late Cenozoic subsidence and uplift, and Quaternary geology. Necessary geoinformatics capabilities include visualization and cartography, file format and database conversion, geostatistics, comprehensive data query capability, data capture, field data capture, and physics-based deformation and landscape development models. Collaborations will improve earth scientists' ability to create knowledge, while providing domain challenges for information technologists.

Keywords: geomorphology, structural geology, neotectonics, paleoseismology, geoinformatics.

*E-mail: ramon.arrowsmith@asu.edu

Arrowsmith, J R., 2006, Active tectonics, tectonic geomorphology, and fault system dynamics: How geoinformatics can help, *in* Sinha, A.K., ed., Geoinformatics: Data to Knowledge: Geological Society of America Special Paper 397, p. 131–139, doi: 10.1130/2006.2397(09). For permission to copy, contact editing@geosociety.org.

INTRODUCTION

Motivations

Earth science interpretations employ the fundamental understandings of the physics, chemistry, and biology of natural systems. Those fundamentals are usually coupled with a broad spatial and historical data gathering and analysis approach. One of the major research themes with great basic science interest and broad earthquake and volcanic hazard applicability has been plate tectonics, the theory of Earth processes posits that Earth's solid surface can be broken into a series of mobile, and largely coherent and rigid, blocks that move independently. As these plates move relative to each other, new solid material (lithosphere) may be created, or plates may be pushed into the underlying asthenosphere, or the plates may move laterally past each other. Most deformation occurs at these plate boundaries. However, oceanic and continental lithospheric plates have different material properties, and interest has begun to focus on the active deformation within more diffuse plate boundary zones and in continental interiors. Active tectonics, tectonic geomorphology, and fault system dynamics are subdomains of plate-tectonics investigations that relate to continental crustal deformation over late Quaternary to annual time scales.

Active tectonics is the study of crustal motions of interest to society (Wallace, 1986). Active tectonics research questions include: What are spatial and temporal scales of continental lithospheric deformation? What are the roles of faulting versus distributed deformation? What drives the deformation? What is the role of preexisting heterogeneity? A related area of research is tectonic geomorphology, which is the study of landforms and landscapes that record a measurable tectonic signal (Burbank and Anderson, 2001). Typical research questions are: How do surficial processes respond to deformation (and what are the possible feedbacks between them)? How can we use the signal of deformation in the landscape (and the late Quaternary) to bridge the gap between annual and million-year time scales? Finally, fault system dynamics refers to the seismic and aseismic processes operating within a network of many faults over time scales from seconds to days to millennia. Example research questions in this area with emphasis on the longer time scales include: What are the spatial and temporal distributions of earthquakes in a region over the (late) Holocene, and how do they reflect the regime and the modulation of deformation? What is the effect of mechanical interaction?

Answering these questions requires interdisciplinary collaborations and data and model integration. Geoinformatics provides a powerful suite of tools to enable such integrative efforts. Outside of North America, "geoinformatics" commonly is synonymous with Geographic Information Science, which is the discipline that brings together surveying, remote sensing, geography, cartography, computer science, and statistics to develop tools for the gathering, manipulation, archiving, processing, and visualization of geospatial data. In the United States, earth scientists and database information technology practitioners have co-opted the name "geoinformatics" to refer more to extracting knowledge from the rock record through construction of disciplinary databases and developing a fully integrated geosciences information network (e.g., http://www.geongrid.org/about.html).

EarthScope—A Motivating Initiative

Many of the above questions are at the heart of EarthScope—a major equipment and science initiative in the solid earth sciences using modern observational, analytical, and telecommunications technologies to investigate the structure and evolution of the North American continent and the physical processes controlling earthquakes and volcanic eruptions (http://www.earthscope.org). It has several components: (1) USARRAY—a continental-scale seismic observatory of fixed and mobile seismometers arrayed across North America, (2) SAFOD (San Andreas Fault Observatory at Depth)—a deep borehole designed to measure the conditions under which earthquakes occur, (3) PBO (Plate Boundary Observatory)—a geodetic observatory measuring the active deformation of the crust and volcanoes of western North America, and ultimately (4) InSAR (interferometric synthetic aperture radar)—a space-based platform designed to provide a synoptic and high-precision measure of surface displacements. GeoEarthscope is a geologic component in which the value of the temporal and spatial context provided by geomorphic and geologic studies through high-resolution topography and enhanced Quaternary geochronological capability is emphasized (http://www.unavco.org/pubs_reports/proposals/pbo/geo_pbo_wp.pdf). While EarthScope provides a rallying point, the studies of active tectonics, tectonic geomorphology, and fault system dynamics will continue beyond EarthScope, both in time and beyond the North American continent.

How Geoinformatics Can Help

In order to assess how the tools of information technology and geoinformatics in particular may enhance our ability to address the questions identified above, I present three information integration scenarios and related data system and modeling tool descriptions: (1) earthquake geology site context, (2) basic digital elevation model analysis, and (3) tectonic geomorphology. In addition, I identify important geoinformatics tools, analysis methods, and data sets for addressing questions in active tectonics, tectonic geomorphology, and fault system dynamics. These are not comprehensive or definitive, but they do illustrate the kinds of analyses, integrations, and visualizations that are commonly employed. These scenarios and their associated data system and modeling tools indicate areas of potential collaboration with our information technology colleagues. Successful collaborations will improve the ability of earth scientists to create knowledge, while providing useful domain challenges for information technologists.

INFORMATION INTEGRATION SCENARIOS

Earthquake Geology Site Context

Extending the earthquake record to prehistory requires focused investigations of the effects of past earthquakes on the ground surface. By documenting the local geomorphic and stratigraphic record of ground rupture, and dating the deformed units, it is possible to produce information about the fault's activity that can be used as input for seismic hazard investigations or to address active tectonics and fault systems dynamics questions. Such documentation is usually done at a place where the sedimentary record is easily accessible by excavation. To complement and inform such research, contextual information is necessary (e.g., Fig. 1).

The Hog Lake paleoseismic site is located along the San Jacinto fault in an area of broad regional shearing of the Pacific–North America plate boundary (Rockwell et al., 2003). Figure 1A shows the Southern California Earthquake Center Horizontal Deformation Velocity Map version 2 (http://www.scecdc.scec.org/group_e/release.v2). Figure 1B shows a 1990 Landsat image of the San Andreas fault system that includes three main faults: the San Andreas fault, San Jacinto fault, and Whittier-Elsinore fault. The deformation is partitioned among them. Each of the faults traverses southern California mostly through a series of discontinuous fault-parallel troughs. Figure 1C shows the USGS Digital Orthophoto Quarter Quadrangle (DOQQ) images of the Hog Lake site. This image is detailed enough to show the fault trough and geomorphic setting. Whereas the images in Figures 1A–1C are raster-based, vector data such as topographic and other features we surveyed using a total station (Fig. 1D) are also important. These data show the trench locations, field-defined fault traces, and the geomorphic context for the earthquake geology investigation. For example, the alluvial fan to the north dams flow from the southeast to actually produce the low-energy sedimentation environment conducive for preservation of the earthquake record (Fig. 1F). In the field, we also used a balloon aerial photography system to document the excavated site and microgeomorphology (Fig. 1E). I georeferenced the balloon photographs, and their joint review with the topographic information is essential for understanding the environment of formation implied by the sediments exposed in the trench (Fig. 1F). Note that the balloon photography is ~10 times higher resolution than the 1 m/pixel DOQQs, illustrating the wide range of scales from which critical information can be derived with a geoinformatics approach.

Basic Digital Elevation Model Analysis

Earth's topography is a result of the combined effects of surface and tectonic processes. The configuration of the landscape also controls the types and rates of surface processes within a given geologic and climatic regime. Digital elevation models (DEMs) are available at various scales, ranging from 1 km with nearly global coverage to tens of centimeters for local studies. Their analysis is routine for active tectonics, tectonic geomorphology, and fault system dynamics (see Pike, 2002, for a recent bibliographic review). The USGS (U.S. Geological Survey) has made available a seamless DEM for the conterminous United States called the National Elevation Dataset (NED; 10–30 m/pixel) at http://seamless.usgs.gov.

Manipulation of the NED DEM of the Carrizo Plain of south-central California illustrates some of the basic requirements necessary for DEM analysis (Fig. 2). Here, the northwest-southeast-trending San Andreas fault cuts across the northeast side of the plain. It is a closed basin perched 300–800 m above the adjacent valleys (the Cuyama and San Juan Creek drainages to the west, and the San Joaquin Valley to the east). A perspective view of the elevation-coded DEM (Fig. 2A) and the topographic contours (Fig. 2B) both show the Carrizo Plain, the bounding ranges (Temblor to the northeast and Caliente and La Panza to the southwest), and the surrounding valleys. Strata in both sets of bounding ranges are deformed into broad arches with smaller and tighter folds superimposed (e.g., Dibblee, 1974). Figures 2C and 2D show examples of simple raster- or grid-based calculations that are essential for DEM analysis: slope (first derivative of elevation with respect to horizontal distance) and aspect (facing direction of local slope), respectively. They both enhance the small-scale features of the topography in the Carrizo Plain and show that the flat Carrizo Plain and San Joaquin Valley contrast with high slopes in the adjacent ranges. These calculations also bring out some of the finer-scale tectonic landforms along the San Andreas fault (Figs. 2C and 2E inset) (e.g., Arrowsmith, 1995). For example, the topographic aspect (Fig. 2D) highlights the San Andreas fault–parallel fabric in the northwestern portion of the Carrizo Plain where numerous folds were mapped by Dibblee (1974). Finally, interactive drawing of arbitrarily oriented or stepped topographic profiles further illustrates some of the physiographic patterns described above (Fig. 2E). Both profiles show the perched Carrizo Plain, bounding ranges, and even important active secondary structures such as the several-hundred-meter-high Elk Hills fold on the northeast end of the southern profile.

Along with local slope (e.g., Fig. 2E), determination of the drainage network and the associated drainage area above a given point in the landscape (using DEM data and geoinformatics tools) is critical for understanding controls on geomorphic processes (e.g., Whipple, 2004; Dietrich et al., 2003, and numerous references in both). Local channel gradient can be related to upstream drainage area as a power law in which the coefficient of proportionality is the steepness index and the power is the concavity index. These two parameters are most often computed from DEMs, and they have been shown to be sensitive to rock uplift and incision rate (among other things; Whipple, 2004). Dietrich et al. (2003) (building on much previous work) and Hilley (2001) show how numerous landscape thresholds (across which the surface processes may be driven by surface uplift) are best depicted in terms of local slope and contributing area.

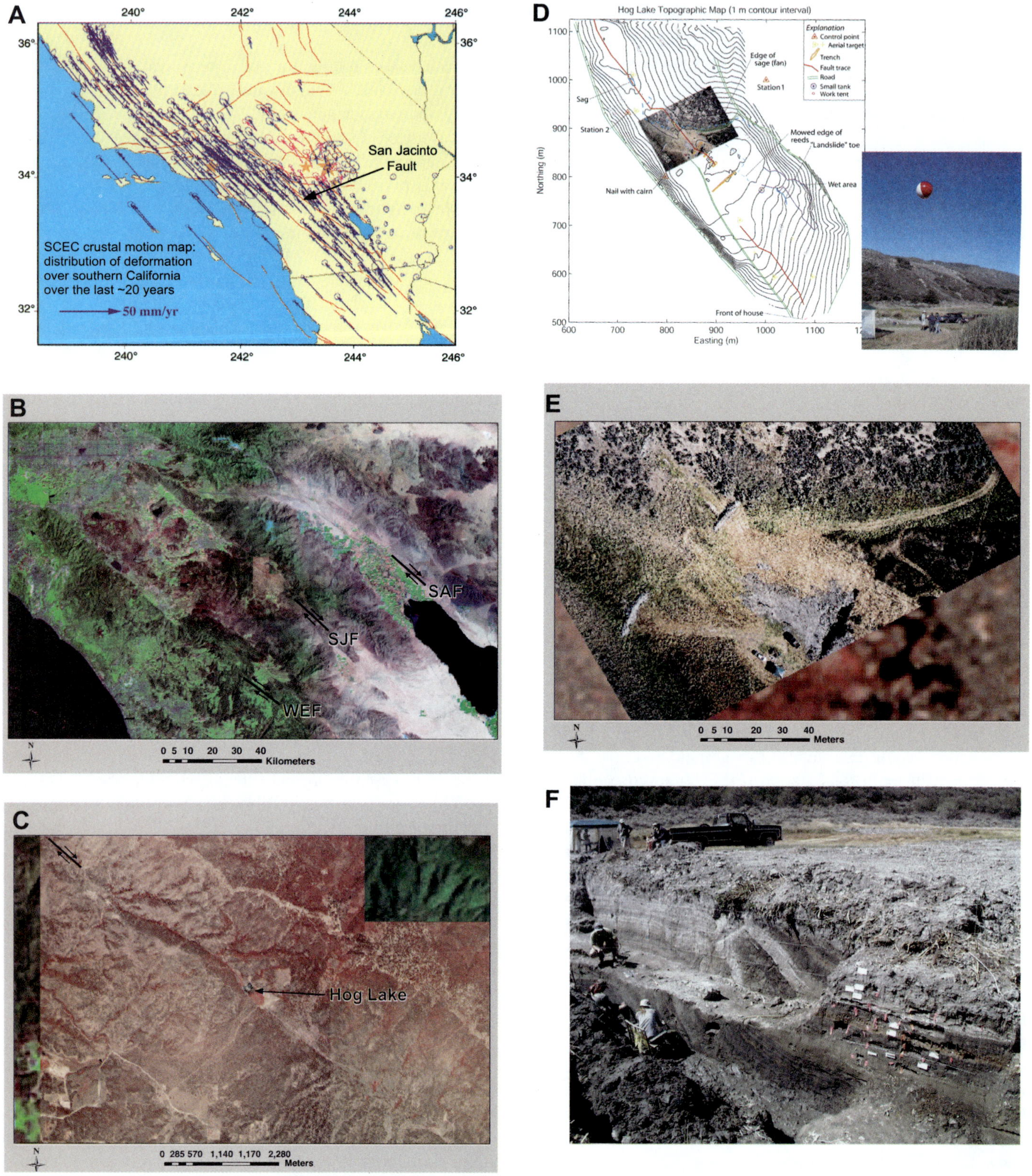

Figure 1. Data integration scenario for paleoseismic application at Hog Lake along the San Jacinto fault in southern California. The images span scales from regional overview to the site level. A: Southern California Earthquake Center Horizontal Deformation Velocity Map version 2, showing the several-hundred-kilometer-wide shear zone of the southern San Andreas fault system (from http://www.scecdc.scec.org/group_e/release.v2/). B: 1990 Landsat image showing the physiographic setting of the major faults of the region. SAF—San Andreas fault, SJF—San Jacinto fault, WEF—Whittier-Elsinore fault. C: USGS Digital Orthophoto Quarter Quadrangle (DOQQ) images with red colors indicating vegetation near Hog Lake. D: Surveyed topography and other features at Hog Lake with the georeferenced balloon photograph shown in Figure 1E. Balloon photography in action at lower right. E: Georeferenced balloon photography (15 cm resolution) over DOQQ (1 m/pixel). F: Photograph of one of the paleoseismic excavations at Hog Lake. The stratigraphy records recurrent ground deformation of interfingered, fine-grained, organic-rich marsh and gravel-rich alluvial-fan deposits (Rockwell et al., 2003).

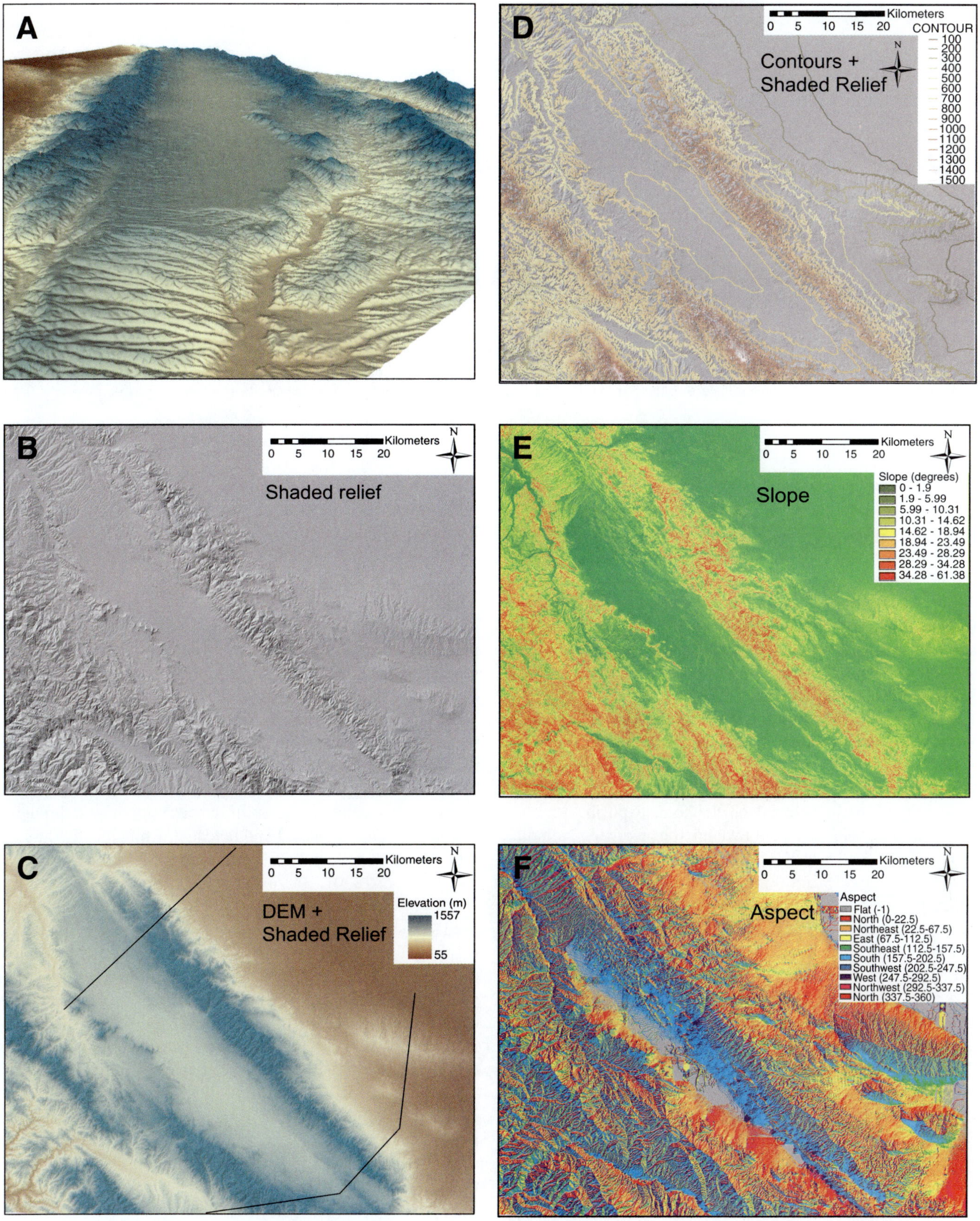

Figure 2 (*continued on following page*). Digital elevation model analysis scenario for the Carrizo Plain area, central California. The San Andreas fault zone cuts through the center of this area from northwest to southeast. A: Oblique view of the USGS National Elevation Dataset (NED) topography toward the southeast, showing the perched Carrizo Plain bounded on the left (northeast) by the Temblor Range and on the right (southwest) by the Caliente Range. Shaded relief (B), elevation and shaded relief (C), and contours and shaded relief all depict the regional geomorphology. Simple raster processing products from these data include slope (E) and aspect (F).

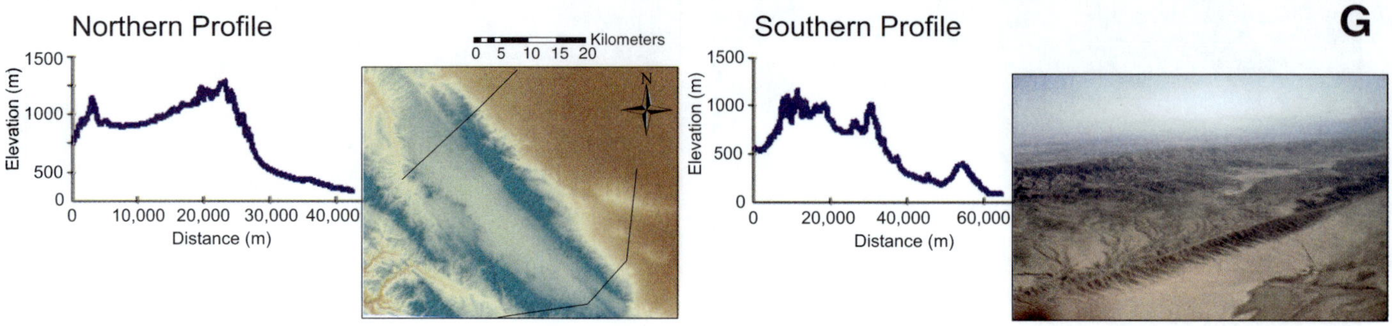

Figure 2 (*continued*). G: Topographic profiles (SW-NE) across the NED (locations marked on the inset colored shaded relief map and C) crossing the Caliente Range, the Carrizo Plain (the San Andreas fault is on the right or northeast side), and the Temblor Range. The inset photograph shows the Dragon's Back pressure ridge along the San Andreas fault in the Carrizo Plain just northwest of the southern profile.

Tectonic Geomorphology

Building on the simple DEM processing and visualization presented above, the tectonic geomorphology scenario I present illustrates data integration for the study of the interaction between tectonic and surface processes. Typical tectonic geomorphology studies include geologic mapping of young (mostly Quaternary) rock units and landforms, topographic analyses (such as those described in the previous section), and they may include model calculations to aid interpretation.

Wheeler Ridge is a quintessential example of a tectonic landform. It is an active fold, growing toward the southeast, that has progressively narrowed, deflected, and defeated drainages that once crossed the piedmont it deforms (Fig. 3). Drainage networks have developed on the fold, exposing its core. This feature has been analyzed by numerous workers, including Shelton (1966), Keller et al. (1998), Mueller and Talling (1997), Brozovic et al. (1995), and Burbank and Anderson (2001). Aerial photography (see perspective view of Fig. 3A) shows the main features of the eastern part of the fold: the wind and water gaps where paleodrainage cuts through the growing fold, the drainage network development and incision in the older western portion of the fold, and the deformed original piedmont surface in the east. Keller et al. (1998) used relative and numerical age indicators of soil development and drainage patterns to map five main geologic units that document the eastward fold growth (Fig. 3C). The angular relationship between the relatively shallow dip of the geomorphic units and the steeply dipping fold core and the development of the wind and water gaps indicate that the fold was actively cut and beveled by piedmont processes prior to isolation from further erosion and uplift (Mueller and Talling, 1997).

Burbank and Anderson (2001) use analysis of Wheeler Ridge to highlight the importance of integrated structural, geologic, and geomorphic analyses in studies of fault and fold growth and landscape development and surface process response. These determinations are made by employing a substitution of space for time, or ergodic, hypothesis. To demonstrate an analysis of this type, I calculated the topographic residual, or local relief, over the eastern portion of the ridge (Figs. 3D and 3E). The topographic residual is the arithmetic difference between an enveloping topographic surface and the current topography (Fig. 3B) (e.g., Bürgmann et al., 1994; Hilley, 2001; Burbank and Anderson, 2001). To define the envelope, I used the extents and elevations of the mapped deformed geomorphic surfaces over the fold (from Keller et al., 1998) and prominent ridgelines. I interpolated linearly between those elevations (cutting off some intervening higher topography—thus the negative residuals) and subtracted raster representations of the enveloping surface and the actual topography. The maximum residual is ~127 m where incision into the growing eastern portion of the fold is greatest on its northern side and in the water gaps. As the fold propagates eastward, deformation and uplift increase slopes, and throughgoing channels are deflected and defeated, while locally drainages initiate and remove material, exhuming the fold core (Burbank and Anderson, 2001; Brozovic et al., 1995).

To illustrate how numerical models of deformation might be used to interpret the deformation at Wheeler Ridge in our tectonic geomorphology analysis, I used an elastic dislocation (boundary element) model (Okada, 1985). This tool allows for the specification of arbitrary displacement discontinuities along arbitrarily oriented and located faults embedded in an elastic half-space. It is useful for short time period crustal deformation modeling (see discussions, for example, in King et al., 1988; Stein et al., 1988; Taboada et al., 1993; Roering, 1995; Arrowsmith et al., 1996; Burbank and Anderson, 2001). Figure 3F shows the vertical displacements calculated on a roughly 200 m grid resulting from 1 m of slip along a 65° south-dipping blind thrust fault whose tip is ~450 m below the crest of the fold. Peak surface displacements of ~40 cm result along the fold axis. In addition, the model simulates the asymmetry of the fold with its gentle south limb. This slip and uplift signature might indicate what would happen in a single earthquake ($M \approx 7$) and what would be repeated many times to build the overall topography of the fold (see Mueller and Suppe, 1997; Mueller and Talling, 1997; Medwedeff, 1992; and

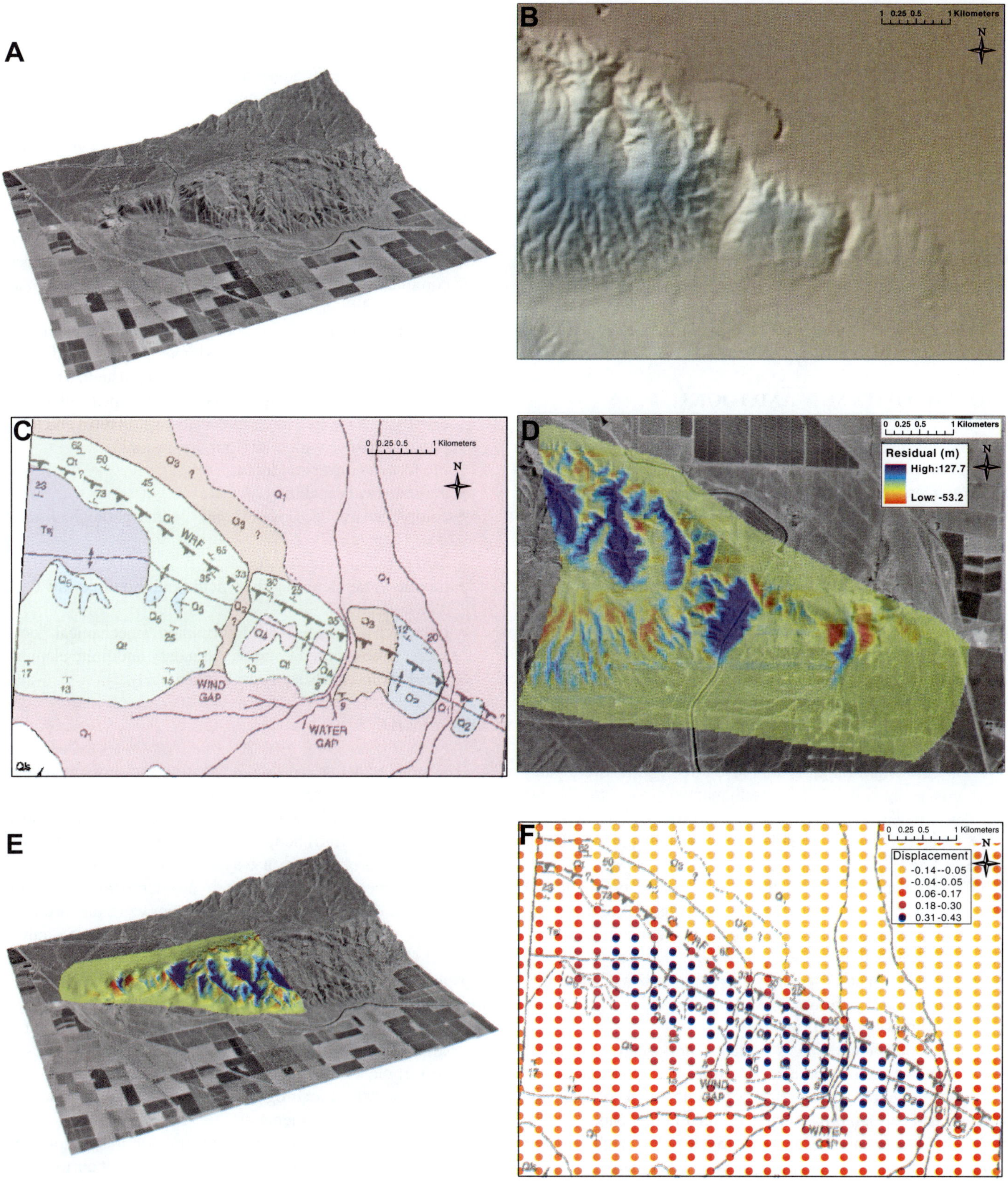

Figure 3. Tectonic geomorphology analysis of Wheeler Ridge, central California. A: Perspective view of DOQQ imagery draped on NED topography looking south-southwest over the eastward-growing fold. B: DEM and shaded relief of Wheeler Ridge. C: Geologic map with polygons digitized and semitransparent over georeferenced geologic map from Keller et al. (1998). From youngest to oldest, the geologic units are Tsj—Tertiary San Joaquin formation, Qt—Pliocene-Pleistocene Tulare formation, Qls—Quaternary landslide deposits, WRF—buried Wheeler Ridge thrust fault, and Q1–Q5—progressively older geomorphic surfaces (Keller et al., 1998). D: Topographic residual (e.g., Bürgmann et al., 1994; Hilley, 2001) produced by subtracting the current topography (Fig. 3B) from an enveloping surface defined partially by the geomorphic surfaces in Figure 3C and the tops of prominent ridgelines. Note the improved information display with the transparency overlay on the DOQQ. E: Drape of residual (Fig. 3D) and DOQQ on topography to better illustrate the relationship of residual to landforms. F: Vertical displacements (cm) calculated from 1 m slip along 65° southwest-dipping blind thrust fault with a tip 450 m below the crest of the fold. The extent of this map is the same as Figures 3B, 3C, and 3D. Note the geologic map from Keller et al. (1998) (Fig. 3C) underneath.

Keller et al., 1998, for more discussion). Stress transfer modeling of slip along an adjacent fault (such as the San Andreas fault) and its loading on the Wheeler Ridge fault would be a simple extension of the boundary element model (e.g., Stein and Toda, 2002). The direct integration of this elastic boundary element model result with the other analyses simply demonstrates the interpretive power that such tools add to a study of this type. The elastic approach is presented here for illustrative purposes; no doubt permanent aseismic and inelastic deformation contribute to the growth of Wheeler Ridge. Other modeling tools that simulate landscape development might be brought to bear within such an interpretive environment as well (e.g., Densmore et al., 1998; Tucker et al., 2001).

REQUIRED DATA SETS AND TOOLS

The information integration scenarios presented above show the kinds of investigations that geoinformatics tools can enhance. The examples are not comprehensive, however. Given the breadth of typical geologic investigations, it is difficult to a priori identify all of the data and analysis that are necessary. Nevertheless, below, I identify a number of data sets and geoinformatics functions of potential use in active tectonics, tectonic geomorphology, and fault system dynamics studies.

- seismicity (instrumental, historic, paleoseismic)
- geodetically determined decadal velocity field (e.g., Fig. 1A)
- stress measurements (e.g., World Stress Map—http://www.world-stress-map.org/)
- topography (DEMs at all scales; e.g., Figs. 1D, 2, and 3), in particular at high spatial resolution and possibly gathered using airborne laser swath mapping (e.g., http://www.ncalm.ufl.edu/)
- imagery (Landsat/ASTER/aerial photos; e.g., Figs. 1B–1E and 3A)
- Quaternary faults (locations, sense of motion, rates of movement, recurrence interval; e.g., Figs. 1A and 3C)
- Late Tertiary–Quaternary subsidence and uplift (basins and exhumation)
- geological maps, especially those emphasizing Quaternary geology and geomorphology (e.g., Fig. 3C)
- geochronologic information

Information integration is evident with visualization tools. However, some integration is simply impeded by format differences between typical tools used. For example, all of the visualization and cartography I have presented in this paper is done in ArcGIS 8.3 from Environmental System Research Institute (ESRI; http://www.esri.com). However, many earth scientists use the Generic Mapping Tools (GMT; http://gmt.soest.hawaii.edu/). Both include some capability for Boolean and mathematical raster overlay and geostatistics. The largest difference is that the ArcGIS tools allow attributes to be assigned to spatial objects, which can be queried. GMT does not allow this fundamental GIS (geographic information system) functionality. For the earth science community to take advantage of databases and tools, the databases must be populated, and information about the data must be recorded in a coherent way. Ideally, these software tools may even be brought into the field so that existing data and model results may guide further observation. In addition, field-based digital data collection reduces recording and transcription errors, and limits the need for subsequent off-site digitizing. This capability significantly improves the quality of community data sets. My tectonic geomorphology example discussed the importance of the combination of models and data. Necessary geoinformatics tool functionality includes:

- visualization/cartographic tools for combined mapping of raster and vector data sets with simple grid math and interpolation capabilities, and the capability to visualize beyond the 2.5 dimensions of digital topography draped with imagery (Fig. 3A) to true three-dimensional structures and their possible change with time (four dimensions).
- file format conversion tools
- geostatistical capability
- comprehensive geological and geomorphological data query capability
- data population/upload tools
- mobile systems to augment field reality and permit field data collection
- interpretive models for deformation (mechanical tools such as those based upon the boundary and finite element method) and landscape development

CONCLUSIONS

The study of active crustal deformation and surface process response requires integration of diverse data and model results. New insights may be gained (and required of the imminent flood of data from EarthScope) if we can present our domain (earth science) problems in a way that our information technology colleagues can recognize them as interesting challenges with common themes in other domains. For example, are bioinformatics applications portable to the geoinformatics environment? Can the computational successes of astronomy and atmospheric science inspire and become possible for earth scientists? The GEON project (http://www.geongrid.org) is a large collaboration between earth scientists and information technology experts largely from San Diego Supercomputer Center. The IT effort clearly leverages tool development from multiple similar collaborations (ecology, bioinformatics, hydrologic sciences, and atmospheric sciences).

In this paper, I have presented information integration scenarios that show how some workers might pursue their research and that identify where integration enhances understanding. Necessary data sets will have to be built, imported, manipulated, analyzed, visualized, and compared with model results using geoinformatics tools. Building and deploying such tools can change the way that earth scientists produce knowledge, promoting a cycle of knowledge creation, observation, and tool refinement.

ACKNOWLEDGMENTS

This paper is largely based on a presentation delivered at the Society for Industrial and Applied Mathematics Conference on Mathematical and Computational Issues in the Geosciences, March 2003, in Austin, Texas, at the Minisymposium on Geoinformatics: Creating Cyberinfrastructure for the Geosciences organized by Chaitan Baru and Vladik Kreinovich. Discussions with GEON colleagues Nathan Toké, Dallas D. Rhodes, Tom Jordan, and Lela Prashad have been helpful. Susan Selkirk assisted with the figure production. Amy Rector helped with the bibliography. Mike Zoldak taught me the raster analysis used in Figure 3. I used the implementation of Okada (1985) for Matlab by Brandon Meade. Online data from the California Spatial Information Library (http://www.gis.ca.gov/), the USGS National Map Seamless Data Distribution System (http://seamless.usgs.gov), and the NASA Earth Science Applications Directorate Landsat MrSid Image server (https://zulu.ssc.nasa.gov/mrsid/) made these demonstrations feasible and are important examples of necessary open data access. Thank you to José Hurtado and Jim Spotila for helpful and constructive reviews. This work was supported by two National Science Foundation grants: (1) ITR/IM+AP (GEO) Collaborative research: Creation of a geospatial data system for the transition between the Colorado Plateau and the Basin and Range provinces (Geoinformatics in Action)—EAR-0112960; and (2) ITR Collaborative research: GEON: A research project to create cyberinfrastructure for the geosciences—EAR-0225543.

REFERENCES CITED

Arrowsmith, J R., 1995, Coupled tectonic deformation and geomorphic degradation along the San Andreas fault system [Ph.D. thesis]: Stanford University, 356 p.

Arrowsmith, J R., Pollard, D.D., and Rhodes, D.D., 1996, Hillslope development in areas of active tectonics: Journal of Geophysical Research, v. 101, p. 6255–6257. (Correction: 1999: Journal of Geophysical Research, v. 104, p. 805.)

Brozovic, N., Burbank, D.W., Fielding, E.J.F., and Meigs, A.J., 1995, The spatial and temporal topographic evolution of Wheeler Ridge California: New insights from digital elevation data: Geological Society of America Abstracts with Programs, v. 27, no. 6, p. 396.

Burbank, D.W., and Anderson, R.S., 2001, Tectonic geomorphology: Malden, Massachusetts, Blackwell Science, 274 p.

Bürgmann, R., Arrowsmith, J R., Dumitru, T.A., and McLaughlin, R.J., 1994, Rise and fall of the southern Santa Cruz Mountains, California, from fission tracks, geomorphology, and geodesy: Journal of Geophysical Research, v. 99, p. 20,181–20,202, doi: 10.1029/94JB00131.

Densmore, A.L., Ellis, M.A., and Anderson, R.S., 1998, Landsliding and evolution of normal-fault-bounded mountains: Journal of Geophysical Research, v. 103, p. 15,203–15,219, doi: 10.1029/98JB00510.

Dibblee, T.J., 1974, Regional geologic map of San Andreas and related faults in Carrizo Plain, Temblor, Caliente, and La Panza ranges and vicinity, California: U.S. Geological Survey Miscellaneous Geological Investigations Map I-757, scale 1:62,500, 1 sheet.

Dietrich, W.E., Bellugi, D.G., Sklar, L.S., Stock, J.D., Heimsath, A.M., and Roering, J.J., 2003, Geomorphic transport laws for predicting landscape form and dynamics, *in* Wilcock, P.R., and Iverson, R.M., eds., Prediction in geomorphology: Washington, D.C., American Geophysical Union Geophysical Monograph Series 135, p. 103–132.

Hilley, G.E., 2001, Landscape development and fault behavior in tectonically active areas [Ph.D. thesis]: Tempe, Arizona State University, 191 p.

Keller, E.A., Zepeda, R.L., Rockwell, T.W., Ku, T.L., and Dinklage, W.S., 1998, Active tectonics at Wheeler Ridge, southern San Joaquin Valley, California: Geological Society of America Bulletin, v. 110, p. 298–310, doi: 10.1130/0016-7606(1998)110<0298:ATAWRS>2.3.CO;2.

King, G.C., Stein, R.S., and Rundle, J.B., 1988, The growth of geological structures by repeated earthquakes, 1: Conceptual framework: Journal of Geophysical Research, v. 93, p. 13,307–13,318.

Medwedeff, D.A., 1992, Geometry and kinematics of an active, laterally propagating wedge thrust, Wheeler Ridge, California, *in* Mitra, S., and Fisher, G.W., eds., Structural geology of fold and thrust belts: Baltimore, Johns Hopkins University Press, p. 3–28.

Mueller, K., and Suppe, J., 1997, Growth of Wheeler Ridge anticline, California: Geomorphic evidence for fault-bend folding behavior during earthquakes: Journal of Structural Geology, v. 19, p. 383–396, doi: 10.1016/S0191-8141(96)00112-5.

Mueller, K., and Talling, P., 1997, Geomorphic evidence for tear faults accommodating lateral propagation of an active fault-bend fold, Wheeler Ridge, California: Journal of Structural Geology, v. 19, p. 397–411, doi: 10.1016/S0191-8141(96)00089-2.

Okada, Y., 1985, Surface deformation due to shear and tensile faults in a half-space: Bulletin of the Seismological Society of America, v. 75, p. 1135–1154.

Pike, R.J., 2002, A bibliography of terrain modeling (geomorphometry), the quantitative representation of topography: Supplement 4.0: U.S. Geological Survey Open-File Report 2003-018937.

Rockwell, T.K., Young, J.J., Seitz, G.G., Meltzner, A.J., Verdugo, D.M., Khatib, F., Ragona, D., Altangerel, O., and West, J., 2003, 3000 years of ground-rupturing earthquakes in the Anza Seismic Gap, San Jacinto fault, Southern California: Time to shake it up?: Seismological Research Letters, v. 74, p. 236–237.

Roering, J.J., 1995, Near-surface and secondary deformation associated with blind thrust faults [M.S. thesis]: Stanford University, 85 p.

Shelton, J.S., 1966, Geology illustrated: San Francisco, W.H. Freeman and Company, 434 p.

Stein, R.S., and Toda, S., 2002, Response of the San Andreas fault to the 1983 Coalinga-Nuñez earthquakes: An application of interaction-based probabilities for Parkfield: Journal of Geophysical Research, v. 107, 2126, doi: 10.1029/2001JB000172.

Stein, R.S., King, G.C., and Rundle, J.B., 1988, The growth of geological structures by repeated earthquakes, 2: Field examples of continental dip-slip faults: Journal of Geophysical Research, v. 93, p. 13,319–13,331.

Taboada, A., Bousquet, J.C., and Philip, H., 1993, Coseismic elastic models of folds above blind thrusts in the Betic Cordilleras (Spain) and evaluation of seismic hazard: Tectonophysics, v. 220, p. 223–241, doi: 10.1016/0040-1951(93)90233-A.

Tucker, G.E., Lancaster, S.T., Gasparini, N., and Bras, R., 2001, The channel-hillslope integrated landscape development (CHILD), *in* Harmon, R.S., and Doe, W.W., III, eds., Landscape erosion and evolution modeling: New York, Kluwer Academic/Plenum Publishers, p. 349–388.

Wallace, R.E., editor, 1986, Active tectonics: Studies in geophysics: Washington, D.C., National Academy Press, 266 p.

Whipple, K.X., 2004, Bedrock rivers and the geomorphology of active origins: Annual Review of Earth and Planetary Sciences, v. 32, p. 151–185, doi: 10.1146/annurev.earth.32.101802.120356.

Manuscript Accepted by the Society 18 June 2005

Geological Society of America
Special Paper 397
2006

Fusion of Landsat ETM+ and radar data to enhance the extraction of surface and near-subsurface information

Hongjie Xie*
Department of Earth and Environmental Science, University of Texas, San Antonio, Texas 78249, USA

G. Randy Keller*
Department of Geological Science and Pan-American Center for the Earth and Environmental Sciences, University of Texas, El Paso, Texas 79968, USA

ABSTRACT

With the introduction of many new types of remotely sensed imagery in recent years, opportunities to compare and integrate these data abound. In this paper, we present a hue-saturation-value (HSV) transformation approach for the fusion of surface roughness and shallow subsurface information from Airborne Synthetic Aperture Radar (AIRSAR) data with surface spectral reflectivity from Landsat 7 Enhanced Thematic Mapper Plus (ETM+) data. The fused image is informative and brings out new features that are not evident in the original images, and it also helps to identify many features that are not clear in the original images. In this paper, we also present a combined principal components analysis (CPCA) approach to effectively remove the banding noise that degrades the radar image quality. We examine the differences in surface roughness, texture, and penetration ability at different radar frequencies, the differences between radar and optical (ETM+) images, and the utility of radar images in mapping morphologically defined structures such as fault scarps. The earthquake-related active faults in the Franklin Mountains and Valentine region in West Texas are presented as examples of successful applications of data fusion.

Keywords: data fusion, remote sensing, image processing, mapping fault scarps, radar, Landsat ETM+.

INTRODUCTION

Many scientific communities have come to recognize the value of integrating a wide range of geoscience data. Data integration can be in many forms and various procedures and have been referred to as *data fusion*. Data fusion can be thought of as quantitatively combining or merging data from multiple sources in an effort to maximize the extractible information from the data in a quantitative scheme. Data fusion is used to merge remote sensing data of different spatial and/or spectral resolutions to create a higher spatial and/or spectral resolution result (Gross and Schott, 1998). Such data fusion improves the reliability of image interpretation and increases the efficiency of information and feature extraction (Shi et al., 2003). Many research efforts on data fusion have been conducted and many fusion methods have been developed and tested. These include but are not limited to hue-saturation-value (HSV), intensity-hue-saturation (IHS) or

*E-mails: hongjie.xie@utsa.edu; corresponding author (Keller)—keller@utep.edu.

Xie, H., and Keller, G.R., Fusion of Landsat ETM+ and radar data to enhance the extraction of surface and near-subsurface information, 2006, *in* Sinha, A.K., ed., Geoinformatics: Data to Knowledge: Geological Society of America Special Paper 397, p. 141–151, doi: 10.1130/2006.2397(10). For permission to copy, contact editing@geosociety.org.

hue-lightness-saturation (HLS) color transforms (Carper et al., 1990; Zhang, 1999), Brovey transform (Gillespie et al., 1987), smoothing filter-based intensity modulation (SFIM) (Liu, 2000), principal component analysis (PCA) (Shettigara, 1992), high-pass filtering (Filberti et al., 1994), band ratio methods (Price, 1987; Munechika et al., 1993), two-band wavelet transformation (Zhou et al., 1998; Li et al., 1999), and multiband wavelet transformation (Bi et al., 1999; Wisutlmethangoon and Nguyen, 1999; Shi et al., 2003). In general, data fusion can be used to (1) integrate low-spatial-resolution multispectral data with high-spatial-resolution panchromatic data to increase the spatial resolution; (2) integrate low-spatial- and high-spectral-resolution (hyperspectral) data with high-spatial- and low-spectral-resolution data to improve both spatial and spectral resolution of the output image, in which high spectral resolution allows identification of materials, while high spatial resolution locates those materials (Gross and Schott, 1998; Robinson et al., 2000); and (3) integrate two complementary but completely different types of remote sensing data sets.

Many studies have been made to fuse the complementary information from different sensors such as synthetic aperture radar (SAR) and optical sensors. The SAR's ability to image through rain, hail, smoke, fog, and clouds makes it a particularly viable alternative or complement to optical remote sensing techniques where data acquisition can be hindered by the weather conditions (Foody, 1988; Lombardo et al., 2003). The energy-matter interaction in radar imaging is a function of surface roughness and geometry of the imaged target in contrast to the reflectivity recorded from optical imagery. Hence, SAR and optical imagery measure different surface characteristics, and their fusion will enhance the interpretation capability of the integrated image (Lombardo et al., 2003). Van der Sanden et al. (2000) fused Advanced Very High Resolution Radiometer (AVHRR) data depicting sea surface temperature and RADARSAT data giving the roughness and texture of the sea surface to produce an output image that can detect relationships between temperature and roughness. Abdelsalam et al. (2002) utilized fused Spaceborne Imaging Radar-C/X SAR and Landsat ETM+ images to map volcanic flows in Ethiopia. Michelson et al. (2000) studied European Remote Sensing satellite (ERS-1) SAR and Landsat TM images for agricultural and forest classification in Sweden, and found that the SAR data contained more separable land-cover information than did the TM data and that the merged SAR and TM image achieved the highest level of separation. Haack et al. (2002) utilized principal components analysis (PCA) to fuse RADARSAT and Landsat TM images for urban landscape classification, and found that the fused image achieved higher classification accuracy than the original ones. Chen et al. (2003a) utilized the intensity-hue-saturation (IHS) color transformation technique to fuse the Airborne Visible/Infrared Imaging Spectrometer (AVIRIS) and Airborne Synthetic Aperture Radar (AIRSAR) images to better distinguish between urban features in Park City, Utah.

In this study, we used the AIRSAR data provided by the National Aeronautics and Space Administration's (NASA) Jet Propulsion Laboratory (JPL) and Landsat 7 ETM+ data as a part of the Pan-American Center for Earth and Environmental Studies' (PACES) efforts to study geological hazards and water resources in western Texas. We focus on the AIRSAR data processing, banding noise removal, radar polarimetric signature extraction, radar and ETM+ image comparison, radar and ETM+ image fusion, and active fault mapping. We apply the HSV transformation approach to integrate the AIRSAR and ETM+ images. The fused image provides significant improvement in the visual interpretation of structural features.

DATA AND METHODS

Study Area and Background

Our study area in western Texas (Fig. 1) is a semiarid region where the annual mean precipitation is ~230 mm. The metro-

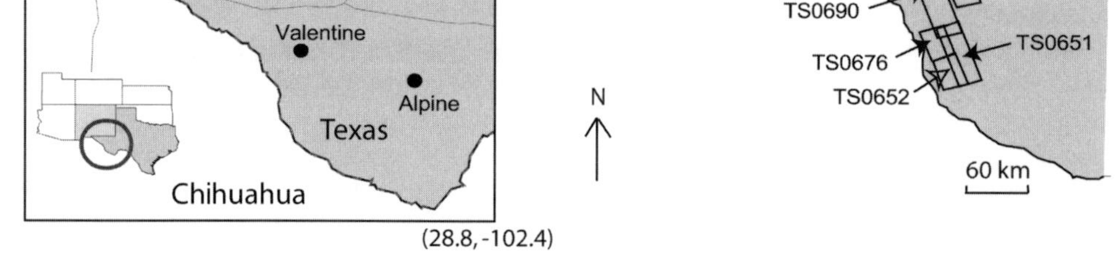

Figure 1. Study area showing the AIRSAR/TOPSAR flight lines with numbers indicating JPL data set identifiers. In the El Paso area, the flight direction was south to north, and the look direction was to the west. In the Valentine area, the flights were flown from southeast to northwest, and the look direction was to the southwest.

politan area of El Paso, Texas, Las Cruces, New Mexico, and Juarez, Mexico, has a population of more than 2.5 million and is growing fast. Thus, the increasing demand for water resources is a concern. Western Texas is also part of the tectonically active Rio Grande rift that triggered a number of earthquakes in this region (Keller and Baldridge, 1999; Dumas et al., 1980; Doser, 1987). For example, Valentine, Texas, experienced a magnitude 6.4 earthquake on 16 August 1931 (Dumas et al., 1980; Doser, 1987). The largest paleoearthquake experienced in the El Paso metropolitan area had an estimated magnitude of 6.9 and is dated by Keaton and Barnes (1996) as occurring 12,000–5000 yr ago. This earthquake was related to the East Franklin Mountains fault that extends north-south for ~45 km. It is a normal fault with slope angles of 18°–25° to the east (Machette, 1987).

Data and Preprocessing

The AIRSAR system, developed and operated by NASA/JPL, flew its first scientific mission in 1988. It is capable of imaging sites in the P (68 cm, 0.45 GHz), L (24.5 cm, 1.22 GHz), and C (5.7 cm, 5.26 GHz) bands simultaneously in polarimetric mode or in the L and C bands in along-track interferometric (ATI) mode, cross-track interferometric (XTI) mode, and bistatic mode (Lou et al., 2001). Both L-band and C-band XTI mode have been successfully used to generate topographic maps (in this case, AIRSAR is also called TOPSAR).

In this study, we utilize AIRSAR data of C-band XTI mode acquired by JPL/NASA's DC-8 aircraft over the Valentine (17 April and 19 April 1998) and El Paso (21 May 1999), Texas, areas (Fig. 1). Table 1 shows the main parameters for these data acquisitions.

The AIRSAR/TOPSAR data sets contain both P and L bands in compressed Stokes matrix format (one pixel is ten bytes), C-VV (transmitting vertically polarized C band energy and receiving only vertically polarized C band energy) data and C-band-derived digital elevation model (DEM) in integer × 2 (signed 16-bit) format, a local incidence angle map derived from the digital elevation model in which each pixel is one byte, and a correlation map containing the normalized correlation coefficient between the two C-band interferometric channels in which each pixel is one byte. The local incidence angle map was used to geometrically and radiometrically correct the C, L, and P bands by taking the topography into account and resampling to ground range of 10 m by 10 m pixel spacing (http://airsar.jpl.nasa.gov/). Before using these data, however, preprocessing was conducted to (1) convert the C-band integer × 2 data to C-VV sigma zero (floating point data); (2) convert integer × 2 DEM to elevation meters (floating point data); (3) convert local incidence angle from byte value (0–255) to degrees (0°–180°); (4) convert correlation coefficient from byte value (0–255) to 0–1; and (5) decompress the L- and P-band polarimetric data and synthesize them into HH (horizontal transmitted and horizontal received), VV (vertical transmitted and vertical received), and HV (horizontal transmitted and vertical received), and TP (total power) images (floating point data).

Speckle removal algorithms have been applied to all resulting images (C-VV, L-VV, L-HH, L-HV, L-TP, P-VV, P-HH, P-HV, and P-TP). The gamma maximum a posteriori filter (Lopes et al., 1990) with a moving window of 7 × 7 pixels provided the best performance in terms of removing the speckle noise while preserving the polarimetric properties, radiometric information, and edge features (Xie et al., 2004).

The two Landsat 7 ETM+ images employed in this study are path 33 / row 38 of 14 October 1999 and path 31 / row 39 of 30 September 1999 in the El Paso and Valentine regions, respectively. We first georeferenced the ETM+ images using 1:24,000 digital line graph (DLG) data available from the U.S. Geological Survey through the Web site http://edc.usgs.gov/geodata/. These DLG data are in the Universal Transverse Mercator (UTM) projection system with the North American Datum (NAD) of 1927 (NAD27). Subsequently, we coregistered the radar images to the same coordinate system using the ETM+ panchromatic band 8 (15 m resolution) as a base image. The root-mean-square (RMS) errors of all registrations are less than one pixel.

Destriping

Striping (banding) noise was also commonly found in most AIRSAR images. Thus, in order to fuse the radar image with the Landsat ETM+ images, we attempted to remove the striping noise. Generally, (1) there is little or no banding in the C-band imagery; (2) there is always banding in the L-HV and/or L-TP images that is parallel to the azimuth direction, located in the

TABLE 1. AIRSAR/TOPSAR FLIGHT LINE PARAMETERS

Flight line ID	Fight elevation (m)	Flight direction	Look direction	Area covered (km)	Near look (°)	Far look (°)
TS0637	8547	SE → NW	SW	21.3 × 63.2	21.5	71.2
TS0651	8419	SE → NW	SW	20.6 × 63.5	28.1	71.8
TS0652	8380	SE → NW	SW	20.7 × 63.6	26.6	71.8
TS0676	8380	SE → NW	SW	20.6 × 33.3	27.5	71.9
TS0690	8419	SE → NW	SW	20.8 × 63.4	26.6	71.6
TS1114	7732	Near S → N	Near W	20.1 × 61.8	31.2	73.0
TS1117	7732	Near S → N	Near W	20.1 × 49.35	31.5	73.0
TS1118	7762	Near S → N	Near W	20.1 × 62.1	31.2	73.0

near-range side of the image, and is manifested as dark stripes (similar to weaker backscatter) compared with surrounding pixels; and (3) there is always banding in the P-band images that is nearly perpendicular to the azimuth direction, located in the far-range side of the image, and is manifested as bright stripes (similar to stronger backscatter) compared with surrounding pixels. Figure 2 shows an example of banding in the P-TP image (left) and L-HV image (right).

Other important observations about banding are that it is not perfectly parallel to either the azimuth or the range direction and that the intervals between adjacent stripes are variable (Fig. 2). Thus, the banding is less structured other than truly systematic noise. Baghadai et al. (2001) suggest that the banding is caused by nadir ambiguity and by uncertainties in the range antenna pattern's extreme incidence angles (near range and far range). We tested some commonly used destriping methods, including low-pass filtering, moment-matching filtering (Gadallah et al., 2000; Wegener, 1990), radiometric equalization filtering (Algazi and Ford, 1981; Corsini and Diani, 2000; Poros and Peterson, 1985), finite-impulse response filtering in the frequency domain (Chen et al.,. 2003b), and line-based cross-smoothing (Albani and Klinkenberg, 2003). However, none of these approaches were successful because of the random nature of the stripes.

We present a new method that we refer to as the combined principal component analysis (CPCA) to remove the stripes. The principal component analysis (PCA) method reduces the amount of data by eliminating redundancy, resulting in fewer, more-effective components. The PCA components created have no physical meaning but statistically describe the total variance. The first principal component (PC1) describes most of the variance (Goyal et al., 1999) and is usually noise-free. In this study, we first performed principal component analysis (PCA) on nine bands (C-VV, L-VV, L-HH, L-HV, L-TP, P-VV, P-HH, P-HV, P-TP). We visually checked banding in the nine principal components (PCs). We found that the first two PCs contain nearly 95% of the critical information contained in the nine bands, and PC1 does not contain any striping noise. Then, we applied PCA to the four L bands and four P bands separately. We found that only PC1 of the L-band data does not contain striping noise. Thus, we created a red-green-blue (RGB) false-color image by using the PC1 of the nine bands, PC1 of the L-band data, and the C-VV band, which does not contain banding noise. The result is a very useful image for visual interpretation. The PC1 image from the nine AIRSAR bands contains ~90% of their information and was used for the data fusion discussed below.

Data Fusion

Data fusion was conducted in three steps. The first step involves principle component analysis of the six 30 m bands (bands 1, 2, 3, 4, 5, and 7) of the Landsat ETM+ image covering the same area as the radar image. We then created a false-color image using the first three principal components (PC1, PC2, and

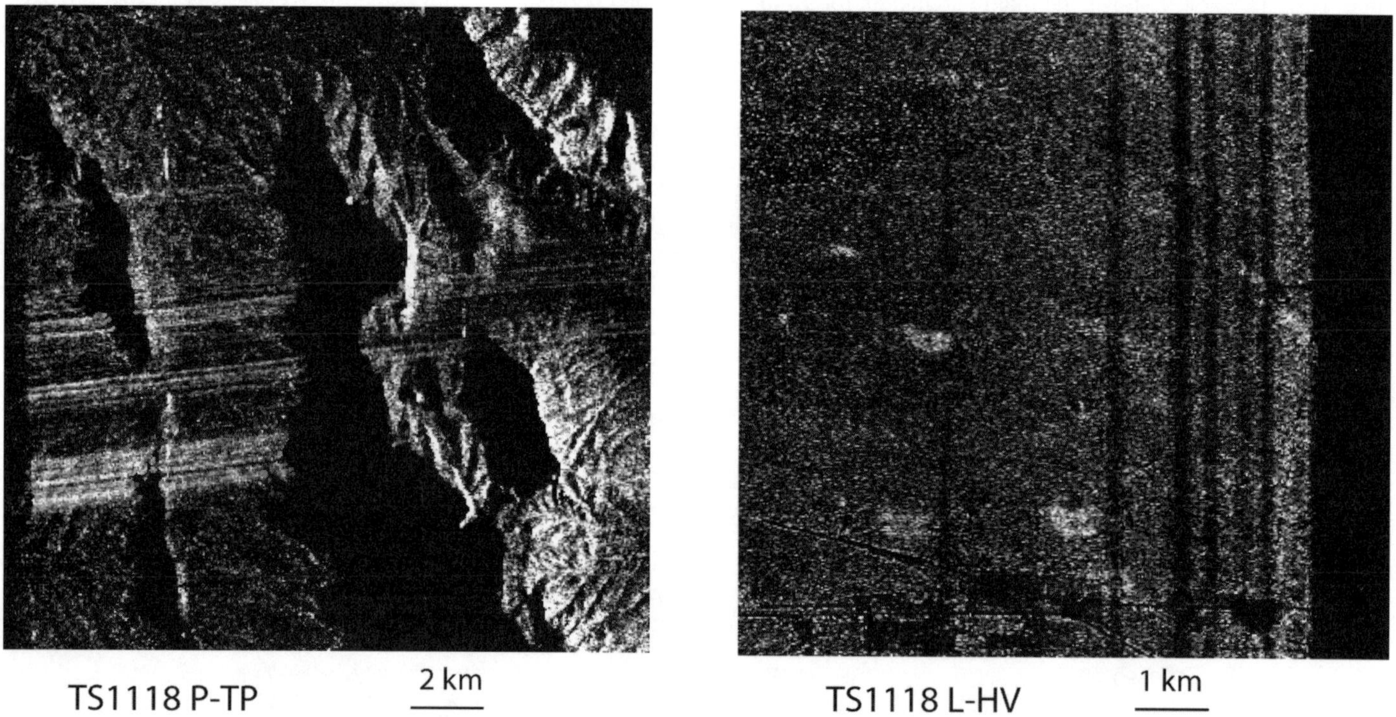

Figure 2. Banding in TOPSAR P-TP band (left) where the banding is located in far range nearly perpendicular to the azimuth direction, and L-HV band (right) where the banding is located in near range nearly parallel to the azimuth direction.

PC3). In the second step, we (1) transformed the above false-color image to HSV (hue, saturation, value) space, (2) replaced the V (value) component with the 15 m panchromatic band, and then (3) transformed back to RGB color space and obtained a fused (routine-sharpened) ETM+ image with 15 m spatial resolution. In the third step, we (1) transformed the sharpened ETM+ image to HSV space, (2) replaced the V (value) component with the first principal component (PC1) previously generated from nine AIRSAR data (10 m resolution), and then (3) transformed the result back to RGB color space and obtained a fused AIRSAR/ETM+ image with 10 m resolution.

RESULTS AND DISCUSSION

We utilized the preprocessed AIRSAR and ETM+ images, destriped false-color radar image, and fused radar/ETM+ image to (1) study the radar polarimetric properties of ground objects, (2) explore the penetration depth of different radar bands, (3) compare the differences between AIRSAR, ETM+ imagery, and fused imagery, and (4) map fault scarps.

Penetration Abilities and Roughness Indications

It is well known that the penetration depth increases as the wavelength of radar bands increases from the C band to L band to P band. The penetration of the C band is very limited. The L band typically has penetration depth ranging from several centimeters on sandy and small pebble alluvial surfaces to 2 m in dry, wind-blown sand (McCankey et al., 1982; Schaber, 1999; Schaber and Breed, 1999). The P band has more penetration ability than the L band. Our radar image (TS0690) in the Valentine area (Fig. 1) illustrates the different penetration abilities and surface roughness at different bands (Fig. 3). In Figure 3, the top panel of C-VV, L-VV, and P-VV images shows the town of Valentine and its surroundings; the bottom panel shows another area ~6 km northwest of Valentine. Flags identify two prominent linear features whose signature brightens from the C- to L- to P-band images, while the linear feature is totally invisible in the C-VV image of the bottom panel. However, the linear features are totally invisible in the C-VV image of the bottom panel. This suggests that the linear objects are buried, because the C band has very limited penetration ability. Our field investigation of the area confirmed that these features are buried water pipelines (~2 m deep).

The radar image in Figure 3 also manifests the surface roughness. Radar backscattering is a function of surface roughness and the wavelength of the radar signal. The surface roughness is measured in centimeters and is determined by the dimensions of the textural features relative to the signal wavelength (Sabins, 1996). Whether a surface is seen as smooth, intermediate, or rough in a radar image completely depends on the radar beam depression angle and radar wavelength (equations 1–3):

Smooth: $$h \le \frac{\lambda}{25 \sin\gamma} \quad (1)$$

Intermediate: $$h = \frac{\lambda}{8 \sin\gamma} \quad (2)$$

Rough: $$h \ge \frac{\lambda}{4.4 \sin\gamma} \quad (3)$$

where λ is the radar wavelength (in cm), γ is the radar beam depression angle, and h (in cm) is the local height of objects or local surface relief.

From the above equations, it is clear that, for a specific feature height and depression angle, the shorter the wavelength of the radar signal, the more likely it is that the feature will appear in a radar image. For example, the linear feature in the images (Fig. 3) indicated as crosses is the Union Pacific railroad, which has the property of a corner reflector producing high signal backscattering in the C band and L band (rough) but produces little or no backscattering on the P band (intermediate or smooth). The stars indicate the town of Valentine, where the buildings act as corner reflectors, and small streets are smooth and thus dark. The buildings are very clear on the C band, moderately clear on the L band, and barely visible on the P band. This shows that the C-band image, due to its shorter wavelength, contains more spatial detail than the P-band image. This is also very clear in the bottom panel of Figure 3, in which the C-band image clearly shows the rough Quaternary alluvial-fan sediments (bright), while the young, smooth axial stream sediments are dark (lower middle of the image). However, only the northern part of the area is rough (bright) in the L-band image, where some details in the drainage pattern could be subsurface features. The P-band image shows some of the features but is otherwise smooth (dark). These comparisons of three images in the bottom panel of Figure 3 show that the surface is rougher in the northern portion of the scene than that in the south. The northern portion consists of the proximal parts of alluvial fans with boulders at the surface, in contrast to the small cobbles to the south.

Comparison of Radar and ETM+ Images

Figure 4 shows the coregistered false color of radar (left) and ETM+ (right) images of the same areas as in Figure 3. The radar images are subscenes of the despeckled and destriped image produced previously, and the ETM+ images (path 31 / row 39, 30 September 1999) are subscenes using bands 7, 4, and 2 as red, green, and blue, respectively. Obviously, the radar and optical images are very different. The buried water pipelines (flags) are clearly seen on the radar image and are not visible on the ETM+ image; the Union Pacific railroad (crosses) is clearly seen in both radar and ETM+ images; and U.S. Highway 90 is seen well in the ETM+ image and fairly well in the town area on the radar image due to the distinct contrast of dark (roads) and bright (buildings). However, outside the town, the highway is not seen in the radar images (Figs. 3 and 4) at all. The highway is very smooth and produces no or little backscattered signal to the radar

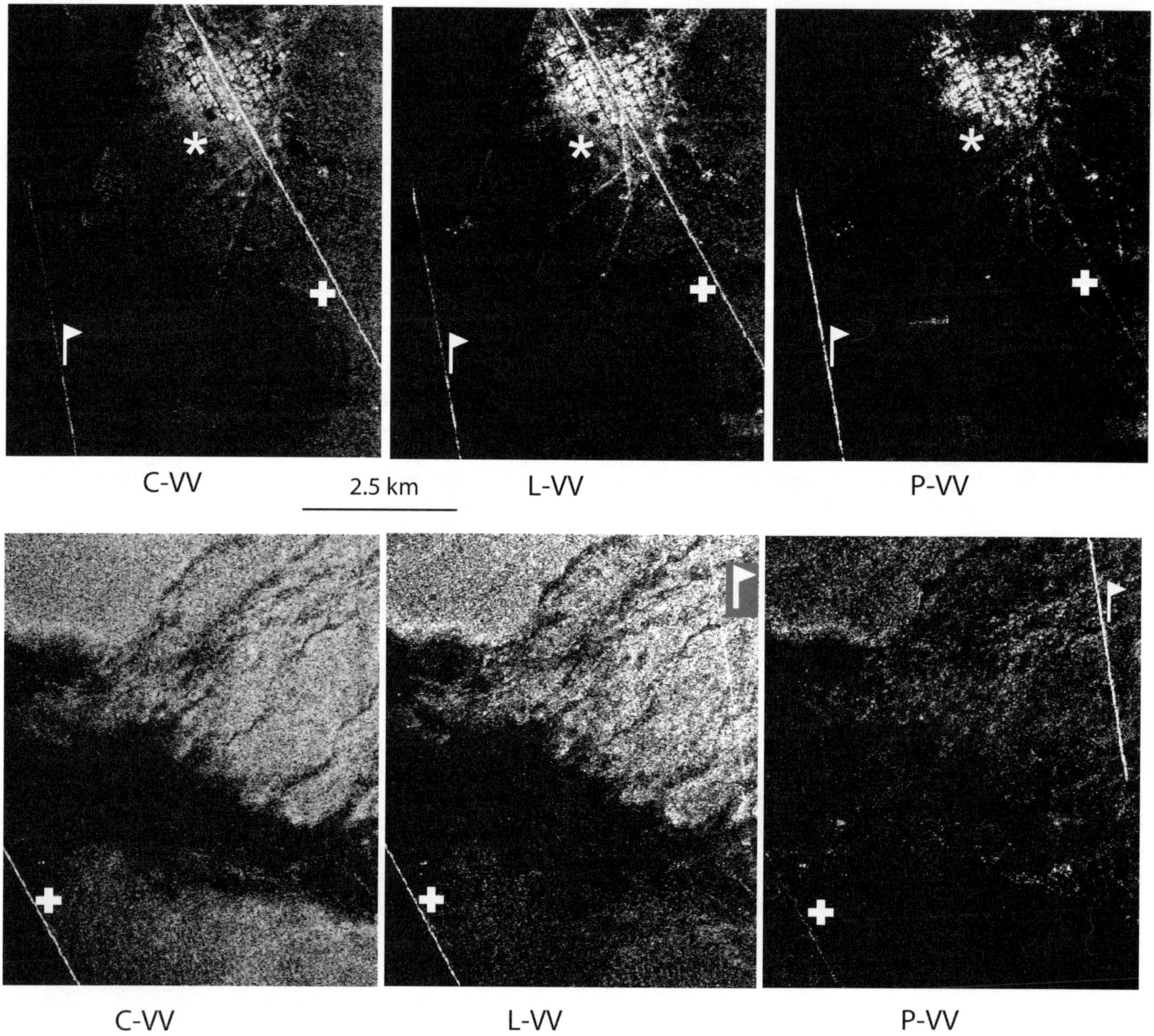

Figure 3. Portions of TS0690 image showing the differences in penetration ability and feature roughness. The top panel includes the town of Valentine, Texas, and the bottom panel is located ~6 km northwest of Valentine.

sensor (dark), while the railroad acts as a corner reflector and has strong backscatter (bright) in the C and L bands.

Mapping Active Faults

We also used AIRSAR images to map the East Franklin Mountains fault and relative small normal faults within the Hueco bolson (basin) near El Paso, Texas (Fig. 1). Figure 5 shows the georeferenced C-VV image of flight line TS1118 (Fig. 1 and Table 1). Northeast El Paso and the Franklin Mountains are visible in the bottom portion of the image, and White Sands Missile Range headquarters and the Organ Mountains are visible in the top portion of the image. The trace of the East Franklin Mountains fault interpreted from fault scarps is identified on the image. The fault scarps face toward the east and have a very bright signature on the image; thus one can easily identify them. The East Franklin Mountains fault is the boundary between tilted fault block of the mountains to the west and the Hueco bolson (graben) to the east, which is major groundwater source for the El Paso–Juarez region. There are a number of small active

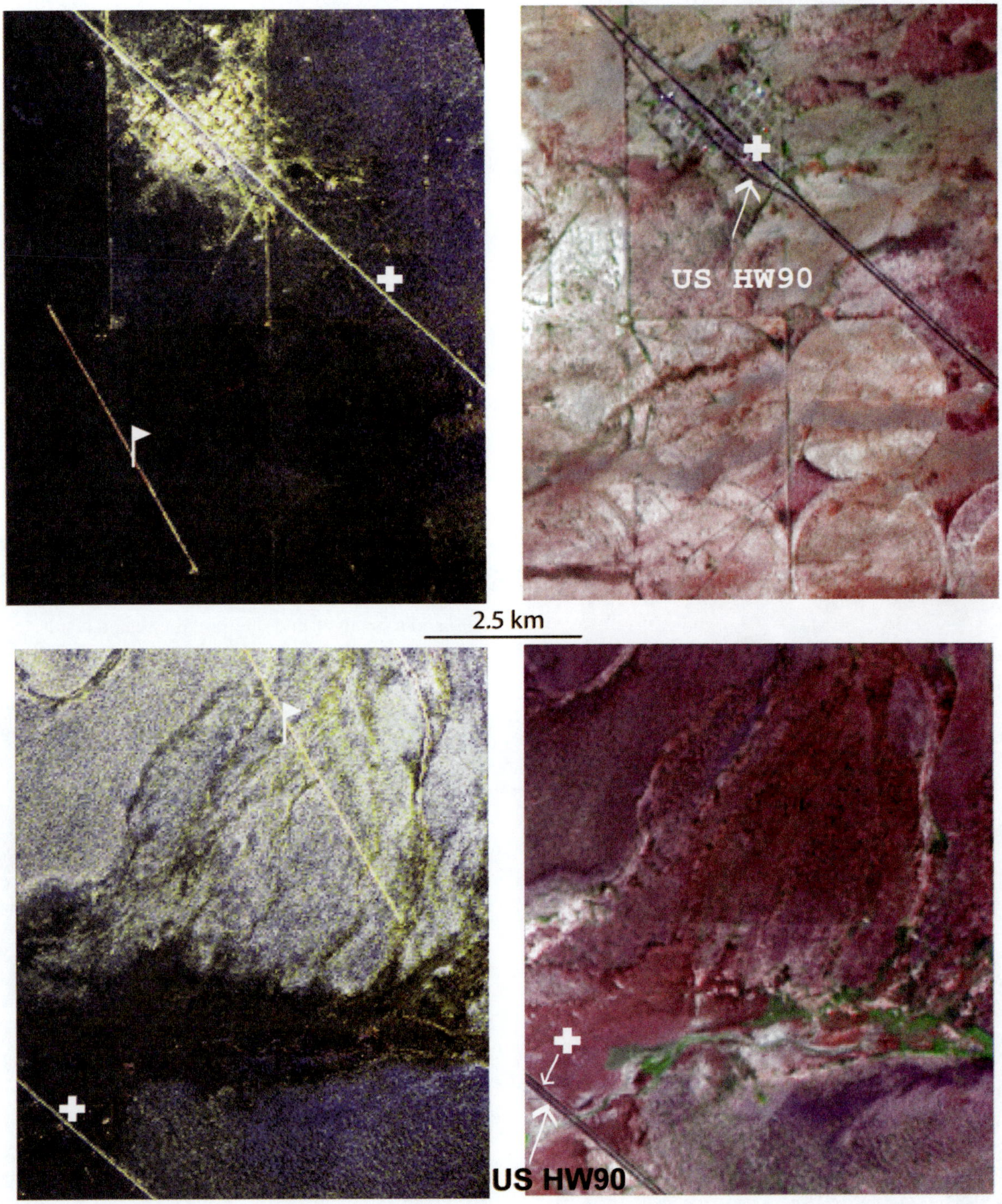

Figure 4. Comparison of coregistered AIRSAR image (left) and ETM+ image (742 of path 31 / row 39 of 30 September 1999). Features are buried water pipeline (flags), railroad (crosses), and highway. The images are of the same area as Figure 3. USHW90—U.S. Highway 90.

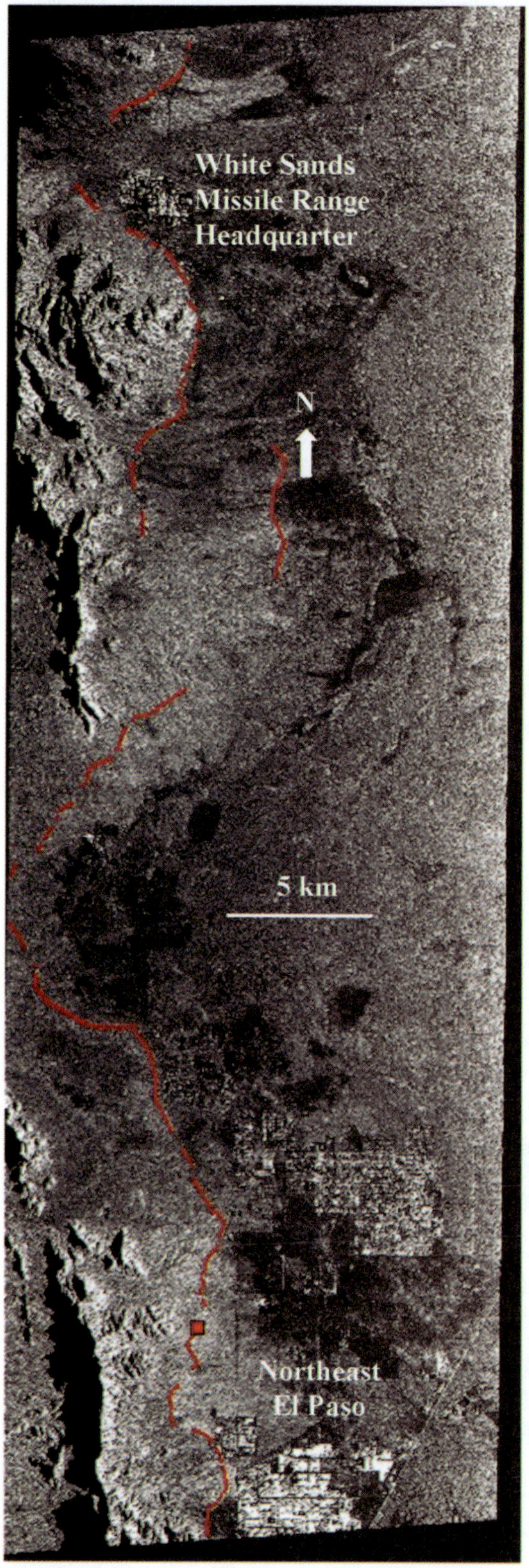

Figure 5. East Franklin Mountains fault interpreted from the AIRSAR C-VV image (TS1118, 20.1 km × 60.1 km). The square in the bottom portion of the image is the location of the trench studied by Keaton and Barnes (1996).

normal faults inside the Hueco bolson that attest to the continuing tectonic activity associated with the Rio Grande rift. These small normal faults have been identified using both AIRSAR and ETM+ images (Fig. 6) by Xie (2002). The combination of AIRSAR and ETM+ data allows us to easily map the faults in this region and produce images that clearly depict their locations to decision makers.

Our field investigations proved that the interpretations from the remote sensing data closely match the position of the East Franklin Mountains fault. However, the Hueco bolson area is relatively flat, and it is difficult to see and trace the fault scarps in the field, but they are very clear in the images (Fig. 6). Our interpretations of remote sensing images were also compared with published maps. Figure 7 shows that the position of the East Franklin Mountains fault located by remote sensing data almost exactly matches the geologic mapping of Scherschel (1995).

Interpretation and Analysis of the Fusion Result

The examples discussed above underscore the fact that both radar and optical images provide distinct and valuable information. However, visually comparing images from two different sensor systems does not reveal spatial relations with precision and does not quantitatively integrate the data. This was the key reason to drive us to undertake the data fusion effort presented here and to analyze and interpret the fusion result.

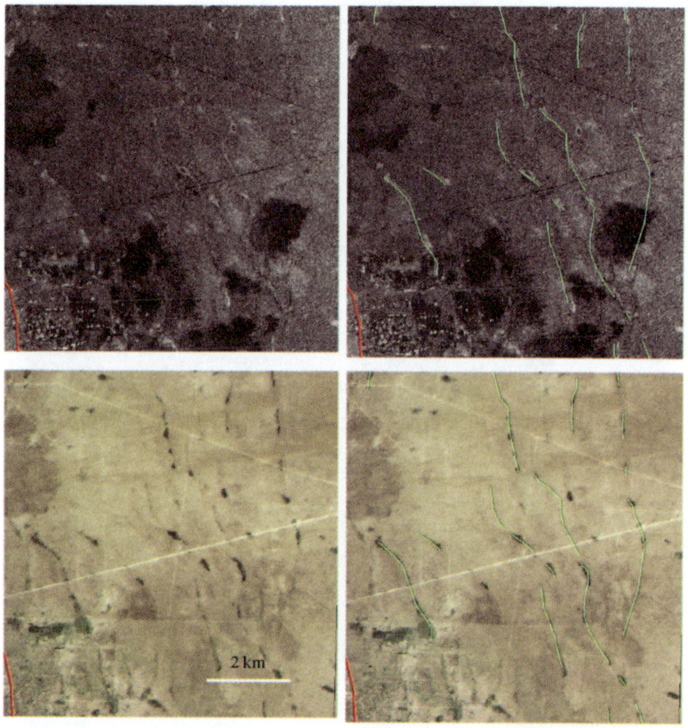

Figure 6. Active normal faults within the Hueco bolson identified by both AIRSAR (upper panel, C-VV of TS1118) and ETM+ images (lower panel, path 33 / row 38 of 14 October 1999).

To evaluate the data fusion results, we visually compared the false-color AIRSAR image, the sharpened ETM+ image, and the fused AIRSAR/ETM+ image. Figure 8 presents an example showing three images in the left panel—false-color radar (top), sharpened ETM+ (middle), and fused radar/ETM+ (bottom)—and two subscenes (A and B) in the middle and right panels, respectively. Firstly, the sharpened ETM+ image is colorful compared with the false-color radar image, on which it is not easy to interpret vegetation cover and surface features. The features manifested in different colors in the ETM+ image are transferred to the fused radar/ETM+ image. For example, the green color in the ETM+ image indicates vegetation cover, which is retained in the fused image. Secondly, the surface roughness, textural features, and even subsurface features recorded by the radar image are also shown in the fused image. Thus, the fused image includes land-cover information as well as the texture and subsurface information. This makes the fused image more informative and more interpretable than either the ETM+ or the radar image alone. For example, the buried water pipelines, the roads and railway, and roughness and textural manifestations of the terrain and exposed geologic features are all integrated into the fused image.

The advantages of the fused result are even better demonstrated by detailed examination of subscenes A and B. The magenta region in subscene A of the fused image, indicated by a green dot, was not identifiable before data fusion. This region matches the outcrop of the Tertiary Chambers Tuff, and the magenta is probably due to a mixture of the lithologic texture and vegetation associated with the outcrop. In fused subscene B, additional detail and terrain differentiation in the large area shaped like the human heart (identified by the red star) that dominates the central portion of the subscene is clear as a result of the data fusion (Fig. 8). The rocks of this heart-shaped area are Tertiary basalt. There is a small and bright hill (also Tertiary basalt) that produces a feature shaped like an "8" located just west of the heart-shaped feature, indicating NE-facing high relief that produces a strong radar signal return. However, on the ETM+ image, this area is just a dark spot. Furthermore, there are many linear features (fence lines) marked by red flags that appear on subscene B of the radar and fused images, but not on the sharpened ETM+ image. Of most interest to us is the linear feature marked by two arrows in subscene A that is best seen on the fused image. This linear feature is the fault scarp of the middle section of the Mayfield fault (Collins and Raney, 1994; Doser, 1987) that is probably associated with the 6.4 magnitude Valentine earthquake that occurred on 16 August 1931 (Dumas et al., 1980; Doser, 1987). This earthquake is by far the largest known historical earthquake to occur in Texas.

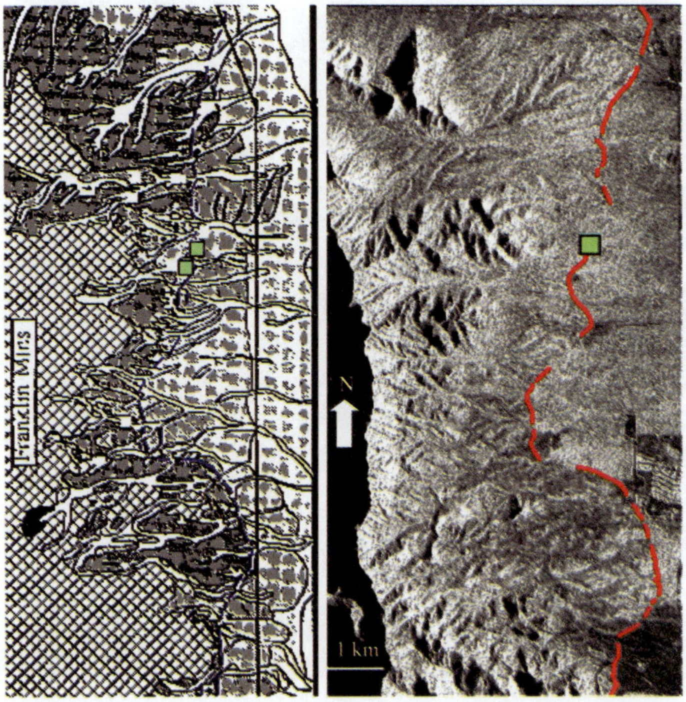

Figure 7. Comparison of a scanned published map (left) from Scherschel (1995) and the interpretation result from the AIRSAR image (right). Squares are the trench study sites (Keaton and Barnes, 1996).

CONCLUSIONS

This paper describes a processing methodology for NASA/JPL's AIRSAR data that includes evaluating the distribution of banding noise and removing it effectively using our combined principal components analysis (CPCA) method. We examined the importance of radar polarimetric properties as a function of texture, surface roughness, and penetration abilities. Our results indicate, as expected, that the C band has the finest spatial detail and the P band has the greatest penetration. Our data fusion results showed that while both radar and optical images contain distinct and valuable information, the fused radar/optical image is even more informative and more interpretable than either the ETM+ or the radar image alone. In particular, the combination of AIRSAR and ETM+ data provides a method for fast, effective mapping of fault scarps.

ACKNOWLEDGMENTS

This work was supported by NASA under cooperative agreement NCC5-209 and through acquisition of the AIRSAR/TOPSAR data. We would like to thank Rebecca Dodge and Mohamed Abdelsalem for their helpful reviews.

REFERENCES CITED

Abdelsalam, M.G., Thurmond, J.B., and Thurmond, A.K., 2002, Resolving lithologic and temporal contrasts of lava flows using optical-radar fusion: ERTE' ALE volcano, Ethiopia: Geological Society of America Abstracts with Programs, v. 34, no. 6, p. 451.

Albani, M., and Klinkenberg, B., 2003, A spatial filter for the removal of striping artifacts in digital elevation models: Photogrammetric Engineering and Remote Sensing, v. 69, p. 755–765.

Algazi, V.R., and Ford, G.E., 1981, Radiometric equalization of nonperiodic striping in satellite data: Computer Graphic Image Processing, v. 16, p. 287–295.

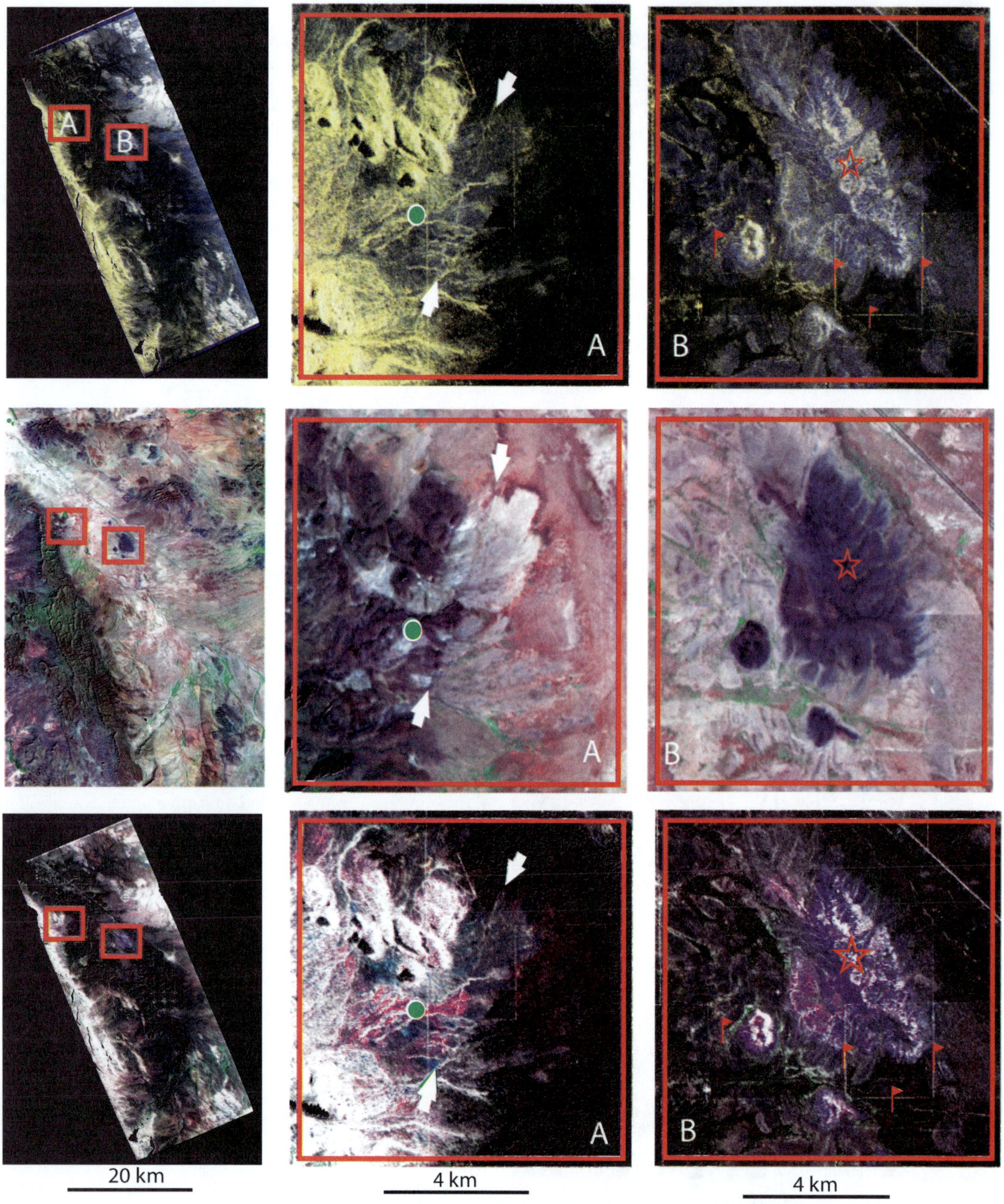

Figure 8. False-color radar image of TS0690 (top), sharpened ETM+ image (middle), and fused radar/ETM+ image (bottom) in the left panel. Two subscenes (A and B) in the middle and right panels, respectively. Symbols correspond to features discussed in the text.

Baghadai, N., Berneir, M., Gauthier, R., and Neeson, I., 2001, Evaluation of C-band SAR data for wetlands mapping: International Journal of Remote Sensing, v. 22, p. 71–88.

Bi, L., Dai, X.R., and Sun, Q.Y., 1999, Construction of compactly supported M-band wavelet: Applied and Computational Harmonic Analysis, v. 6, p. 113–131.

Carper, W.J., Lillesand, T.M., and Kiefer, R.W., 1990, The use of intensity-hue-saturation transform for merging SPOT panchromatic and multispectral image data: Photogrammetric Engineering and Remote Sensing, v. 56, p. 459–467.

Chen, C.M., Hepner, G.F., and Forster, R.R., 2003a, Fusion of hyperspectral and radar data using HIS transformation to enhance urban surface features: International Society for Photogrammetry and Remote Sensing Journal of Photogrammetry and Remote Sensing, v. 58, p. 19–30.

Chen, J., Shao, Y., Guo, H., Wang, W., and Zhu, B., 2003b, Destriping CMODIS data by power filtering: IEEE Transactions on Geoscience and Remote Sensing, v. 41, p. 2119–2124.

Collins, E.W., and Raney, J.A., 1994, Tertiary and Quaternary tectonics of Hueco bolson, trans-Pecos, Texas, and Chihuahua, Mexico, *in* Keller, G.R., and Cather, S.M., eds., Basins of the Rio Grande rift: Structure, stratigraphy, and tectonic setting: Geological Society of America Special Paper 291, p. 265–283.

Corsini, G., and Diani, M., 2000, Stringing removal in MOS-B data: IEEE Transactions on Geosciences and Remote Sensing, v. 38, p. 1439–1446.

Doser, D.I., 1987, The August 16, 1931, Valentine, Texas, earthquake: Evidence for normal faulting in west Texas: Bulletin of the Seismological Society of America, v. 77, p. 2005–2017.

Dumas, D.B., Dorman, H.J., and Latham, G.V., 1980, A reevalution of the August 16, 1931, Texas earthquake: Bulletin of the Seismological Society of America, v. 70, p. 1171–1180.

Filberti, D.P., Marsh, S.E., and Schowengerdt, R.A., 1994, Synthesis of imagery with high spatial and spectral resolution from multiple image sources: Optical Engineering, v. 33, p. 2520–2528.

Foody, G., 1988, Crop classification from the airborne synthetic aperture radar data: International Journal of Remote Sensing, v. 9, p. 655–668.

Gadallah, F.L., Csillage, F., and Smith, E.J.M., 2000, Destriping multidetector imagery with moment matching: International Journal of Remote Sensing, v. 21, p. 2505–2511.

Gillespie, A.R., Kahle, A.B., and Walker, R.E., 1987, Color enhancement of highly correlated images, II: Channel ratio and "chromaticity" transformation techniques: Remote Sensing of Environment, v. 22, p. 343–365.

Goyal, S.K., Seyfreind, M.S., and O'Neill, P.E., 1999, Correction of surface roughness and topographic effects on airborne SAR in mountains rangeland areas: Remote Sensing of Environment, v. 67, p. 124–136.

Gross, H.N., and Schott, J.S., 1998, Application of spectral mixture analysis and image fusion techniques for image sharpening: Remote Sensing of Environment, v. 63, p. 85–94.

Haack, B.N., Solomon, E.K., Bechdol, M.A., and Herold, N.D., 2002, Radar and optical data comparison/integration for urban delineation: A case study: Photogrammetric Engineering and Remote Sensing, v. 68, p. 1289–1296.

Keaton, J.R., and Barnes, J.R., 1996, Paleoseismic evaluation of the East Franklin Mountains fault, El Paso, Texas: Final report to the USGS for grant number 1434-94-G-2389, 45 p.

Keller, G.R., and Baldridge, W.S., 1999, The Rio Grande rift: A geological and geophysical review: Rocky Mountain Geology, v. 34, p. 131–148.

Li, J., Zhou, Y.U., and Li, D.R., 1999, Fusion of high-resolution panchromatic and multi-spectral images by using wavelet transform: China Journal of Remote Sensing, v. 3, p. 103–107 (in Chinese, with English abstract).

Liu, J.G., 2000, An evaluation of Landsat-7 ETM+ panchromatic band for spectral preserve image fusion to improve apparent spatial resolution of reflective spectral bands: Proceedings of the 14th Environmental Research Institute of Michigan (ERIM) Conference on Applied Geological Remote Sensing, Las Vegas, November 6–8, 2000, p. 151–158.

Lombardo, P., Oliver, C.J., Pellizzeri, T.M., and Meloni, M., 2003, A new maximum-likelihood joint segmentation technique for multitemporal SAR and multiband optical images: IEEE Transactions on Geoscience and Remote Sensing, v. 41, p. 2500–2518.

Lopes, A., Nezry, E., Touzi, R., and Laur, H., 1990, Maximum a posteriori filtering and first order models in SAR images: Proceedings of IEEE International Geoscience and Remote Sensing Symposium, v. 28, p. 2409–2412.

Lou, Y., Kim, Y., and van Zyl, J., 2001, The NASA/JPL airborne synthetic aperture radar system: http://airsar.jpl.nasa.gov (June 2005).

Machette, M.N., 1987, Preliminary assessment of paleoseismicity at White Sands Missile Range, southern New Mexico: Evidence for recency of faulting, fault segmentation, and repeat intervals for major earthquakes in the region: U.S. Geological Survey Open-File Report 87-444, 46 p.

McCankey, J.F., Schaber, G.G., Breed, C.S., Grolier, M.J., Haynes, C.V., Issawi, B., Elachi, C., and Blom, R., 1982, Subsurface valleys and geoarchaeology of the eastern Sahara revealed by shuttle imaging radar: Science, v. 218, p. 1004–1020.

Michelson, D.B., Lilijeber, B.M., and Pilesjo, P., 2000, Some comparison of algorithms for classifying Swedish landcover using Landsat TM and ERS-1 SAR data: Remote Sensing of Environment, v. 71, p. 1–15.

Munechika, C.K., Warnick, J.S., Salvaggio, C., and Schott, J.R., 1993, Resolution enhancement of multispectral image data to improve classification accuracy: Photogrammetric Engineering and Remote Sensing, v. 59, p. 67–72.

Poros, D.J., and Peterson, C.J., 1985, Method for destriping Landsat Thematic Mapper image—A feasibility study for an online destriping Landsat Thematic Mapper image processing system: Image Processing, v. 10, p. 69–93.

Price, J.C., 1987, Combing panchromatic and multispectral imagery from dual resolution satellite instruments: Remote Sensing of Environment, v. 21, p. 119–128.

Robinson, G.D., Gross, H.N., and Schott, J.R., 2000, Evaluation of two applications of spectral mixing model to image fusion: Remote Sensing of Environment, v. 71, p. 272–281.

Sabins, F.F., 1996, Remote sensing principals and interpretation (3rd edition): New York, W.H. Freeman and Company, 494 p.

Schaber, G.G., 1999, SAR Studies in the Yuma Desert, Arizona: Sand penetration, geology, and the detection of military ordnance debris: Remote Sensing of Environment, v. 67, p. 320–347.

Schaber, G.G., and Breed, C.S., 1999, The importance of SAR wavelength in penetrating blow sand in northern Arizona: Remote Sensing of Environment, v. 69, p. 87–104.

Scherschel, C.A., 1995, The Quaternary geologic and geomorphic framework for neotectonic analysis of the northeastern Franklin Mountains, El Paso, Texas [M.S. thesis]: College Station, Texas A&M University, 75 p.

Shettigara, V.K., 1992, A generalized component substitution technique for spatial enhancement of multispectral images using a higher resolution data set: Photogrammetric Engineering and Remote Sensing, v. 58, p. 561–567.

Shi, W., Zhu, C., Zhu, C., and Yang, X., 2003, Multi-band wavelet for fusing SPOT panchromatic and mulitspectral images: Photogrammetric Engineering and Remote Sensing, v. 69, p. 513–520.

Van der Sanden, J.J., Vachon, P.W., and Gower, J.F.R., 2000, Combing optical and radar satellite image data for surveillance of coastal waters: Proceedings of the 6th International Conference on Remote Sensing for Marine and Coastal Environments, May 1–3, 2000, Charleston, South Carolina, 8 p.

Wegener, M., 1990, Destriping multiple detector imagery by improved histogram matching: International Journal of Remote Sensing, v. 11, p. 859–875.

Wisutmethangoon, Y., and Nguyen, T.Q., 1999, A method for design of Mth-band filters: IEEE Transactions on Signal Processing, v. 47, p. 1669–1679.

Xie, H., 2002, Application of integrated remote sensing and GIS technologies to geological, agricultural, water and environmental issues in far west Texas [Ph.D. thesis]: El Paso, University of Texas, 248 p.

Xie, H., Keller, G.R., and Langford, R.P., 2004, AIRSAR/TOPSAR Image Processing and Interpretation: An application example in far west Texas, USA: Geocarto International, v. 19, p. 33–41.

Zhang, Y., 1999, A new merging and data fusion by means of discrete two-dimension wavelet transform: Journal of the Optical Society of America, v. 12, p. 1834–1841.

Zhou, J., Civco, D.L., and Silander, J.A., 1998, A wavelet transform method to merge Landsat TM and SPOT panchromatic data: International Journal of Remote Sensing, v. 19, p. 743–757.

Manuscript Accepted by the Society 18 June 2005

Geological Society of America
Special Paper 397
2006

Data fusion in geophysics: Seismic tomography and crustal structure in Poland as an example

Matthew G. Averill
G. Randy Keller
Kate C. Miller
Department of Geological Sciences, University of Texas, El Paso, Texas 79968, USA

Piotr Sroda
Institute of Geophysics, Polish Academy of Sciences, Warsaw, Poland

Tiffni Bond
Aaron Velasco
Department of Geological Sciences, University of Texas, El Paso, Texas 79968, USA

ABSTRACT

The Trans-European Suture zone in eastern Europe is a complex region that marks the suture of Phanerozoic western European terranes with the Precambrian East European craton. Because of sedimentary cover and the Baltic Sea, the nature of this suture is known primarily from geophysical studies. Seismic velocity and gravity models from earlier experiments indicate changes of crustal thickness from 28–35 km to 42–47 km across the Teisseyre-Tornquist zone from the Paleozoic platform of western Europe to the East European craton. Tectonic models suggest the presence of a Precambrian–early Paleozoic passive margin beneath the Teisseyre-Tornquist zone. The Holy Cross Mountains in southeastern Poland represent an anomalous crustal block whose origin and interaction with the East European craton is unknown. We have employed a new approach to quantitatively integrate (i.e., data fusion) industry-acquired drilling and seismic reflection data, as well as refraction data from the CELEBRATION 2000, POLONAISE '97, and Teisseyre-Tornquist zone experiments in a tomographic inversion. We find that it is feasible to integrate data of different resolutions in a tomographic inversion. Secondly, we constrained upper-crustal velocities within our study area to create a more completely resolved velocity structure. Thirdly, we modeled a uniform gravity data set to validate all available data and construct models of crustal structure across the Holy Cross Mountains region.

Keywords: Trans-European Suture zone, East European craton, Holy Cross Mountains, tomographic inversion, data fusion.

Averill, M.G., Keller, R.G., Miller, K.C., Sroda, P., Bond, T., and Velasco, A., 2006, Data fusion in geophysics: Seismic tomography and crustal structure in Poland as an example, *in* Sinha, A.K., ed., Geoinformatics: Data to Knowledge: Geological Society of America Special Paper 397, p. 153–168, doi: 10.1130/2006.2397(11).

INTRODUCTION

The term *data fusion* has been used to describe many types of data integration efforts. In this study, we present a new approach for quantitative fusion of three fundamentally different types of seismic data in a joint tomographic inversion to determine the velocity variations in Earth in three dimensions. We applied our data fusion technique to a region in Poland where we had access to a variety of data sets. Our study area was part of the Trans-European Suture zone. This geologically complex and interesting feature extends 2000 km from the North Sea to the Black Sea (Fig. 1) and has evolved as the result of several phases of continental-scale deformation (Pharaoh, 1999). In our case, the inversion produced a volume in which seismic velocity is determined for blocks whose dimensions are on the order of a few kilometers, but smaller-scale investigations should work without difficulty. Our ultimate integrated analysis also employed gravity data and a variety of geologic observations. Thus, we will briefly illustrate the final geologic implications of the tomographic result that we obtained.

The data incorporated into this study include (1) results from a series of large-scale seismic experiments (Fig. 2) that targeted the deep structure of the Trans-European Suture zone: POLONAISE '97 (POlish Lithosphere ONsets—An International Seismic Experiment) (Guterch et al., 1999), CELEBRATION 2000 (Central European Lithospheric Experiment Based on Refraction) (Guterch et al., 2001), and the Teisseyre-Tornquist zone experiment (Grad et al., 1999), (2) industry reflection seismic profiles with stacking-velocity tables acquired from Apache Oil Company, (3) industry well log data including check-shot surveys, downhole velocity surveys, and stratigraphic columns acquired from the Apache Corporation, (4) gravity from the Polish onshore gravity database provided by Stanislaw Wybraniec at the Polish Geological Institute, and (5) geologic data from previous regional studies and the Geologic Atlas of Poland. The goals of this study are to integrate the available seismic (travel-time) data into a three-dimensional tomographic inversion to create a model for the velocity structure within this region and to provide a better understanding of the origin and tectonic evolution of this and adjacent crustal blocks in relation to the eastern Trans-European Suture zone and East European craton. In turn, the seismic data are correlated with gravity and geologic data to constrain structural models across the study area.

Geologic Background

The Trans-European Suture zone is composed of a series of sutures created during the amalgamation of Phanerozoic European terranes with Proterozoic eastern European terranes (Guterch et al., 1999). This zone extends 2000 km from the North Sea to the Black Sea (Fig. 1) and has evolved as the result of Caledonian, Variscan, and Alpine composite orogenic cycles (Pharaoh, 1999). The Caledonian and Variscan orogenies resulted in accretion of Phanerozoic European terranes to the rifted southwestern margin of Baltica (East European craton) from 490 Ma to 290 Ma (Keller and Hatcher, 1999). Phanerozoic Europe consists of an unknown number of microcontinent-sized crustal blocks possibly derived either from tectonic transport around the margin of Baltica (having Baltica affinity) or from the Gondwanan supercontinent margin (Pharaoh, 1999). Terrane boundaries are generally poorly exposed oceanic sutures, primarily marked by geophysical lineaments and defined using faunal affinities (Pharaoh, 1999). The Teisseyre-Tornquist zone marks the eastern boundary and southeastern extent of the Trans-European Suture zone adjacent to Baltica (East European craton), as well as the coalescence of several tectonic features (Fig. 1). Trending northwest-southeast from the Baltic Sea to the Black Sea, the Teisseyre-Tornquist zone is also thought of as the contact zone between the eastern and western European platforms (Lamarche et al., 1999). This zone and its adjacent terranes are marked by distinct upper-crustal features as well as deeper structure constrained by geophysical techniques such as refraction seismic, gravity, magnetic, and heat flow (e.g., Guterch et al., 1999).

Large-Scale Seismic Experiments

The Teisseyre-Tornquist zone experiment targeted a key portion of the Trans-European Suture zone adjacent to the axial part of the Polish trough (Grad et al., 1999) (Fig. 2). Along this profile, a thick (up to 20 km) layer of sediments and metasediments was found, and the thickness of Earth's crust increases from northwest to southeast from 35 to 41 km. In turn, profiles from the POLONAISE '97 project were designed to parallel and cross this previous experiment to further constrain the crustal structure southeastward along the Trans-European Suture zone.

In May of 1997, the POLONAISE '97 (Guterch et al., 1999) seismic refraction / wide angle reflection experiment took place predominantly in Poland and included extensions of the P4 profile into Germany and Lithuania. This experiment utilized 800 recording instruments to record along five profiles (Fig. 2). The layout of the four main interlocking profiles was designed to produce significant three-dimensional coverage, and tomographic analysis has been undertaken on these data in both two (Guterch et al., 1999) and three dimensions (Sroda et al., 2002). These results indicate the presence of a deep, up to 20 km, asymmetric basin consisting of sedimentary to metasedimentary or volcanic rocks with velocities of <6.1–6.2 km/s overlaying a higher-velocity (6.7–7.5 km/s) lower crust with the base varying significantly in depth across the Trans-European Suture zone.

The goals of CELEBRATION 2000 (Guterch et al., 2001) were to (1) investigate the deep structure of the southwestern margin of the stable continental interior of Europe, (2) delineate the major terranes that compose western Europe in the region, and (3) construct a three-dimensional model of the lithospheric structure and evaluate and develop geodynamic models for the evolution of the area. This experiment produced a very large data set by recording along a series of interlocking profiles whose total length was ~8900 km. These profiles were designed to form

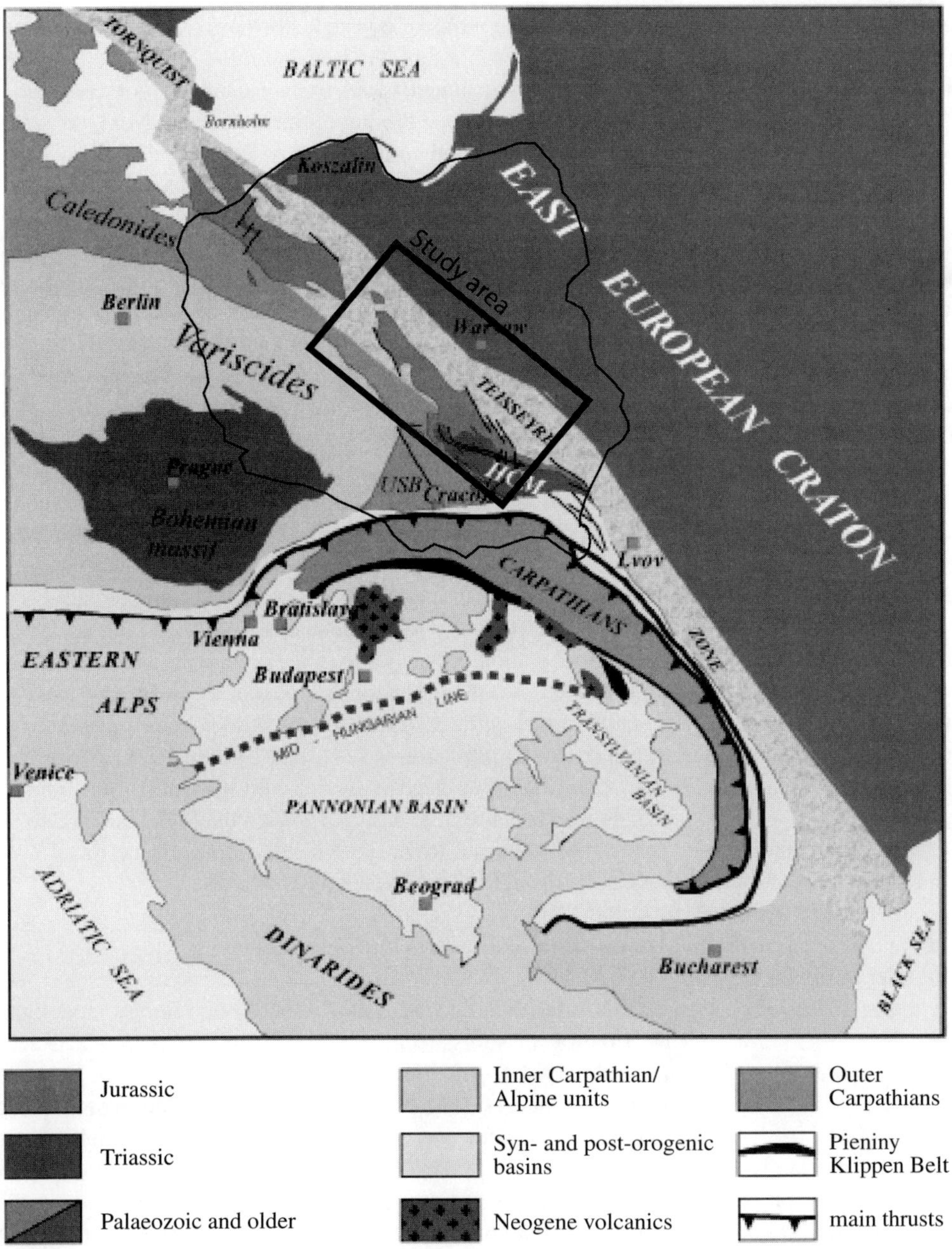

Figure 1. The Tornquist-Teisseyre zone and adjacent tectonic features of the eastern European Trans-European Suture zone. Rough outline of Poland in bold. HCM—Holy Cross Mountains; USB—Upper Silesian block. (Modified from Guterch et al., 1999.)

a three-dimensional array to augment the 5400 km of traditional profiles (Fig. 2). The analysis of data from this experiment and fusion of these data with those from other sources was the primary motivation of this study.

TOMOGRAPHIC INVERSION

Seismic tomography is a relatively new approach to determine the velocity structure of the earth (e.g., Nolet, 1987). In this approach, velocity anomalies are detected by analyzing the travel times of many seismic rays through a region of the earth. Usually some reference earth model (velocity structure) is assumed, and variations from this model are determined and typically expressed as percent change in velocity and displayed in color. This approach is similar to that used in medicine to form images from X-rays or ultrasound, such as the CAT scan (computerized axial tomography), and is substantially different from the traditional approaches used in seismic reflection and refraction studies (e.g., Yilmaz, 2001). The situation in the earth is more complex because seismic rays bend as they travel and because the distribution of seismic recorders and seismic sources is not uniform. As in most other geophysical techniques, tomography can be applied on either a local or global scale. Thus, seismic tomography is an active area of research that is producing interesting new images of the velocity variations in the earth at a variety of scales.

The basic concept behind seismic tomography is that the travel time of a body wave traveling through the earth is

$$T = \int ds \,/\, V(s)$$

where T is the travel time, $V(s)$ is the functional expression for the velocity variation along the travel path, s is the travel path or raypath, and the integral is along s. It is important to note that the above equation is nonlinear due to its dependence on the slowness field (reciprocal of velocity) and is generally linearized by inverting travel-time residuals (dt) for changes in slowness (du).

The essence of seismic tomography is to use a large number of observations for waves that have traveled a variety of paths in order to mathematically determine a velocity model. The most commonly employed approach is to approximate the earth by a three-dimensional mesh of blocks (e.g., Hole, 1992). The travel-time integral above is used to sum the travel times through each block intersected by a raypath. Many observations then lead to a large system of linear equations where the unknowns are the velocities of the blocks. The quality of the solution depends on the number of observations and their accuracy, the distribution of raypaths, and the degree to which the block model can actually approximate the true velocity structure.

In this study, we were in the unique position of having access to a considerable amount of additional data in the form of borehole check-shot surveys and stacking velocities derived from the processing of seismic reflection data. Thus, we devised a method to include these data in the tomography in order to obtain better constraints on the velocity structure of the upper 20 km of crust in the study area.

Seismic Tomography Algorithm

The tomographic approach that was used for this study was a three-dimensional seismic travel-time tomography or back-projection method for determining three-dimensional velocity structure from first-arrival travel-time data (Hole, 1992). It uses a linear tomographic inversion algorithm that minimizes computational time, provides a relatively high-resolution model, and can handle structures with large velocity contrasts (Hole, 1992). It is roughly a two-step process: (1) First-arrival travel times are calculated for each node in the grid by solving the eikonal equation (an approximation to the wave equation) with a finite-difference method (Vidale, 1988), and (2) travel-time residuals are inverted for changes to the velocity model. As a starting model, we used a three-dimensional expansion of a simple one-dimensional initial velocity distribution for the region that was based on known geology and previous seismic studies. This model is then expanded to a uniform grid with equal spacing in all three dimensions.

Once travel times have been calculated through the initial model, the second or inverse step begins. The model is parameterized in terms of constant slowness (reciprocal of velocity) cells. The goal of this step is to invert the differences between observed and calculated travel times (residuals) for changes in slowness that minimize the travel-time residuals. Next, the slowness of each cell in the initial model is determined by the average of the slowness of the cell's eight bounding nodes (Zelt and Barton, 1998). The computations produce two sums for each cell that are needed in order to update the slowness. The first is the raypath-averaged slowness perturbation for each ray passing through the cell, which equals the travel-time residual (observed minus calculated) of the ray divided by the ray's total length through the cell. The second sum is equal to the number of rays passing through each cell. Once these calculations have been performed for each ray that traverses the model space, the first sum is divided by the second to obtain the slowness update for each cell through which at least one ray passes. The update is then added to the model and the whole process starts again once a moving average filter is applied to stabilize or smooth the updated model. The moving average filter is input into the program as a starting parameter and works most efficiently when its size is decreased after each run. For a more complete discussion of the computational method, see Hole (1992). A typical run consists of four to seven iterations. As a result, models subject to smaller smoothers define progressively smaller wavelength features (Zelt and Barton, 1998). Since it is not reasonable to update the model until the travel-time residual equals zero because of picking errors, a model is considered to be acceptable when the RMS (root-mean-square) of the travel-time residuals is roughly equal to the picking error.

Description of Travel-Time Data Sets

In order to define the boundaries of our velocity model, the distribution of shots, receivers, check-shot surveys, and stacking-velocity determinations were plotted relative to the tectonic features of interest (Fig. 3). Based on the compilation of all these data, a model space was constructed to maximize the data coverage with respect to the Trans-European Suture zone and Polish trough (rectangle in Fig. 3).

The refraction data set consisted of observations of seismic arrival times for receivers and sources that lie within the bounds of the model space. This large data set included 500 arrival times from the Teisseyre-Tornquist zone experiment (from six shots and 155 receivers), 1520 arrival times from the POLONAISE '97 experiment (from 16 shots and 253 receivers), and 5806 arrival times from the CELEBRATION 2000 experiment (from 33 shots and 616 receivers). Since the Teisseyre-Tornquist zone and POLONAISE '97 data within our model space have been analyzed in several previous studies (Grad et al., 1999; Guterch et al., 1999; Jensen et al., 1999; Sroda et al., 1999, 2002), the first-arrival times (picks) were well established. As a result, these data were obtained directly from the Institute of Geophysics of the Polish Academy of Sciences in a format appropriate for input to the inversion code. By contrast, the CELEBRATION 2000 data set has only been analyzed along individual profiles, and the results

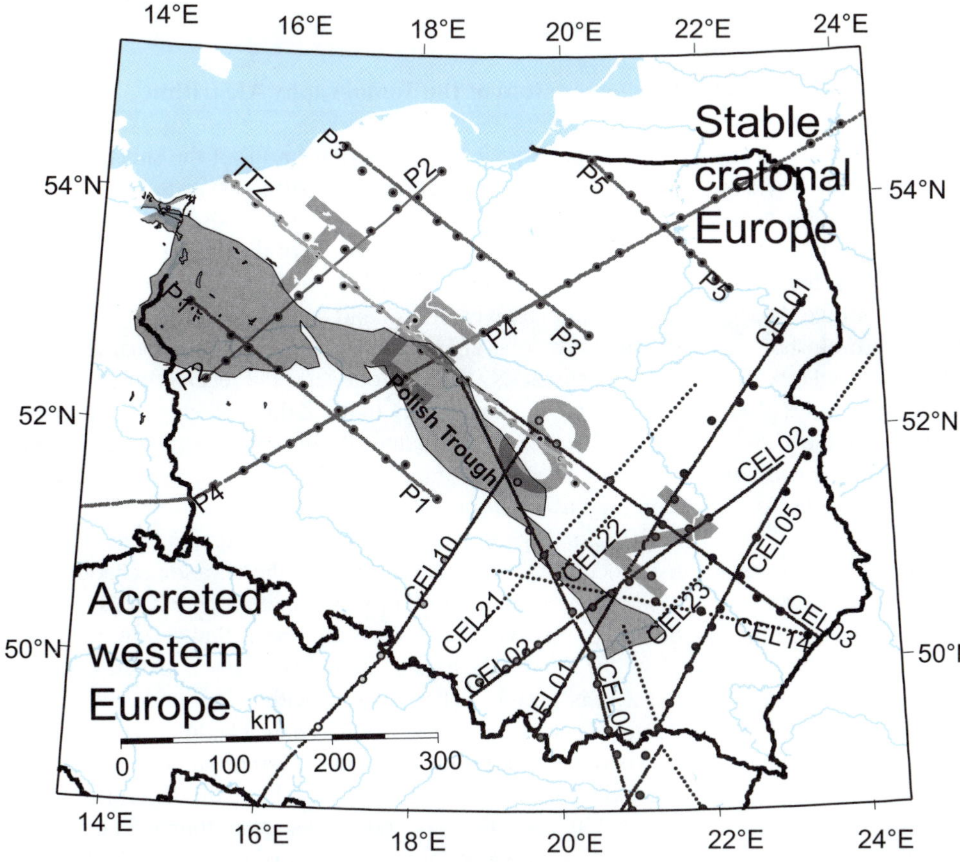

Figure 2. Index map of available transects for this study. TESZ—Trans-European Suture zone; TTZ—Tornquist-Teisseyre zone profile; P—POLONAISE '97 profiles; CEL—CELEBRATION 2000 profiles.

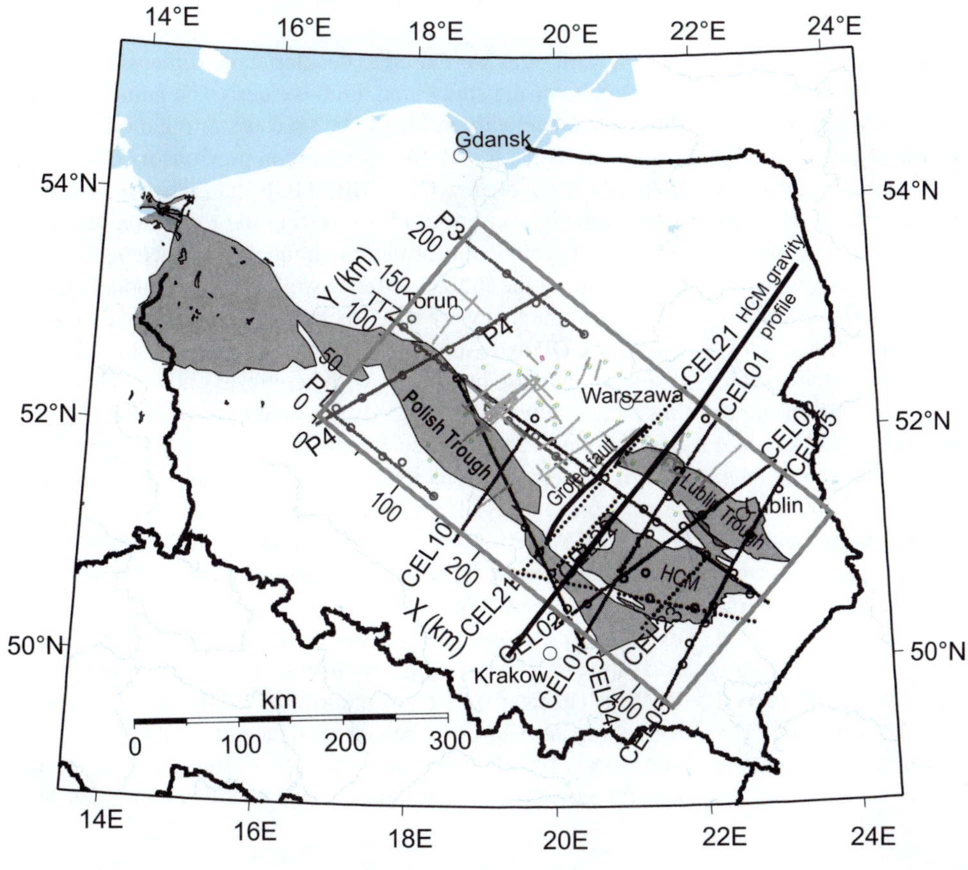

Figure 3. Index map of data used for the tomographic inversion along with tectonic features of interest taken from the Geologic Atlas of Poland (Kotanski, 1997). HCM—Holy Cross Mountains; TTZ—Tornquist-Teisseyre zone profile; P—POLONAISE '97 profiles; CEL—CELEBRATION 2000 profiles. Light gray profiles show location of common depth point data; small circles indicate location of well check-shot surveys.

are preliminary. Therefore, for consistency in this study, we picked first arrivals both for individual profiles and for fan shots within the established model space, and formatted them for use in the inversion. An example of the seismograms for a profile is shown in Figure 4. Dashed lines mark the approximate location of the first arrivals that were input into the inversion code on this figure.

The "nonrefraction" data set is a compilation of well data and velocity data from reflection profiles acquired from the Apache Corporation. The well data were in the form of check-shot surveys in which a seismometer is lowered into the well and records a source that is located near the well. This technique provides a very accurate measurement of the velocity distribution in the vicinity of the well. However, this type of survey is expensive and thus relatively rare. The processing of seismic reflection data involves the summing of seismograms from shot-receiver pairs that share the same approximate subsurface reflection point. This process is called common depth point (CDP) stacking. Before the stacking can be done, the seismograms must be aligned so that the reflections will add in a constructive fashion at the expense of random noise. This alignment requires the estimation of seismic velocity as a function of depth called stacking velocities (e.g., Yilmaz, 2001).

Check-shot surveys from individual wells consisted of either depth-time pairs or depth-velocity pairs for the position within the well. For purposes of the inversion, these data were converted to a one-way travel time from a shotpoint at the surface to a receiver at depth. The geometry for each well location was prepared by assigning a shotpoint number to the well. In addition, a receiver number was assigned to each depth for which there was a travel time. The travel-time picks available were for very short depth intervals (30–50 m), given the large spatial scale of our study. Thus, these data were decimated to consist of picks at only 2 km intervals in order to avoid numerical problems within the 2 km model cells.

The travel-time data that we determined from the processing of seismic reflection data consisted of travel times derived from the tabulation of stacking velocities available at intervals along the seismic profiles. These data were provided by the staff of Geophysika Torun (2003, personal commun.). Although it would have been ideal to completely reprocess the reflection data and thus have direct knowledge of the reliability of these velocity determinations, the original data were not available. Since stacking-velocity data take the form of time-velocity pairs, they had to be converted to depth-time pairs at each CDP location for input to the inversion. Once this was accomplished, each CDP location was assigned a shotpoint number. Each depth-time pair at each location was then assigned a receiver number in a fashion similar to the approach taken with the check-shot data. As with the well data, travel-time picks derived from stacking velocities were decimated to 2 km intervals. In addition, CDP data were cut off at 2.5 s one-way time due to the decreasing resolution of stacking velocities with increasing depth.

The total "nonrefraction" data set consisted of 213 arrival times from check-shot surveys conducted in 72 wells, and 525 arrival times derived from stacking-velocity determinations made at 234 CDP locations (Fig. 3).

Implementation of the Tomography Algorithm

The geographical information for all of the shot points and receivers of each data set (Teisseyre-Tornquist zone, POLONAISE '97, CELEBRATION 2000, wells, and CDPs) was transformed into the X and Y coordinate system of the model space via shell scripts. These scripts are used to produce a series of output files that contain receiver offsets and travel-time picks derived from interpretative first-arrival picking, receiver locations, and shotpoint locations (X, Y, Z). In our case, the receiver offsets for the well and CDP data are only in the Z dimension, with the surface location treated as the source. Therefore, velocities are only calculated along vertical raypaths below the source. These files are then used as input into the inversion.

The main parameters set during the inversion process include the model dimensions, cell size, and the iteration schedule. The model dimensions were set to 440 km along the X axis (perpendicular to the axis of the Trans-European Suture zone), 240 km along the Y axis (the axis of the Polish trough), and 70 km along the Z axis (depth). The cell size within the model was set to 2 km for each dimension. The iteration schedule was set to iterate six times at each of five moving average smoothing windows decreasing in dimension and incorporating all data for receivers within 350 km of the corresponding seismic source. The smoothing windows were set by the number of cells in the X, Y, and Z dimensions at $64 \times 64 \times 16$, $32 \times 32 \times 8$, $16 \times 16 \times 4$, $8 \times 8 \times 4$, and $8 \times 8 \times 2$.

Results of Tomographic Inversions

Several steps were used to develop three-dimensional velocity models for the study area. First we derived a number of one-dimensional velocity models to be used as starting models for the inversion, which took into consideration previous models for the area as well as the new CELEBRATION data. These one-dimensional models were tested by inverting the refraction data only. This effort provided information on the dependence of the inversion result on the initial model as well as control models for the refraction data that were used to observe the effects of adding the industry CDP and well data. Based on these tests, a model with a smooth increase in velocity with depth was chosen. This model has a generally smooth velocity gradient that steepens at ~22 km depth, followed by a gradual decrease in the velocity gradient to 62 km. A negative velocity gradient at the base of the model is used to prevent rays from escaping at its base.

An RMS error was calculated and reported for each shotpoint based on travel-time residuals—differences between calculated and observed travel times—for first-arrival picks. The overall RMS error for a given iteration is the sum of the RMS for each shotpoint scaled by the number of picks for the shot. Initial RMS values near 0.75 converge to final values of 0.14 s for the refraction data, 0.14 s for refraction data with well data included, and 0.18 s for all the data. In order to monitor the inversion, two-dimensional vertical slices through the three-dimensional

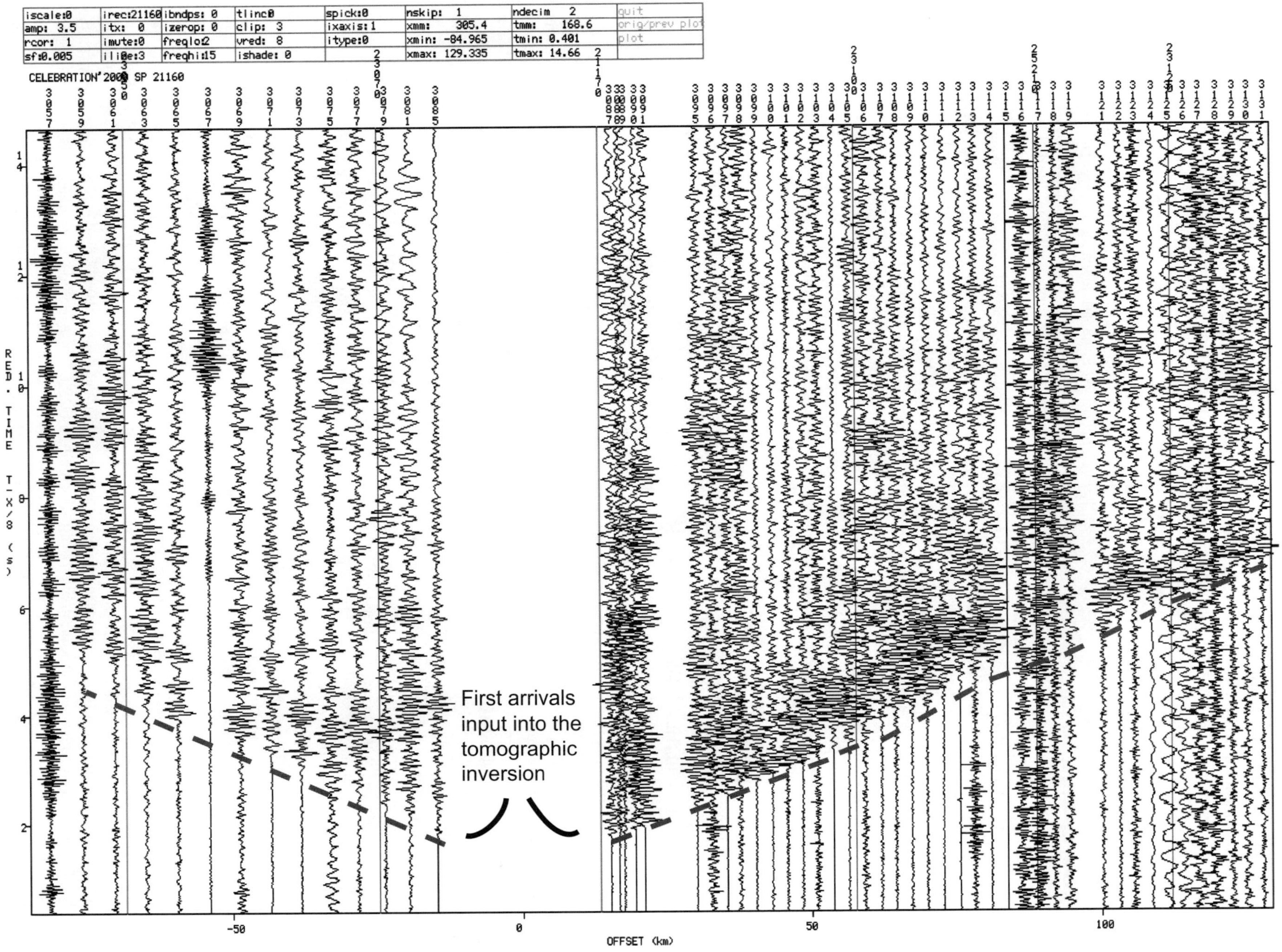

Figure 4. Fan record section along CELEBRATION profile 03 from shot number 21160. From left to right shows a northwest-southeast profile. Travel times reduced to 8.0 km/s.

velocity model and horizontal slices at various depth intervals were examined for individual runs. The vertical slices were chosen based on the data density and geologic considerations and correspond to profiles at X = 300 km, which is perpendicular to the axis of the Trans-European Suture zone and crosses the Holy Cross Mountains and Lublin Trough area, and at Y = 100 km, which follows the axis of the Polish trough (Fig. 3).

Results Using Only the Refraction Data

The first step in the process was to undertake a conventional inversion of first-arrival picks for the combined Teisseyre-Tornquist zone, POLONAISE '97, and CELEBRATION 2000 refraction experiments. After some experimentation, a starting model with a smooth increase of velocity was employed because it produced the most stable inversion. A vertical slice (X = 300 km) of the result of this inversion is shown in Figure 5. This slice crosses the Teisseyre-Tornquist zone and, in the upper 20 km of the crust, shows distinct changes along the profile. The spatial distribution of these velocity variations is shown in a horizontal slice through the model at a depth of 15 km (Fig. 6). These views of the velocity models show velocities between 5.0 and 6.0 km/s forming a layer that quickly thickens into a deep (~20 km) basin from the southwest (at Y = 0) toward the northeast into the Trans-European Suture zone. In Figure 5, this basin thins northeastward and develops a secondary shallower basin (~12 km) before thinning to less than 10 km at the northeastern end of the profile (at Y = 240 km). In turn, a thin (0–5 km) cover of velocities less than 5 km/s develops midway along the profile and then thins to the northeast. The interval with velocities between 6.0 and 6.8 km/s is clearly thickest (10–20 km) to the southwest and dramatically thins to the northeast along the profile.

Results with Industry Data Included

Industry well and reflection data sets were independently added into the inversion with the refraction data. Well data, consisting of check-shot surveys, were added to the refraction data set, and the inversion procedure was repeated. In turn, CDP data, based on industry reflection lines, were added to the refraction and well data, and the inversion procedure was repeated again. Results for inversions including CDP data showed instabilities, and model RMS values did not decrease below 0.18 s, compared to values of 0.14 s for both refraction data and the combined refraction and well data. Although this result is disappointing, we attribute some of the errors to inherent difficulties in calculating stacking velocities from reflection seismic data. We suspect that careful velocity analysis, rather than the production mode approach used when all that is sought is a good stacking result, would have given us a good result, but we cannot quantify this assertion. Another consideration is that the CDP data were very closely spaced and thus could contain travel-time anomalies from local geologic features that were too small to be resolved by the much more widely spaced refraction and check-shot data (Fig. 3).

We calculated the differences between the different inversion results and found that some groupings of CDPs with high RMS values showed strong correlation with areas where drilling and seismic reflection data demonstrate the presence of large bodies of salt, which have high seismic velocities. This indicates that some of the problems encountered by including the CDP data are not due to erroneous data but are due to our attempt to fuse data with spatial resolutions that are too different to be compatible.

The inclusion of the check-shot surveys, however, improved our inversion results. These well data, limited to the uppermost crust (<6 km), provide valuable and reliable velocity information for sedimentary rocks within the Polish trough. Plots showing the differences between the velocity models for the refraction data alone and for the refraction and well data combined show the effects of adding the well data (Fig. 7). In this figure, positive values indicate a decrease in velocity when the well data are added. This corresponds to better resolution of the low-velocity sediments in the upper crust (<5 km). In turn, velocities at deeper levels are affected by changes in the model in the uppermost crust, and velocity gradients are steepened beneath the basins. The difference plots helped delineate low-velocity regions as possible basin structures and demonstrated the positive effects of constraining the upper crust in addition to refraction data.

Vertical slices (X = 300 km; Y = 110 km) through our final velocity model resulting from the fusion of the refraction and well data are shown in Figure 8. These diagrams show that while this study includes a very a large amount of data, the extensive size of the study area provides only limited three-dimensional tomographic ray coverage. As a way of displaying only those areas of the model with good resolution, areas lacking "ray coverage," i.e., that are not traversed by raypaths, were set to a null value, "masked," in this model. As discussed in Averill (2003), this model has many interesting geologic implications that are beyond the scope of this paper. However, we will show how it can be qualitatively fused with a totally different type of geophysical data in the section below.

GRAVITY DATA AND MODELING

Measurements of Earth's gravitational field are a very cost-effective source of information about the structure of the lithosphere. These very precise measurements are processed so that known variations in the gravity due to factors such as elevation and latitude are removed. The remaining anomalies are the manifestation of geologic structures and are mapped and modeled to determine density variations. For many years, the scientific community has cooperated to compile gravity measurements into databases on a global and regional basis (e.g., http://paces.geo.utep.edu). The database for Poland is particularly good, and the gravity map of Poland (Fig. 9) contains several major anomalies of interest (Wybraniec, 1999) that can be compared with our velocity models. These include an anomalous block of high gravity values in the southeastern portion of our study area and the northwest-southeast-trending series of highs and lows that extend

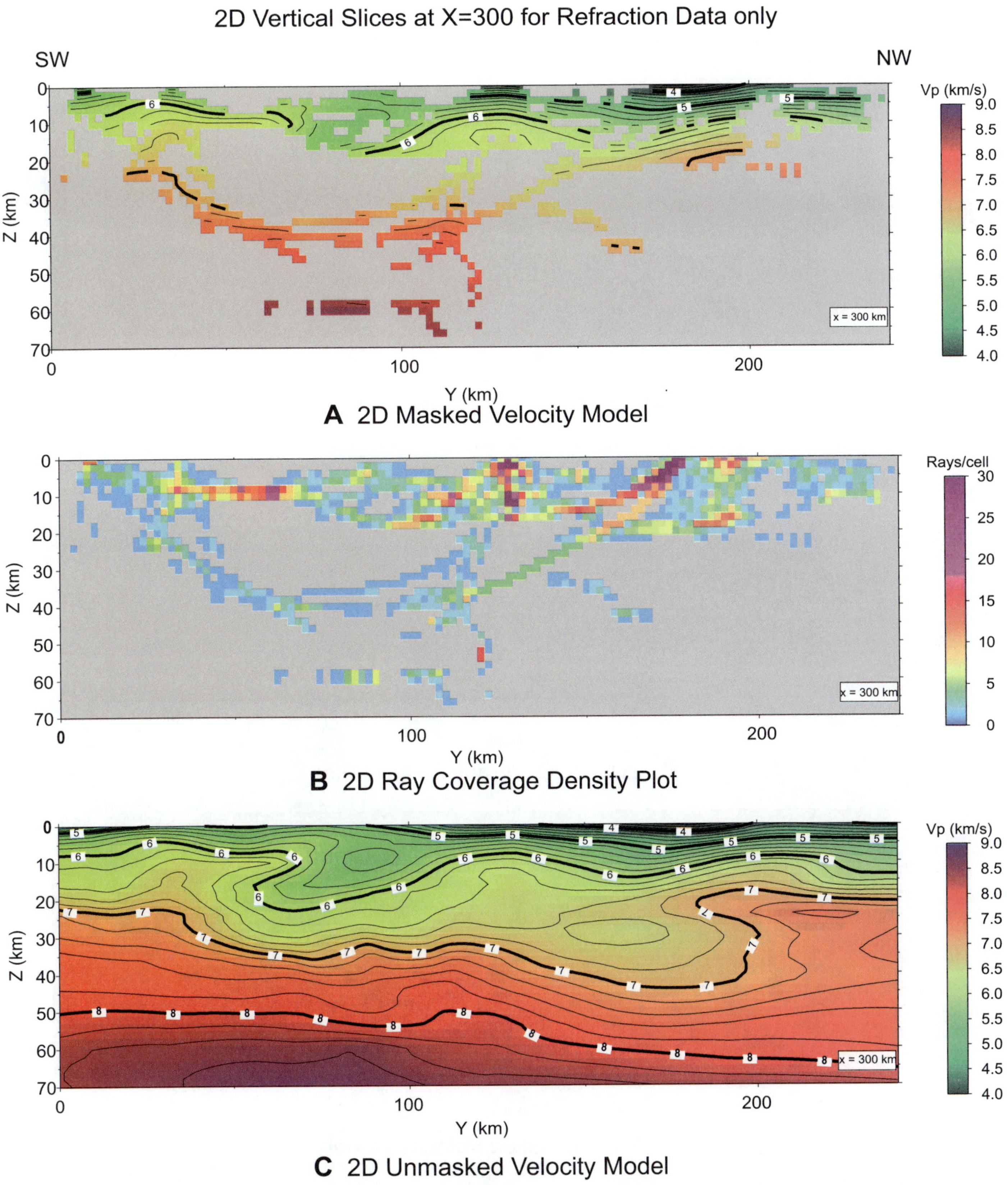

Figure 5. Two-dimensional vertical tomographic slices at X = 300 km perpendicular to the Teisseyre-Tornquist zone axis. A: Two-dimensional masked velocity model (areas without ray coverage are set to null value, e.g., "masked"). B: Two-dimensional ray coverage density plot. C: Two-dimensional unmasked velocity model. Vertical exaggeration is ~0.9.

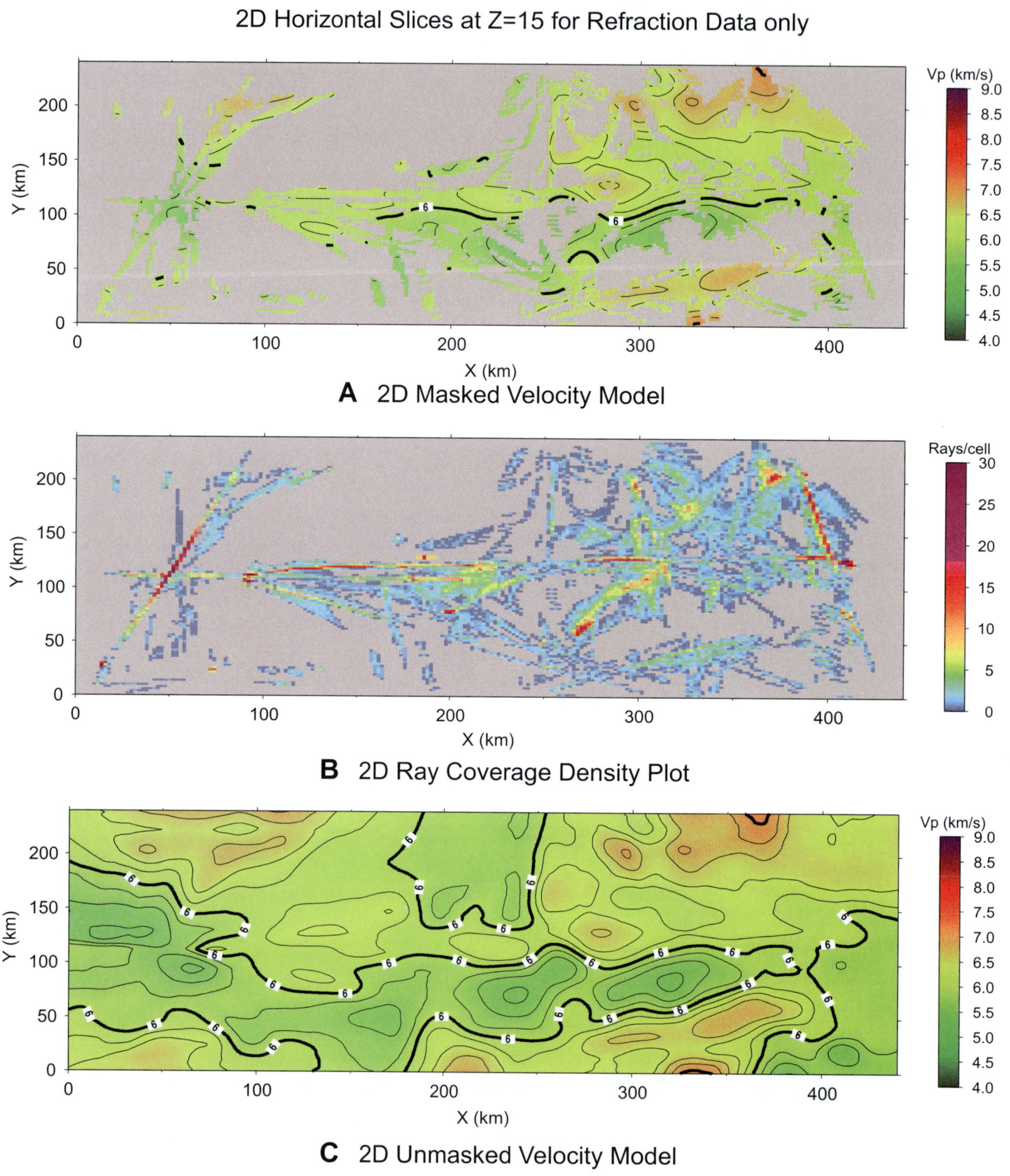

Figure 6. Two-dimensional depth slices at Z = 15 km. A: Two-dimensional masked velocity model. B: Two-dimensional ray coverage density plot. C: Two-dimensional unmasked velocity model.

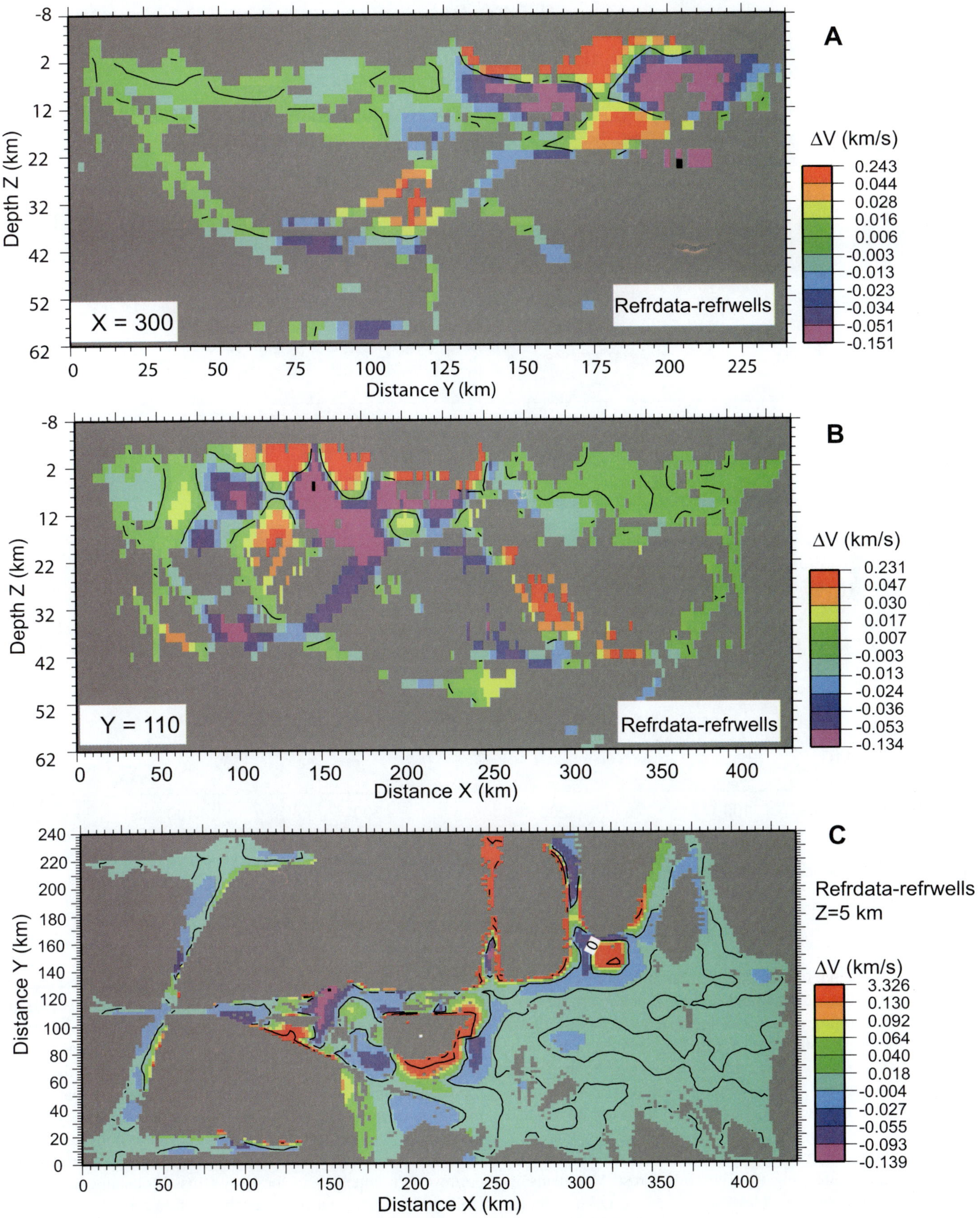

Figure 7. Plots of model velocity differences between refraction data alone and refraction data with wells. Models are masked to show only areas with ray coverage. A: Two-dimensional vertical slice at X = 300 km. B: Two-dimensional vertical slice at Y = 110 km. C: Two-dimensional depth slice at Z = 5 km.

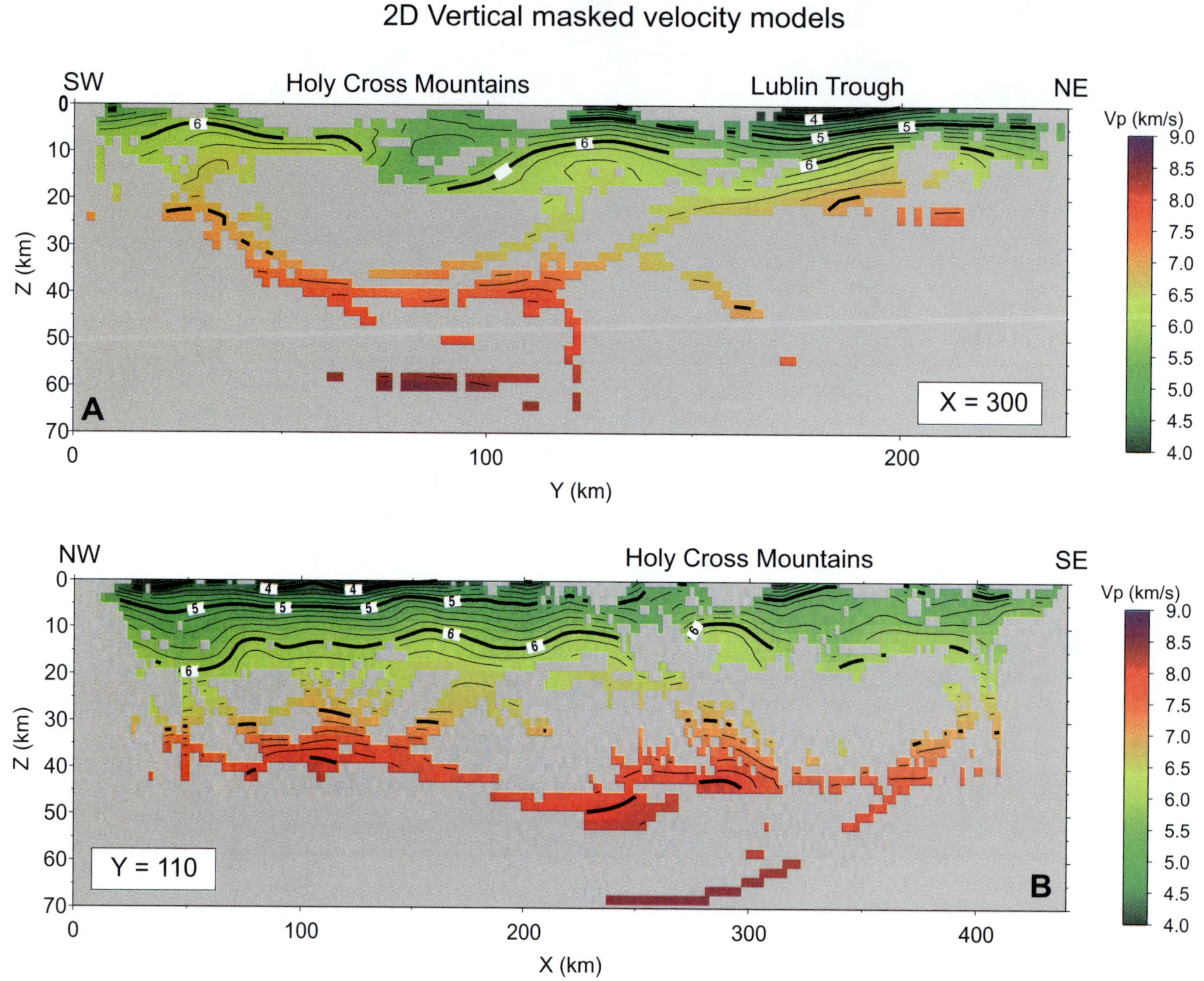

Figure 8. Two-dimensional masked velocity models for the integrated refraction and well data at X = 300 km (A) and Y = 110 km (B). Vertical exaggeration is ~0.9 for the X profile and ~0.5 for the Y profile.

across central Poland. In fact, examination of the gravity anomalies helped to determine the bounds for our tomographic analysis. In unison with tomographic modeling, our goal was to jointly constrain the crustal structure of this region with gravity data. The gravity data used for this study were provided by the Polish Geological Institute (S. Wybraniec, 2003, personal commun.) in the form of a 5 km spaced gridded data set that was reformatted for input into our gravity modeling and mapping programs.

The gravity anomalies clearly delineate several tectonic features in the region (Fig. 9). The Holy Cross Mountains in the southeastern region of Poland as well as the Lublin Trough clearly coincide with gravity highs, shown by the hotter colors. A northwest-southeast–trending gravity low corresponds to the deep (>10 km) sedimentary deposits of the Polish trough. The Sudetes region in southwestern Poland is the core of a crystalline uplift in the Variscan terrane in this region and is clearly marked by a gravity high. Features such as the Odra fault zone and the Holy Cross dislocation (a fault that is a major tectonic boundary in the Holy Cross Mountains [Valverde-Vaquero et al., 2000]) are marked by lineations in the gravity anomalies and truncation of gravity highs. The Grojec fault (a subsurface feature that trends northeast across the study area at approximately x = 230 km; Fig. 9) marks the northwestern boundary of the Holy Cross Mountains block (the large gravity high that occupies the southeastern portion of the study area) and is clearly a major feature associated with the sharp truncation of the gravity high in that region.

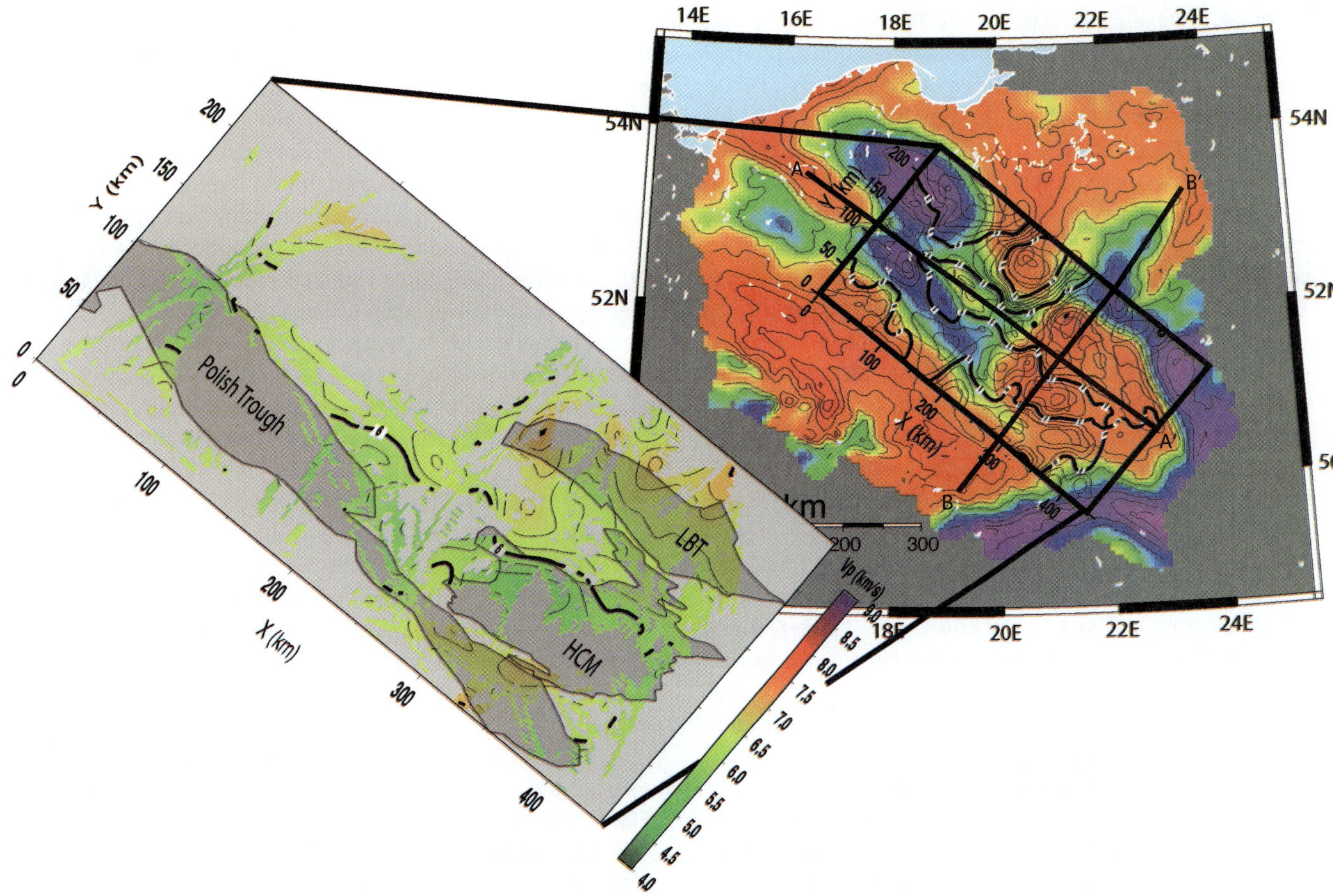

Figure 9. Bouguer gravity map with an overlay of a two-dimensional tomographic slice at Z = 15 km. Overlay shows contours of the unmasked velocity model. Inset projection shows the masked velocity model indicating the constrained portion of the model. LBT—Lublin Trough; HCM—Holy Cross Mountains. Lines A–A′ and B–B′ show locations for modeled gravity profiles.

By overlaying two-dimensional velocity depth slices onto our gravity map, we were able to relate velocity structures to the observed gravity anomalies. For example, at a depth of 15 km (Fig. 9), a northwest-southeast-trending low-velocity (<6.0 km/s) zone correlates with the trend of the Polish trough that is marked by low gravity anomaly values and represents an accumulation of sedimentary strata. A high-velocity (6.4–7.0 km/s) anomaly located over the Lublin Trough (northeastern half of the Holy Cross Mountains block) clearly accounts for the high gravity anomaly and suggests the presence of dense mafic material. The steep velocity gradient bounding the Holy Cross Mountains block on the northwest correlates with the Grojec fault, showing that it is a major, deep-seated boundary.

Gravity Modeling

Gravity modeling requires plotting a profile of gravity values against distance, and a profile was constructed along the line shown in Figure 3. During modeling, the goal is to match this gravity profile of observed values with a calculated gravity profile produced by forward modeling of crustal features. Inverse methods can also be used to model gravity anomalies, but we chose forward modeling order to be able to integrate a wide variety of constraints. As this is a nonunique process, it is necessary to start with a well-constrained model, given known geologic relationships along the profile and independent geophysical constraints.

To constrain our models, we included locations of geologic and tectonic features, well and seismic data provided by industry, interpreted seismic refraction lines, and slices from tomographic modeling discussed above. The first step was to correlate available data with their location along the strike of the profile line. In turn, these data were used to create an initial density model that honors the constraints at any designated location. This initial model is then adjusted, honoring the constraints to best fit the observed profile. The result is an integrated structural model along the strike of the profile that jointly satisfies the seismic, gravity, and geologic data available.

In order to prepare the model, crustal structure is represented by individual polygons within the model. These polygons are defined by nodes in model space and are assigned a density value based on constraints such as velocity and lithology from wells, seismic surveys, and known velocity-density relationships. Our computer code, which is based on the method of Cady (1980), uses the dimensions and density of these polygons to produce a calculated profile for the model structure. Due to the integrated process used to derive these models, they should be considered as crustal-scale cross sections, as opposed to crustal models derived from gravity anomalies alone.

The gravity model across the Holy Cross Mountains (Fig. 10) clearly shows a transition from a Paleozoic platform terrane (Upper Silesian block) to the East European craton. At this transition, the thin (31–33 km) crust of the Paleozoic terrane is in juxtaposition with the three-layer, 43–45 km thick crust of the East European craton. A large mafic body marks the edge of the rifted cratonal margin and accounts for the large gravity high over the ~15 km deep Paleozoic basin within the Holy Cross Mountains block. A thickened portion of the mafic body underlies and accounts for the Lublin Trough gravity high. A thickened carbonate Paleozoic section (10–15 km) within the transition zone underlies the Holy Cross Mountains region. These features are suggestive of a volcanic continental margin formed during the breakup of Rodinia. These bodies were modeled with lateral variation in and out of the model, corresponding to the limited lateral extent of the associated gravity anomalies. The wedge-shaped feature modeled within the upper crust of the East European craton margin is interpreted to correspond to deformation of the margin. Deformation of the Paleozoic section along the East European craton margin corresponds to uplift of the Holy Cross Mountains. The lack of large-scale orogenic features such as major basement deformation and uplift suggests an incomplete or indirect collision during the Variscan orogeny.

The lateral extent of some of the features is shown in Figure 9, where a horizontal slice of the velocity model is shown along with the gravity anomaly map. A region of high velocities (~7.0 km/s) coincides with the thickened mafic body underlying the Lublin Trough and marks the northeastern edge of the Teisseyre-Tornquist zone. A corresponding velocity high to the southwest corresponds to the southwestern edge of the transition zone and the edge of the Paleozoic European platform. A trough of velocities less than 6.0 km/s corresponds to the thickened sedimentary strata. The 6.0 km/s contour roughly coincides with the top of the crystalline basement across the transition, and the Grojec fault shows up as a discontinuity in the velocity structure at 15 km, confirming that it is a major crustal feature. The result of this integrated modeling thus sheds additional light on the deep structure of this region and represents a form of data fusion in which a two-dimensional gravity model is used as the medium for the integration of a wide variety of data.

CONCLUSIONS

The results of this study include a successful demonstration of the benefits and drawbacks of quantitative fusion of small-scale industry-acquired well and reflection data with large-scale refraction data as well as an integrated interpretation of the structure and evolution of a portion of the Trans-European Suture zone and the surrounding region in southeastern Poland. Using the results of this study and previous work we can conclude the following:

1. The integration of travel times from small-scale data sets, such as well and reflection data, into a large-scale tomographic inversion can provide additional constraints on the structure of the uppermost crust and in turn more confidence about the resolution of deeper structure. Our approach has the advantage of including all travel-time data in one inversion.

2. The presence of a mafic lower crust, indicated by a thickened layer of high-velocity (6.8–7.4 km/s) material and a distinct high gravity anomaly is consistent with observations from modern rifted passive continental margins (e.g., Lizarralde and Holbrook, 1997).

3. In the Holy Cross Mountains block, the presence of a high-velocity zone (6.6–7.2 km/s), modeled as a high-density (3.1 g/cm^3) mafic body, accounts for the observed gravity anomaly (greater than +30 mgal) located beneath an ~15 km deep basin with velocities <6.0 km/s. These velocities are consistent with Paleozoic sedimentary rocks of East European craton margin affinity. In turn, they are part of the Holy Cross Mountains block.

4. Across the Trans-European Suture zone, there is a transition from a thick (>40 km), three-layer crust to a thinner (31–33 km) crust that is consistent with previous work.

5. Lack of large-scale orogenic features across the transition and the observed deformation of Paleozoic marginal sediments in the Holy Cross Mountains suggest indirect or incomplete Variscan collision/suturing along the East European craton margin.

6. Our integrated gravity modeling is a quantitative but less formal type of data fusion in that the various constraints are used in a qualitative fashion. We are presently investigating ways in which other types of seismic data can be used in a mathematically formal fusion via tomographic inversion.

ACKNOWLEDGMENTS

We would like to particularly acknowledge the assistance of John Hole, who helped us understand his tomography program and provided a helpful review. This work is part of an ongoing collaboration between the University of Texas, El Paso, and the Institutes of Geophysics of the Polish Academy of Sciences and the University of Warsaw. Apache Petroleum provided most of the industry data used in this study. Our participation in the seismic data collection was funded by grants from the National Science Foundation (NSF Celebration: OISE-0001356) and the State of Texas Higher Education Coordinating Board. Our research on data fusion is funded by the GEON project (NSF grant EAR-0225670) and by NASA through PACES (cooperative agreement NCC5–209). IRIS/PASSCAL is supported by the National Science Foundation (NSF) and provided the majority (~700 of 1200) of the instruments for this experiment, and most of these instruments were purchased through grants to the University of

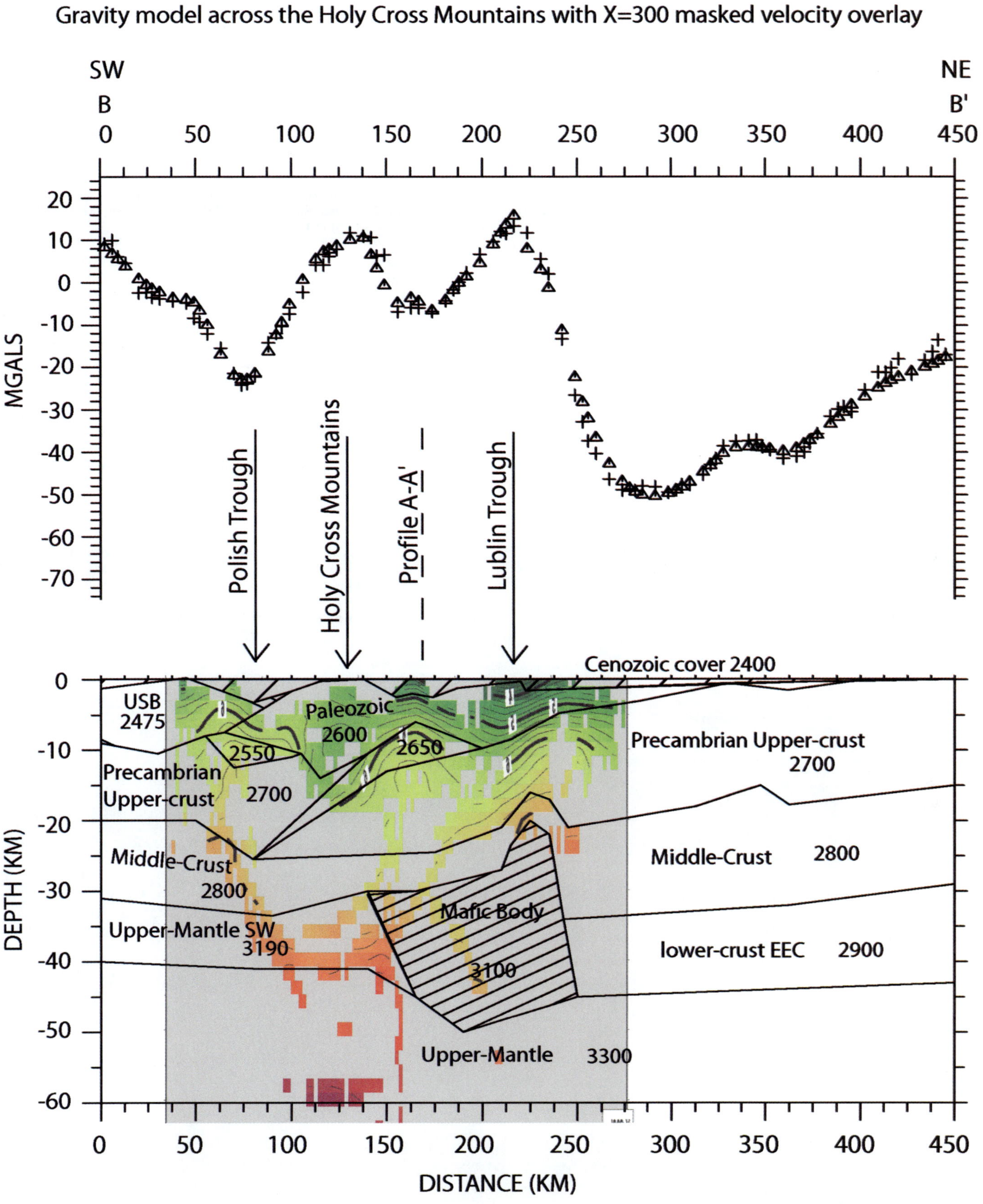

Figure 10. Two-and-a-half-dimensional model constructed for gravity profile B–B′ that crosses the Holy Cross Mountains along with an overlay of the masked velocity model for X = 300 km. Density units in kg m^{-3}. USB—Upper Silesian block; EEC—East European craton

Texas, El Paso (State of Texas Higher Education Coordinating Board, NSF/MRI, and U.S. Department of Defense).

REFERENCES CITED

Averill, M., 2003, An integrated geophysical study of the Holy Cross Mountains region of the Trans-European Suture zone in Poland [M.S. thesis]: El Paso, University of Texas, 172 p.

Cady, J.W., 1980, Calculation of gravity and magnetic anomalies of finite-length right polygonal prisms: Geophysics, v. 45, p. 1507–1512, doi: 10.1190/1.1441045.

Grad, M., Janik, T., Yliniemi, J., Guterch, A., Luosto, U., Tiira, T., Komminaho, K., Sroda, P., Höing, K., Makris, J., and Lund, C.E., 1999, Crustal structure of the Mid-Polish Trough beneath the Teisseyre-Tornquist zone seismic profile: Tectonophysics, v. 314, p. 145–160, doi: 10.1016/S0040-1951(99)00241-3.

Guterch, A., Grad, M., Thybo, H., and Keller, G.R., 1999, POLONAISE '97—An international seismic experiment between Precambrian and Variscan Europe in Poland: Tectonophysics, v. 314, p. 101–121, doi: 10.1016/S0040-1951(99)00239-5.

Guterch, A., Grad, M., and Keller, G.R., 2001, Seismologists celebrate the new millennium with an experiment in central Europe: Eos (Transactions, American Geophysical Union), v. 82, p. 529, 534–535.

Hole, J.A., 1992, Nonlinear high-resolution three-dimensional seismic travel time tomography: Journal of Geophysical Research, v. 97, p. 6553–6562.

Jensen, S.L., Janik, T., and Thybo, H., POLONAISE Working Group*, 1999, Seismic structure of the Paleozoic platform along POLONAISE '97 profile P1 in northwestern Poland: Tectonophysics, v. 314, p. 123–143.

Keller, G.R., and Hatcher, R.D., Jr., 1999, Some comparisons of the structure and evolution of the southern Appalachian-Ouachita orogen and portions of the Trans-European Suture zone: Tectonophysics, v. 314, p. 43–68, doi: 10.1016/S0040-1951(99)00236-X.

Kotanski, Z., 1997, Geologic atlas of Poland: Geologic maps of horizontal cutting: Warsaw, Poland, Polish Geological Institute, p. 15–37.

Lamarche, J., Mansy, J.L., Bergerat, F., Averbuch, O., Hakenberg, M., Lewandowski, M., Stupnick, E., Swidrowska, J., Wajsprych, B., and Wieczorek, J., 1999, Variscan tectonics in the Holy Cross Mountains (Poland) and the role of structural inheritance during Alpine tectonics: Tectonophysics, v. 313, p. 171–186, doi: 10.1016/S0040-1951(99)00195-X.

Lizarralde, D., and Holbrook, W.S., 1997, U.S. mid-Atlantic margin structure and early thermal evolution: Journal of Geophysical Research, v. 102, p. 22,855–22,875.

Nolet, G., editor, 1987, Seismic tomography: Dordrecht, Netherlands, D. Reidel Publishing Company, 386 p.

Pharaoh, T.C., 1999, Paleozoic terranes and their lithospheric boundaries within the Trans-European Suture zone (TESZ): A review: Tectonophysics, v. 314, p. 17–41, doi: 10.1016/S0040-1951(99)00235-8.

Sroda, P., and POLONAISE Profile 3 Working Group, 1999, P- and S-wave velocity model of the southwestern edge of the Precambrian East European craton, POLONAISE '97: Profile P3: Tectonophysics, v. 314, p. 175–192.

Sroda, P., Czuba, W., Grad, M., Guterch, A., Gaczynski, E., and POLONAISE Working Group, 2002, Three-dimensional seismic modeling of crustal structure in the TESZ region based on POLONAISE '97 data: Tectonophysics, v. 360, p. 169–185, doi: 10.1016/S0040-1951(02)00351-7.

Valverde-Vaquero, P., Dorr, W., Belka, Z., Franke, W., Wiszniewska, J., and Schastok, J., 2000, U-Pb single grain dating of detrital zircon in the Cambrian of central Poland: implications for Gondwana versus Baltica provenance studies: Earth and Planetary Science Letters, v. 4, p. 225–240.

Vidale, J., 1988, Finite-difference calculation of travel times: Bulletin of the Seismological Society of America, v. 78, p. 2062–2076.

Wybraniec, S., 1999, Transformations and visualization of potential field data: Polish Geological Institute Special Papers, v. 1, p. 1–88.

Yilmaz, O., 2001, Seismic data analysis: Processing, inversion, and interpretation of seismic data: Tulsa, Oklahoma, Society of Exploration Geophysicists, 2027 p.

Zelt, C.A., and Barton, P.J., 1998, Three-dimensional seismic refraction tomography: A comparison of two methods applied to data from the Faeroe basin: Journal of Geophysical Research, v. 103, p. 7187–7210, doi: 10.1029/97JB03536.

Manuscript Accepted by the Society 18 June 2005

Geological Society of America
Special Paper 397
2006

Schema to ontology for igneous rocks

A.K. Sinha
A. Zendel
Department of Geosciences, Virginia Tech, Blacksburg, Virginia, USA

B. Brodaric
Geological Survey of Canada, 615 Booth Street, Ottawa, Canada

C. Barnes
Department of Geosciences, Texas Tech University, Lubbock, Texas

Jihane Najdi
Department of Geosciences, Virginia Tech, Blacksburg, Virginia, USA

ABSTRACT

Extracting knowledge from the rock record stored in databases is one of the primary goals of the information-oriented geoscientist. This activity requires well-designed organizational structures to facilitate queries, and ultimately cyber-aided geological research. Such structures need to encompass information about geologic objects and the processes that affect or produce the objects. Therefore, our goal is to create a prototype of a computer-based knowledge environment that specifically reflects the reasoning used by a geoscientist, with the recognition that his/her primary interest lies in understanding processes that affect the rock record. In order to start development of such capabilities, we have utilized an organization of attributes and their definitions to construct a database schema for field-based igneous rocks, and show that its conversion into a knowledge base requires the application of both object and process ontologies.

Keywords: cyberinfrastructure, petrology, geoinformatics, ontology.

INTRODUCTION

It is well recognized that earth science is both an observational and computationally oriented science. In order to better identify and characterize the natural processes affecting earth through time, it is a common practice to combine observational information from the field with sophisticated numerical models. However, a geologist's dealing with the natural world is often based on sparse data, the interpretations for which necessarily lead to multiple underdetermined models (e.g., Schumm, 1991). In an attempt to select the optimal model or search for new models within this environment, emphasis on integration of data from disparate sources and disciplines utilizing knowledge discovery methods must be more fully implemented (e.g., Brodaric and Gahegan, this volume). As geologists become more adept in extracting new knowledge from databases, they become more effective in comparing and contrasting information through query-based search engines. This practice of comparison and integration is most effectively supported via a geoinformatics cyberinfrastructure: the application of informatics technologies such as development of knowledge bases, logic frameworks, and ontologies to geosciences. In this paper, we describe the

Sinha, A.K., Zendel, A., Brodaric, B., Barnes, C., and Najdi, J., 2006, Schema to ontology for igneous rocks, *in* Sinha, A.K., ed., Geoinformatics: Data to Knowledge: Geological Society of America Special Paper 397, p. 169–182, doi: 10.1130/2006.2397(12). For permission to copy, contact editing@geosociety.org.

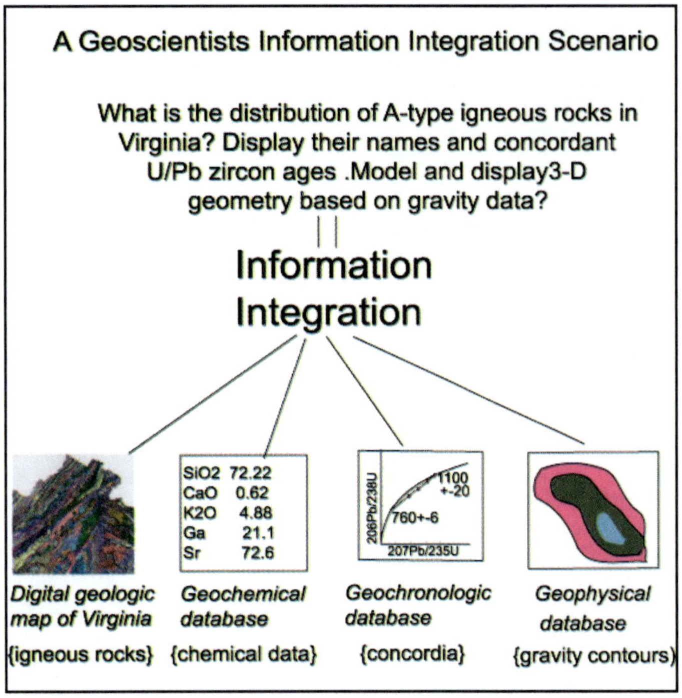

Figure 1. A schematic representation of a geoscientist's information integration problem. Note that multiple databases need to be accessed, analyzed, and displayed. To meet the science need of the user, an ontologic framework for integration as well as access to databases is necessary.

development and application of a prototype Web-based environment for accessing and querying data from igneous rocks through the application of knowledge discovery methods. We suggest that similar activities will be useful for research, education, and communication with laypersons and decision makers, within all disciplines of geology.

As examples of this approach, we use two motivating scenarios, each of which asks "Why do this exercise?" and "How may a result be accomplished?" The first scenario involves an example of query-based data retrieval for finding regularities in the data, specifically a search for A-type plutons of a certain age in a defined area, followed by estimation of their shape and volume. The second scenario involves hypothesis testing using known processes over existing data. We use the Tuolumne Intrusive Series (Bateman and Chappell, 1979) as an example of a complex plutonic system for which multiple petrogenetic hypotheses have been proposed. Essential to both examples and the approach in general is representation of geological knowledge, in the form of classification systems and hypotheses, and its subsequent use in discovering additional knowledge.

Scenario I

In our first proposed scenario (Fig. 1), the scientist's query of locating, accessing, modeling, and visually displaying the results requires logical access to distributed spatial and aspatial databases, and to tools that must be interlinked using existing catalogs of explicitly represented knowledge (ontologies) and workflows. Although reliant on several sophisticated geological data-processing tools and methods, this scenario is relatively simple, as it is designed to access data and perform sequential searches and functions utilizing only a few tools and their products. It does, however, require access to a hierarchical ontology containing rock and mineral classifications, as well as definitions of terms, e.g., concordant age.

Table 1 includes a summary of the logical flow of information, including tasks and requirements, needed to complete the retrieval and computational processing posed in the query. These logical functions represent decomposition of the primary query into several subqueries, each of which uses Web-based access to data and tools. Use of ontologies is limited, and information regarding locations of databases and tools, as well as mechanisms for either automated or user-controlled application of tools, must be in place (see Ludäscher et al., this volume).

Scenario II

In this scenario, the objective is to demonstrate how one might use ontology-based tools to represent and augment our understanding of the multiple possible process models for the genesis of an object. For example, the Tuolumne Intrusive Series (Fig. 2), which has been studied for several decades, provides a geologic framework associated with the formation of a zoned pluton. In order to explain the origin of the observed zoning, multiple explanations for its evolution have been proposed. Fractional crystallization of a single magma involving the process of crystal-liquid separation (Bateman and Chappell, 1979) can be readily contrasted with the hypothesis that mixing of mafic magmas controlled the observed element variability in each zone (Reid et al., 1983). Alternatively, based on coupling of geochemical data with Sr and Nd isotopic studies, Kistler et al. (1986) suggested that the Tuolumne Intrusive Series was the result of multiple pulses of heterogeneous magma, originating from multiple sources. More recently, Glazner et al. (2004) have argued for incremental emplacement of magmas over several million years to form the Tuolumne Intrusive Series. If such diverse explanations exist for this well-exposed, intensively studied body, it is clear that new techniques of data and knowledge processing are required for a synoptic view to depict the Tuolumne Intrusive Series magma source and emplacement processes.

The Tuolumne Intrusive Series scenario is fundamentally related to the concept of heterogeneity or homogeneity and the geologic processes that produce these conditions. Heterogeneity is defined as a state (e.g., pluton) consisting of dissimilar constituents and can be attributed to both open- and closed-system processes. For example, fractional crystallization can be defined as a closed-system process, yet it results in highly heterogeneous rocks. In contrast, many open systems, acting either individually or in complex combinations, lack sufficient conditions to uniquely identify processes that have acted on the object (magma).

TABLE 1. REQUIREMENTS FOR COMPLETING THE TASKS FORMULATED IN THE QUERY SHOWN IN FIGURE 1

Task	Requirement
(1) Locate state of Virginia.	GIS database with gazetteer
(2) Identify all igneous rocks from geologic maps.	rock classification schema; distinguish between "igneous," "plutonic," "tonalite," etc.
(3) Access databases of igneous rocks.	data access functions to access distributed databases; formalized knowledge of the characteristics of igneous rocks
(4) Apply discriminant functions to classify.	classification functions as igneous rocks as A-type applied to igneous rocks; specification of classification types
(5) Access radiometric age databases.	temporal ontology; distributed data access
(6) Use zircon as mineral for identifying U/Pb concordant age.	temporal ontology; inference links to method ontology; mineral classification
(7) Access gravity databases.	gravity databases
(8) Overlay gravity data over polygons of A-type granite in Virginia.	See (2), (7).
(9) Deploy tool for identifying shape of pluton.	knowledge of regional gravity field; knowledge of appropriate rock densities; numerical tools
(10) Display the three-dimensional geometry of the igneous rocks.	Web-based three-dimensional visualization functions

Examples of open magmatic systems include wall-rock reaction, assimilation, and magma mixing, and in all cases materials have been introduced from sources external to the system. Therefore, evaluation of open-system processes requires recognition, development, and application of explicitly represented knowledge associated with how multiple processes and materials interact to impact the genesis of a particular object, because rocks in a pluton may have different sources and different contaminants.

Similarly, when a range of ages is observed in igneous rocks, implementation of time ontology is required. By necessity it has to be earth-science-specific, e.g., drawing temporal relationships between computed ages and closure temperature processes (Dodson, 1973; Spear, 1993). Although this activity is not the primary focus of this paper, constructing an earth science time ontology can readily utilize concepts already developed for a more general (calendar) ontology of time. Each measurement of age (time) may represent an instance of an event within the history of the pluton such as the Tuolumne Intrusive Series, so that its relationship to other processes occurring in the magma or solidified pluton becomes a key in understanding the heterogeneous geologic record.

Therefore, based on the two scenarios identified in earlier sections, there is a need to develop multiple, generic process ontologies that encompass space and time, each of which may represent a distinct interpretation. Although different interpretations were made to explain the origin and geologic history of the Tuolumne Intrusive Series, covering both closed- and open-system behavior, the petrogenetic processes are probably not independent of each other in nature. The same is true, in a way, of generic process ontologies: Conceptualization of processes can be seen as independent entities in an ontology, such as independent

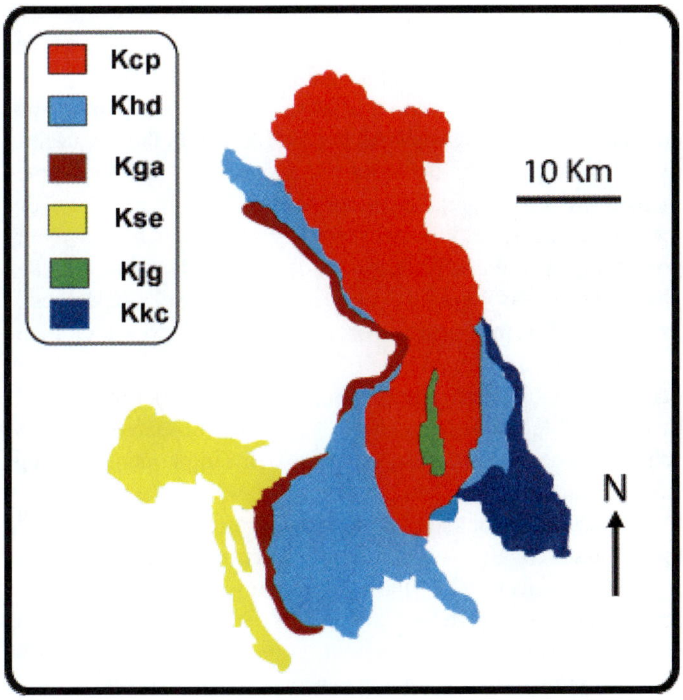

Figure 2. Geologic map of the Tuolumne Intrusive Series showing the multiple igneous bodies, which range in composition and age (Paterson and Vernon, 1995; Bateman and Chappell, 1979). See Coleman et al. (2004) for information on rock types and ages.

characterization of foundational processes (e.g., transport in general), but act in concert with representations of geologic events occurring in a specific space-time region (transport in this region). While it is true that plutons such as the Tuolumne Intrusive Series may become heterogeneous by subsolidus processes, these processes are secondary and can be identified on the basis of alteration mineralogy, veining, and isotopic data. In scenario II, the need to construct alternate conceptual models and explanations through testing against existing data for goodness of fit can facilitate forward computational models.

In order to provide the capabilities to engage in testing of new hypotheses based on goodness of fit to existing data, we present an ontologic framework for accessing and utilizing igneous rock databases.

INTRODUCTION TO DATABASE SCHEMAS AND ONTOLOGIES

In this paper, we use the term *database schema* in a very narrow sense to refer to an explicit representation of a database structure. By *structure* we refer to the database organization of entities in a domain, e.g., rocks, minerals, and plutons, and include associated characteristics such as attributes, potential vocabulary, relations, and constraints. We also follow convention and refer to three types of database schemas: conceptual, logical, and physical. Conceptual schema reflect the natural organization of entities in the domain, e.g., igneous rocks, whereas logical and physical schemas increasingly impose structures devised for computer processing, e.g., tables for igneous rocks.

Conceptual schemas in this sense are therefore domain-specific and as technology-neutral as possible within the constraints of the paradigm in which they are expressed. They are implemented by logical and physical schemas, each progressively more tied to a specific technology; e.g., logical schemas incorporate object-oriented or relational database constructs, whereas physical schemas implement these constructs in specific database systems such as the ESRI (Environmental System Research Institute) geodatabase, Oracle, and Microsoft Access. For example, because our schema for igneous rocks, discussed below, is stated in terms of relational databases, it can be considered a logical schema—one that is, however, physically implemented in the ESRI geodatabase system.

Conceptual database schemas and ontologies are typically differentiated on the basis of the generality of contents and on the rigor (expressiveness) of the representation. Thus, a conceptual schema can be less general and expressive than an ontology, or equally general or expressive, but it cannot be less general and equally expressive, nor equally general and less expressive.

Conceptual schemas, although technologically neutral, are typically not general, as they are designed to fit a specific suite of data, often from a specific scientific perspective and for a specific scientific purpose. For example, two conceptual database schemas for igneous rocks might be quite different depending on the designer's needs, goals, purposes, observations, measurement methods, etc., without necessarily implying that the knowledge those designers hold of igneous rocks is dissimilar. Of course it is also quite possible that the schemas diverge because of differences in theoretical or cognitive orientation, education, and experience, but this is not always the case. Conceptual database schemas are thus inherently partial and purposeful.

Ontologies, on the other hand, are normally intended to be rigorous representations of broadly applicable domain knowledge shared by a community such as geologists (Gruber, 1993; Guarino, 1995). There is obviously a continuum between ontologies (general) and conceptual database schemas (specific), as it is debatable the degree to which any knowledge representation, such as ontology, can ultimately be free of purpose and end use. Nonetheless, in practice ontologies can often be fruitfully distinguished from conceptual database schemas based on their level of generality, and conceptual, logical, and physical database schemas are distinguished based on their commitment to specific technologies. Another distinguishing criterion involves how entities are defined, in that ontologies tend to rigorously define their components using some axiomatic language with a broad capacity for inference and complex structure, whereas database schemas are often expressed in semicomputable formalisms such as notations based on diagrams (e.g., UML; Rumbaugh, et al., 1999) or declarative languages (e.g., SQL) with a relatively narrow capacity for inference and structure. In short, we use ontologies to mean expressive representations of broad domain knowledge intended to be applicable across many conceptual database schemas. We also therefore distinguish between ontologies and related collections of definitions such as lexicons, glossaries, taxonomies, and classification schemes (e.g., Struik et al., 2002), when the latter are not rigorously represented, nor highly structured, and do not address a general scope of content. When these latter collections do meet the two ontologic criteria of generality and expressiveness, we consider them a form of ontology.

Contents of the Proposed Pluton Schema

Identification of entities and their attributes is required prior to development of a logical database schema for igneous rocks. This involves isolating key entities and attributes that act as indices for connecting tables in a relational database and aiding subsequent searches—for example, a location, sample number, and pluton name. These are typically present in the many well-established pluton schemas (e.g., see Walker et al., this volume; Lehnert et al., 2000; Pearce, 1987), though the schema presented here integrates both analytical and field observations and has a hierarchical organization that can be extended easily beyond the information listed in Table 2.

Based on the attributes we think are commonly utilized by geologists to describe a pluton, we have developed a spatial database schema that organizes data describing the geochemistry, mineralogy, and radiometrics of igneous rock samples, and the geological bodies, enclaves, fractures, fabrics, textures, and

TABLE 2. FIELD-BASED AND LABORATORY-BASED ATTRIBUTES THAT DESCRIBE THE OBJECT: PLUTON

PLUTON
Pluton name
Quadrangle name for location
UTM coordinates (center of pluton)
Latitude/longitude
Section/township/range
Location (state, country)
FIELD AND GEOSPATIAL OBSERVATIONS
Sample number (primary key attribute)
Sample location
Surface sample
Elevation (in meters)
Core: depth from surface (in meters)
Fabric: foliated-magmatic/metamorphic
Strike, dip
Unfoliated (= massive)
Texture: equigranular, porphyritic
Seriate, cumulate
Color index (CI)
Alteration: major, minor
Distribution of enclaves
MAFIC ENCLAVES (those with CI > host)
Size: coarse, medium, fine
Orientation: dip, strike
Shape: discoid, lenticular, angular
Reaction state: reacted, unreacted
OTHER MAGMATIC ENCLAVES
Size: coarse, medium, fine
Orientation: dip, strike
Shape: discoid, lenticular, angular
Reaction state: reacted, unreacted
XENOLITHS
Size: coarse, medium, fine
Orientation: dip, strike
Shape: discoid, lenticular, angular
Reaction state: reacted, unreacted
Fracture: orientation-dip, strike, aperture sealed, open, spacing
Soil: thickness, type
Shape: Concordancy and discordancy of contact
Contact aureole: present or absent
Width of contact aureole

(*continued*)

TABLE 2. (*continued*)

ANALYTICAL DATA FOR WHOLE ROCK SAMPLES
Sample number (key attribute; same as used in Field geospatial observation category)
Image (hand sample)
Petrographic: modal data, image
Geochemistry: Major and trace element abundance
Isotope measurements
For radiogenic systems
U/Th/Pb
K/Ar
Rb/Sr
Lu/Hf
Sm/Nd
Others as needed
For stable isotopes: oxygen, deuterium, sulfur
Physical properties:
Magnetic susceptibility
ASTM test data (density, compressive strength, permeability, storage capacity, seismic velocity, resistivity, thermal conductivity)
ANALYTICAL DATA FOR MINERAL SAMPLES
Mineral chemistry
Ore-grade major and trace elements, image
Non-ore grade major and trace elements, images
Isotope measurements
For radiogenic systems
U/Th/Pb, images
Sm/Nd, images
Ar/Ar, images
Rb/Sr, images
Fission track, images
U/He ages
Stable isotope ratios: oxygen, deuterium, sulfur, nitrogen
Fluid inclusions, images, composition
Melt inclusions, images, composition
SOME CALCULATED/MODELED/OBSERVED INFORMATION AND LINKS
Crystallization sequence for pluton
Calculated intensive variables
Evolution of volatiles (related to magma batches)
P-T-X of melting in source (related to pluton)
Emplacement rate
Cooling rate
Paleomagnetic data
Exhumation

Note: The sample number is the primary key.

inclusions to which they are geologically and geographically related. Our pluton database system also accesses images of geologic samples taken at varying scales (Fig. 3). Hand samples, thin sections, and individual mineral images are linked to spatially documented samples and provide the end user a means to explore igneous rocks from outcrop to microscopic level. Because this logical database schema contains spatial features and over two dozen tables with relationships of varying cardinality, we have physically implemented it in the Environmental System Research Institute's (ESRI) geodatabase format. The geodatabase is essentially a spatially enabled relational database management system that adheres to the structure of Microsoft's SQL database software, thus enabling us to exploit the strengths of both systems.

In our proposed schema (Fig. 4), the Sample table contains the ssID primary key representing a sample's identifier,

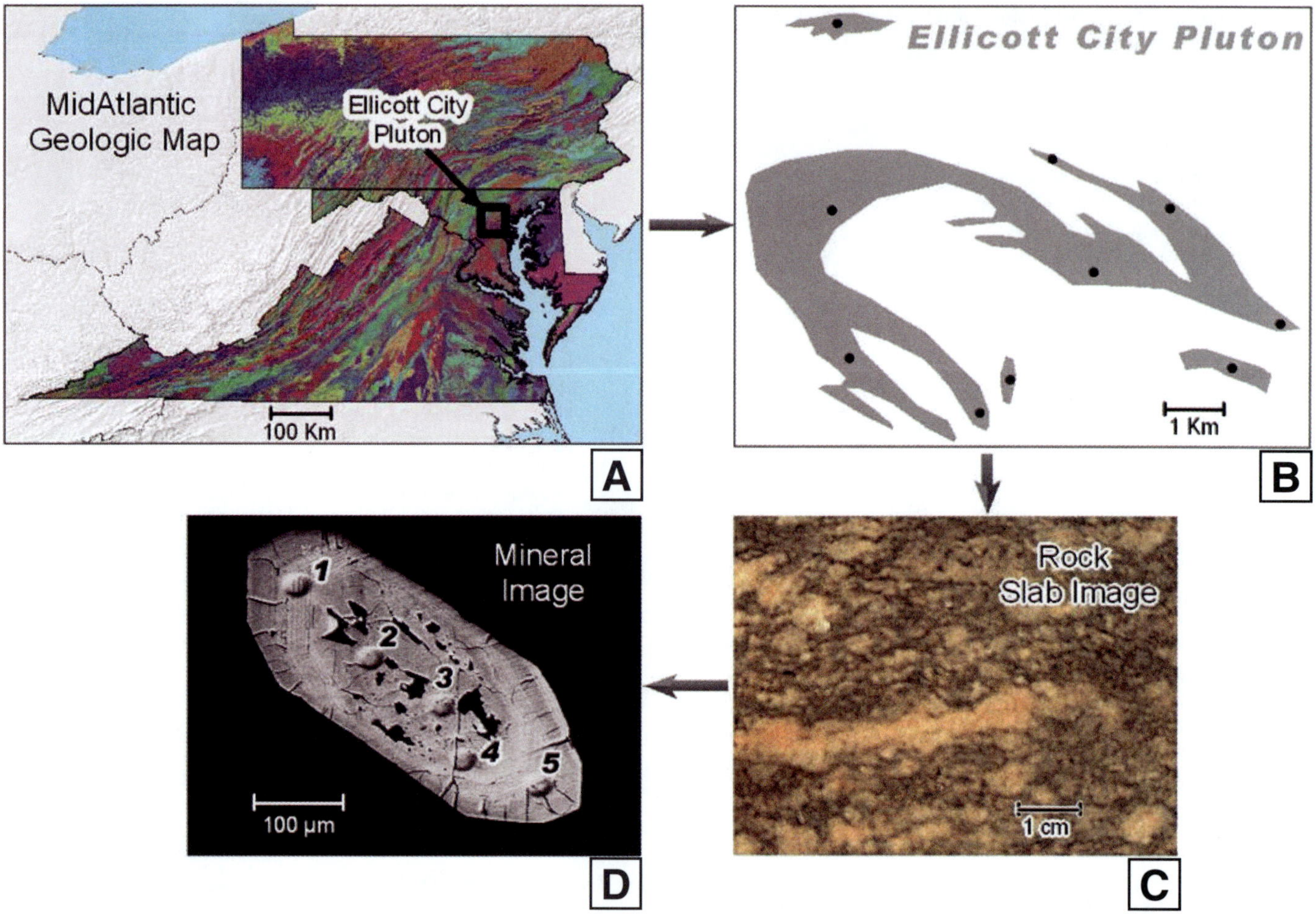

Figure 3. Graphical representation of scales of observations and associated data in a pluton database. Images are spatially referenced by ssID (primary key) over 11 orders of magnitude in scale, e.g., 10^9–10^2 km). A: GIS-enabled geologic map of the mid-Atlantic region of the Appalachian Mountains. B: Shape representation of Ellicott City pluton in Maryland that was mapped at a scale of 1:24,000. The ssID numbers for specific locations are the primary key in the schema. C: Sample image showing texture and fabric of a stained rock slab, information for which is given as a text file attached to ssID. D: Backscattered electron image of a mineral (zircon) from the sample showing micron-scale zoning and craters that mark the spatial location of analyses conducted by SIMS (secondary ion mass spectrometry) for measurement of U/Pb ages. Individual spots carry the same minID (unique mineral identifier number) but are cross-referenced from the image to the isotope analysis number in the radiometrics database.

described later in the section. This identifier links all tables that contain data describing a sample's modal, geochemical, isotopic, and mineral compositions, as well as the fractures, fabrics, textures, and inclusions to which they are related. The Sample table is also linked to the georeferenced SamplePoints entity that enables many of the tables in the schema to be displayed in GIS format. Because geologists prepare maps of plutons, including sample locations, at various scales, the spatial data include representation of pluton geometry at multiple scales. These table linkages (relations) also allow the database user to compare any number of pluton features. For example, pluton shapes may be compared at multiple scales, or amphibole composition(s) in a single sample may be compared with whole-rock geochemistry of the host rock. Visual features of rocks and minerals are also included, as they facilitate a better understanding of textures, structures, and sites of in situ microanalyses (e.g., U/Pb ages using the SIMS technique). The extension of information from the map scale to the micron scale is a key component of the schema.

All the analytically generated mineral and bulk rock data link to metadata regarding authors, citations, and analytical methods in the References field. For example, relationships exist between mineral, whole-rock geochemistry, U-Th-Pb isotope mineral and the AnalyticalMethods entity. As its name implies, the AnalyticalMethods entity contains information on the methods used to determine the geologic age of the mineral, such as SIMS (secondary ion mass spectrometry) and TIMS (thermal ionization mass spectrometry).

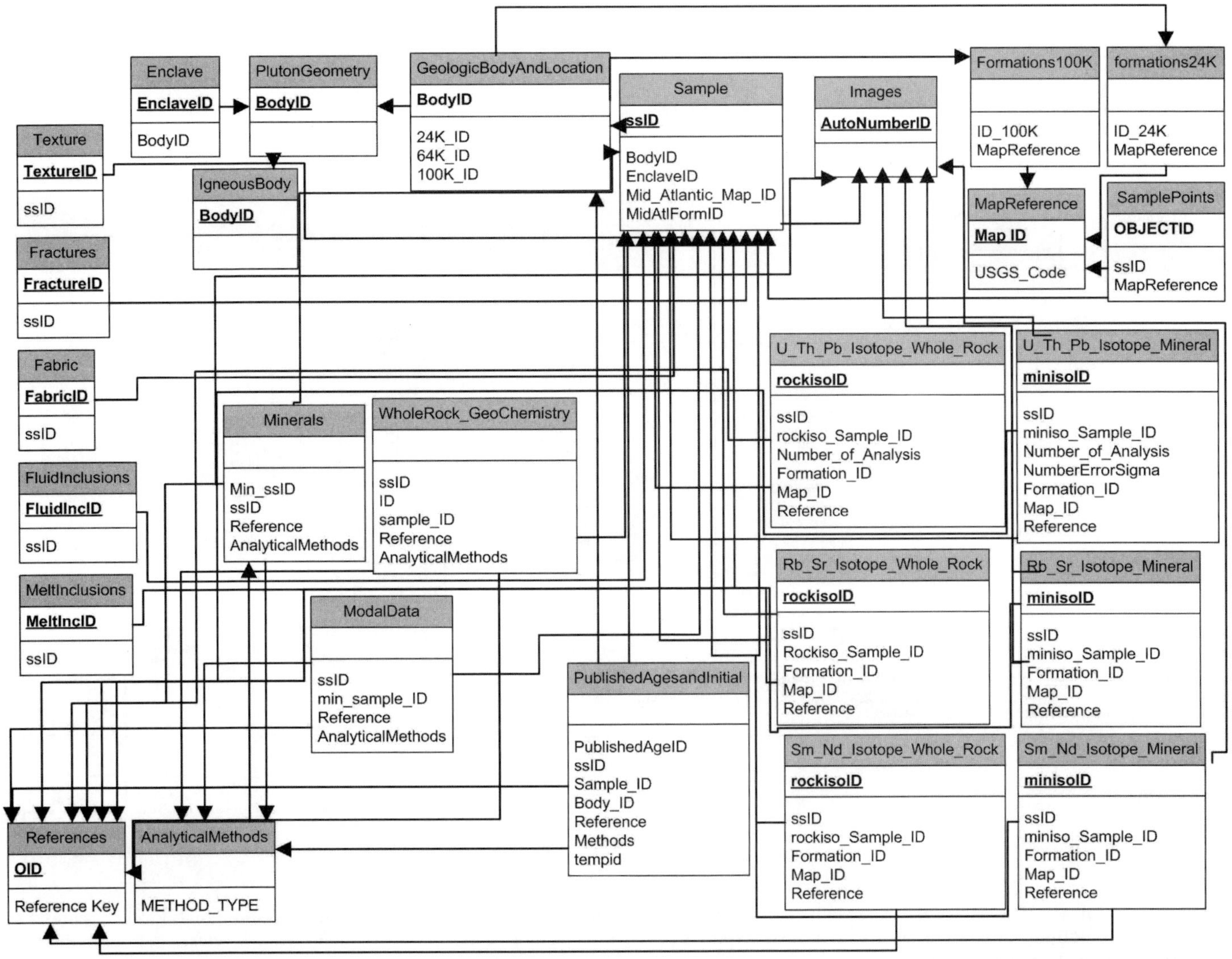

Figure 4. A simplified (logical) relational database schema for accessing information from plutons. The primary key that uniquely describes the geologic content of a sample is its ssID number. The arrows shown in the diagram are links from one table to another and demonstrate, for example, how ssID as a primary key is related to data in all tables.

INTRODUCTION TO OBJECT AND PROCESS ONTOLOGIES

Developing an ontology for igneous rocks requires differentiating between two types of entities, called endurants/perdurants or continuants/occurrents by philosophers and computational ontologists, but which we will simply call *objects* and *processes* (Grenon and Smith, 2004; Gangemi, et al., 2002). Earth science objects refer to physical features and materials that retain their identity over time and sometimes through physical changes. Examples of objects include plutons, rock materials, etc. Processes operate on objects to create the physical changes, and examples of processes include melting, transport, segregation, solidification, etc. In discrete processes, the relatively instantaneous transition points from one physical state to another are called events, whereas continuous processes have a temporal duration, with beginning and end points, and contain progressive states of transition (Sowa, 1999, p. 213–218)

Because the beginning and end points of continuous processes mark a noticeable change of physical state, these process are often also called events in geology. Consequently we use the term "event" to refer to a transition point in discrete processes, or wholly to a continuous process. An object ontology for igneous rocks could therefore contain entities (classes) for types of igneous rocks, their constituent elements, and minerals (see Fig. 5), whereas a process ontology for igneous rocks could contain the types of processes and events that operate on igneous objects, either creating them or operating on them (see Figs. 6 and 7).

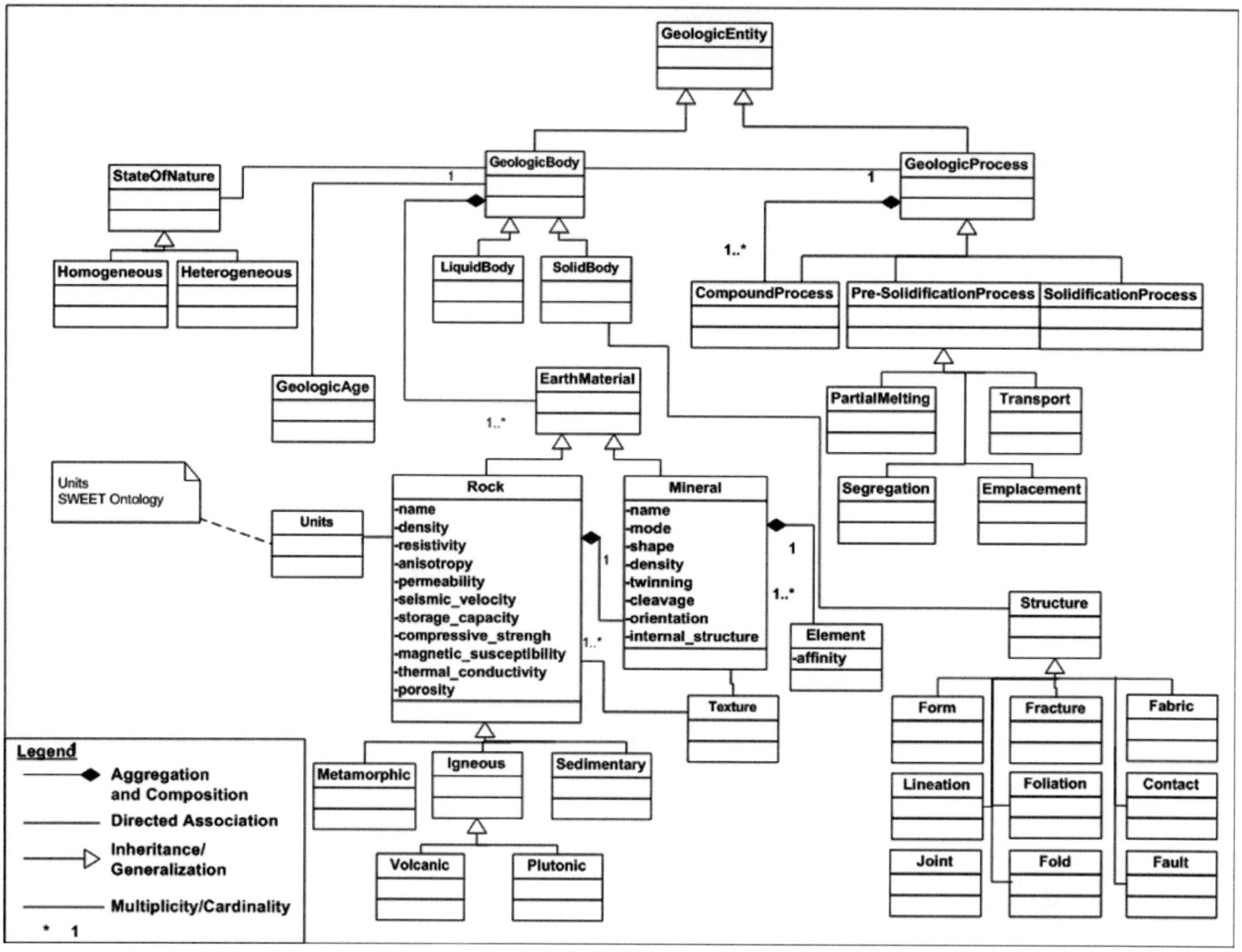

Figure 5. A UML (Unified Modeling Language) class diagram representing an informal pluton object ontology. At a high level of representation, the GeologicBody class is an entity, with attributes (properties), which includes subclasses with their own attributes. These properties can represent analytical data (e.g., geochemical, radiometric) presented in the database schema for plutons.

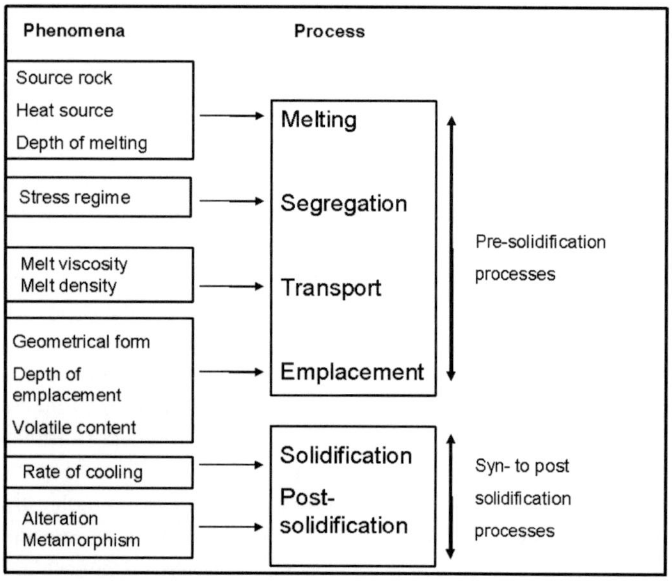

Figure 6. A graphic view of multiple processes that are linked together through process, time, tectonics, and observational relationships. Extracting knowledge from the igneous rocks requires the integration of the entire chain of processes.

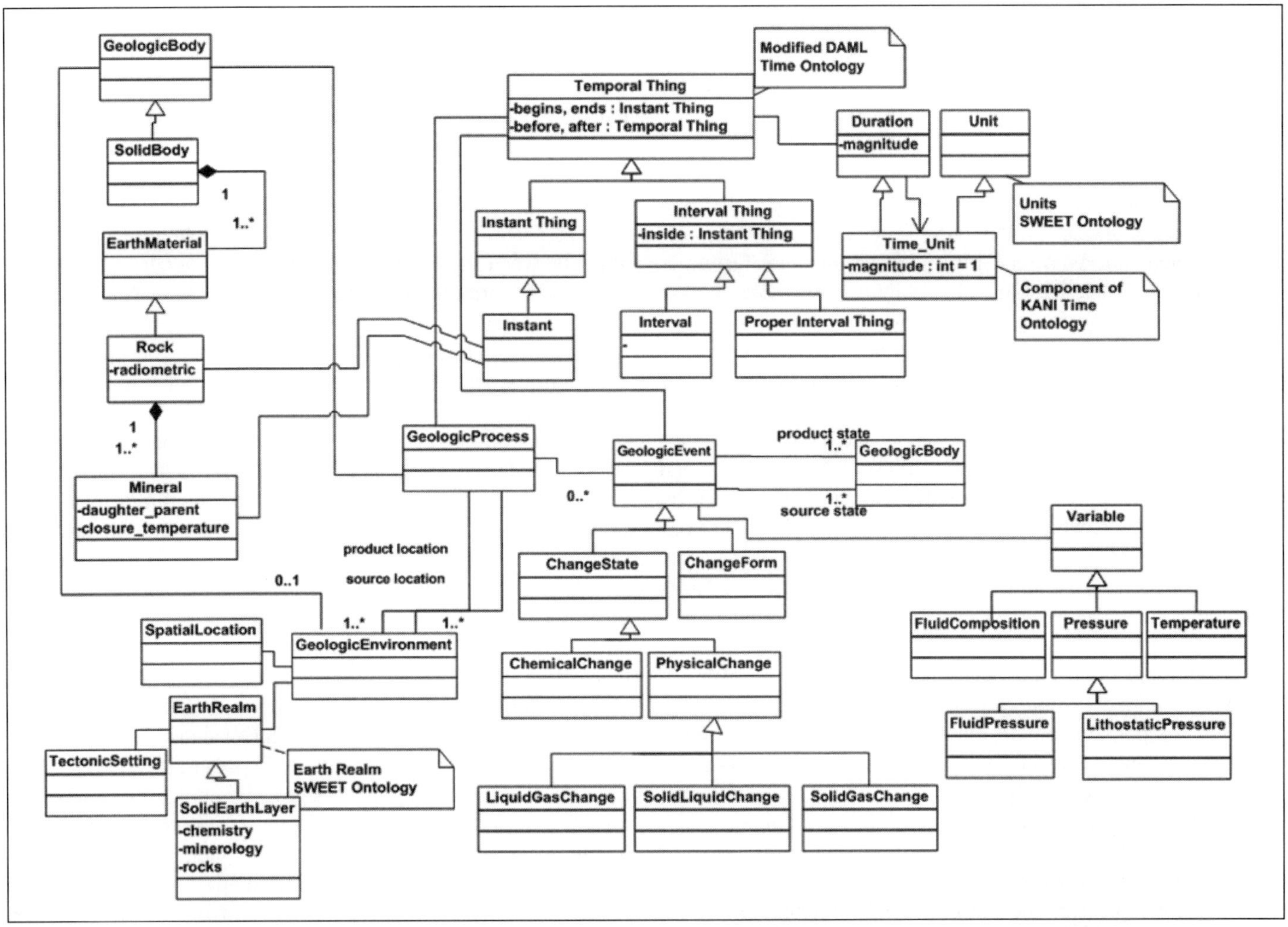

Figure 7. Class diagram representing the relationships between the object (plutonic rock) and the process that operates on the object. The GeologicProcess class has three subclasses: solidification process (includes syn- to postsolidification events), presolidification process, and compound process (Figs. 5 and 6). Presolidification process represents transport, segregation, emplacement, or partial melting. Compound process, on the other hand, consists of one or more geologic processes. For instance, a process that starts with partial melting and ends with solidification is a compound process. The concepts of temporal relations are also represented in the diagram. The geologic process operates on objects over a time interval. For instance, solidification process occurs over a period of time (cooling period). The GeologicProcess class also contains the GeologicEvent class, which represent the change of state or form of a source object.

These two types of ontologies represent fundamental categories of knowledge that are applicable to the igneous domain.

It is also important to further distinguish here between theories and models in these two types of ontologies (see Brodaric and Gahegan, this volume)—theories refer to abstractions that apply to many places and times, whereas models are situation-specific. For example, a distinction can be made between a type of process, represented theoretically, and its instances in a specific space-time situation, represented concretely in a model to explain the origins of objects in the situation, e.g., "melting as a type of process" and "the melting process and related events that yielded the object Pluton X." Likewise for objects we can consider "Pluton X" as summarized on a map legend versus its spatial representation as a series of polygons on the map face.

Significantly, research questions are structured around gaps between theoretical and situational representation. For example, geologists commonly use prior knowledge about types of processes and objects (theoretical) to help derive models consisting of interpreted processes and objects that explain observations and measurements in an area (situational). In transitioning this method to digital environments, we can envision theoretical knowledge to be stored in ontologies whereas situational knowledge, typically about objects (e.g., plutons), can be stored in databases, thus allowing possible process models for an area to be constrained by (1) theoretical knowledge of how processes might interact with the observed or measured objects, and (2) comparisons with other models (situational). An integrated object-process/theoretical-situational ontology is therefore necessary for computer-aided research into igneous processes and events.

Object Ontology

Development of ontologies is a multistage activity. With few available ontologies in the solid earth sciences, e.g., PetroGrapher (Abel et al., 2002), ParaDIS (Schmatz et al., 1995), and SWEET (Semantic Web for Earth and Environmental Terminology) (Raskin, 2004; Raskin, this volume), new research programs will develop prototype ontologies through iterative processes. Although many methods for developing ontologies have been identified, we follow the general methodology described by Uschold and Gruninger (1996). The first stage is the use of informal techniques that identify the concepts and their relationships within a domain (e.g., Sinha et al., 2003) (Table 2). These can then be represented by the design and construction of a high-level conceptual schema, which depicts the different classes of entities within a domain and their relationships (Elmasri and Navathe, 2004). To reiterate from above, the conceptual schema organizes a geologist's knowledge about the entity without details associated with retrieval, computations, and implementation. It is also important to recognize that conceptual relationships within a domain are unlikely to be uniformly represented by experts, as details of subdomains within the larger disciplinary framework are still unknown. The second stage involves transforming the conceptual schema into a rigorous ontology by representing it in some formal logical language such as OWL (McGuinness and van Harmelen, 2003). We restrict our discussion in this paper to the first step of this process.

Figure 5 represents the object ontology for plutonic rocks as a UML class diagram (Rumbaugh et al., 1999), which is a representation that describes the relationships between different classes of entities and their attributes, which in turn can be used to fully develop formal ontologies expressed in some logic-based language. In Figure 5, the Plutonic class is a subclass of the Igneous class, which is a subclass of the Rock class, which in turn is a subclass of the EarthMaterial class. The Plutonic class inherits from these superclasses attributes and relations such as mineral and texture. An attribute can also represent a concept. The Structure class is an attribute of the SolidBody class, one that has further subclasses. Specific measurements of attributes (single or multivalued) taken in the field or laboratory are also interrelated to each other through relationships and rules of cardinality. For example, the GeologicBody class is related to the EarthMaterial class through a composition relationship, as well as an explicit cardinality expression indicating that a geologic body is composed of one or more earth materials (i.e., rocks and minerals). Similarly, the Rock class consists of minerals, which contain one or more elements. The class diagram also provides a useful framework to categorize the common concepts in the domain. This framework will also enable other scientists to identify and contribute additional conceptual schemas toward the development of more formal ontologies.

Process Ontology

Because of the emerging nature of ontologic research in earth sciences, no complete ontologies for geologic processes are available for reference or refinement. Therefore, as a start, this paper presents a high-level conceptual schema for geologic processes that produce or modify igneous rocks. As such, we emphasize the need to develop a logical understanding of the relationships between processes at the subdomain level, and attendant vocabularies prior to the creation of formal process ontology. For example, as shown in an overview diagram (Fig. 6), it is easy to recognize that there are many interrelated concepts that are woven in the logic of interpreting igneous rocks. It is clear that in the category of presolidification events, the process of partial melting is the key prerequisite for the formation of igneous rocks. Associated with this process are issues of sources, depth of melting, and the tectonic setting of the melting event. Additional processes represented in the presolidification hierarchy are segregation, transport, and emplacement. Syn- to postsolidification processes include phenomena such as the rate of cooling or alteration. It is clear that many phenomena act as operators on any given process, and the cumulative effect of such actions leads to the final product, i.e., igneous rock. The recognition of the characteristic contribution of each process to the final physical form (i.e., shape of pluton, or its chemical signature) is often difficult to assess and leads to multiple hypotheses as discussed for scenario II. The development of process ontologies requires the integration of phenomena and events associated with the entire hierarchy, as represented in Figure 6. As such, linking processes and concepts within an ontologic framework provides the full relationships that exist during formation of an igneous rock, from the partial-melting event to the solidification event, and is similar in concept to the feedback loop applied to granitic plutons (Petford et al., 1997). The physical and chemical properties contained in databases are the result of the interplay of all the processes that affect the nature of the rocks; therefore, the presence of certain geochemical patterns or textures may be related to such things as degree of partial melting, composition of source rocks, pressure and temperature of melting or differentiation, and rate of cooling associated with transport.

The most primitive concept included within igneous rocks is the change of state that occurs from a solid to a magma (melt plus solids = partial melting event) under various physical conditions. The concept of a source region is intimately related to its location within the earth, and can be readily located within earth realms as represented in the SWEET ontology (Raskin, 2004). However, the source *region* is much less important to petrologists than the source *rock*. We can readily extract the nature of the source rock from the data, but it is not nearly so simple to determine the source region; it requires knowledge from other disciplines and databases. For example, the high-level concept of earth realm (contained in SWEET ontology) includes phenomenon, property, substance, space, and time, which translate conceptually to include the solid earth with the crust as part of the realm, and igneous rocks as substances. Because of the nature of the research activity, the emphasis of the SWEET project emphasizes the entire earth, including the oceans and atmosphere, whereas this paper details the subclasses that are required for pluton-specific research. For example, a robust ontology of source rocks (crust, mantle, core) must include (1) knowledge about representative rocks as a function of depth and pressure (located within reference data sets) and (2) access to results of experimental melting studies that provide compositional and physical information on the liquids (melts) and solids that may be produced from a given source during partial melting.

It is well recognized that the geochemical, isotopic, and mineralogic properties of igneous rocks are in many ways reflections of both the source and the environment of melting. Although the compositional or isotopic signatures of igneous rocks can be used to infer possible compositions of source rocks, it is commonly difficult to identify unique source conditions. For example, application of the concept of aluminum saturation index may be used to infer metasedimentary sources. However, this type of logic must be modified by the realization that peraluminous tonalitic magmas may also be derived by partial melting of metabasic source rocks. Therefore, knowledge that is required to interpret the processes that yield magmas and subsequently plutons, and whose data can be modeled to infer such processes, requires a conceptual schema that links presolidification processes to product state, composition, or a measurable event (Fig. 8). For example, among the simplest types of granites (e.g., low-*T* leucogranites), experimental data on the compositions of liquids formed by partial melting of a metasedimentary source at various pressures and temperatures (contained within a reference database) can be compared to observed compositions to infer both the degree of melting and the intensive variables associated with the event (e.g., Patiño Douce, 1999). For more complicated but essentially closed igneous systems, iterative application of partial melting algorithms can result in a close match between computed liquid compositions and observed values stored in databases. Such activities require the explicit representation of concepts of compatible and incompatible elements to permit the selection of appropriate partitioning coefficients of elements involved in the partial melting event. During partial melting, many of the properties of magma are a reflection of the source rocks. Therefore, if two magmas with independent properties physically mix, the mixture has hybrid properties. Therefore, the concept of heterogeneity represents a complex scenario in which multiple processes are able to mediate changes in objects, such as melts and magmas. The design of a high-level process ontologic framework for igneous rocks requires distinctions between states and their properties, transition points between states (events), and the multiple processes that operate the states (magma).

Still more complicated closed and open magmatic systems owe their chemical compositions to differentiation during some combination of processes that involve segregation, transport, emplacement, and solidification; all these processes are in addition to partial melting. Segregation, transport, and emplacement processes are extremely complex, as they are mostly inferred via a combination of field data, textural observations, geophysical information, and numerical simulations. Segregation processes are likely to involve concepts of convection, contiguity, compaction, filter pressing, and buoyancy (Wickham, 1987; Miller et al., 1988). Transport processes involve concepts of multiphase flow, changes in viscosity as a function of temperature and crystal/melt proportions, ascent velocity, and internal shear (Petford et al., 1997). Each of these processes is capable of imposing a chemical/mineralogic signature on the magma but may be very difficult to evaluate because of possible overprinting by other processes. Emplacement processes, although equally complex, may provide

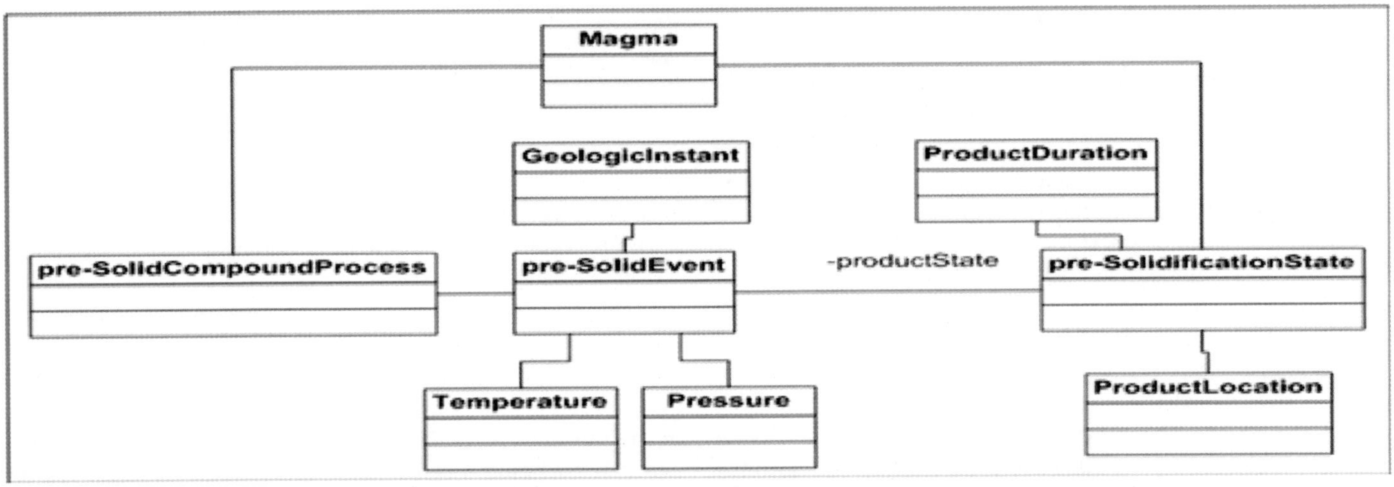

Figure 8. Representation of a presolidified state of igneous rock as an instantiation of the process and object ontologies for plutons. It describes the potential action of a presolidification process, including subprocesses (partial melting, segregation, transport, emplacement), operating on a state of magma at a certain place and time.

physical evidence of late-stage magma movement in the form of magmatic fabrics which include alignment of crystals and flow features (also measured and recorded in data tables as magnetic susceptibility measurements) and provide conceptual links to chamber convection and intensive variables (pressure, temperature, fluid content).

Time Ontology and Its Application

Processes that impart a geologic signature to the pluton are also associated with rates and the duration of individual processes. The aggregate of events that make up the geologic history of the pluton is well suited for the application of time ontology. Although this subject has received a great deal of attention from both the artificial-intelligence and business community, as well as those who perceive time as a philosophical concept (Hobbs, 2002; Zhou and Fikes, 2000; Schreiber, 1990), its application to earth science processes is limited (e.g., NADM, 2004). The temporal aspects of the properties of entities are well documented for the business community through the need to query text narrative in a Web environment (Hobbs and Pustejovsky, 2003). Issues of representation of "(1) identifying an event and anchoring it in time, (2) ordering events with respect to each other, (3) reasoning with contextually underspecified temporal expressions, and (4) reasoning about the persistence of events" are well suited for implementation in the geologic community. We utilize the information from ongoing research (e.g., Zhou and Fikes, 2000) to develop a high-level representation of time and its application to igneous rocks (Fig. 7). We suggest that the concepts of temporal relations that are applicable to the earth science domain include instants and intervals of time. Instants of time are considered to be time points that correspond to real numbers on a time line. The reference value is the present day (time = 0), and granularity is in years. Time units will vary on the basis of experiments; our emphasis on recording time encapsulated in igneous rocks is easily represented by the year or its multiples. The interval of time represents recognition of beginning and end points. For example, a melting event (formation of a magma) occurs over an interval of time, and magmas crystallize (solidification event) over an interval of time. The melting event is represented by the process of fusion of solids of varying enthalpies of fusion, and necessarily at different times, induced by changes in intensive variables (pressure, temperature, fluids). The time required to produce a melt can vary from seconds (meteorite impacts) to several thousands of years and is commonly modeled rather than analytically measured. However, when possible, the characteristic representation of this event in earth science is contained in the application of isotope geochronology, where solids formed from the melt retain a distinctive radioactively stable parent-daughter relationship (e.g., Faure, 1986). The ratio P/D (parent/daughter) coupled with its isotopic abundances is utilized to compute values of absolute time (in years). Ontologically, it is important to recognize that time measured for the same event is represented as a continuum of ages caused by the physical separation of solids (minerals) as a function of temperature, and to some extent the chemical properties of the radioactive parent. Although the dependency of time on temperature and the chemical nature of the solid phase appear to be a derived property of time, it is integral to the representation of time in the geologic context. As such, our proposed ontology includes a relationship to the "kind of solid" to yield useful measure of time.

It is possible to visualize that a solidification event can be represented as multiple instants of time related to incremental change in state (liquid to solid) as physical conditions warrant. At the same instant, based on nucleation kinetics, solid phases (e.g., minerals) begin to form and become the recorders of time associated with that instant event, i.e., change of state from liquid to solid. However, the isotopic clocks that measure time in different minerals are not activated at the same time, even though the minerals themselves were formed at the same instant. A separation in measured time (a function of blocking temperatures) for the same event of crystal formation requires a new way to represent the relationship between time and event. The present concept that an instant of time is related to a specified event is unable to represent time-solid-temperature relationships. We suggest that geologic measurements of absolute time be coupled with knowledge that it represents a temperature history, as much as a record of an event. For example, if the measurement of absolute time associated with the different phases of the Tuolumne Intrusive Series had not been undertaken utilizing the same mineral and the P/D system (Coleman et al., 2004), it is unlikely that the model implying multiple batches of magma associated with the construction of the complex would be acceptable to the community. The implication is that time measured in geologic materials is subject to the concept of closure temperatures, as well as the composition of the host solid containing the P/D system.

SUMMARY AND FUTURE DIRECTIONS

The use of disciplinary object ontologies is a requirement for navigation across databases with varying content and concept. Our paper shows the need to expand this capability to one in which scientists can examine the relationship of external factors from other disciplines that may influence our understanding of a disciplinary process. For example, we need to link tectonic processes (e.g., rifting, subduction, and delamination) to igneous rock processes, or to metamorphic processes as well as the logical transition from high-grade metamorphism to partial melting. We also recognize the need to go from analysis of numerical data to connection of concepts and data inherent in thermodynamics, kinetics, heat flow, mass transfer, etc.

We also consider whether modeling of underdetermined systems can provide directions for future research. Deficiencies in existing models potentially highlight areas of new research and understanding. For example, the multiple explanations for petrogenesis, emplacement, and duration of the Tuolumne Intrusive Series imply highly underdetermined models. This activity works on at least two levels—the conceptual-theoretical level (ontology/theory development) and the physical-situational level (data-driven modeling)—and therefore can help reveal gaps in understanding at the two levels. In our view, use of IT infrastructure is an important part of this activity. Making these gaps apparent through explicit representation and computation mechanisms such as those envisioned in cyberinfrastructure can improve our ability to more rigorously model geologic processes. It then can be used to stimulate and direct future research.

The product of the research activity as summarized in this paper includes mostly a midlevel granularity of ontology that may be linked to those at higher levels, e.g., concepts outlined in the SWEET project. We emphasize the newness of such research in the earth science community and present some basic concepts of integration that are well recognized by geologists, but may be unfamiliar to the information technology scientists or not yet well represented formally. Issues of time and space are of fundamental significance to geology as they represent the evolutionary history of earth. We also emphasize that accessing data, although important, is only the first step in understanding the data for testing a known hypothesis or presenting a new one. The real challenge lies in developing capabilities that utilize process ontology as the key for developing the emerging field of geoinformatics

ACKNOWLEDGMENTS

The authors acknowledge the support of the National Science Foundation through grant NSF-EAR 0225588 awarded to A.K. Sinha. We are also grateful to Art Snoke, Bertram Ludäscher, Clinton Smyth, Dogan Seber, Chaitan Baru, and Murray Journeay for stimulating discussions on entity relationships for igneous rocks. We also acknowledge the input from the Advanced Petrology class at Texas Tech University, Lubbock. Thoughtful reviews by Nancy McMillan, Nathan Greene, and Hassan Babaie greatly improved the content and readability of the paper.

REFERENCES CITED

Abel, M., Silva, L.A., De Ros, L.F., Mastella, L., Campbell, J., and Novello, T., 2002, PetroGrapher: Managing petrographic data and knowledge using an intelligent database application, Intelligent Computing in the Petroleum Industry, ICPI, 12 p.

Bateman, P.C., and Chappell, B.W., 1979, Crystallization, fractionation, and solidification of the Tuolumne Intrusive Series, Yosemite National Park, California: Geological Society of America Bulletin, v. 90, p. 465–482, doi: 10.1130/0016-7606(1979)90<465:CFASOT>2.0.CO;2.

Coleman, D.S., Gray, W., and Glazner, A.F., 2004, Rethinking the emplacement and evolution of zoned plutons: geochronologic evidence for incremental assembly of the Tuolumne Intrusive Suite, California: Geology, v. 32, p. 433–436, doi: 10.1130/G20220.1.

Dodson, M.H., 1973, Closure temperature in cooling geochronological and petrological systems: Contributions to Mineralogy and Petrology, v. 40, p. 259–274.

Elmasri, R., and Navathe, S., 2004, Fundamentals of database systems: New York, Addison Wesley and Benjamin Cummings Publication, 1004 p.

Faure, G., 1986, Principles of isotope geology: New York, John Wiley and Sons, 589 p.

Gangemi, A., Guarino, N., Masolo, C., Oltramari, A., and Schneider, L., 2002, Sweetening ontologies with DOLCE, *in* Proceedings of the 13th International Conference on Knowledge Engineering and Knowledge Management: London, UK, Springer-Verlag, v. 2473, p. 166–181.

Glazner, A.F., Bartley, J., Coleman, D., Gray, W., and Taylor, R., 2004, Are plutons assembled over millions of years by amalgamation from small magma chambers?: GSA Today, v. 14, no. 4/5, p. 4–11, doi: 10.1130/1052-5173(2004)014<0004:APAOMO>2.0.CO;2.

Grenon, P., and Smith, B., 2004, SNAP and SPAN: Towards dynamic spatial ontology: Spatial Cognition and Computation, v. 4, no. 1, p. 69–103, doi: 10.1207/s15427633scc0401_5.

Gruber, T., 1993, A translation approach to portable ontology specifications: Knowledge Acquisitions, v. 5, p. 199–220, doi: 10.1006/knac.1993.1008.

Guarino, N., 1995, Formal ontology, conceptual analysis and knowledge representation: International Journal of Human-Computer Studies, v. 43, p. 625–640, doi: 10.1006/ijhc.1995.1066.

Hobbs, J.R., 2002, A DAML ontology of time: http://www.cs.rochester.edu/~ferguson/daml.

Hobbs, J., and Pustejovsky, J., 2003, Annotating and reasoning about time and events: Proceedings of American Association for Artificial Intelligence, Spring Symposium on Logical Formalization of Commonsense Reasoning, Stanford, California, 9 p.

Kistler, R.W., Chappell, B.W., Peck, D.L., and Bateman, P.C., 1986, Isotopic variation in the Tuolumne Intrusive Suite, central Sierra Nevada, California: Contributions to Mineralogy and Petrology, v. 94, p. 205–220, doi: 10.1007/BF00592937.

Lehnert, K., Su, Y., Langmuir, C.H., Sarbas, B., and Nohl, U., 2000, A global geochemical database structure for rocks: Geochemistry, Geophysics, Geosystems, paper number 1999GC000026.

McGuinness, D., and van Harmelen, F., editors, 2003, OWL Web ontology language overview: http://www.w3.org/TR/owl-features/#s1.

Miller, C., Watson, E.B., and Harrison, T.M., 1988, Perspectives on the source, segregation and transport of granitoid magmas: Transactions of the Royal Society of Edinburgh: Earth Sciences, v. 79, p. 135–156.

NADM, 2004, North American Geologic Map Data Model Steering Committee, NADM Conceptual Model 1.0, A Conceptual Model for Geologic Map Information: U.S. Geological Survey Open-File Report 2004-1334; Geological Survey of Canada, OPEN FILE 4737, (CD-ROM in press).

Paterson, S.R., and Vernon, R.H., 1995, Bursting the bubble of ballooning plutons: A return to nested diapirs emplaced by multiple processes: Geological Society of America Bulletin, v. 107, p. 1356–1380, doi: 10.1130/0016-7606(1995)107<1356:BTBOBP>2.3.CO;2.

Patiño Douce, A.E., 1999, What do experiments tell us about the relative contributions of crust and mantle to the origin of granitic magmas?, *in* Castro, A., et al., eds., Understanding granites: Integrating new and classical techniques: Geological Society [London] Special Publication 168, p. 55–75.

Pearce, J.A., 1987, An expert system for the tectonic characterization of ancient volcanic rocks: Journal of Volcanology and Geothermal Research, v. 32, p. 51–65, doi: 10.1016/0377-0273(87)90036-9.

Petford, N., Clemens, J.D., and Vigneresse, J., 1997, Application of information theory to the formation of granitic rocks, *in* Bouchez, J.L., et al., eds., Granite: From segregation of melts to emplacement fabrics: Boston, Massachusetts, Kluwer Academic Publishers, p. 3–10.

Raskin, R., 2004, Semantic Web for Earth and Environmental Terminology (SWEET): http://sweet.jpl.nasa.gov/ontology/.

Reid, J.B., Evans, O.C., and Fates, D.G., 1983, Magma mixing in granitic rocks of the central Sierra Nevada, California: Earth and Planetary Science Letters, v. 66, p. 243–261, doi: 10.1016/0012-821X(83)90139-5.

Rumbaugh, J., Jacobson, I., and Booch, G., 1999, The Unified Modeling Language reference manual: Reading, Massachusetts, Addison-Wesley, 550 p.

Schmatz, D.R., Engi, M., and Lieberman, J.E., 1995, ParaDIS: A relational database for the consistent documentation and analysis of metamorphic mineral assemblages: Computers and Geosciences, v. 21, p. 1031–1041, doi: 10.1016/0098-3004(95)00039-B.

Schreiber, F.A., 1990, A breviary to time concepts for information systems: Revista di Infomatica, v. 20, p. 59–68.

Schumm, S.A., 1991, To interpret the Earth: Ten ways to be wrong: New York, Cambridge University Press, 133 p.

Sinha, A.K., Ludäscher, B., Brodaric, B., Baru, C., Seber, D., Snoke, A., and Barnes, C., 2003, GEON: Developing the cyberinfrastructure for the earth sciences: http://geon.geol.vt.edu/geon/pubreps/workshop.pdf.

Sowa, J.F., 1999, Knowledge representation: Logical, philosophical, and computational foundations: New York, Brooks Cole Publishing, 594 p.

Spear, F.S., 1993, Metamorphic phase equilibria and pressure-temperature-time paths: Mineralogical Society of America Monograph 1, 799 p.

Struik, L.C., Quat, M.B., Davenport, P.H., and Okulitch, A.V., 2002, A preliminary scheme for multihierarchical rock classification for use with thematic computer-based query systems: Geological Survey of Canada, Current Research 2002-D-10, 9 p.

Uschold, M., and Gruninger, M., 1996, Ontologies: Principles, methods and applications: Knowledge Engineering Review, v. 11, 63 p.

Wickham, S.M., 1987, The segregation and emplacement of granitic magmas: Geological Society [London] Journal, v. 144, p. 281–297.

Zhou, Q., and Fikes, R., 2000, A reusable time ontology, Knowledge Systems Laboratory: http://www.isi.edu/~pan/damltime/time-entry.owl.

Manuscript Accepted by the Society 18 June 2005

Printed in the USA

Geological Society of America
Special Paper 397
2006

GeoNet: Use of grid technologies in geoinformatics for the transition zone between the Colorado Plateau and the Basin and Range province

Leonardo Salayandia
Yuan Huang
Ann Q. Gates
Steven Roach
University of Texas, Pan-American Center for Earth and Environmental Studies, El Paso, Texas 79968, USA

ABSTRACT

Geospatial Network (GeoNet) is a distributed information system accessible through the Internet that will permit users from the geoscience community to share and contribute knowledge from the transition zone between the Colorado Plateau and the Basin and Range province. Specifically, the goals of GeoNet are to facilitate the access, analysis, and visualization of geospatial data for use by scientists, policymakers, educators, and the general public. The types of geospatial data provided through GeoNet include satellite imagery, gravity, heat flow, digital elevation models, remote sensing, seismicity, active faults, and basic geographic data. This paper, written as a primer, describes how state-of-the-art technology is being used to realize GeoNet's goals. It also discusses GeoNet's layered architecture and shows how this architecture supports software applications deployed as service-oriented components. Ontologies based on the emerging OWL Web ontology language standard are used to enhance knowledge discovery. Use-case scenarios describe the features of GeoNet and the expected sequences of interaction between GeoNet and different classes of users. GeoNet's design will permit its functionality and data sets to be integrated with the GEON (Geosciences Network) grid and, thus, will contribute to the larger geoscience community.

Keywords: geoinformatics, geospatial, Internet technologies, ontology, Basin and Range province, Colorado Plateau, GEON, GeoNet, grid services, OWL, southwest United States, use cases, Web services, XML.

1. INTRODUCTION

The geoscience community has been acquiring and storing large amounts of geologic data. There have been recent efforts, e.g., EarthScope [8], GEON [11], and NEES [13], that have focused on advancing science and the understanding of Earth by (1) making geospatial data available to scientists, policymakers, educators, and the general public, and (2) providing tools that facilitate and promote the integration and analysis of the data. To contribute to these efforts, the departments of Geological Sciences and Computer Science at the University of Texas, El Paso, and Geological Sciences at Arizona State University are collabo-

Salayandia, L., Huang, Y., Gates, A.Q., and Roach, S., 2006, GeoNet: Use of grid technologies in geoinformatics for the transition zone between the Colorado Plateau and the Basin and Range province, *in* Sinha, A.K., ed., Geoinformatics: Data to Knowledge: Geological Society of America Special Paper 397, p. 183–193, doi: 10.1130/2006.2397(13).

rating to establish an integrated geospatial information system for the transition zone between the Colorado Plateau and the Basin and Range province. This system, called the Geospatial Network (GeoNet), will consist of distributed data repositories and applications that are accessible through a portal on the Internet.

GeoNet will provide access to geospatial data sets, storage and validation of new data sets, consolidation of data from multiple sources, and integration of the aforementioned types of data. In addition, it will provide a toolkit for data analysis and modeling, which can generate data visualization products (e.g., maps, cross sections and three-dimensional views). The tools included in GeoNet target users and contributors of all levels and are categorized as:

1. input, storage, and extraction applications that enter new data into the system, clean and validate data, access existing data and metadata, and extract data on a geographic, temporal, band, or layer basis

2. related applications that have the ability to modify and reorient existing data to develop new data sets

3. geographic applications that georeference and georectify data and recast data into standard data formats where applicable

4. processing applications that perform data manipulation specific to geospatial data types, e.g., reduction of gravity data

5. interface-support software needed to communicate with commercial software packages

From an applied viewpoint, GeoNet will enable study in areas such as urbanization, agricultural productivity, natural change, and water resources. By facilitating the integration of geospatial data, it also will enable interdisciplinary research. GeoNet will eventually become integrated into the GEON grid [11], a geoinformatics [9] initiative that centers on managing vast amounts of earth science data.

This paper discusses how GeoNet is being designed to achieve an open and flexible information system through which members of the geoscience community can populate and maintain data sets and ensure their quality. In addition, it describes the technologies and approaches that are being adopted to address problems that have hindered development of effective geoscience information systems in the past, e.g., the lack of: standard formats for storing geospatial data, orderly distribution of available geospatial data, documentation about data, and accessible tools for managing, modeling, and analyzing data. Section 2 provides an overview of GeoNet. Section 3 describes some of the supporting technologies that enable the creation of GeoNet. In section 4, the data organization, services, and communication structure that constitute the GeoNet framework are discussed.

2. OVERVIEW OF GEONET

This section describes GeoNet's geographical area of study, data sets, and its relationship to the GEON project. It also provides use-case scenarios of the system, which describe GeoNet from a user's perspective. The section concludes by providing a description of the general architecture of GeoNet.

2.1. Description of Area and Data

GeoNet focuses on the transition zone between the Colorado Plateau and the Basin and Range province, primarily Arizona, New Mexico, northern Mexico, and portions of California, Nevada, Utah, Colorado, and Texas, as shown in the shaded relief map in Figure 1. This region is significant in understanding the geologic evolution of southwestern North America. The project leverages considerable support from several federal and state agencies including NASA, the U.S. Geological Survey, the National Imaging and Mapping Agency, and the Arizona Geological Survey.

The Colorado Plateau is the broad higher-elevation region of the Four Corners area that is separated from the relatively lower-elevation Basin and Range province by the transition zone. In New Mexico and Colorado, the Basin and Range province is manifested as the narrow Rio Grande rift, whereas it is broader and variably active in Arizona, Utah, Nevada, Texas, and Mexico. The westernmost portion of the study area touches the southern San Andreas fault system in California and the Gulf of California.

The data sets that will be accessible through GeoNet are gravity, aeromagnetic, electromagnetic, seismic reflection, broadband seismic, seismicity/earthquake, heat flow, remote sensing (multispectral), drill hole data, geologic maps, faults, geochemistry/petrology, crustal stress and strain, digital elevation models, and remote sensing (e.g., SAR). The sources of the data include University of Washington, Incorporated Research Institutions for Seismology (IRIS), University of Nebraska, U.S. Geologic Survey (USGS), Arizona State University (ASU), and the University of Texas, El Paso (UTEP). Each data set will have associated metadata that include information such as person and organization responsible for collecting data, the date data were collected, specifications of the instruments used to collect the data (if appropriate), person and organization responsible for the validation of data and date when this process was completed, as well as other information deemed pertinent to each specific data set.

2.2. Relationship to GEON

In order to study the dynamics of earth science, geoscientists must integrate and share multidisciplinary data sets. The NSF-funded Geosciences Network (GEON) project represents a collaborative effort among researchers from a broad cross section of information technology and earth science disciplines [11]. The information infrastructure created by GEON will allow scientists to share databases and tools, enabling the interdisciplinary research required to discover relationships among earth science disciplines, e.g., the relationship between tectonics and the evolution of sedimentary basins, broader predictive understanding and modeling capabilities of geologic hazards such as earthquakes and volcanoes, the four-dimensional reconstruction of Earth through time, and the management of natural resources. The need to manage the growing amounts of diverse earth science data was

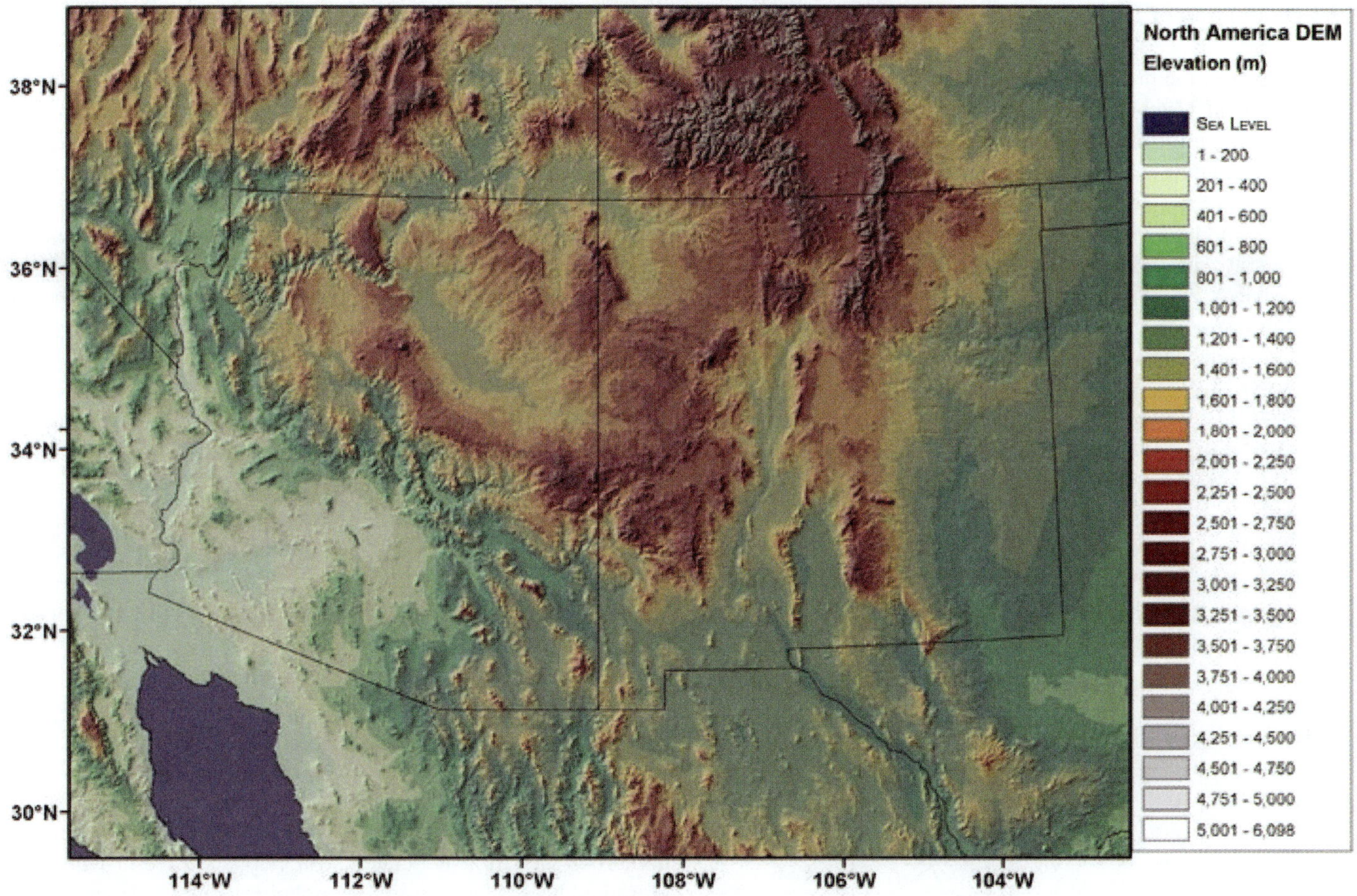

Figure 1. Shaded relief map of the study area [10]. A 1 km U.S. Geological Survey (USGS) digital elevation model (DEM) of the study region: Arizona and New Mexico. The image was plotted in ArcGIS by draping the color scheme over shaded relief.

recognized during a series of NSF-sponsored meetings on geoinformatics [9]. The vision of GEON is to develop a cyberinfrastructure that will integrate multiple database systems and allow for tool sharing across multiple computing systems, facilitating the geoscientists' ability to understand the dynamics of earth science. GeoNet will be integrated into the GEON grid.

The GeoNet effort will develop the toolkits to use and integrate the geospatial data sets. GeoNet is being designed to allow communication, distribute computation resources, and share data among heterogeneous systems by adhering to grid computing standards proposed by the Global Grid Forum (GGF) and described in the Open Grid Services Architecture (OGSA) documentation (discussed in section 3.4).

The information technology challenges for GeoNet and GEON relate to the distributed data and computational grid techniques required to meet the science goals of GEON. These include the formulation, design, and implementation of data modeling and the semantic integration of data sets distributed across heterogeneous computing platforms and across various subdisciplines within the research community. Other challenges include the formalization of various ontological blueprints and configuration of multiple computing systems. The research products and services arising from GEON will be available to the entire scientific community and will provide opportunities for research and collaboration.

2.3. User Perspective of GeoNet

Use cases describe the services of a software system that provide a visible result for an actor. Actors represent external entities that interact with the system, such as other software systems, and classes of human users, such as registered users who use the system for a particular purpose. Furthermore, use-case scenarios are utilized to describe sequences of interaction between the system and actors. The use-case diagram [5] shown in Figure 2 presents the behavior of GeoNet from an external perspective. The diagram depicts use cases as ovals, actors as stick figures, and relationships between use cases and actors as lines.

2.3.1 Description of Use Case Diagram

In GeoNet, the principal actors include:

- **General User**—user who accesses the system to obtain general information about GeoNet, requests data from selected data sets, and runs applications within the system
- **Registered User**—user who has registered with GeoNet and who logs on to the system to access or contribute geospatial data, ontologies, software applications and other services
- **Validator**—geoscience expert who validates submitted geospatial data, ontologies, software applications, and other services before making them available to GeoNet users
- **Distributed Geoscience Database**—set of database management systems that maintain geospatial data and ontology repositories
- **Application Software**—software that interacts with GeoNet to provide additional functionality, such as ontology editors, integration tools, and visualization tools

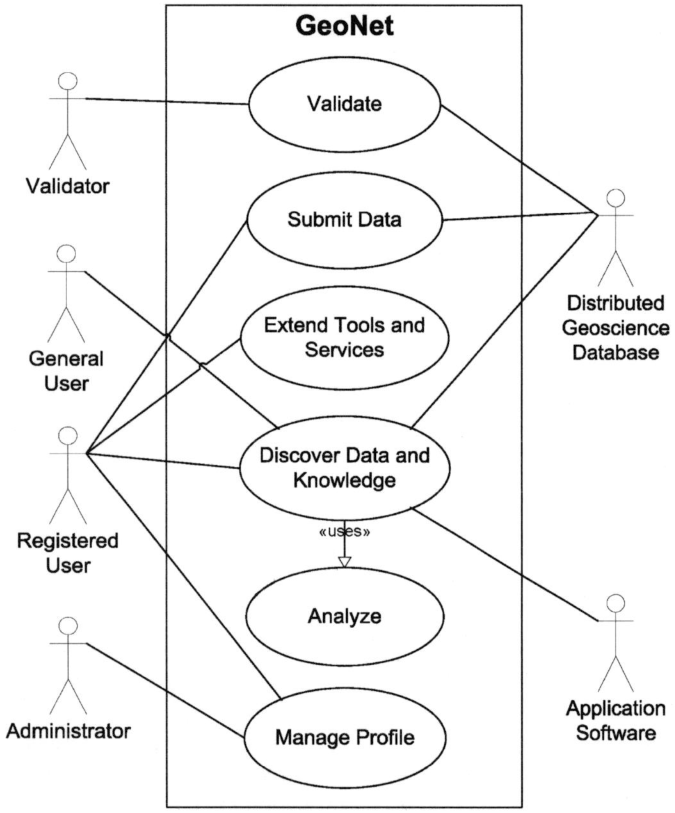

Figure 2. Use-case diagram for the GeoNet system. The use-case diagram shows a user perspective of the system. The ovals represent the different uses of the system; the stick figures represent the external entities that interact with the system.

- **Administrator**—manager of user profile information that includes user name, password, and access level for a particular class of user

The principal use cases for GeoNet include:

- **Validate**—services that aid a geoscience expert in validating geospatial data, ontologies, software applications, tools, and other services that are submitted by registered users and intended to be shared among the geoscience community
- **Submit Data**—services that allow registered users to submit geospatial data with the intention of making the data available to the community
- **Extend Tools and Services**—services that allow registered users to register software applications, ontologies, tools, and other services with the intention of making them available to the community
- **Discover Data and Knowledge**—services that allow users to locate geospatial data and information associated with the data
- **Analyze**—services that allow users to study and explore geospatial data
- **Manage Profile**—services that allow the administrator to manage user profile information and allow registered users to maintain their own

Each use case contains one or more scenarios, and each scenario may contain alternatives. Alternatives give variations to the main sequence of steps described in a scenario. There are numerous scenarios associated with GeoNet, three of which are presented here. In addition, optional alternatives are enumerated after scenario steps where appropriate. The steps that are replaced by an alternative are given in parentheses.

2.3.2. Scenario 1: Retrieve Gravity Data

This scenario describes the Discover Data and Knowledge use case given in Figure 2. The actors are the Distributed Geoscience Database, Application Software, and either a General User or a Registered User.

Discussion: In this scenario, a user wants to retrieve gravity data for a given geographic area. The gravity data will be presented in a table of longitude, latitude, and gravity data. When the data needed to satisfy the request are distributed on multiple database management systems, the retrieval service orchestrates the process of obtaining the data, integrating the results, and presenting the results to the user in the format requested. Furthermore, the data are presented in such a way that the user knows whether the data have been corrected by the U.S. Geological Survey (USGS).

Steps

1. The user selects the *retrieve data* option from the menu.
2. The system displays a map of the southwestern United States.

3. The user draws a rectangular footprint on the displayed map by holding down the right mouse button and dragging the mouse.
4. The system displays a footprint as a transparent box outlined in yellow as the user moves the mouse from the upper left-hand to the lower right-hand corner of the desired area on the map.
5. The footprint is finalized once the user releases the right mouse button.
6. The latitude and longitude values of the footprint are displayed in decimal degrees on the interface.
7. The user submits the request for the footprint area.
8. The corresponding gravity data are displayed in a grid box by default.

Alternatives

1. The user enters latitude and longitude values using the degrees and decimal-minutes format (replaces steps 1–4).
2. The user requests that the results be presented on a gridded map (replaces step 6).
   a. The user selects *gridded map* for displaying the gravity data.
   b. The gravity data are displayed in a gridded map.
3. The user downloads the results into a file (extends the use case after step 6).
   a. The user selects the *download* option.
   b. The user enters the local path in which to save the data.
   c. GeoNet saves the results at the requested location when the service is completed.
   d. The user receives an acknowledgment that the file has been downloaded.

2.3.3. Scenario 2: Submit Gravity Data

This scenario outlines the Submit Data use case and the interaction of the Registered User and Distributed Geoscience Database actors as illustrated in Figure 2. The validation process that takes place after this scenario involves the Validate use case that is part of a different scenario.

Discussion: In this scenario, a Registered User actor wants to upload data to the data repository. Once the submission process is completed, the data are flagged as temporary until they are reviewed by a user with validator privileges who approves and incorporates the data into GeoNet's main data sets. Validation is further described in the Validate use case. When the corresponding metadata are completed by adding the date of validation and the identification number of the validator, the data become available to registered users. Also, the data are further submitted for validation to the USGS. Once the data are approved, the source identification number provided by USGS and its corresponding validation date are entered into the metadata file. The submitter is notified when the submission has been validated and is given a reference number for citation purposes.

Steps

1. The user chooses to submit gravity data.
2. GeoNet displays a metadata form.
3. The user completes the metadata form by entering, e.g., the identification of the person submitting the data, date data were collected, base station identification, and instrumentation used.
4. The system validates the metadata and generates a source identification number.
5. The user selects the directory in which the gravity data are stored.
6. GeoNet uploads the gravity data file from the designated directory and stores the data in a temporary location.

Alternatives

1. The user runs a data reduction algorithm on the newly submitted data (after step 6).
2. The user edits the data (after step 6).

2.3.4. Scenario 3: Request for a Three-Dimensional Rendition of a Remotely Sensed Image Overlaid on a Digital Elevation Map

This scenario refers to the Discover Data and Knowledge and Analyze use cases as given in Figure 2. The following actors are involved: Distributed Geoscience Database, Application Software, and either a General User or a Registered User.

Discussion: In this scenario, a user wants to display digital elevation data and remotely sensed data simultaneously. These data are displayed using three-dimensional imaging.

Steps

1. The user selects the option to display remotely sensed and digital elevation data.

2. The system presents a list of previously defined footprints. The footprints are displayed as outlines on a map of the southwestern United States. Each footprint represents the area covered by a remotely sensed image.

3. The user selects a footprint by clicking the region where the footprint outline is displayed.

4. The system displays thumbnail versions of the available images for the footprint area, and it displays metadata about the image, e.g., date of the image, type of satellite, and scene identification number.

5. The user selects the desired image.

6. GeoNet displays a three-dimensional rendition of the image integrated with a digital elevation map.

7. The system provides an interface for rotating the graphic.

8. The user rotates the graphic to get different views.

9. The user chooses to save the image.

10. The system prompts the user to select a graphical format in which to save the image and a file location.

11. The user selects an image format and enters a file location.

12. The system initiates the download of the image.

13. The user receives an acknowledgment that the file has been downloaded.

2.4. High-Level Architecture

Figure 3 shows the different layers of GeoNet's general architecture: the data layer, the functionality layer, and the presentation layer. The data layer contains data repositories that may be distributed across multiple servers and spread across different geographical locations. There are no restrictions on the platforms used for data repositories. The data layer will typically be distributed across organizations according to their corresponding data set specialty. As an example, University of Texas at El Paso may host and maintain gravity data sets about the transition zone between the Colorado Plateau and the Basin and Range province, while Arizona State University may host and maintain seismicity data sets about the same area. The result is a heterogeneous data layer. This layer must provide the necessary database connectivity and is mainly oriented toward the extraction from and submission to the data repositories.

The functionality layer implements the logistics of the GeoNet system, which is delivered in the form of services that are consumed by client applications. Data querying, retrieval, integration, and processing services are part of the functionality layer. Services are distributed, with different organizations implementing and hosting services on their own servers. Some of these services interact with the data layer in order to extract or submit data, while other services provide data integration or processing by combining other services within the functionality layer. The services provided at the functionality layer are guided by a set of ontological rules that are themselves enforced through services. These rules are necessary to maintain the integrity and consistency of data being accessed across different data sets, and therefore all services provided in the functionality layer must conform to them. The ontological rules are validated by the GeoNet community.

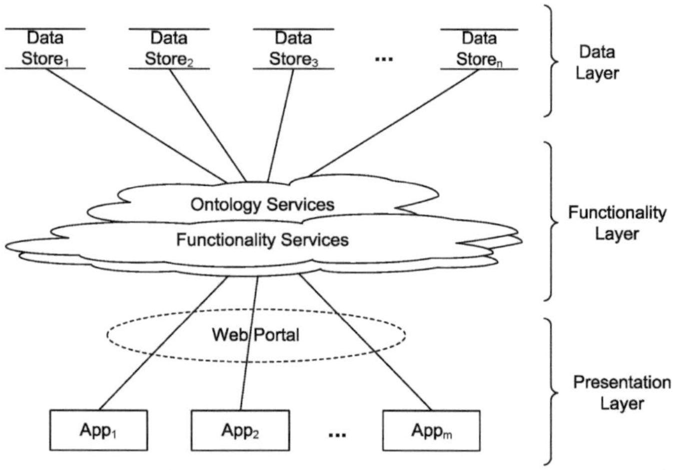

Figure 3. High-level view of GeoNet. GeoNet's architecture consists of a layered design. The data layer is composed of data repositories that are geographically distributed among participating institutions; the functionality layer represents the logistics of GeoNet, and its implementation is based on grid services; the presentation layer consists of client applications that represent the point of interaction for the end user.

A *client* in GeoNet refers to an application that consumes services within GeoNet. The presentation layer consists of Web-based clients and regular clients (i.e., applications that install and execute on the user's platform). This layer provides the user with an interface to the functionality services offered by GeoNet. Although client applications may provide additional functionality beyond that available at the functionality layer, typical applications at the presentation layer will only serve as user interfaces and will pass the functionality logistics to the corresponding services at the functionality layer. This approach results in low-complexity applications at the presentation layer, which have less-demanding platform requirements. As a result, these "thin-client" applications can potentially be deployed on a wide range of end-user platforms, ranging from desktop computers to handheld devices.

Client applications interact with the GeoNet system by invoking functionality services, which are referenced through a registry on the GeoNet portal. The registry directs the client application request for service to an available hosting platform, which executes the service. The service will satisfy the client's request by performing some computation on input data, interacting with other services that may complement the client's request, or interacting with a data store to get or submit data. The services will then respond to the client's application request by sending back the proper data in XML format. The client application will then transform the XML format to an appropriate format for presentation to the user. For example, if the client is a Web-based application, the application may transform the XML format to an HTML format, which can be processed by the client's Web browser.

Finally, the GeoNet portal provides an interface for both geoscientists and public users to access, download, and validate shared resources, interact with other GeoNet users, and create and manage user accounts. Another important role it plays is to allow geoscientists and institutions to extend GeoNet's functionality by registering their services on the system. The profile information maintained for each user can be used to customize GeoNet resources and services.

The GeoNet portal provides the following features:

- **My Profile**—Allows a user to register as a member of GeoNet and to maintain his or her registration information.
- **Tool Center**—Allows users to access tools that interact with GeoNet's functionality layer to access, integrate,

manipulate, and analyze geospatial data. These tools can be accessed directly through the portal or by downloading applications that have the ability to interact with GeoNet's functionality layer. Documentation and tutorials for the use of such tools are available through the Tool Center.

- **Registry Center**—Allows users to contribute toward the development of GeoNet by registering data, ontologies, and software applications to GeoNet's registry.
- **Validation Center**—Allows a designated domain expert to (1) approve submitted geospatial data, ontologies, and software applications based on set procedures, and (2) categorize entries based on level of validation, e.g., a distinction will be made for gravity data that have been inspected by a domain expert and data approved by USGS, or a tool that is of value to the community but has not been thoroughly tested.
- **Discovery Center**—Allows users to explore geospatial data and ontologies, locate tools, and access other services provided within GeoNet. General users can search for tools that provide the necessary functionality for their needs, while software developers can find software components that can be combined and plugged into their projects.
- **GeoNet Forum**—Allows the GeoNet community to discuss issues, pose questions, contribute solutions, and define strategies for the GeoNet system and its relationship to the geoinformatics initiative.

3. SUPPORTING TECHNOLOGIES

GeoNet is built upon Internet-based technologies. Some of these technologies have been standardized, while others are nascent and are in the process of achieving recommendation status by their respective sponsor organizations. These technologies can be described as a layered model, where one category builds on top of another and extends functionality toward a more specific use. These categories include the Extensible Markup Language (XML) and related technologies, the OWL Web ontology language, Web services, and grid services. Each is described next.

3.1. Extensible Markup Language (XML)

XML [17] is a flexible markup language developed by the World Wide Web Consortium (W3C). The main purpose of this language and its related technologies is to satisfy the need for exchanging a wide variety of data over the Web. XML achieves this goal by focusing on the following characteristics:

- **Extensibility**—Like other markup languages, an XML document encapsulates data between tags, which also serve as descriptors for a particular section. While the XML specification indicates a basic syntax for the language, it does not put any restriction on the tags used. This feature allows a particular user to define tags to be used, providing flexibility in the description of data. For example, consider an XML document that contains a section of data delimited by the begin and end tags <GravityData> and </GravityData>. It should be expected that the data contained in this section are in fact gravity data. To validate the format of the data contained in a GravityData section, supporting XML technologies also allow the description of schemas or document types that define the fields that are expected to be present on a particular section. For example, a schema can be created that defines a gravity data element as consisting of a latitude, longitude, and gravity data fields. The schema would indicate that in a well-formed document, each gravity data element delineated by the GravityData tags must include each of these subfields. The schema can be used to verify that these fields are present within the GravityData section.
- **Separation of semantics and presentation**—XML focuses on the description of data and ignores the data's presentation to the user. Tools are available for parsing and transforming XML documents. For example, an XML document may contain data that are to be presented as synthetic speech over a telephone or data that are to be displayed on a Web page and should be presented as an HTML document to a user's Web browser. The speech program and the browser are responsible for the presentation of the data to an end user. The separation of semantics and presentation in XML allows custom messaging systems to be created based on XML. Special communication protocols can be specified using XML technologies, and corresponding parser tools can be developed to adapt the presentation of data between computer systems in order to achieve communication. Some of these XML-based communication protocols (e.g., the SOAP protocol [2]) are also being standardized and play an essential role in Web services technologies, described in section 3.3.
- **Platform independence**—XML is based on a simple text format and, therefore, can be transmitted on a wide variety of network connections operating at different levels of reliability, performance, and security. The simplified text format also eliminates any issues that arise between different computer platforms and their particular interpretation of binary formats. Furthermore, XML allows the use of a wide variety of alphabet character sets, which permits data and metadata to be used by an international audience.

3.2. OWL Web Ontology Language

An ontology is a "formal, explicit specification of a shared conceptualization" [7]. In addition, it precisely defines concepts, attributes, precision requirements of each attribute, mappings between standard controlled vocabularies (synonyms), constraints, and relationships (between concepts and attributes).

OWL [16] is an XML grammar that defines an ontology language designed to process information content. OWL is a standard proposed by W3C and is divided into three increasingly expressive sublanguages. OWL Lite is a subset of OWL DL

(Description Logic), which is a subset of OWL Full. The suitability of a sublanguage for a given ontology definition depends on the complexity of the ontology; guidelines for choosing the appropriate sublanguage are given by W3C.

OWL is used by applications to enhance content interpretation. The idea is to describe the meaning of data by defining the concepts that the data describe. OWL provides the following desired features in GeoNet:

- **Data integration**—Applications that are able to interpret content can more easily reconcile data from different sources.
- **Knowledge discovery**—By permitting users to query content instead of data, users can create queries that are more helpful toward knowledge discovery. For example, once the concept of location is defined, a user could request a gravity map of New Mexico even if the underlying database stores the gravity data using latitude and longitude.
- **Metadata verification and validation**—By having a formal ontology definition, applications can monitor a data set submission to ensure that its accompanying metadata conform to the ontology, thus setting a common base for interaction.

3.3. Web Services

Web services are an Internet-based technology that use XML-based messaging to enhance the sharing of computing resources over a distributed environment. It proposes an application-centric model, where applications communicate among each other without human interaction to share resources over the Web. This view varies from the traditional human-centric model of the Web, where the human user interacts with a Web browser to request resources.

Web services are based on the service-oriented paradigm [3]. The service-oriented approach proposes that resources be exposed to application programs as consumable services. The result is the interaction of three parties:

1. The service provider is the party responsible for hosting a resource. It hides all implementation details of the resource and only exposes a public interface that represents the point of interaction.

2. The service registry is the party responsible for maintaining a directory of available services. It serves as the initial intermediary between the service provider and the service consumer.

3. The service consumer finds the desired service through the service registry and starts interaction with the resource through the service's public interface.

Although there are various alternatives to implement Web services, there are current efforts to standardize the technology. The main candidate organization to achieve this standard is W3C, which has a working group focused toward this end [4].

Web services are characterized by the following properties that are essential for sharing resources over a distributed, application-centric environment:

- **Web services are self-describable.** The description of Web services is generally done using the Web Services Description Language (WSDL) [4], an XML grammar that is readable by both human users and applications. A WSDL document describes the grammar used to access a Web service's public interface. Programs can extract the service interaction requirements from a WSDL document and tailor applications that use the service.
- **Web services are discoverable.** Web services require mechanisms for publishing and finding services. On the developer side, a service provider must be able to access a public registry to add references to the service. On the other hand, search mechanisms must be provided for consumers to locate and access such services. Mechanisms for publishing and searching for Web services vary widely, and standards are still being developed. There are two main alternatives: a centralized and a decentralized approach [1]. The Universal Description, Discovery, and Integration (UDDI) registry provides a "yellow-pages" approach to publish and find services. Service providers can access this global registry system to add references and information about their services. Consumers can access the registry to make queries to find them. The decentralized alternative is the Web Service Inspection Language (WSIL), which provides a "business-card" approach and uses Web conventions (e.g., HTTP GET) to allow users to locate the service.

In order for a service to be consumed, the client has to locate the service using a discovery mechanism, and it has to access its public-interface description as prescribed by its WSDL documentation. Based on the description, a proxy module is created on the client's platform. This proxy class is usually created automatically with the aid of a development environment such as Microsoft Visual Studio .Net. The proxy module executes on the user's platform and takes care of the communication between the user application and the Web service. As a result, the proxy module handles the formatting of data to conform to the appropriate protocol, which is also specified in the WSDL documentation. As shown in Figure 4, the user application sends requests to the local proxy module, which takes care of the interaction with the service over the Internet. The proxy module then receives a response from the service and sends the results back to the user application.

Figure 4 shows that the user application is not concerned with network communication issues. The application sends requests to a proxy module that is executing on the same local platform. On the other side, the Web service is not concerned with presentation details. A Web service expects a request from a proxy module and sends an appropriate message in response. It is the responsibility of the consumer application to present an appropriate interface for the results. The Web service architecture also makes it possible for application programs to discover and access services at run time (as opposed to discovering services at design time). This means that it is possible to construct an appli-

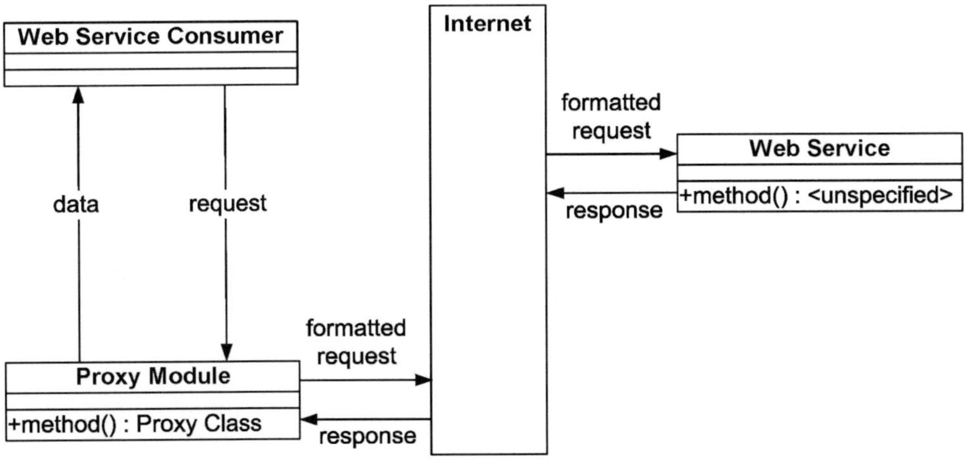

Figure 4. Web services proxy communications. The communication between a consumer application and a Web service is handled by a proxy module, which is generated automatically with the use of the WSDL description provided by the Web service.

cation that utilizes different services based on service attributes such as availability.

3.4. Grid Services

Grid services are an emerging technology for distributed computing that offers an alternative to other approaches, e.g., CORBA (Common Object Request Broker Architecture) [14]. Grid services incorporate the service-oriented approach and the XML-based communication protocols used by Web services, making it more flexible than its previous counterparts. Grid services extend Web services by adhering to a specific set of behaviors and conventions [6].

Grid services are currently undergoing standardization efforts, with the main effort led by the Global Grid Forum (GGF) [12]. This effort has resulted in the creation of the Open Grid Services Architecture (OGSA) [6], which describes the features of grid services. Main features included in OGSA are:

- **Discovery**—Like Web services, grid services also support mechanisms to publish and discover services (i.e., UDDI and WSIL). However, grid services extend this feature to allow a consumer to discover new services based on public data elements that reflect the current state of the service.
- **Standard, stateful, dynamic service creation**—Web services are stateless. This means that a Web service does not keep information about the state of a transaction, such as knowing if a user has been authenticated. Grid services, on the other hand, are designed to maintain state and exist for a finite period of time. Consequently, there can be different instantiations of the same service at any given time, each of which is differentiated by their current state. Grid services define a specialized class of services that have the responsibility of creating new service instances. These types of services are called *factories*, and a standard interface is defined for them. This allows clients to access a consistent interface for instantiating new services across heterogeneous platforms. Factory services create new services and set their initial state and lifetime period.
- **Lifetime management**—Component failure is inevitable in any loosely coupled, distributed system, and mechanisms must be provided to deal with such situations. Grid services offer a soft-state lifetime management approach, where services depend on clients to sustain continued interest in the resources offered in order for the corresponding services to remain active. If a client application or other component of the distributed system chain fails, the affected service will eventually terminate automatically. Explicit termination of the service from the client side or from the server side is also possible.
- **Notification**—In order to enhance collaboration in a dynamic, distributed environment, services must be able to notify each other of changes in their states. Grid services specify mechanisms that allow services to send notifications of their state to other services that have requested to be notified.

Grid services, like Web services, assure interoperability among distributed, heterogeneous systems and offer transparency to user applications in terms of physical location and invocation of service. Furthermore, grid services offer the opportunity to create and maintain transient, stateful services that can be created and destroyed dynamically. Consequently, grid services provide a richer framework than Web services for the development of complex, loosely coupled distributed systems and allow features such as one-time log-in to resources, data set replication across nodes, and transaction computing to be implemented uniformly across heterogeneous platforms.

There are concurrent efforts within the GGF community to refine the standards and to address higher-level services. These services are described as follows:

- Distributed data management services support access to and manipulation of distributed data that are stored in databases.

- Workflow services support the coordinated execution of multiple application tasks on multiple distributed grid resources.
- Auditing services support the recording of usage data. This includes secure storage of data, analysis of data for purpose of fraud, and intrusion detection.
- Instrumentation and monitoring services include the discovery of sensors in a distributed environment, collection and analysis of information from these sensors, and generation of alerts when unusual conditions are detected.
- Problem determination services for distributed computing include dump trace and log mechanisms with event tagging and correlation capabilities.
- Security protocol mapping services enable distributed security protocols to be transparently mapped onto native platform security services.

4. GeoNet Framework

This section describes the GeoNet framework in more detail from three perspectives: data, services, and communication. The data perspective describes the general guidelines for organizing geospatial data within GeoNet and the use of ontologies to describe attributes, constraints, and relationships within and between data types. The services perspective describes general services of GeoNet that enable an open, heterogeneous implementation. The communication perspective illustrates the transmission of data through the layered architecture proposed for GeoNet.

4.1. Data

As described in section 2.4, "High-Level Architecture," geospatial data within GeoNet will typically be hosted and maintained across geographically distributed organizations. The challenge is managing the data and providing access to the metadata that ensure consistency. This section discusses these challenges.

4.1.1. Guidelines for Organizing Data

The data layer of GeoNet consists of database servers that are maintained by organizations within the scientific community. In addition to reading data sets, the GeoNet community will be able to submit new data sets and update data sets, which is done through a well-defined process that requires validation of data by expert scientists.

GeoNet's architecture is designed to be service-oriented. The service-oriented paradigm proposes that the implementation details of a service be hidden from consumers, exposing only a public interface for interaction. At the data layer of GeoNet, this means that the specific choice of a database management system and the structure (i.e., database schema) used to store the data are irrelevant to the intended client. In order to maintain interoperable data sets from different sources, metadata must be provided to preserve the integrity of the data set integration. For example, GeoNet's data layer services must provide, within their public interface, methods that extract data precision information of a data set to make sure that the quality of the data is compatible with that of another data set being integrated. Furthermore, in order to maintain the quality of data, metadata must be provided to impose minimum requirements of precision and valid ranges expected of new records submitted for an existing data set.

Given that the data layer of GeoNet provides access to the data sets, along with corresponding metadata, the functionality layer of GeoNet can then provide the midlevel services necessary to transform and adapt the data sets. For example, services at the functionality layer may be able to convert data sets between different units of measure.

4.1.2. Ontology

An ontology of geospatial data and physical properties that are relevant to GeoNet is needed to facilitate communication among services independent of individual system technologies, information architectures, and application domains [15]. The geospatial ontology in GeoNet is being created using the OWL Web ontology language described in section 3.2. The ontology facilitates three primary functions. First, since the ontology categorizes information, the ontology can serve as a navigation tool that facilitates the identification of existing services. Second, the ontology can be used to assist users in composing queries that are not data-specific. For example, a user can create a query requesting data about New Mexico, even though the data in the database are stored by latitude and longitude. Third, since GeoNet is an open system, users will be developing and contributing new tools. The ontology provides guidelines for tool developers that will enable new tools to be integrated into GeoNet as they are developed. Furthermore, GeoNet will provide the ability to extend the ontology by allowing the geoscience community to contribute new concepts and relationships. As with data set contribution, ontology contribution is guided by a well-defined process that requires review by domain experts.

A long-range goal is to provide the user with perspectives for integrating and viewing the data. This will permit filtering of data based on the type of the user. For example, a K–12 educator would want data presented in a different format than a geologist acquiring the same data.

4.2. Services

The geospatial ontology provides information to developers that can help address interoperability of data sets. As discussed in section 2.4, "High-Level Architecture," GeoNet provides access to data sets and corresponding metadata through services at the data layer. Services built on top of these data services correspond to the functionality layer of GeoNet as shown in Figure 3. The knowledge stored in the ontology maintains consistency within these services.

The GeoNet portal discussed in section 2.4 categorizes services by functionality into the following groups: Discovery Center, Validation Center, Registry Center, GeoNet Forum, Tool Center, and My Profile. Besides offering an organization for services, the purpose of the GeoNet portal is to provide the user with a graphical user interface (GUI) that does not tax the user's computer system and does not require a savvy programmer to access a service. Since GeoNet is to be an open system hosted and maintained by the geoscience community, the organization of the GeoNet portal allows the community to contribute tools and services in an orderly, well-defined fashion. It is expected that the portal organization of GeoNet will change as the community contributes new services, resulting in the creation of new categories and the refinement of existing categories.

As with data set and ontology contributions, the contribution of services by the community is also guided by a well-defined process that includes validation by domain experts. In order to maintain consistency of the system and data set interoperation, services contributed to GeoNet must be consistent with the geospatial ontology. If a service is contradictory to the guidelines dictated by the ontology or if the service does not fit into the current ontology, either the ontology has to be updated or the service has to be redefined to achieve consistency.

4.3. Communication

As services are contributed to the system and organized into the defined structure of the GeoNet portal, users can access the services by navigating the structure through a front-end graphical user interface (GUI). GeoNet's navigation GUI, which is also enhanced by the guidelines dictated by the geospatial ontology, facilitates the discovery of knowledge by allowing users to find and utilize the services offered through GeoNet.

As users call services through GeoNet's navigation portal, services perform in the background and offer results to the user. The user can then download the results from the service interaction or continue navigating to call additional services, with the obtained results as input. This forms a pipe-and-filter mechanism, which if saved and parameterized, can be considered as a recipe or workflow to obtain complex results. Such workflows may be helpful in automating common tasks that otherwise may be tedious or time-consuming.

Finally, given the layered architecture of GeoNet, the user interacts with a centralized front end to access distributed service-oriented modules that offer tools and expose access to distributed data repositories. For the user, this means access to a potentially vast resource of data and tools at his/her disposal from a centralized point of interaction over the Internet.

5. SUMMARY

GeoNet is being developed to establish an integrated geospatial data system that collects and stores data from the southwestern region of the United States. The effort is supported by the departments of Geological Sciences and Computer Science at the University of Texas, El Paso, and Geological Sciences at Arizona State University. GeoNet consists of distributed data repositories that are accessible through services over the Internet. It provides a set of tools that allow the creation of three-dimensional models for visualizing the data and that allow scientists, as well as public users, to retrieve, update, validate, and analyze data sets.

The Internet-based technologies used to design and implement GeoNet are described, including XML, Web services, grid services, and the ontology language OWL. The issues that must be considered in defining such a network are presented, as well as a basic framework based on the service-oriented paradigm that supports the development of GeoNet. Specifically, the exposure of metadata at the data layer to facilitate the data set integration from distributed sources is presented, and a geospatial ontology to ensure consistency is discussed. Furthermore, to facilitate the extensibility of GeoNet by the community, a defined process of submission is proposed, which maintains quality of data and tools by requiring the review of contributions by domain experts before they are accepted.

REFERENCES CITED

[1] Appnel, T., An Introduction to WSIL, http://www.onjava.com/pub/a/onjava/2002/10/16/wsil.html, October 16, 2002.

[2] Box, D., Ehnebuske, D., Kakivaya, G., Layman, A., Mendelsohn, N., Nielsen, H., Thatte, S., and Winer, D., The Simple Object Access Protocol (SOAP), Version 1.1, http://www.w3.org/TR/SOAP, May 8, 2000.

[3] Cerami, E., Web Services Essentials: Sebastopol, California., O'Reilly & Associates, February 2002.

[4] Christensen, E., Curbera, F., Meredith, G., and Weerawarana, S., Web Services Description Language (WSDL), Version 1.1, http://www.w3.org/TR/wsdl, March 15, 2001.

[5] Cockburn, A., Writing Effective Use Cases: Boston, Addison-Wesley, 2001.

[6] Foster, I., Kesselman, C., Nick, J., and Tuecke, S., The Physiology of the Grid, An Open Grid Services Architecture for Distributed Systems Integration, http://www.globus.org, June 22, 2002.

[7] Gruber, T.R., 1993, A Translation Approach to Portable Ontology Specifications: Knowledge Acquisition, v. 5, p. 199–220, doi: 10.1006/knac.1993.1008.

[8] EarthScope, Exploring the Structure and Evolution of the North American Continent, http://www.earthscope.org, October 2004.

[9] Geoinformatics, A Defining Opportunity for Earth Science Research, http://www.geoinformatics.vt.edu, May 2004.

[10] Geoinformatics in Action!, http://www.geoinformaticsnetwork.org/swgeonet, October 2004.

[11] GEON, Cyberinfrastructure for the Geosciences, http://www.geongrid.org, October 2004.

[12] The Global Grid Forum, http://www.ggf.org, October 2004.

[13] The NEES Cyberinfrastructure Center (NEESit), http://it.nees.org/about/index.html, October 2004.

[14] The Object Management Group, http://www.omg.org, October 2004.

[15] Ontology, Org Forum, Enabling Virtual Business, http://ontology.org, August 2000.

[16] Web Ontology Language Overview, OWL, W3C Recommendation, http://www.w3.org/TR/owl-features, February 2004.

[17] The World Wide Web Consortium, http://www.w3.org, October 2004.

Manuscript Accepted by the Society 18 June 2005

Printed in the USA

Geological Society of America
Special Paper 397
2006

Development of ontologies for earth system science

Rob Raskin
Jet Propulsion Laboratory, California Institute of Technology, Pasadena, California 91109, USA

ABSTRACT

The Semantic Web for Earth and Environmental Terminology (SWEET) is a NASA-funded prototype to explore the technological challenges in making the semantic Web a reality for earth system science data and information. SWEET includes an extensive collection of ontologies written in the Ontology Web Language (OWL). These ontologies encompass both faceted (decomposed) concepts and integrative (unifying) concepts. Taken together, these enable a representation of the two science processes of reductionism and synthesis. SWEET stores ontology elements optionally in a DBMS (database management system), with a two-way translator between the XML and DBMS representations. SWEET also includes wrappers for large external online databases (gazetteers, earthquakes, etc.) to enable their contents be virtual components of the ontologies (http://sweet.jpl.nasa.gov).

Keywords: ontology, metadata, earth system science, knowledge representation.

SEMANTIC INTEROPERABILITY FOR EARTH SYSTEM SCIENCE DATA

Earth system science spans many scientific disciplines, and its data are diverse enough to be characterized by multiple data formats, data models, community data standards, and terminologies describing common concepts. Discovery, access, and usage of data across disciplines is not a seamless process. As Web services for location and integration of data products (e.g., over a data grid) become a reality, there is a need for a common semantic framework for expressing science and data concepts. *Semantic interoperability* is the next level of interoperability beyond data interoperability, in which a machine understanding of data content is achieved during data exchanges.

Semantic understanding has a more fundamental application in the process of data discovery, as data providers (experts) and data consumers (often novices) may have a semantic mismatch. Such a mismatch occurs when putting in the "wrong" search term into a search engine and receiving too many or too few search results.

In the early days of computing, an initial level of data interoperability resulted when data structures (arrays) created on one computer system were readable by another computer (given a common transfer medium). This type of interoperability was possible, in part, through standards for representing integers, floating-point values, etc. Scientific data formats such as HDF emerged later to extend this level of interoperability to more complex data structures and across arbitrary vendor platforms. Standard metadata representations enabled humans to interpret some properties of the data set. The Internet later brought on protocols such as OPeNDAP/DODS (Cornillon, et al., 2003), which supported modification of the data structure (subset extraction) during the transfer. All of the exchanges described say nothing about the scientific interpretation of the data on the receiving end. A variable name is assigned to a data structure, but human intervention is required to make sense of it.

These examples demonstrate *syntactic interoperability* for machines (to read the data) and semantic interoperability for humans (humans are still needed to interpret the data). How do we enable machines to have a semantic understanding of science data concepts?

The HDF-EOS format (an extension for HDF for earth science data) remedied the semantic interoperability problem for

Raskin, R., 2006, Development of ontologies for earth system science, *in* Sinha, A.K., ed., Geoinformatics: Data to Knowledge: Geological Society of America Special Paper 397, p. 195–199, doi: 10.1130/2006.2397(14). For permission to copy, contact editing@geosociety.org.

independent variables by standardizing the spatial and temporal terminologies used to query and extract data subsets. The Open Geospatial Consortium (OGC) specifies similar standards for spatial/temporal extents in its Web Mapping Service (WMS) and Web Coverage Service (WCS) protocols.

Semantic interoperability for dependent variables has generally meant the use of controlled keyword lists. For instance, the NASA Global Change Master Directory (GCMD) (Olsen, 2001) defines ~1000 controlled keywords, each with a dictionary definition. Such a representation, in the form of a single hierarchy for science keywords, was developed through input and support of the earth system science community. This type of semantic interoperability has proved successful for classification and discovery of data resources, using manual classification by data providers.

Common keyword hierarchies fall short when called on to support more challenging scientific discovery and usage tasks, including applying computer reasoning to respond to general queries, to chain services together. The deficiency occurs because the representation does not support inheritance of concepts for knowledge reuse, does not provide a rich expression of the relationship between the keywords, and is not directly extendable by the user. This paper describes a more scalable solution to semantic interoperability in the context of earth system science, which supports the representation of commonsense knowledge. Central to this type of representation is the *ontology*, a term that conveys a constructed "world view" or "shared understanding."

INTRODUCTION TO ONTOLOGIES

An ontology is a formal representation of concepts and their mutual relationship in a form that captures domain knowledge. Ontologies are integral to semantic Web and artificial-intelligence applications where a knowledge base assists processes of search, classification, automated decision making, electronic commerce, machine-to-machine communication, etc.

Ontologies are a form of controlled terminology that differ fundamentally from taxonomies, thesauri, and other controlled hierarchical or linear lists of domain terms commonly adopted by organizations.

Ontologies enable child terms to inherit all properties of their parents, rather than being only subcategories of their parents. This fundamental assumption enables knowledge to be reused and supports scalable knowledge construction, as it is not necessary to redefine higher-level concepts previously defined. Ontologies provide the mechanism for articulating how a child concept differs from its parent, using the ontology concepts themselves. Furthermore, ontologies support multiple inheritance, so that compound concepts are easily generated. Use of open standards for ontology representation enables community members to extend existing concepts and to declare equivalence of terms across ontologies.

As an example, the troposphere is a subclass of an atmospheric layer, which itself is a subclass of a spatial object and part of the atmosphere. The troposphere differs from its parent atmospheric layer based both on the property of height above Earth's surface and on the temperature distribution with height. But a meteorologist would not be included as a subclass of atmospheric science; that concept would be an instance of a person, declared further with an occupation specialty.

Ontologies support formal representation of knowledge, so that complex relationships between concepts can be described, intelligent search and reasoning can be applied, and a *semantic Web* can be realized. The semantic Web (Fensel, et al., 2003) is an ambitious extension to the existing Internet environment, coordinated by the World Wide Web Consortium (W3C). The semantic Web vision includes commonsense knowledge integrated into Web pages, via XML tags around Web page terms that reference ontology elements, where the terms are "defined." A search tool that can interpret these metadata tags can distinguish multiple uses of the same term (e.g., "fall" as a season versus "fall" as a downward motion) and can find semantically equivalent resources without the need for an exact word match.

ONTOLOGY DEVELOPMENT

The Semantic Web for Earth and Environmental Terminology (SWEET) (Raskin and Pan, 2003) was an initial NASA-funded prototype to develop a comprehensive set of ontologies of the earth system to improve discovery and use of earth system science data. We started with the GCMD keywords and manually populated the ontologies using the OilEd and Construct ontology editing tools. The GCMD keyword collection was developed with an extensive community, although we decomposed and normalized the concepts into a scalable structure. Other domain-specific controlled keyword lists provided additional terms. Currently, over 2000 science-related terms are included.

As a result of the normalization of the concept space, complex concepts such as sea surface temperature are decomposed into their components (sea as a substance, surface as a spatial concept, and temperature as a physical property). The design principles in the ontology construction process are given in Table 1. The Earth Science Information Partners (ESIP) Federation (Raskin, et al., 2002) currently maintains these ontologies.

A schematic of the ontologies is shown in Figure 1. The boxes in the figure represent ontologies, and the arrows represent major pathways in which concepts are defined in terms of other concepts. Thus, the arrow from Process to Phenomena indicates that physical, chemical, and biological processes are significant factors in defining large-scale phenomena.

Basic Building Blocks

Substances (Nonliving)

This ontology consists of the physical and chemical building blocks of nature, such as particles, chemical compounds, and electromagnetic radiation. Substances may be declared to have particular properties, such as radioactive and organic.

Substances (Living)

This ontology includes plant and animal species, imported from the GCMD "biosphere" taxonomy.

Processes

This ontology describes processes (generally small-scale) that substances undergo, such as diffusion, conduction, respiration, and absorption.

Environment

Earth Realm

The layers of the earth system constitute an earth realm ontology, based upon physical or chemical composition. First-order layers such as atmosphere, ocean, and geosphere, have associated subrealms (such as ocean floor and atmospheric boundary layer). The subrealms generally are distinguished from their parent classes, based upon altitude and physical/ chemical properties. Earth realms may be declared to have properties such as primary substances, boundaries, reference heights, and associated ecosystem types.

Observables

Property (Physical, Chemical, Biological)

This ontology describes properties of elements in other ontologies (e.g., temperature, pressure, height, and albedo). Properties may have values, default units, and actual units. They are grouped by property type (radiative, thermal, solid mechanical, dynamical, gravitational, etc.).

TABLE 1. ONTOLOGY DESIGN SPECIFICATIONS

Machine readable	Software must be readily parsable
Scalabl	Design must be capable of handling very large vocabularies
Orthogonal	Compound concepts should be decomposed into their component parts, to support recombining concepts in new ways
Extendable	Easily extendable to enable specialized domains to build upon existing general ontologies
Application-independent	Structure and contents should be based upon the inherent knowledge of the discipline, rather than on how the domain knowledge is used
Natural language-independent	Structure should provide a representation of concepts, rather than of terms
Community involvement	Community input should guide development

Ancillary Ontologies

Units

Units are defined using Unidata's UDUnits (http://www.unidata.ucar.edu/software/udunits). The resulting ontology

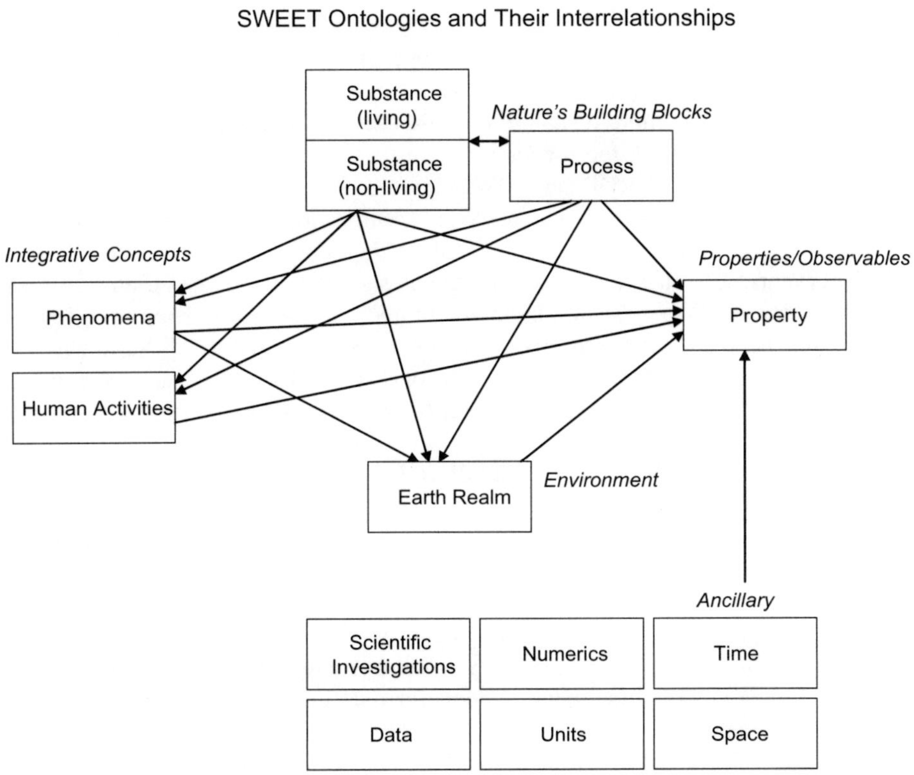

Figure 1. Semantic Web for Earth and Environmental Terminology (SWEET) ontologies and their interrelationships.

includes conversion factors between various units. Prefixed units were added, such as km, defined as a special case of m with appropriate conversion factor.

Numerics

Numerical extents include interval, point, 0, R^2, etc. Numerical relations include greater than, max, etc. We defined multi-dimensional concepts such as coordinate systems, mathematical operators, and functions.

Time

Time is essentially a numerical scale with terminology specific to the temporal domain. We developed a time ontology in which the temporal extents and relations are special cases of numeric extents and relations, respectively. Temporal extents include duration, season, century, 1996, etc. Temporal relations include after, before, etc.

Space

Space is essentially a three-dimensional numerical scale with terminology specific to the spatial domain. We developed a space ontology in which the spatial extents and relations are special cases of numeric extents and relations, respectively. Spatial extents include country, Antarctica, equator, etc. Spatial relations include above, north of, etc. Spatial extents may have properties such as coordinates, directions, locations, and administrative jurisdictions.

Data

The data ontology provides support for data set concepts, including representation, storage, modeling, format, resources, grid computing, and distribution.

Scientific Investigations

This ontology provides support for instrumentation, measurement devices, and missions.

Unifying Concepts

Phenomena

A phenomena ontology is used to define transient events. A phenomenon crosses bounds of other ontology elements. Examples include hurricane, earthquake, El Niño, volcano, and terrorist event, and each has associated time, space, planetary realms, nonliving elements, living elements, etc. We also include specific instances of recent phenomena.

Human Activities

This ontology is included for representing impacts of environmental phenomena such as commerce and fisheries.

ONTOLOGIES AS A UNIFYING KNOWLEDGE FRAMEWORK

The first several ontologies listed above represent orthogonal concepts (or dimensions), often called *facets*. Traversing down a tree associated with a facet follows the scientific path of reductionism by adding additional details and specialization to more general concepts.

The unifying ontologies (phenomena and human activities) represent the opposite approach, as these concepts are synergetic rather than mutually orthogonal. These entries define holistic, synthesizing concepts using elements from the faceted ontologies (e.g., a hurricane is associated with particular coastal areas, and is characterized by high winds, rainfall, flood impacts, etc.).

Taken together, these two complementary approaches mirror the scientist's dual methods of reductionism and synthesis, and support a comprehensive framework for capturing scientific knowledge and investigation.

ONTOLOGY LANGUAGES

The World Wide Web Consortium (W3C) has adopted a pair of XML languages as its recommended standard for ontology representation: the Resource Description Framework (RDF) (W3C, 2004a) and the Web ontology language OWL (W3C, 2004b). XML is human- and machine-readable and platform-independent, but its tags have no agreed-upon semantics. RDF specializes XML by standardizing meanings for class, subclass, property, subproperty, domain, range, etc. (Table 2). OWL is the language used in SWEET. OWL is a further specialization of RDF, adding meanings for concepts such as functional properties, set algebra, and cardinality (Table 3). Both RDF and OWL statements can be visualized outside of their usual XML form as collections of "triples." For example, the assertion that shear is a subset of the class of mechanical processes can be expressed as

Shear isSubsetOf MechanicalProcess

a more readable (but less usable) form than the corresponding XML representation.

RDF and OWL have representations for numbers only through a separate W3C specification (W3C, 2001) that defines number types (e.g., real number, unsigned integer) and intervals of these types (e.g., the closed interval between 0 and 1). This specification is limited in that it contains no notion of operators or relations on these numbers (e.g., "greater than," "max") and no notion of a multidimensional space $\mathbf{R}^n$. SWEET defined numerical relations and vector concepts for use in its ontologies.

ONTOLOGY STORAGE

XML-based languages such as OWL are well suited to data and model exchange but are less practical for storage and query of large ontologies such as SWEET, which currently holds over 3000 terms. Existing database management systems provide the needed functionality in storage and indexing of robust ontologies, including support for data integrity, concurrency control, etc. Consequently, we adopted the PostGRES object-oriented DBMS as an alternative representation of ontology elements.

Two-way translators were created between the internal DBMS representation and the standard XML representation. By placing all term declarations in the DBMS, searches are very rapid.

To represent instances of spatial concepts, we used bounding polygons to describe regions, where possible. Polygons are a native data type in PostGRES.

Many large knowledge bases are dynamic and remotely stored. We executed the concept of virtual ontologies using OWL wrappers around remote, dynamic data repositories. This enabled SWEET to include earthquake events available through the USGS database interface and gazetteer information from multiple sources.

APPLICATION

SWEET can be used to aid a search by integrating the ontologies with a search engine. A prototype search tool was developed that accepts user search requests, consults the ontologies for alternate definitions, then submits the union of these terms to the GCMD search engine. Such a mechanism eliminates the need for exact keyword matches. Currently, this tool returns links to GCMD metadata entries known as Directory Interchange Formats (DIFs). Matches can be made using synonym, parent, or child terms. Ranking of results is based on a heuristic that places lower weights on those from "distant relatives." A better approach is to cluster the results on any alternate terms used. Such a tool is currently under development for the ESIP Federation using the Nutch public-domain search engine. Similar search functionality is available through commercial products such as Seamark from Siderean Software (http://siderean.com) and TAP (http://tap.stanford.edu).

DISCUSSION

Ontologies are not intended to replace controlled vocabularies developed by specific communities. Rather, communities can use these ontologies as common building blocks to declare that term X in ontology A is equivalent to term Y in SWEET. If term Z in ontology B is also declared equivalent to Y in SWEET, then X and Z are equivalent by transitivity, without the need to provide a direct declaration. In this sense, SWEET provides a common semantic framework. It is possible that a large number of small ontologies will win out over fewer larger ones. In such cases, the *scope* of the ontology should be selected minimally. SWEET includes a *substance* ontology, but ultimately this concept can be defined in a more generic science ontology, when available.

The next step in the ontology development process is to coordinate SWEET ontologies with those developed for the solid Earth community in GEON (Sinha, this volume) as well as planetary science. The resulting ontologies represent a generic description of any planetary body.

Discussions are under way with the Earth System Grid (ESG) and Earth System Modeling Framework (ESMF) to maintain consistency with grid and modeling efforts.

TABLE 2. CAPABILITIES OF RDF/RDFS

Boolean combinations	Union, complement, intersection
Functional properties	Inverse, transitive, symmetric
Equality and inequality	Same as, different from
Property restrictions	Has some, has all
Cardinality restrictions	Minimum, maximum, equal

Note: RDF—Resource Description Framework.

TABLE 3. ADDED CAPABILITIES OF ONTOLOGY WEB LANGUAGE

Building blocks	Class, subclass, URL, literal value
Set properties	Subclass, subproperty, domain, range, instance
Assertions	Subject, predicate, object, value
Information	Comment, see also, label
Groupings	Bag, list, sequence, first, rest, nth, nil

A specific application where ontologies will prove useful is in formal definitions of data-mining operators. Currently, no language exists for a complete definition to the extent that others can replicate the procedures. Ontologies enable users to formally define a phenomenon in terms of its underlying characteristics.

What is the value of a scientific knowledge representation system that supports semantic interoperability? It is assumed that semantic Web services will open up a wide range of new data services and new users, and result in a reduced burden on the knowledge seeker. All of this is possible if the scientific knowledge (embodied in ontologies) is there.

REFERENCES CITED

Cornillon, P., Gallagher, J., and Sgouros, T., 2003, OPeNDAP: Accessing data in a distributed, heterogeneous environment: Data Science Journal, v. 2, no. 5, p. 164–174.

Fensel, D., Hendler, J., Lieberman, H., and Wahlster, W., editors, 2003, Spinning the semantic Web: Cambridge, Massachusetts, MIT Press, 479 p.

Olsen, L., 2001, Supporting interagency collaboration through the Global Change Data and Information System (GCDIS): Proceedings of the 17th International Conference on Interactive Information and Processing Systems (IIPS) for Meteorology, Oceanography, and Hydrology, American Meteorological Society, Albuquerque, New Mexico, January 14–18, 2001, p. 187–190.

Raskin, R., Burrows, H., Conover, H., Gallagher, J., Major, G., and Rhyne, T., 2002, Discovering and accessing information from the Federation of Earth Science Information Partners (ESIP): Eos (Transactions, American Geophysical Union), v. 83,, no. 47, p. 19.

Raskin, R.G., and Pan, M., 2003, Ontology development in the Semantic Web for Earth and Environmental Terminology (SWEET): Geological Society of America Abstracts with Programs, v. 35, no. 6, p. 368.

W3C, 2001, XML schema part 2: Datatypes: http://www.w3.org/TR/xmlschema-2 (June 2005).

W3C, 2004a, RDF: http://www.w3.org/RDF (June 2005).

W3C, 2004b, OWL: http://www.w3.org/TR/owl-ref (June 2005).

Manuscript Accepted by the Society 18 June 2005

Geological Society of America
Special Paper 397
2006

*The case for transforming the geological survey knowledge system: Where digital geoscience spatial models meet the semantic Grid**

T.V. Loudon†
British Geological Survey, West Mains Road, Edinburgh EH9 3LA, UK

ABSTRACT

As the Internet develops into a more powerful cyberinfrastructure (the Grid), geological survey knowledge will enter a framework where users assess its value by its relevance and evaluated reliability rather than its source, and where artificial boundaries of place and discipline lose significance. Within geological surveys, much effort has been devoted to the development of the geological map through digital cartography. Consequently, a firm base already exists for more radical change to the geoscience knowledge system. The geoscience map is prepared from field records and structured data and is given meaning through the illustrated text narratives of map explanations. A more comprehensive view of this knowledge system is needed to gain the full benefits of the Grid. The flexibility of markup languages can link narrative to database, spatial modeling, and semantic representations, and thereby combine the insights of human understanding with the power of computation. Rationalizing and extending the scope of existing ontologies (specifications of conceptualizations) to align their structure and vocabularies could provide a more comprehensive conceptual view. Thus, a major challenge for geological surveys will be to transform their knowledge system to conform to a new service-oriented infrastructure devised by others, with its future evolution determined by users.

Keywords: geoscience survey, digital cartography, spatial model, markup language, semantic Grid, cyberinfrastructure.

1. THE OPPORTUNITY

Recent advances in geoinformatics suggest that it is timely to review the system for handling information and knowledge. Many of those involved in geoscience surveying may question the need for further change. This paper attempts an answer, directed at geoscientists, by indicating some possible directions and expected benefits. In this paper, the term *geological surveys* usually refers to official surveying organizations.

The acceptance of the concept of the digital geoscience spatial model (DGSM) as an approach to geological surveying (Smith, 2001), and the deployment of the semantic Grid as infrastructure for e-science, combine to offer an unusual opportunity, namely, to transform part of the geoscience knowledge system. The work and proposals recorded at the GEON Web site on cyberinfrastructure for the geosciences (GEON, 2004a, 2004b) indicate the radical changes envisaged for the broader field of geoscience. The ultimate aim for a geological survey knowledge

*This paper is published by permission of the director of the British Geological Survey (NERC). It reflects a personal, not a corporate, viewpoint.
†Email: v.loudon@bgs.ac.uk

Loudon, T.V., 2006, The case for transforming the geological survey knowledge system: Where digital geoscience spatial models meet the semantic Grid, *in* Sinha, A.K., ed., Geoinformatics: Data to Knowledge: Geological Society of America Special Paper 397, p. 201–209, doi: 10.1130/2006.2397(15). For permission to copy, contact editing@geosociety.org.

system may be to meet user requirements and expectations, but the results will be more effective if they emerge from careful consideration of the subject matter. A successful outcome therefore depends on geological input at scientific, technical, and management levels. Some potential benefits to geoscience surveying are reviewed in section 2, some underlying concepts in section 3, and prospects for extending the vision of future geoscience surveying in section 4, with conclusions in section 5.

The Grid

The revolutionary impact of information technology on society (such as the railway, road, electricity grid, or telephone system before it) may occur when it ceases to comprise differentiated, proprietary technologies and becomes a ubiquitous infrastructure shared by all: a commodity, no longer offering a strategic advantage over competitors, but a cost to be managed like office supplies. This is seen as a likely consequence of the current evolution of the Internet into a new infrastructure of information and communications technology referred to as the cyberinfrastructure or Grid (Foster and Kesselman, 2003).

The Report of the (U.S.) National Science Foundation Advisory Panel on Cyberinfrastructure states in its summary (Atkins et al., 2003, p. 81), "The scope and scale of the ACP [Advanced Cyberinfrastructure Project] will require an annual budget of about $1 billion over and above the current PACI and infrastructure programs in NSF.... Nevertheless, our recommendations should be considered as only a beginning." Geoscientists are involved, for example, through the GEON project—a prototype national geosciences cyberinfrastructure network. An aim of GEON is helping to weave the separate strands of the solid Earth disciplines and data into a unified fabric. Initiatives elsewhere include European Commission funding, which supports the EGEE project (Enabling Grids for E-science in Europe), a two-year project of a four-year program, with links to government initiatives and funding (EGEE, 2004; e-Science Institute, 2004; Research Councils UK, 2004).

The Grid is bringing a large increase in the capacity and power of computer networks, thus allowing more ideas to be expressed, shared, and structured to extend human knowledge. It aims to provide facilities that, among other things, enable scientists to generate, analyze, share, discuss, and communicate their investigations in a more effective manner. Hinke (2004) describes how Grid services "could revolutionize the ability of developers to easily construct extensible data handling and processing systems, by providing seamless access to various types of resources including computational resources, data archives, scientific sensors and the services to process data into some desired product, which itself could be formulated as a service." The Grid might, for example, change the emphasis in a geological survey's objectives, from delivering published end products, to maintaining information resources from which end users could select services that would respond more flexibly to their specific needs.

The DGSM

Geoscience surveying operates at the interface of the solid Earth and geological knowledge, linking them through spatial location. It thus has important roles in collecting information and testing interpretations. The digital geoscience spatial model extends beyond conventional map-based surveying. The model overcomes constraints of the map, such as its static two dimensions, fixed scale, inflexible information density, limited topics, rigid sheet boundaries, separation of spatial depiction from text explanation, and laborious and time-consuming drafting and revision procedures.

The model can take a flexible view of the geology as a set of interrelated things of interest (objects), which exhibit attributes such as properties, behavior, and relationships. Its structure accepts classifications by distinguishing between object classes and instances (described in section 3) and recognizing class hierarchies (as in stratigraphy) within which attributes can be inherited. Potentially, it can relate observed objects to, say, the outcome of simulated processes, and track reasoning processes from observation to explanation or vice versa (Voisard, 1999). Being scale-free, the model can accommodate multiscalar concepts such as fractal representations. Objects described in markup languages, such as XML, can combine text, data, and images readable by human beings with tags readable by computer, and thereby connect their respective strengths. The results of a DGSM can be stored and maintained in a hypermedia knowledge repository or object store, where the semantic Web records relationships and reasoning.

The Semantic Web

Extensions to the Web aim to provide a language expressing both information and rules for reasoning about it (including rules in existing knowledge representation systems), enabling people and computers to work in closer cooperation (CEOS, 2004). Sharing and processing information by automated tools as well as by people should lead to more effective discovery, integration, and reuse of the information (Gahegan and Brodaric, 2002). The Extensible Markup Language (XML) allows users to extend their documents by embedding structures in them that computers can interpret (W3C, 2004; Cox, 2004; POSC, 2004). The Resource Description Framework (RDF) gives form to such a structure, as a triple of three linked items, like a sentence expressing the idea that an object has an attribute with a value. A Universal Resource Identifier (URI) can identify each of the three items, thus linking them to unique definitions available on the Web. The value can be another object, and the RDF statement can itself be regarded as an object or attribute.

There is thus scope to record, as part of an object's description or at the appropriate points in a document, inference rules that define (1) object behavior (its response to types of process), such as a class of surfaces being appropriately modeled in a specific way; (2) relationships, such as that the W, X, and Y formations lie in a conformable sequence in the south, but the W lies unconformably on the Y in the north; and (3) reasoning, such as linking the form of a surface model to that of process simulations and thence drawing conclusions about its origin. A URI can point to anything, including a physical object, thus bridging the interface from the virtual to the real world.

Ontologies also play a part in the semantic Web, each a published specification of a conceptual framework for a particular topic (NASA, 2004; W3C, 2004). They typically provide taxonomies that define classes of objects, the relationships between them, and sets of inference rules. They aim to be consistent and unambiguous and to integrate diverse viewpoints, facilitating sharing of content and services. "Human endeavor is caught in an eternal tension between the effectiveness of small groups acting independently and the need to mesh with the wider community.... An essential process is the joining together of subcultures when a wider common language is needed.... The Semantic Web, in naming every concept simply by a URI, lets anyone express new concepts that they invent with minimal effort. Its unifying logical language will enable these concepts to be progressively linked into a universal Web" (Berners-Lee et al., 2001).

The Semantic Grid

The semantic Grid brings together the ideas of the semantic Web and the Grid. It supports the vision of e-science, helping scientists to work and collaborate more effectively by providing a language that expresses both data and rules for reasoning about the data. "Our vision is of a generically usable e-science infrastructure, comprised of easily deployed components whose utility transcends their immediate application, providing a high degree of easy-to-use and seamless automation and in which there are flexible collaborations and computations on a global scale. The key to this is an infrastructure where all resources, including services, are adequately described in a form that is machine-processable, i.e., knowledge is explicit—in other words, the infrastructure provided by the technologies of the Semantic Web. The goal is semantic interoperability" (De Roure, 2003).

The semantic Grid adopts a service-oriented view (De Roure et al., 2001) "based upon the notion of various entities providing services to one another under various forms of contract (or service level agreements)." Numerous software systems or agents that perform useful tasks on behalf of a user would interact in this setting. A service provider, such as a geological survey, would therefore have to ensure long-term digital preservation of trusted sources of evaluated, up-to-date knowledge (PADI, 2004). In due course, individuals and organizations must evolve to meet their changing roles within a more comprehensive and flexible system supporting the repository of scientific knowledge. For geological surveys, there is an opportunity to gain many benefits, mentioned next, provided essential features of the present system are not neglected.

2. THE BENEFITS

Worldwide, the work of geological surveys converges on geoscience maps and their accompanying explanations. These conventional products follow consistent standards and give extensive coverage at several fixed scales. The large format and high resolution of the maps enable users to focus on points of specific interest within a context supplied by peripheral vision. They generally correlate with topographic and other related maps. They are robust, highly portable, and can be used without supporting technology. They are assembled with care and consistent quality control. The maps are typically supplemented by text explanations that provide an illustrated background account of the geology of the area as a whole and of salient details, illustrated where appropriate by detailed maps and diagrams. Because they are formal publications, they provide a fixed reference point for geoscientists. They are widely available at reasonable cost, and copies are held indefinitely in several secure libraries. Revised versions are published when appropriate, generally after many years. The qualities of these conventional publications make it unlikely that they will be supplanted in the foreseeable future. But the end products constrain the approach and procedures of investigation.

Information technology (IT) has a growing role in the processes of geoscience surveying and in the preparation of maps and associated documents. Digital cartography can make the legacy of old documents available in new forms, which subject experts can enhance for inclusion in a system of digital spatial models. Field mapping can generate digital records directly. Modern information and communications technology introduces models that may lead to more effective methods of surveying and presentation. Adherence to standards makes it easier to find and exchange information (FGDC, 2004; USGS, 2004a, 2004b). New systems can offer many benefits but require careful design to retain existing valuable qualities and to ensure that they tie in with related systems of wider scope (Brodaric, 2000).

An IT-based system can model spatial and other aspects of geological knowledge, including concepts, objects, and their properties, relationships, and processes. Anticipated long-term gains for users of such a geoscience knowledge system include the following.

Costs

The costs of preparing, storing, finding, and using conventional documents are opaque, making cost comparisons difficult. However, at each stage the corresponding long-term costs of an IT system appear highly competitive, with more transparent costing leading to more efficient procedures.

Flexible Structure

Maps have fixed scales, rigid content, and sheet boundaries unconnected to the significance and extent of geological objects. Digital spatial models can be seamless with no artificial boundaries of area, detail, or subject matter. Unlike maps, models can handle complex systems and multiscalar views of the geology, as in the self-similar dissipative-system models of sand bodies (van Wagoner et al., 2003).

Timeliness

Conventional publication is slow, and decades may elapse between data collection and publication. Much information is now collected digitally and could be made instantly available, with access controlled by authorization procedures rather than by the mechanics and economics of printing and publishing.

Ease of Access

Discovering and accessing relevant information from conventional sources is slow and laborious, particularly for users lacking the facilities of a major survey. Web access, on the other hand, can be made available anywhere. Web search engines, metadata, and browsers help the user to discover, filter, and retrieve spatial and related information from many sources.

Connected Information

Information from a single investigation is inconveniently segregated by form of presentation, such as maps; memoirs; journal articles; textbooks; databases; archives of field records, samples, and specimens; photographs; sketches; preliminary results; and human knowledge. Multimedia and markup languages can link and combine information types at any level of detail.

Flexible Presentation

Multimedia records can be filtered for relevance and assembled and presented to meet a wide range of purposes and contexts. For example, a map of a rock body could be shown onscreen alongside text accounts of the processes of its formation and photographs of its outcrop, with options for, say, superimposing fitted diagrams of its origin and development, or displaying the consequences of cut-and-fill for road building.

Flexible Visualization

The map is a single, rigid visualization. Spatial models provide information that can be visualized as maps, sections, fence diagrams, block diagrams, video sequences, correlation diagrams, and so on, for user-selected areas, content, scales, and projections. Thematic maps and multiple visualizations can overcome the information overload of the single map, clarifying the interpretation.

Formal Structure

The object-oriented model separates metadata from detail, distinguishes between object classes and instances, and creates object-class hierarchies with multiple inheritance of attributes and behavior. It conforms well to patterns of geological thought and provides a clear framework for uniform standards on what to call things, what they mean, how they relate, where to put them, and where to find them.

Reusing Objects

By routinely identifying the properties and behavior of object classes, it should be possible to reuse object instances, such as the boundary of a rock body, for various purposes, such as thematic studies for land use, geological risk assessment, resource estimation, and other products custom-built for specific user needs. The Web and, in due course, the Grid will enable experts in any field to assemble objects and software from diverse sources, and record their methods in scripts that in turn may be reused in other contexts.

Reasoning

Object-oriented methods can link field evidence and conclusions to an ontology (for consistent terminology) and to objects describing aspects of theory and specific chains of reasoning. These could provide the evidence justifying the conclusions, help to evaluate the findings, and identify the wider consequences of new observations or ideas.

Integration

The interpretation depends on many aspects of geological reasoning, such as depositional, structural, and metamorphic models, which may have to be considered together to explain even a small feature. The rich connections within and among hypertext records can help to clarify links between the separate lines of reasoning at any level of detail.

Reconciling Viewpoints

Object-oriented techniques should enable information from many sources to be combined. Multiple sources reduce uncertainty and narrow the range of plausible hypotheses. But the original sources may have adopted viewpoints with different objectives and emphasis, leading to conflicting results. Reconciling them requires the judgment of human experts, and the computer can record their reasons and justification.

The motive for transforming the information systems of geological surveys is to support better science more efficiently. Survey records could be connected in more detail with the historical configurations of geological objects and processes and their underlying physics, chemistry, and biology, while positioning geoscience appropriately in the mainstream of development of the larger, unbounded knowledge system. It is subject specialists (not computers) that have the background understanding of geoscience and of their own ways of thinking that enabled them to build the existing knowledge system. They are best placed to reorganize that system by formalizing a framework within which IT can help users to record, store, find, relate, select, assemble, communicate, and visualize their knowledge. Deploying these mechanical aids effectively calls for some unaccustomed introversion by geoscientists about their working practices and the consequences of change made possible by the emerging methods of IT.

3. ASPECTS OF THE GEOSCIENCE KNOWLEDGE SYSTEM

Two streams of IT development contribute to the potential benefits listed in section 2. The first is knowledge representation (Sowa, 2000), including abstraction and modeling; the second is communication and its infrastructure of Web-based technologies. The two streams converge in the hypermedia repositories and knowledge systems supported by the semantic Web and Grid. It may be useful here to clarify some terms, and to highlight some approaches and suggest how they fit together. Their combined relevance in an integrated system for geological surveying remains largely unexplored.

Standards

International standards extend the ability to communicate within a comprehensive knowledge system. Various information communities (such as CGI, 2004; ISO, 2004; OGC, 2004; POSC, 2004; W3C, 2004) are concerned with procedures and standards in specific areas, including easier access to information (GILS, 2004; INSPIRE, 2004; National Geological Surveys Committee, 2004).

Systems

A system is a collection of elements or components that are organized for a common purpose. System boundaries are somewhat arbitrary, but the systems approach brings together widely used concepts, vocabulary, and insights from many fields of study. The system of geological surveying observes and measures features of the real world. Its procedures represent, record, interpret, explain, and communicate the results, extending the system of scientific knowledge with conclusions that can be tested against reality.

Abstraction

The system that makes scientific knowledge comprehensible to the human mind requires a process of abstraction; that is, reducing the volume of information while retaining salient features and relationships. Geoscience surveying progressively reduces the full detail of the real world to observations, then to field records, then (by a process of interpretation, reasoning, and explanation) to general conclusions drawn from specific instances, and to summarization and generalization. It aims (by feedback from higher levels to detailed levels of abstraction) to extend or modify a shared paradigm of a parsimonious, unified account of the observed phenomena.

Reconciliation

The paradigm is shared at a general level, but detail may be specific to the investigation. The objectives, methods, and opinions of the investigators determine which features are regarded as salient, and influence how the abstraction proceeds. Using information from one study in another context therefore requires filtering and adjustment, usually under human control, to reconcile information from one investigation with that of another, for the purposes at hand (Kent, 1978; Sowa, 2000, Chapter 6 therein).

Models

A model is an approximate representation, for a specific purpose, of some limited aspects of the real or imagined world. It aims to reduce their inconceivable complexity to a manageable representation. Hierarchical models can represent phenomena at several levels of complexity or detail. Geoscience survey models interface (through spatial coordinates) with the real world, where observations are made and deductions tested.

Object-Oriented Analysis

The computer-based system should work in harmony with human thought patterns. Object-oriented analysis (Budd, 2000; Coad and Yourdon, 1991) aims to build on the methods commonly used to organize our observations of the real world. It differentiates the observed world into things of interest (objects) and their attributes and relationships, distinguishes between the whole object and its component parts, groups individual occurrences (instances) of objects into classes with shared properties and behavior, and provides for hierarchies of classes and inheritance of properties within them. The objects can potentially be reused in various applications. Object-oriented design and implementation (Brodaric and Hastings, 2002), as opposed to analysis, raise many issues not considered here.

Spatial Models

Geoscience spatial models aim to represent the spatial properties and relationships of geoscience objects within a consistent mathematical framework (Mallet, 2002). The behavior of the objects can be defined, relating them to reasoning from process models and tying the interpretation to database management systems and geographic information systems (to discover, filter, and retrieve relevant information from the object store), interpolation systems (for filling gaps and estimating uncertainty), and visualization systems (to display the results).

The General Geoscience Spatial Model

The observations and interpretations of geoscience are linked to a more general conceptual model. This refers at all levels of detail to the three-dimensional disposition and configuration (where things are and how they are arranged) of the present-day observable and conceptual models of geoscience, to their observed and interpreted properties and composition, and also to their history throughout geological time, including the processes that created or transformed them and are crucial to their interpretation. It is too vast and inaccessible for full representation, and so remains an abstract concept. But it provides the essential conceptual framework for relating and reconciling various facets of its fragmentary representation.

The Challenge

The multiplicity of views in geoscience might be reflected in the infrastructure by a variety of big and little ontologies, with working connections established through linkages, in a system that encourages their rationalization and reconciliation for specific purposes, in response to real user needs. The interests and viewpoints of the end users of geoscience knowledge differ from (and may be unknown to) the collectors of the information, and the ultimate users of geological knowledge are not generally geologists. The future infrastructure will develop to offer links to all areas of general and scientific knowledge. There is a need for geoscientists and their information communities to adapt their knowledge system to a new infrastructure devised by others, and to keep pace with changing user needs.

This points to the need for planning ahead to meet the challenges of geoinformatics. Portals, browsers, and selection, interpolation, and visualization systems will be subsumed in the infrastructure and accessed through agents (programs that act in a self-interested manner on behalf of their users, while interacting with many other agents in the computer network). Collecting, finding, and using geological information will be undertaken in a multitude of different contexts, not just on the terms set by one cohesive geoscience information community. The challenge is therefore, not to prejudge the end result, but to influence and participate in the development and exploitation of an infrastructure that can offer the mechanisms for diversity, evaluation, selection, and inheritance that are required for the beneficial evolution of knowledge.

The infrastructure offers no competitive advantage from selective ownership of new technology. On the contrary, as with purchasers of a camera-phone, the utility of each owner's investment increases as the standard technology is more widely adopted. Early adopters might benefit by subsidy from bodies anxious to maintain a country's or region's long-term competitive position. But the main advantages of an organization's early involvement lie in gaining time to evaluate the scenarios, analyze the processes within the organization, determine priorities, and achieve a smooth adjustment to the developing knowledge system without unnecessary disruption to staff and customers.

For example, the British Geological Survey (2004) currently provides a portal and browser for access to its information. But in the outside world, enquirers dealing with, say, epidemiology or insurance risks might overlook the possibility that the portal, or indeed geology, could be relevant. In the longer run, customers may wish to access geoscience information through a general search engine, where it could be integrated with information from other sources, rather than through an entry point specific to one organization. If so, a flexible plan for transition is required some years in advance. Time is essential to plan ahead, explore alternatives, and gain experience in unfamiliar fields such as establishing trusted, service-oriented, digital systems.

Long-term adjustments may need immediate attention, such as reviewing the structure and boundaries of organizations and information communities, establishing tentative pathfinder standards, defining areas of priority research, and reconsidering field methods, staff training, and balance of staff expertise. Decisions may also be required on the assignment and fulfillment of long-term responsibilities, such as digital preservation and maintenance (PADI, 2004; Fedora, 2004) and procedures for rationalizing geoscience ontologies, maintained in the light of the current general geoscience model.

4. EXTENDING THE VISION

The vision of IT in geoscience surveying is still being adjusted to the concepts of database, computer-aided design, geographic information systems, spatial models, object-oriented analysis, and the Internet. Projects based on these concepts have evolved step-by-step, and are now incorporated in the procedures of many geological surveys, although not yet bringing the full benefits listed in section 2. Maps can be scanned, stored, and reproduced; geological features can be vectorized and described in a database; map sheets can be edge-matched; GIS-based spatial indexes can record and retrieve a wide range of spatial features; metadata can be standardized for geoscience records; spatial models can record and visualize the three-dimensional form of geological features; object-oriented methods help to rationalize the design; digital field recording integrates the system with electronic surveying tools, the global positioning system, and (potentially) augmented reality; the Web can select and deliver the results. Not only is a vision defined, but also many parts of the system are implemented and doing useful work. Many standards, such as those referred to in section 3, are already in place. They give a firm base for planning the next steps.

There is now a clear need for extending this vision to match the developments in the cyberinfrastructure. For example, a seamless geological map should allow the user to specify the required area, type of content, scale, projection, and format. But simply edge-matching existing sheets will not meet these needs. Map explanations and marginalia, such as stratigraphic tables, cross sections, and generalized vertical sections, prove difficult to keep in step with depiction of the map face for an arbitrary area.

The underlying problem is that the map is a static two-dimensional projection of incomplete fragments of the general geoscience model mentioned in section 3. The surveying process organizes the fragmentary information by long-standing procedures designed for paper documents. They hold back the introduction of user map design and other concepts, such as georeasoning (Gahegan and Brodaric, 2002; Pshenichny, 2003), or a view of Earth processes sharing the common physics of complex dissipative systems (van Wagoner et al., 2003). The problem cannot be resolved by technical advance alone: ideas are shaped by the manner of their representation, and geoscientists must be involved.

Linking models of spatial objects to an account of the reasoning process is desirable for at least three reasons. First, the surveying model must include evidence and reasoning to fulfill its role as the interface connecting geoscience knowledge and the real solid Earth. Second, all scientists should be able to explain their procedures, so that colleagues can follow their reasoning and evaluate and test their results. Third, the links are needed to establish how the knowledge base must be amended to respond to the inevitable revision of interpretations and observations. Taken together, these issues suggest that the vision of future geoscience surveying may be too narrowly focused on extending the geoscience map as a spatial model, and that the underlying rationale needs reconsideration. The human brain appears to handle spatial and narrative information along different pathways (Loudon, 2000, parts J and K therein). The separation reappears in geoscience surveying in the distinction between maps or spatial models on the one hand, and text-based narrative explanations on the other.

Space has special properties, reflected in methods (topology, geometry, and spatial transformations) for describing spatial pattern and relationships. Spatial thinking takes a comprehensive three-dimensional view, perhaps considering geological time as analogous (in some respects) to a fourth spatial dimension. "We can zoom in to see the detail, zoom out to see the spatial context, pan around to see the situation elsewhere, and compare spatial patterns arising from different topics. Narrative text cannot offer these abilities, but can be intimately linked to a spatial representation" (Loudon, 2000, p. A89).

Information in narrative form may refer to sequences of events, reasoning, and explanations. In a similar way, narrative may describe properties by sequences of words and sentences, perhaps enumerating the characteristics of a rock, and proceeding from general comments to detail. In our memories and in written text or the spoken word, the narrative follows threads of thought, linked and woven together to make sense of our experience by defining objects and processes, drawing analogies, establishing relationships, abstracting, and summarizing.

Markup languages enable surveyors to organize their ideas as a set of narrative accounts (objects), trace the threads of reasoning through hypertext links between the objects, and link the objects with their counterparts in a digital spatial model. All are potentially tied to a shared ontology through interactive templates and are developed together in the field and office. Spatial representations, such as the geological map or spatial model, fill the space depicted, resorting to interpolation, approximation, and generalization where direct knowledge is inadequate. The spatial coordinates provide additional GIS links to, for example, stratigraphic metadata, cross sections, databases, and explanatory text objects.

The knowledge base cannot record the full general geoscience spatial model, which remains only as a concept. However, the overall structure and features of geoscience and its general model are described in many textbooks and implied in bibliographic databases such as GeoRef (AGI, 2004) and metadata such as the Epicentre model (POSC, 2004, see Specifications). They provide the framework, classes, and categories for the indexing and cataloguing of geoscience information.

The same framework guides the surveyors and the users of the completed survey. Within such a framework, the user should be able to search among the diverse incomplete fragments of the general model, select the most relevant, and visualize them as filtered extracts, possibly requiring reconciliation by the user. The user's agent might apply filters to select from such aspects as the area of interest; the business setting; the range of topics and levels of detail in each object hierarchy; the user's objectives and the emphasis placed on particular properties; the sources and evaluation of the information; the levels of resolution and generalization that match the scale of visualization; the levels of confidence, ranging from confirmed observation to tentative explanation; the spatial relationships valid at the selected levels; the appropriate spatial properties, such as existence, location, slope, form, texture, arrangement, and relationships; and the place of the objects and properties in chains of reasoning and process-response models. Visualization of the filtered extracts would then involve geometrical transformations including projection.

The ontology for the digital geological survey model is a component of this more general framework. The model might be thought of as representing a self-consistent, approved, and authorized interpretation and explanation of the disposition and configuration of the rock units and aspects of their properties, composition, relationships, and history, within a defined area or volume. The internal consistency of the model reduces the need for users to reconcile fragments. The model should enable users to view surveying as a process, not just as a product. It must be flexible to accommodate the variety of routes that surveying procedures can follow, and the levels of detail to which they refer. For example, the framework must be able to accept as input not just multimedia streams of marked-up field notes and spatial models, but also existing visualizations (such as the legacy of geoscience maps) and related material, reworked or otherwise. The knowledge system should thus be able to start with conventional information and incorporate more advanced methods as they prove their worth.

The framework must support evolution by modification and replacement of items in the intertwined sets of submodels, objects, and methods for processing them. It must be amended and extended as new topics enlarge its scope. The ontology that specifies this conceptualization must accommodate more specialized ontologies referring to submodels. Likewise, geoscience ontologies must take their place within the much broader framework of the cyberinfrastructure, for geoscience explanations reach out to other branches of science, and users of geoscience information approach it from many different business settings. In the long run, specific markup languages, ontologies, and methods may be subsumed in infrastructure products of greater generality, to the advantage of all concerned.

No committee can assemble the wisdom to rationalize the diversity of the knowledge system. The semantic Grid may therefore be designed to simplify and rationalize the plethora of methods and standpoints by selection through evolutionary processes, perhaps guided by the rituals of scientific discourse and of the marketplace, supported by the shared efforts of consortiums, standards committees, geological surveys, and other information communities dedicated to the common good.

5. CONCLUSIONS

The Grid extends the means of representing, processing, and communicating human knowledge, by greatly increasing the capacity and power of computer networks and infrastructure services. Geological surveying has long focused on cartography, and more recently on its digital equivalents. The large volumes of electronic data from digital spatial models and the larger volumes from exploration seismic data can be more readily transmitted through the increased bandwidth of the Grid. But its ability to accommodate detailed relationships may prove more significant. Spatial models and narrative explanations are built side by side, and the Grid provides the opportunity to maintain their close links at all levels of detail, linking data and observational evidence with reasoning, interpretation, and explanation.

Markup languages have the potential to link text narrative at all levels of detail with spatial models, metadata, Earth processes, experiments, databases, images, material collections, and human experts. Collaborative development of standard frameworks and ontologies brings unified structures, classifications, and vocabularies to this knowledge system and encourages their rationalization, thus bringing together the results of many individual efforts. The Grid will enable geological surveys and similar organizations to provide services from which users (who best understand their own requirements) can obtain appropriate products or create their own.

Basic geology is at the core of the geoscience knowledge system, and geologists now have an opportunity to embrace a technology free of the constraints of pen, paper, and printing press. Geoscientists can move their legacy of existing knowledge into the mainstream of a more comprehensive knowledge system and ensure its appropriate digital preservation. Geological surveys are well positioned to advance the transformation of this part of the knowledge system, while academic research, such as the GEON program, brings new perspectives to geoscience surveying.

ACKNOWLEDGMENTS

I am grateful to the British Geological Survey for making this work possible, in particular the help of Bill Hatton, Ian Smith, and John Laxton, and to Boyan Brodaric of the Geological Survey of Canada for his encouragement in improving this paper.

REFERENCES CITED

AGI, 2004, American Geological Institute: GeoRef Information Services: http://www.agiweb.org/georef/ (September 2005).

Atkins, D.E., Droegemeier, K.K., Feldman, S.I., Garcia-Molina, H., Klein, M.L., Messerschmitt, D.G., Messina, P., Ostriker, J.P., and Wright, M.H., 2003, Revolutionizing science and engineering through cyberinfrastructure: Report of the National Science Foundation Blue-Ribbon Advisory Panel on Cyberinfrastructure, 84 p., http://www.communitytechnology.org/nsf_ci_report/ExecSum.pdf (September 2005).

Berners-Lee, T., Hendler, J., and Lassila, O., 2001, The semantic Web: ScientificAmerican.com, May 17, 2001: http://www.sciam.com/article.cfm?articleID=00048144-10D2-1C70-84A9809EC588EF21 (September 2005).

British Geological Survey, 2004, BGS GeoPortal: http://www.bgs.ac.uk/scripts/GeoPortal/home.cfm (September 2005), BGS GeoIndex: http://www.bgs.ac.uk/geoindex/home.html (September 2005).

Brodaric, B., 2000, Digital geological knowledge: From the field to the map to the Internet: Digital Mapping Techniques '00—Workshop Proceedings: U.S. Geological Survey Open-File Report 00-325: http://pubs.usgs.gov/of/of00-325/brodaric.html (September 2005).

Brodaric, B., and Hastings, J., 2002, An object model for geologic map information (14 p.): International Society for Photogrammetry and Remote Sensing Symposium on Geospatial Theory, Processing and Applications, Ottawa, 2002: http://www.isprs.org/commission4/proceedings/pdfpapers/131.pdf (September 2005).

Budd, T., 2000, Understanding object-oriented programming with Java: Reading, Massachusetts, Addison-Wesley, 420 p.

CEOS, 2004, Committee on Earth Observation Satellites—Semantic Web Link Collection: http://www.op.dlr.de/ceos/dstt/semantic_web.html (September 2005).

CGI, 2004, Commission for the Management and Application of Geoscience Information: http://www.bgs.ac.uk/cgi_web/welcome.html (September 2005).

Coad, P., and Yourdon, E., 1991, Object-oriented analysis: Englewood Cliffs, New Jersey, Yourdon Press, 233 p.

Cox, S., 2004, XMML—eXploration and Mining Markup Language: https://www.seegrid.csiro.au/twiki/bin/view/Xmml/WebHome (September 2005).

De Roure, D., 2003, The semantic Grid: Vision: http://www.semanticgrid.org/vision.html (September 2005).

De Roure, D., Jennings, N., and Shadbolt, N., 2001, Research agenda for the semantic Grid: A future e-science infrastructure: Technical report UKeS-2002-02, UK e-Science Technical Report Series, National e-Science Centre, Edinburgh, UK, 37 p.: http://www.semanticgrid.org/documents/semgrid-journal/semgrid-journal.pdf (September 2005).

EGEE, 2004, Enabling Grids for E-science in Europe: http://public.eu-egee.org/ (September 2005).

e-Science Institute, 2004, UK National e-Science Centre: http://www.nesc.ac.uk/esi/intro.html (September 2005).

Fedora, 2004, The Fedora™ Project: An open-source digital repository management system: http://www.fedora.info/ (September 2005).

FGDC, 2004, U.S. Federal Geographic Data Committee—Geospatial metadata standards: http://www.fgdc.gov/metadata/metadata.html (September 2005).

Foster, I., and Kesselman, C., editors, 2003, The Grid: Blueprint for a new computing infrastructure (2nd edition): Amsterdam, Morgan Kaufmann, 748 p.

Gahegan, M., and Brodaric, B., 2002, Computational and visual support for geographical knowledge construction: Filling in the gaps between exploration and explanation (19 p.): Symposium on Geospatial Theory, Processing and Applications, Ottawa, 2002: http://www.isprs.org/commission4/proceedings/pdfpapers/154.pdf (September 2005).

GEON, 2004a, The Geosciences Network: Building cyberinfrastructure for the geosciences: http://www.geongrid.org/ (September 2005).

GEON, 2004b, Ontologies: http://www.geongrid.org/resources/ontologies.html (September 2005).

GILS, 2004, Global Information Locator Service: http://www.gils.net/ (September 2005).

Hinke, T.H., 2004, Grid technology as a cyberinfrastructure for earth science applications: NASA Earth Science Technology Conference, June 2004, 8 p.: http://esto.nasa.gov/conferences/estc2004/papers/a3p1.pdf (September 2005).

INSPIRE, 2004, Infrastructure for Spatial Information in Europe—European Geo-Portal: http://eu-geoportal.jrc.it/gos (September 2005).

ISO, 2004, International Organization for Standardization: http://www.iso.org/iso/en/ISOOnline.frontpage (September 2005).

Kent, W., 1978, Data and reality: Amsterdam, North-Holland Publishing Company, 211 p.

Loudon, T.V., 2000, Geoscience after IT: Computers and Geosciences, v. 26, no. 3A, 142 p.

Mallet, J.-L., 2002, Geomodeling: Oxford, UK, Oxford University Press, 599 p.

NASA, 2004, Semantic Web for Earth and Environmental Terminology: SWEET ontologies: http://sweet.jpl.nasa.gov/ontology/ (September 2005).

National Geological Surveys Committee, 2004, Canadian Geoscience Knowledge Network: http://cgkn.net/ (September 2005).

OGC, 2004, Open Geospatial Consortium: http://www.opengeospatial.org/ (September 2005).

PADI, 2004, Preserving access to digital information: http://www.nla.gov.au/padi/index.html (September 2005).

POSC, 2004, Petrotechnical Open Standards Consortium: http://www.posc.org (September 2005).

Pshenichny, C.A., 2003, Conference Reports, Georeasoning Workshop: IAMG Newsletter, no. 67, p. 15, 19, http://www.iamg.org/newsletter/newsletter.html (September 2005).

Research Councils UK, 2004, Links to e-Science on the Web: http://www.rcuk.ac.uk/escience/links (September 2005).

Smith, I.F., 2001, The digital geoscience spatial model—The shape of the BGS of the future: Earthwise, issue 15, p. 8: http://www.bgs.ac.uk/magazine/magazines/docs/15/23.pdf (September 2005).

Sowa, J.F., 2000, Knowledge representation: Logical philosophical, and computational foundations: Pacific Grove, California, Brooks-Cole, 594 p.

USGS, 2004a, International Clearinghouse for Geologic Map Databases and Standards: http://ngmdb.usgs.gov/intdb/.

USGS, 2004b, The North American Geologic Map Data Model: http://nadm-geo.org/ (September 2005).

van Wagoner, J.C., Hoyal, D.C.J.D., Adair, N.L., Sun, T., Beaubouef, R.T., Deffenbaugh, M., Dunn, P.A., Huh, C., and Li, D., 2003, Energy dissipation and the fundamental shape of siliciclastic sedimentary bodies: American Association of Petroleum Geologists Search and Discovery, Article #40080, 9 p.: http://www.searchanddiscovery.com/documents/vanwagoner01/images/vanwagoner.PDF (September 2005).

Voisard, A., 1999, Abduction and deduction in geologic hypermaps: Lecture Notes in Computer Science, no. 1651, p. 311–329: http://citeseer.ist.psu.edu/voisard99abduction.html (September 2005).

W3C, 2004, World Wide Web Consortium: http://www.w3.org/ (September 2005); Web Ontology Language Overview, OWL, http://www.w3.org/TR/owl-features/ (September 2005); Resource Description Framework, RDF, http://www.w3.org/RDF/ (September 2005); Semantic Web, http://www.w3.org/2001/sw/ (September 2005); Extensible Markup Language (XML), http://www.w3.org/XML/ (September 2005).

Manuscript Accepted by the Society 18 June 2005

Geological Society of America
Special Paper 397
2006

The concept of geological uncertainties and new ways of their geomathematical evaluation

György Bárdossy*
Hungarian Academy of Sciences, Roosevelt tér 9, H-1051 Budapest, Hungary

János Fodor*
John von Neumann Faculty of Informatics, Budapest Tech, Bécsi út 96/b, H-1034 Budapest, Hungary

ABSTRACT

Geoinformatics documentations, e.g., relational databases, generally consist of single-valued input data, that is, of real numbers. The uncertainties related to these data are not expressed. As a consequence, the results of data set evaluations can be considered as statements, but their reliability is not quantified. The introduction of uncertainties quantification into geoinformatics would be highly useful. One of the main difficulties is that the related concepts are not clarified and properly defined; in some cases they are even contradictory. The aim of this paper is to clarify the basic concepts related to geological uncertainties and to call attention to some new mathematical methods, suitable to quantify the uncertainties at the database level and to express the reliability of evaluation results. In the first part of the paper, the mathematical and the suggested geological classifications of uncertainties are presented. The role of "frequentist" and "Bayesian" information and probabilities are discussed. Limitations of probability theory for uncertainty analysis are shortly outlined. In the second part of the paper, the "uncertainty-oriented" mathematical methods are outlined, such as interval analysis, fuzzy set theory, possibility theory, probability bounds theory, and various hybrid methods. The authors describe their experience in the geological application of some of these methods. Test calculations, preformed by the authors, are listed before presenting some concluding remarks.

Keywords: uncertainty, fuzzy sets, probability, possibility.

INTRODUCTION

Uncertainty is involved in many real applications. Apart from the definition, it is an important matter of decision whether one would like to model uncertainty in an explicit way. If not, then it is typical to use a deterministic model as an approximation of the uncertain phenomenon in question. In addition, the *worst-case analysis* is an approach that acknowledges the presence of uncertainty without modeling it explicitly. It works with upper (or lower) bounds, trying to ensure that no larger (or smaller) value of an uncertain parameter may occur in the given system. In this paper, we shall focus on those cases in which the decision is to model uncertainty explicitly.

Before the mid-1960s, probability theory and statistics were the only tool to model uncertainty. Since that time, additional alternative theories have been suggested for uncertainty mod-

*E-mails: Bárdossy—h4750bar@helka.iif.hu; Fodor@bmf.hu

Bárdossy, G., and Fodor, J., 2006, The concept of geological uncertainties and new ways of their geomathematical evaluation, *in* Sinha, A.K., ed., Geoinformatics: Data to Knowledge: Geological Society of America Special Paper 397, p. 211–215, doi: 10.1130/2006.2397(16). For permission to copy, contact editing@geosociety.org.

eling. However, we share the surprise of Zimmermann (2000), who could not find any general definition of uncertainty. In this paper, we also follow him in the sense that "we shall focus on the human-related, subjective interpretation of uncertainty which depends on the quantity and quality of information which is available to a human being about a system or its behavior that the human being wants to describe, predict or prescribe" (Zimmermann, 2000). The choice of an appropriate uncertainty calculus should depend on the causes of uncertainty, quantity and quality of information available, and type of information processing required by the respective uncertainty calculus.

The paper is organized as follows. In the next section, we summarize some of the fundamental concepts, especially those related to uncertainty. Then we briefly recall traditional methods for handling uncertainty, together with their shortcomings. After that, a summary of some promising alternative uncertainty calculi is presented. Before the concluding remarks, we touch upon our experiences in geological applications of these new methods, performed by us in diverse test calculations. The interested reader should consider the recent book by Bárdossy and Fodor (2004) for more details both on the theory and on the applications.

BASIC CONCEPTS

Geological investigations are characterized by particularly high uncertainties. It is therefore of paramount importance to understand the concept of uncertainty in general and its application to geology. In our opinion, uncertainty is a general term expressing lack of certainty and of precision in describing a geological object, a feature, or a process. Highly competent scientists distinguish the following types of uncertainties (Dubois and Prade, 2000; Zimmermann, 2000):

1. *imprecision* or *inaccuracy*, expressing the deviation of measurements from the true value
2. *vagueness* or *ambiguity*, the uncertainty of nonmeasurable objects or properties
3. *incompleteness*, the uncertainty due to incomplete information
4. *conflicting evidence*, the uncertainty arising from contradicting evidences present in the studied system
5. *presumption* or *belief*, when all available information is subjective (the well-known "expert's opinion" belongs to this group)

The above classification is valid for geological investigations as well. However, the complexity of most geological problems requires a more detailed and specific classification. The one elaborated by us distinguishes two main groups (Bárdossy and Fodor, 2004):

A. uncertainties due to natural variability (also called *aleatory* uncertainties)

B. uncertainties due to human shortcomings and incomplete knowledge (also called *epistemic* uncertainties)

Natural variability is a property of nature, existing independently of us. It can be described and quantified by mathematical methods, but not diminished by further empirical studies (although it may be better characterized). That's why it is also called *irreducible* uncertainty. The term *aleatory* uncertainty intends to emphasize its relation to the *randomness* in gambling and games of chance.

On the other hand, the second kind of uncertainty is the uncertainty that comes from scientific ignorance, measurement uncertainty, inobservability, censoring, or other lack of knowledge. In contrast with aleatory uncertainty, epistemic uncertainty is sometimes called *reducible* uncertainty since, at least in principle, it can generally be reduced by additional empirical effort. This fundamental difference has been often neglected in geological investigations.

The second group has been divided by us for both computational and geological reasons into the following subgroups:

B1. uncertainties in the phase of preparing the input data

B2. uncertainties in the phase of geological evaluation

They are generally evaluated separately for each variable, but in some cases the calculation of the total uncertainty is required.

The sources and characteristics of these uncertainties have been discussed in detail in the book of Bárdossy and Fodor (2004).

TRADITIONAL APPROACHES FOR HANDLING UNCERTAINTIES

As we already mentioned in the Introduction, there exist *deterministic* methods that notice the presence of uncertainty without modeling it explicitly. On the other hand, probability theory and statistics (may be called *stochastic* methods) allow for a certain type of uncertainty in the given system. In this theory, *probability* is considered as a measure of uncertainty. (We suppose that the reader is familiar with the main concepts and methods of probability theory.) Two main approaches can be distinguished. The *frequentist* approach generally applied in geology requires repeated identical experiments to calculate the main characteristics (such as expected value, standard deviation, and/or the probability distribution) of the studied variable. However, this can be fulfilled only in few cases because of technical and financial reasons. For example, how to repeat borehole sets? This represents a significant limitation for the application of the frequentist approach in geological investigations. Neglecting this requirement leads to more or less important bias of the results. Within the stochastic framework, there exists a second approach as well, the *Bayesian* one. This approach works with subjective probabilities, depending not only on the phenomenon itself, but also on the state of knowledge and on the belief of the researcher. Bayesian probabilities can be applied also to unrepeatable phenomena. They change with time as new pieces of information are acquired by the investigations. This concept is based on the famous *Bayes' Theorem*.

In this context, *prior* probabilities refer to the event prior to updating by new pieces of information, and *posterior* probabilities refer to the same event after updating by new pieces of information acquired by new investigations. Geological investigations are characterized by new pieces of information obtained step-by-step during the investigation. This is why they are particularly suitable for the application of the Bayesian approach. Unfortunately, relatively little has been done so far for the geological application of this concept.

For several decades, the methods based on probability theory have been applied successfully for the evaluation of geological investigations. However, some important limitations and difficulties appeared, particularly for the evaluation of geological uncertainties:

1. The input data of the deterministic and stochastic approaches are single-valued (real numbers), also called *crisp* numbers. Only natural variability can be expressed by this type of data and not the uncertainties due to human shortcomings and incomplete knowledge.

2. The basic axioms of probability theory, called *Kolmogorov axioms*, deal only with mutually exclusive subsets. However, in geology disjoint subsets are rare; transitions are much more frequent.

3. Uncertain propositions and statements cannot be evaluated in terms of repeated experiments.

4. Semiquantitative and qualitative (linguistic) variables are not suitable for most statistical evaluations.

For the reasons listed above, we consider that methods based on probability theory are mathematically correct, but they do not offer optimal solutions for several geological problems, particularly for the evaluation of uncertainties.

REVIEW OF UNCERTAINTY-ORIENTED MATHEMATICAL METHODS

New mathematical methods have been developed since the 1960s to efficiently handle the uncertainties, especially to represent what is known about real-valued but uncertain quantities. For situations in which the uncertainty is purely aleatory, probabilistic and statistical methods are usually preferred. When the gaps in our knowledge involve both aleatory and epistemic uncertainty, several competing approaches have been suggested. Let us stress that the frequently used term *uncertainty analysis* has been applied so far only for probabilistic evaluations. Thus, it could not offer the entire evaluation of all types of uncertainties. The most important new methods consist of *interval analysis* (Moore, 1979), *fuzzy set theory* (Zadeh, 1965), *neuro-fuzzy systems* (Fullér, 2000), *possibility theory* (Zadeh, 1978; Dubois and Prade, 1988), *probability bounds theory* (Walley and Fine, 1982; Williamson and Downs, 1990), and different *hybrid methods* (Cooper et al., 1996; Guyonnet et al., 2003). The common feature of these methods is that instead of single-valued (i.e., crisp) input data, they apply different new types of data expressing the amount of uncertainty related to the given input data.

Interval analysis (Moore, 1979) replaces crisp numbers by uncertainty *intervals* (Fig. 1). The topic has become even more important with the advent of computers (Dubois et al., 2000): The motivation is "the quest for rigor in numerical computation on machines." It is assumed that the true value is somewhere within the interval. Interval analysis lacks gradations and is the simplest method to express uncertainty through arithmetic calculations. The method guarantees that the true value will always remain within the interval, but this goal is achieved at the cost of precision. During the calculations, the intervals become wider and wider and the final results become too conservative.

Possibility theory, a generalization of interval analysis, provides a suitable model for the quantification of uncertainty by means of the possibility of an event (Zadeh, 1978; Dubois and Prade, 1988). The theory acknowledges that not all types of uncertainty can be handled by probability distributions. Instead, it uses membership functions to represent nonquantified uncertainty. The membership value of a number, varying between zero and one, expresses the plausibility of the occurrence of that number. The theory has been applied successfully in biology, health, and medicine (Ferson and Ginzburg, 1996; Ferson et al., 1999) and in different branches of industry and economics (Bardossy and Duckstein, 1995; Fodor and Roubens, 1994).

The related *fuzzy set theory* expresses uncertainty very often by the use of *fuzzy numbers*. They represent estimates of uncertainty at different levels of possibility (or membership degree). Membership functions of fuzzy numbers are by definition unimodal, and they have to reach at least in one point the possibility level one, that is, the full possibility. In geology, mainly trapezoidal and triangular fuzzy numbers are applied. They can be both symmetrical and asymmetrical. The smallest and the largest possible values of the given variable represent the lower and the upper bounds of the *support* of the fuzzy number. All values of the variable must be within these boundaries. The values reaching the possibility level one are considered as the most possible estimates of the given variable, and this interval is called the *core* of the fuzzy number. Fuzzy numbers are generalizations of tra-

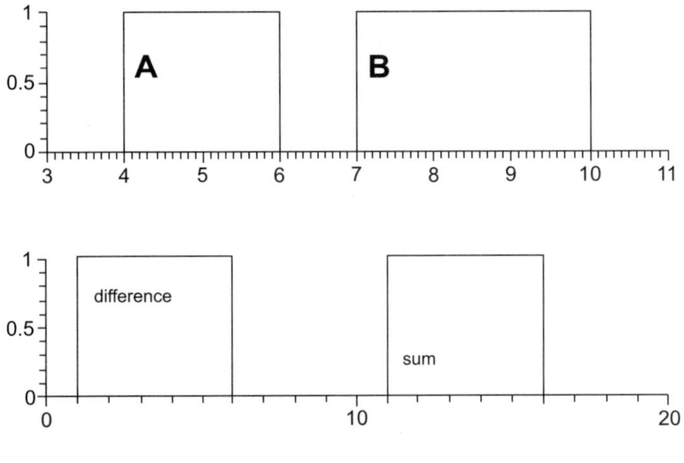

Figure 1. Two intervals (A and B) and their sum (A + B) and difference (B – A).

ditional real numbers, as the latter ones can be regarded as fuzzy numbers with a single-point support.

All usual arithmetic calculations can be carried out with fuzzy numbers. In contrast with probability, one of their great advantages is that they do not require the knowledge of the correlations among the variables and the type of their probability distribution (Takács and Várkonyi-Kóczy, 1999a, 1999b). For the sake of numerical comparisons and ranking, fuzzy numbers can be reconverted into crisp numbers. This calculation is called *defuzzification*. But the main advantage of the fuzzy method is that prior geological experience can be incorporated into the construction of fuzzy numbers. This goal can be achieved by joint constructing of the fuzzy numbers by geologists and mathematicians. The method allows the appropriate evaluation of semi-quantitative and qualitative input data as well. The frequent transitions of the geological populations, as mentioned before, can be also represented by fuzzy numbers (Fig. 2). Cagnoli (1998) showed the application of the fuzzy set theory in the study of volcanic rocks. In recent years, it has found a broad application in the geographical information systems as well (Altman, 1994; Macmillan, 1995; Unwin, 1995).

The way of constructing fuzzy numbers raises the problem of their *robustness*. Imagine that several well-trained and experienced experts are asked to construct fuzzy numbers, based on the

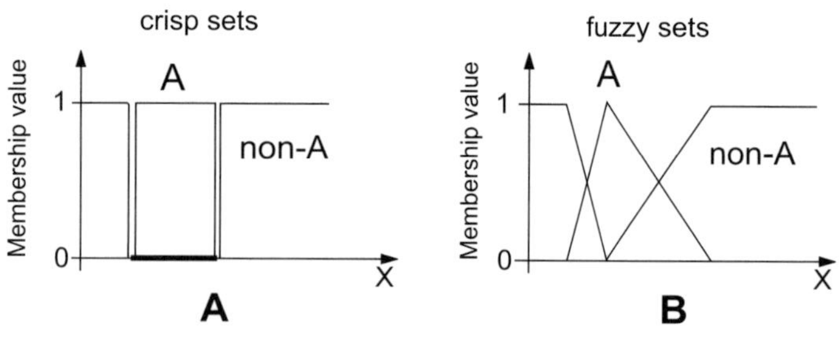

Figure 2. A: Crisp set A and its complement non-A. Their intersection is empty, and their union is the set of all elements of the universe. B: Fuzzy set A and its complement non-A. They overlap.

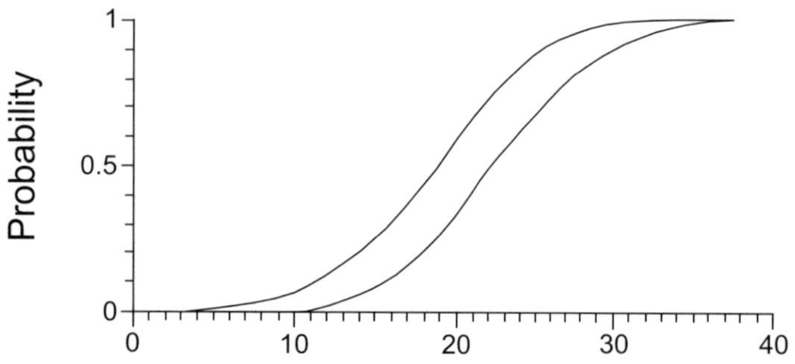

Figure 3. Probability bounds given by two probability distribution functions.

same crisp data. It is certain that the resulting fuzzy numbers will not be exactly identical. However, the differences are expected to be rather small. Luckily, all the mathematical operations one has to carry out with fuzzy numbers are stable, that is, small changes in the input data yield only small changes in the results. As a consequence, the final results are not sensitive to small differences in the initial fuzzy numbers.

The *probability bounds* theory (Walley and Fine, 1982; Williamson and Downs, 1990) is a combination of probability theory with interval analysis, by representing epistemic uncertainty within the context of probability theory. The final idea is that one can work with *bounds* on probability for this purpose. It expresses uncertainty by two cumulative probability distributions, called *probability boxes*, or *p-boxes* for short. The area between the two curves represents the extent of uncertainty of the given variable, as can be seen in Figure 3. Probability bounds are considered as a generalization of crisp numbers, intervals, and probability distributions. Williamson and Downs (1990) also described algorithms to compute arithmetic operations (addition, subtraction, multiplication, and division) on pairs of p-boxes. These operations generalize the notion of *convolution* between probability distributions. The great advantage of this method is that it can apply different probability distributions—e.g., normal, lognormal, and exponential—and correlations for the variables to be studied. But the method works also without making any prior assumptions. The probability bounds get narrower with more empirical information about the given geological object. Its disadvantages are the more complicated calculations to be carried out. Nevertheless it seems for us to be a highly efficient approach in the case of safety assessments, when prior information is abundant. A detailed overview of p-boxes and their links to other representation of uncertainties can be found in Ferson et al. (2002).

The methods of *hybrid arithmetic* (Cooper et al., 1996; Guyonnet et al., 2003) combine probability distributions with intervals, fuzzy numbers, and probability bounds in different ways. These methods allow the use of all kinds of numbers, this being their greatest advantage. These are the newest among the methods of uncertainty analysis, and there are very few publications on its application.

APPLICATIONS OF FUZZY SETS IN GEOLOGY

As outlined in the foregoing sections, membership functions can express the degree of *uncertainty* of any input data collected in the course of a geologic investigation. But Dubois and Prade (2000) pointed out that they can express the degree of *similarity* and the degree of *preference* as well. In our opinion, these two latter meanings can be applied also to geological problems. In a mathematical sense, the *degree of similarity* expresses the proximity of x to the chosen prototype elements of A. In geology, a prototype can be a rock, a lithologic unit, a mineral deposit, a fossil species, etc. Note that probability theory, as defined by the axioms of Kolmogorov, does not admit *transitions*, only full or no membership, that is, only disjoint sets of data. However, in geology transitions are very frequent. When we apply a probabilistic evaluation, we are obliged to draw sharp boundaries and to cut transitional zones into different pieces. Obviously this is a distortion of the natural reality. Membership functions can be applied with success to solve this problem.

The *degree of preference* refers to a set of more or less preferred objects and $\mu_A(x)$ represents the degree of preference in favor of the object x. This case may occur in the exploration of mineral deposits, or of groundwater resources, when we have to choose among several potential regions. The degree of our preference for a given region can be represented by a membership function. The choice between suitable locations for commercial, toxic, or radioactive waste disposal can also be represented by membership functions, instead of the traditional ranking of crisp numbers. Note that when using the degree of preference, the choice of the preferred object is ours.

In the last four years, a number of test calculations were carried out by us—applying the fuzzy set theory—on different geological problems. Parts of the results were published in separate articles or in the book of Bárdossy and Fodor (2004). The main topics of our calculations have been

1. estimation of the resources of solid mineral deposits
2. quantitative mineralogical phase analysis of rocks and ores by X-ray diffractometry and by thermal analysis (separately)
3. safety assessment of radioactive waste disposal
4. application to paleontological biometry
5. measurements of groundwater transmissivity in boreholes
6. evaluation of geochemical transitions in bauxite deposits

CONCLUSIONS

In geology, uncertainty has long been considered a removable adverse circumstance that should gradually disappear with the overall development of the earth sciences. However, one must recognize that a part of this uncertainty is an inherent feature of nature. Therefore, understanding and appropriate handling of uncertainties should be part of all future geological investigations.

Traditional mathematical methods—deterministic and probabilistic—applied so far in geological investigations are mathematically correct, but by far not optimal for the treatment of all kinds of uncertainties.

New mathematical methods summarized in this paper are suitable to evaluate in a mathematically correct way semiquantitative and qualitative ("linguistic") input data and to determine the uncertainties and errors connected with them.

It is important to emphasize that the traditional and the new uncertainty-oriented methods are context-dependent, and complete each other in geological investigations.

A thorough study of the geological objects and processes is indispensable for any mathematical evaluation in geology. Without that, even the most sophisticated method becomes an empty formalism.

REFERENCES CITED

Altman, D., 1994, Fuzzy set theoretic approaches for handling imprecision in spatial analysis: International Journal of Geographical Information Systems, v. 8, p. 271–289.

Bardossy, A., and Duckstein, L., 1995, Fuzzy rule based modeling with applications geophysical, biological and engineering systems: New York, CRC Press, 232 p.

Bárdossy, Gy., and Fodor, J., 2004, Evaluation of uncertainties and risks in Geology: Berlin, Heidelberg, London, New York, Springer-Verlag, 221 p.

Cagnoli, B., 1998, Fuzzy logic in vulcanology: Episodes, v. 21, p. 94–96.

Cooper, J.A., Ferson, S., and Ginzburg, L.R., 1996, Hybrid processing of stochastic and subjective uncertainty data: Risk Analysis, v. 16, p. 785–791.

Dubois, D., and Prade, H., 1988, Possibility theory: An approach to computerized processing of uncertainty: New York, Plenum Press, 263 p.

Dubois, D., and Prade, H., editors, 2000, Fundamentals of fuzzy sets: The handbook of fuzzy sets, series 7: Boston, London, Dordrecht, Kluwer Academic Publishers, 647 p.

Dubois, D., Kerre, E., Mesiar, R., and Prade, H., 2000, Fuzzy interval analysis, *in* Dubois, D., and Prade, H., eds., Fundamentals of fuzzy sets: The handbook of fuzzy sets series 7: Boston, London, Dordrecht, Kluwer Academic Publishers, p. 483–581.

Ferson, S., and Ginzburg, L.R., 1996, Different methods are needed to propagate ignorance and variability: Reliability Engineering and Safety, v. 54, p. 133–144.

Ferson, S., Root, W., and Kuhn, R., 1999, RAMAS Risk Calc: Risk assessment with uncertain numbers: Setauket, New York, Applied Biomathematics, 184 p.

Ferson, S., Kreinovich, V., Ginzburg, L., Myers, D.S., and Sentz, K., 2002, Constructing probability boxes and Dempster-Shafer structures: SAND2002-0835 Technical Report: Albuquerque, New Mexico, Sandia National Laboratories, p. 143.

Fodor, J., and Roubens, M., 1994, Fuzzy preference modeling and multicriteria decision support: Dordrecht, Netherlands, Kluwer Academic Publishers, 272 p.

Fullér, R., 2000, Introduction to neuro-fuzzy systems: Heidelberg, Physica Verlag, 289 p.

Guyonnet, D., Bourgine, B., Dubois, D., Fargier, H., Côme, B., and Chilès, J.-P., 2003, A hybrid approach for addressing uncertainty in risk assessments: Journal of Environmental Engineering, v. 129, p. 68–78, doi: 10.1061/(ASCE)0733-9372(2003)129:1(68).

Macmillan, W., 1995, Modeling: Fuzziness revisited: Progress in Human Geography, v. 19, p. 404–413.

Moore, R.E., 1979, Methods and applications of interval analysis: Philadelphia, Society for Industrial and Applied Mathematics, Studies on Applied Mathematics, Vol. 2.

Takács, O., and Várkonyi-Kóczy, A.R., 1999a, Fuzzy handling of uncertainty in nonlinear systems, *in* De Baets, B., et al., eds., Proceedings of EUROFUSE-SIC '99, May 25–28, 1999, Budapest, p. 22–27.

Takács, O., and Várkonyi-Kóczy, A.R., 1999b, Information processing based on mixed-classical and fuzzy-data models, *in* Institute of Electrical and Electronics Engineers International Workshop on Intelligent Signal Processing, September 4–7, 1999, Budapest, p. 23–27.

Unwin, D., 1995, Geographical information systems and the problem of error and uncertainty: Progress in Human Geography, v. 19, p. 549–558.

Walley, P., and Fine, T.L., 1982, Towards a frequentist theory of upper and lower probability: Annals of Statistics, v. 10, p. 741–761.

Williamson, R.C., and Downs, T., 1990, Probabilistic arithmetic I: Numerical methods for calculating convolutions and dependency bounds: International Journal of Approximate Reasoning, v. 4, p. 89–158, doi: 10.1016/0888-613X(90)90022-T.

Zimmermann, H.-J., 2000, An application-oriented view of modeling uncertainty: European Journal of Operational Research, v. 122, p. 190–198, doi: 10.1016/S0377-2217(99)00228-3.

Zadeh, L.A., 1965, Fuzzy sets: Information and Control, v. 8, p. 338–353, doi: 10.1016/S0019-9958(65)90241-X.

Zadeh, L.A., 1978, Fuzzy sets as a basis for a theory of possibility: Fuzzy Sets and Systems, v. 1, p. 3–28, doi: 10.1016/0165-0114(78)90029-5.

Manuscript Accepted by the Society 18 June 2005

Printed in the USA

Geological Society of America
Special Paper 397
2006

Application of fuzzy arithmetic and prior information to the assessment of the completeness of a mineral exploration program: A case study

György Bárdossy*
Hungarian Academy of Sciences, Roosevelt tér 9, H-1051 Budapest, Hungary

János Fodor*
John von Neumann Faculty of Informatics, Budapest Tech, Bécsi út 96/b, H-1034 Budapest, Hungary

ABSTRACT

The completeness of an exploration project is of crucial importance for making a decision to start or to give up a mining investment, or to continue the exploration to get complementary information. The authors discuss this problem on the example of the Halimba bauxite deposit, Hungary. Two-hundred thirty-seven core drills were executed on a 14 ha area. Resource calculations were carried out in 12 subsequent stages by fuzzy arithmetic with the aim to quantify the uncertainties of ore tonnage and grade. Prior information and prior probabilities were applied to complete the exploration data. Their validity was checked by the subsequent stages. Ranges of influence for the main variables were calculated by geostatistical methods (variograms). Spatial variability and spatial continuity of the orebodies were mathematically evaluated. The authors found that there is no single "overall" value to express the completeness of a mineral exploration program, but the main geological, mining, and economic factors must be evaluated separately and ranked according to their importance. The reliability of the results can be quantified by the application of new "uncertainty-oriented" mathematical methods.

Keywords: mineral exploration, prior information, uncertainty, fuzzy arithmetic.

INTRODUCTION

Exploration of solid mineral deposits is generally an expensive task. Even more expensive and risky is the successive mining investment. It is of paramount importance therefore to optimize the exploration expenses and to minimize the risks of the mining investment. This double task has been considered so far as a purely geological and mining-engineering problem; however, in our opinion, geoinformatics and the application of some different mathematical methods may considerably improve the results.

The aim of this paper is to show the application of these new methods by a case study. After a brief review of the basic concepts, we consider sources and variability of the initial data. Constructions of fuzzy numbers for the deposit area, the ore thickness, and the bulk density are shown, and then the tonnage of the resource is expressed also as a fuzzy number. Then we analyze the completeness of the exploration project by applying prior probabilities. The Halimba bauxite deposit in Hungary has been chosen as example. At the end we draw some conclusions.

*E-mails: Bárdossy—h4750bar@helka.iif.hu; Fodor—Fodor@bmf.hu

Bárdossy, G., and Fodor, J., 2006, Application of fuzzy arithmetic and prior information to the assessment of the completeness of a mineral exploration program: A case study, *in* Sinha, A.K., ed., Geoinformatics: Data to Knowledge: Geological Society of America Special Paper 397, p. 217–224, doi: 10.1130/2006.2397(17).

BASIC CONCEPTS

The completeness of an exploration campaign is generally expressed by the resource assessment (tonnage and grade) and its overall reliability. The spatial distribution and spatial variability of ore grade and the spatial continuity of ore within the deposit are further important aspects (Henley, 2000; Wellmer, 1989; Yamamoto, 1999). However, the well-known traditional methods of resource assessment are not able to quantify the reliability of the estimation results. The fuzzy set theory has been applied recently by the authors for this purpose on some solid mineral deposits with full success (Bárdossy et al., 2003; Bárdossy and Fodor, 2004). Fuzzy sets have been applied for the resource assessment of skarn tin deposits by Luo and Dimitrakopoulos (2003).

A further improvement can be achieved by applying the concept of Bayesian probabilities. It is well known that the so-called *frequentist approach* requires repeated identical experiments. However, this requirement can be fulfilled only rarely in geologic investigations. The *Bayesian approach*, on the other hand, is able to evaluate unrepeatable phenomena as well. Bayesian probability depends only on the (subjective) state of knowledge about the given problem and how it changes with time as new information is acquired (Bárdossy and Fodor, 2004). A mineral exploration program has also this changing character as new information is obtained about the given deposit by drilling new boreholes, etc. For this reason, it is reasonable to suggest applying also prior information and Bayesian probabilities for the evaluation of exploration results and other geoscientific problems (Wood and Curtis, 2004).

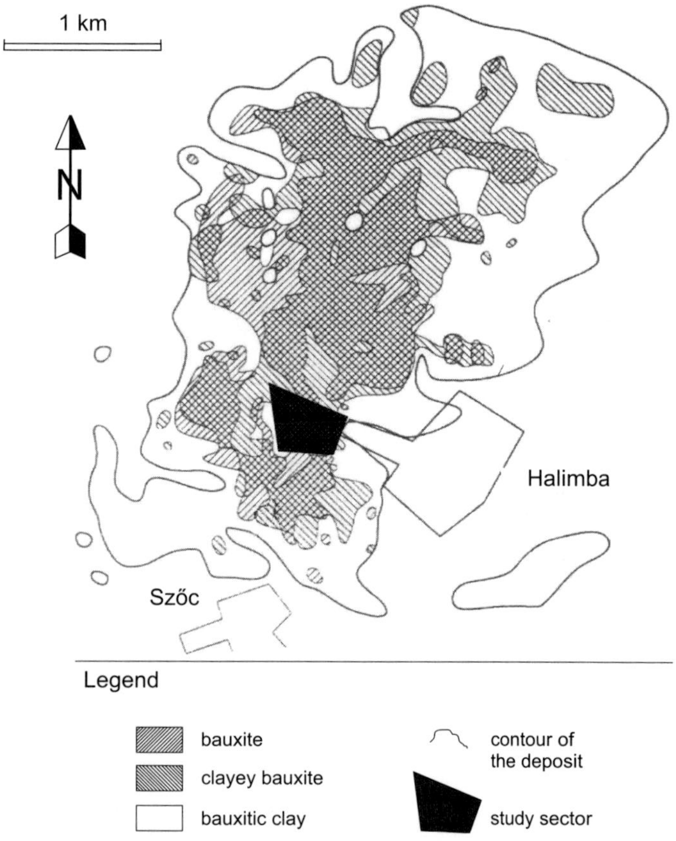

Figure 1. The Halimba bauxite deposit, Hungary.

INITIAL DATA

The bauxite deposit of Halimba, Hungary, selected for this case study has been explored since 1943, and up to the present around 2600 core boreholes have been drilled. Underground mining started in 1950 and is still running. Computerized relational databases have been established (AutoCAD) for the main data obtained about the deposit, particularly for the chemical composition of the ore. The sector of the test calculations has been intensively explored during the last three years. It covers an area of 15 ha with 237 borehole sites, and it is situated in the southern part of the deposit (Fig. 1). A 10–40 m thick bed consisting of bauxite, clayey bauxite, and bauxite clay covers the karstified surface of Upper Triassic dolomite and limestone. The overburden is of Middle Eocene age. The deposit is of fluvial origin. The area of the studied sector is of floodplain facies. The bauxite accumulated during short inundation phases, forming very irregular orebodies within the continuous clayey bauxitic layer. Underground mining operations started in the western part of the study area in 2004. They confirmed the above-outlined deposit model.

First the *spatial variability* of the main variables has been evaluated by us, applying the well-known methods of geostatistics (Goovaerts, 1997). Ranges of influence have been calculated by the Variowin program (see Pannatier, 1996) for the thickness of the bauxitic bed, for the bauxite ore, and for the Al_2O_3, SiO_2, Fe_2O_3, CaO, and MgO contents of the bauxite. Our basic idea was to follow the changes of knowledge that occurred as the exploration progressed. For this reason, resource assessments were carried out by us after every 20 new boreholes were finished. Thus, a growing number of boreholes served as a base of the successive resource assessments. Altogether 12 resource assessments have been performed, by the above-mentioned fuzzy method.

It is well known that the three basic components of any resource assessment of solid mineral deposits are the area of the deposit, the thickness of the ore, and its bulk density. Fuzzy numbers have been constructed for all three components. A fuzzy number is a quantity whose value is imprecise rather than exact as is the case with ordinary ("crisp") numbers. Any fuzzy number is represented by a *membership function* whose domain is the set of real numbers, and whose range is the closed unit interval. The most typical trapezoidal and triangular fuzzy numbers are used also in Figure 2. Each numerical value in the domain is assigned a specific "grade of membership" where 0 represents the smallest possible grade, and 1 is the largest possible grade. The "support" of a fuzzy number is defined to extend from the minimum to the maximum possible value. In the case of the *deposit area*, the minimum value is determined by straight lines connecting the extreme productive boreholes. The maximum possible area is obtained by connecting the closest unproductive boreholes around the productive area. The "core" of the fuzzy number is defined to represent the geologically most likely area, determined by the deposit model and its contour line.

This is a relatively simple and unambiguous task in the case of well-explored deposits. However, in the early stages of exploration, the number of boreholes is often not sufficient for the above-outlined constructions. In these cases, we extrapolated from the given productive borehole the range of influence of the bauxite thickness in all directions, obtaining this way the minimum possible area. The maximum possible area was obtained by taking in all directions twice the range of influence. In this case, an interval has been chosen also for the core of the fuzzy number, expressing the larger uncertainty of the deposit area. The extrapolated resource boundaries are replaced gradually by straight lines connecting the neighboring boreholes, as new boreholes are drilled. According to our experience, exploration should not be finished before replacing all the extrapolated boundaries by the connecting straight lines. Thus, the trapezoidal fuzzy numbers are gradually replaced by triangular ones, as one can see in Figure 2, by comparing the area for stage 2 and stage 11.

The fuzzy numbers representing the *ore thickness* correspond to the averages of the borehole results. Before calculating the averages, the main statistics of ore thickness have been calculated by us, applying the 12.0 version of the SPSS program (see SPSS Base 12.0 User's Guide, 2003). The histograms and the "skewness" values indicated a strong right-asymmetrical distribution. To eliminate the corresponding bias, "maximum-likelihood" estimators have been calculated instead of the common averages. Tukey's biweight estimator was found to correspond best to an unbiased average (Tukey, 1977). It has been applied in all cases when the skewness statistic exceeded 1.0. We have taken minimum and maximum values of the support of the fuzzy numbers corresponding to the endpoints of the corresponding confidence interval, at 95% level of confidence. The core of the fuzzy number is an interval determined by the standard error of the mean.

The *bulk density* of the ore has been measured in the laboratory on borehole cores and in the mine on large samples, several hundred times. The distribution of the results is symmetrical. The mean value is 2.29 t/m^3. The analytical error is less than ten relative percent. The variability of the bulk density is very limited over the test area. For this reason, the same fuzzy number has been applied for all 12 resource assessments. In the same way as for the ore thickness, the support of the fuzzy number corresponds to the confidence interval at 95% level of confidence and the core to the standard error of the mean, plus the analytical error.

The *tonnage of the resource* is the product of the above-discussed three components. Fuzzy multiplication was applied for the three corresponding fuzzy numbers. The uncertainty of the resource assessment is expressed in tons by the length of the support and the core, separately. Additionally, relative deviations from the average values—expressed as percentages—were also calculated.

The *average grade* of the ore has been calculated in a similar way, constructing average fuzzy numbers for all the listed chemical components. To avoid biases due to asymmetrical distribution histograms, skewness values were calculated and robust M-estimators were applied whenever the skewness exceeded the value 1.0. As in the case of tonnage, absolute and relative uncertainties have been calculated for all the evaluated chemical components.

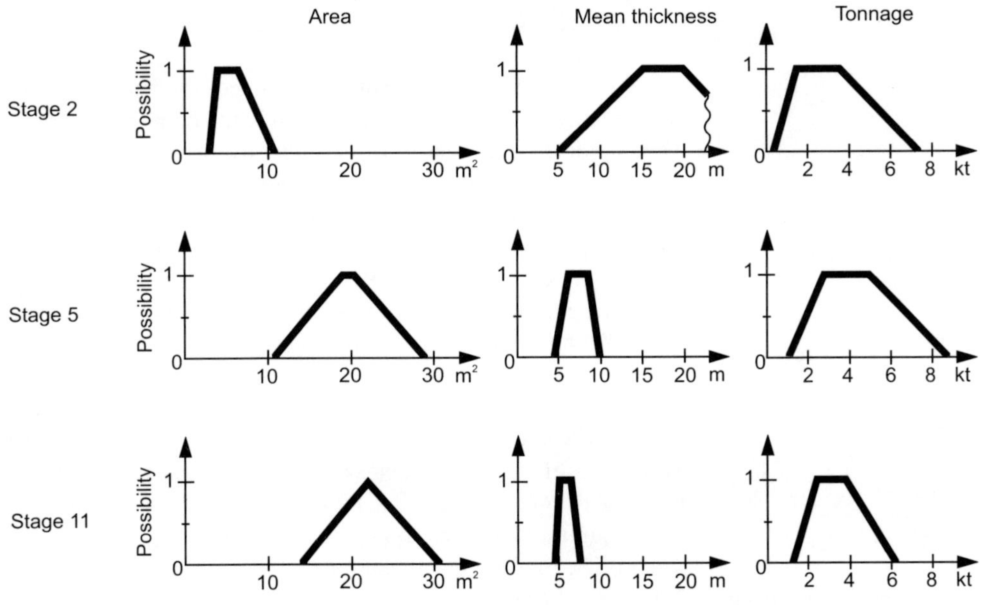

Figure 2. Fuzzy numbers expressing the area, mean thickness, and tonnage for stages 2, 5, and 11.

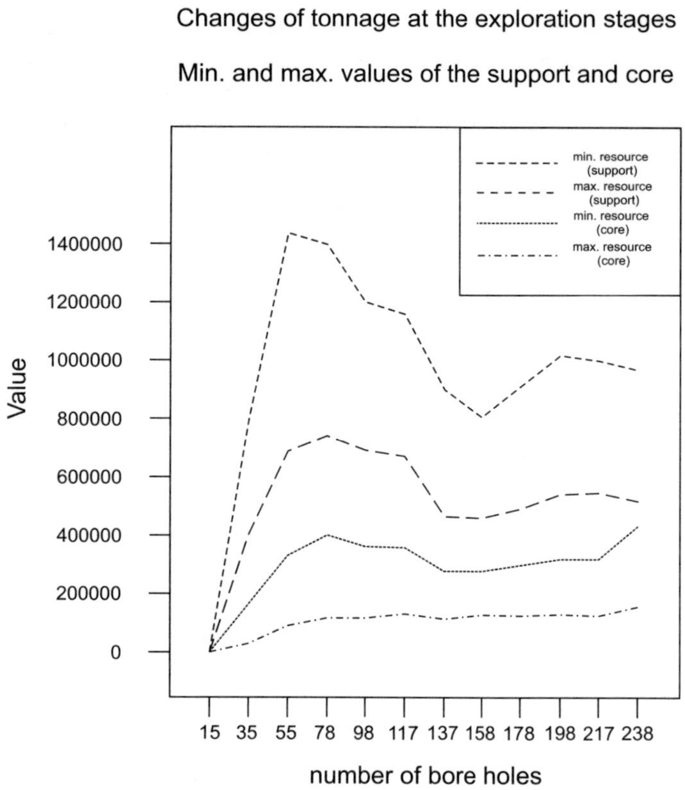

Figure 3. Quantitative evaluation of tonnage: successive changes of the maximum and minimum of the support and the core.

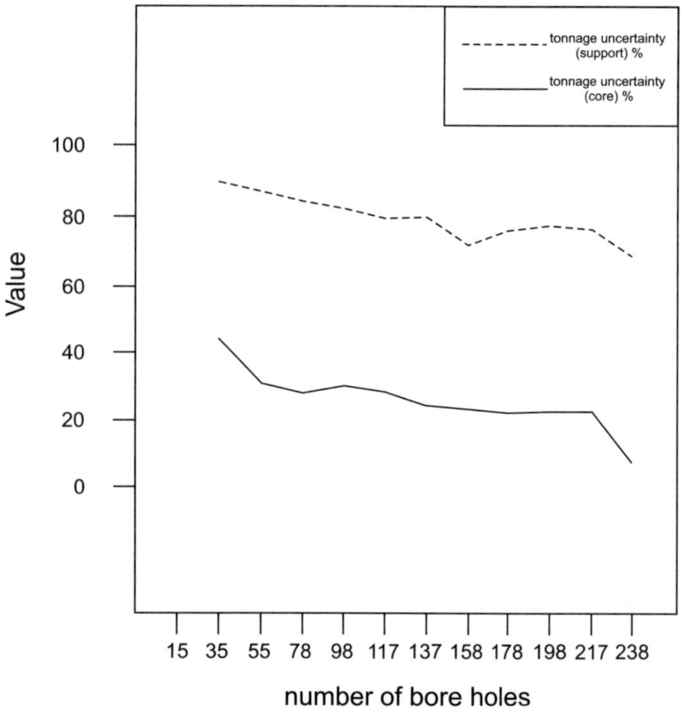

Figure 4. Relative uncertainty of the tonnage for the core and the support.

As mentioned above, all the calculations have been repeated 12 times, starting with 17 boreholes and adding 20 new ones every time. Fuzzy numbers for selected stages are presented in Figure 2.

EVALUATION OF THE COMPLETENESS OF THE EXPLORATION

For the starting situation, that is, before the drilling of the first borehole in the sector, the following *prior probabilities* have been assumed, based on the experiences of the neighboring explored and mined sectors:

- The bauxite–clayey bauxite bed is continuous over the exploration sector: *.8 probability*.
- The bed is not continuous over the exploration sector: *.2 probability*.

Second item of the prior probabilities:

- Commercial bauxite orebodies are situated within the bed: *.6 probability*.
- No commercial bauxite orebodies occur within the bed: *.4 probability*.

The end situation—after the 12th stage—confirmed all the above prior probabilities.

For the first two stages of exploration, the number of productive boreholes was not sufficient to calculate reliable variograms. For this reason, the already calculated ranges of influence of the neighboring sectors were applied, supposing similar values in the study area. At the end of the third stage, variograms could be calculated for the ore thickness. By applying different "lags" and variogram models, ranges of influence from 10 to 20 m length were obtained. With increasing number of boreholes, the variograms became more accurate, and after the 12th stage, a 15 m range of influence was accepted for the entire exploration area. However, locally even smaller ranges of influence exist, as confirmed by the latest mining operations. As outlined later, these changes significantly influenced the results of the successive resource assessments.

At the end of each exploration stage, circles were constructed around each borehole, expressing the range of influence. The boreholes were placed in a "randomly stratified" grid, with the aim to optimize the contouring of the very irregularly shaped orebodies. For this reason, unclassified areas remained between some neighboring boreholes. Prior probabilities have been calculated for each area separately. If a probability exceeded the .5 value, the corresponding area has been included into the resource assessment. In most cases, the prior probabilities have been confirmed by the boreholes of the successive exploration stages, and at the places where mining operations started.

Different variables have been chosen for the quantitative evaluation of the completeness of exploration, first of all, the *tonnage of the resources*. In Figure 3, the successive changes of the minimum and maximum values of the support are represented. The minimum value of the ore tonnage steeply increased in the first four stages of exploration, followed by much smaller increase in the later stages. The fluctuation of the diagram reflects the randomness of the results at some stages. The possible maximum tonnage also increased steeply in the first stages, but it was followed by an unexpected gradual decrease until the eighth stage. The last stages show a slight increase. The peculiar form of this diagram can be explained by the higher uncertainty of the maximum tonnage, influenced by the position of the closest unproductive boreholes and by the extrapolation of the contour line in the first stages of exploration. The peak between the third and fourth stages is clearly a random effect that may occur in the first stages of any exploration campaign. As exploration progressed, the difference between the two diagrams of Figure 3 diminished, as the area between the minimum and maximum contours became narrower.

The tonnage expressed by the core of the fuzzy numbers has a much shorter uncertainty interval, presented also in Figure 3. The random overestimation of the tonnage between the third and fourth stages is clearly visible on both diagrams, but it is gradually equalized in the later stages without reaching a constant value. Theoretically, the exploration is still not complete, but the changes of the tonnage are insignificant. Thus, the tonnage of the resources alone is in favor of finishing the exploration drilling.

A further aspect of the evaluation is the *relative uncertainty of the tonnage*, expressed as a percentage relative to the mean (crisp) tonnage. We calculated it separately for the support and for the core of the corresponding fuzzy numbers. The results are presented in Figure 4. It is obvious that the uncertainty of the tonnage expressed by the support is much larger than that expressed by the core. It decreases in the successive stages until the eighth stage—from ±91% to ±73%. This is followed in the later stages by a fluctuation and a final value of ±69%. On the other hand, the relative uncertainty of the tonnages expressed by the core is much smaller. The starting ±46% relative uncertainty diminishes to ±9%. This indicates a nearly complete exploration result.

The tonnage values of the fuzzy numbers and their relative uncertainties are presented in Table 1. Let us stress that these data represent a significant complement to the single-valued traditional resource estimation results. But even these data are insufficient in our opinion to make a reliable decision on the completeness of an exploration campaign. The *main chemical components* have been evaluated by us, too, according to the successive exploration stages. The resulting main statistics calculated by the SPSS program are presented in Table 2.

The Al_2O_3 content has the smallest relative variance, ±7%. The distribution of this component is almost normal, thus the mean value is unbiased. It diminished from the second to the latest exploration stage from 52.8% to 51.2%, considered by us as a very small change. During the same time, the standard error of the mean diminished from ±1.0% to ±0.3%, indicating a high reliability of the results. It can be concluded that regarding the alumina content the exploration has been complete since the early stages.

The Fe_2O_3 content follows with ±14% to ±16% relative variance. The distribution is symmetrical, and the mean decreased

TABLE 1. MAIN RESULTS OF THE RESOURCE CALCULATIONS

Stages	Number of boreholes	Deposit area				Tonnage				Length of the core interval	Relative uncertainty of the tonnage (%)	Length of the support interval	Relative uncertainty of the tonnage (%)
		a	b	c	d	a	b	c	d				
1	15	0	0	0	0	0	0	0	0	0	0	0	0
2	35	5300	8100	12000	17400	41700	158200	425400	826400	267200	46	784700	90
3	55	11800	26200	29800	46400	104400	342400	675200	1388400	332800	30	1284000	86
4	78	12610	29100	30800	47960	126100	391400	710900	1367200	319500	29	1241100	83
5	98	14090	29300	31000	46600	126600	350000	643000	1200000	293000	30	1073400	81
6	117	15630	30400	32100	47700	136500	344500	626700	1152500	282200	29	1016000	79
7	137	14090	29700	30000	45630	106000	279600	472600	929000	193000	26	823000	79
8	158	16980	30200	30700	44100	130900	281900	473800	843200	191900	25	712300	73
9	178	16700	32200	32400	47870	126400	303300	499100	917500	195800	24	791100	76
10	198	18940	35300	35300	52280	135900	329200	548000	1028500	218800	25	892600	77
11	217	20800	36000	36300	52200	136200	323800	549100	1002700	224100	25	870900	76
12	238	24300	41600	42000	60300	178800	439000	524200	960000	85200	9	781100	69

Legend:
a: lower bound of the support of trapezoidal fuzzy number [a,b,c,d]
b: lower bound of the core of trapezoidal fuzzy number [a,b,c,d]
c: upper bound of the core of trapezoidal fuzzy number [a,b,c,d]
d: upper bound of the support of trapezoidal fuzzy number [a,b,c,d]

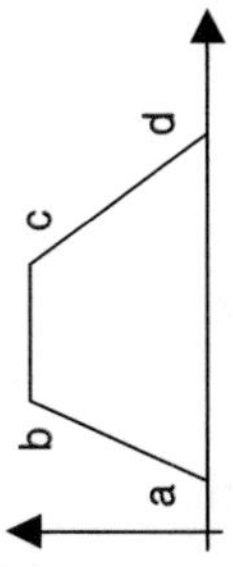

from 25.3% to 24.6%, close to the range of the analytical error. As with the Al_2O_3, the standard error of the mean diminished from ±1.0% to ±0.3%. Thus, the exploration is considered complete also in this respect.

The SiO_2 content of the ore is more variable, the relative variance ranging from ±39% to ±45%. The distribution is almost symmetrical, and the mean remained the same within the range of the analytical error. Only the standard error of the mean diminished from ±0.7% to ±0.2%, indicating a high reliability of the results. The exploration is complete also in this respect.

CaO is one of the main contaminants in the Halimba bauxite. This is the most variable analyzed chemical component, the relative variance ranging from ±83% to ±114%. The distribution is strongly skewed, as indicated by the high positive skewness value. For this reason, Tukey's M-estimator has been applied instead of the normal mean. It increased gradually from 0.6% to 1.0% in the last stage of the exploration. It cannot be predicted whether a further increase would occur with the drilling of new boreholes. The reason for this high variability is that the CaO is present in the form of secondary calcite precipitations, irregularly distributed within the orebodies.

Thus, regarding the evaluated chemical components, the exploration can be considered as completed, except the CaO content.

A further aspect influencing the completeness of the exploration is the detection of the *spatial distribution of the orebodies* and the degree of their variability. The question is, how much was the precision of these predictions increased by the exploration, and can it be regarded complete after the 12th stage? To answer these questions, prior probabilities have been applied. The borehole sites have been ordered into five categories, and a prior probability has been attached to each category, based on the overall exploration experiences of the entire deposit: (1) The site is within the productive area: .3 probability; (2) the site is in an ~1 m wide strip on the border of the productive area: .05 probability; (3) the site is within the possible area: .2 probability; (4) the site is in an ~1 m wide strip on the outer border of the possible area: .05 probability; and (5) the site is within the unproductive area: .4 probability.

The borehole sites situated beyond the range of influence of bauxite thickness have not been categorized, as the prior information cannot be extended to them. In the next step, all existing borehole sites were categorized based on the resource assessment maps of the 12 exploration stages, and the changes of categories were presented in the form of a table. Table 3 shows these changes for 20 selected borehole sites, as the limited extent of this paper does not allow the presentation of all 237 borehole sites. (The uncategorized sites are indicated by question marks). It is obvious that the number of uncategorized sites diminishes in the successive exploration stages. Several boreholes have been drilled at such sites having no prior information. This "haphazard" approach led often to negative results, as illustrated by Table 3. The categories of the borehole sites often changed in positive or negative sense indicating the incompleteness of the exploration. Theoretically, exploration should be considered complete if the site category would not change in the final two or three exploration stages, before the drilling of the given borehole. Unfortunately, this condition was only partly fulfilled even for the last, 12th stage. Thus, in this respect the exploration cannot be accepted as complete.

A more complete evaluation can be obtained if several exploration stages are considered together. Table 4 shows the results of the first seven stages. Numbers below the main diagonal indicate prior categorization, and numbers above show actual data. For instance, 35 boreholes were categorized a priori as "possible"; 16 of these became actually "productive", 17 remained "possible", and only 2 turned out to be "unproductive". Even more interesting results were obtained, when evaluating all stages together, as presented in Table 5. From the 203 prior probabilities, 92 were confirmed by the drilling of the corresponding boreholes. Even more important is that in 97 cases the prior probabilities were changed positively and only in 14 cases negatively. These results underline the effectiveness of the exploration campaign.

A further aspect, important for the planning of a mining investment, is the *completeness of the contouring of the orebodies*. In our case, this means that the orebodies should be surrounded from all sides by unproductive boreholes. The evaluation is simple: The exploration is incomplete at all places where the contour of the orebody is determined only by extrapolation. In the study area, four places remained incomplete in this respect. An overall relative index can be computed when comparing the length of the completely contoured borders with the length of the extrapolated ones.

A further aspect is the *rate of lateral changes in the thickness and altitude of the orebodies*. This aspect is very important in the case of underground mining, as it can be a limiting factor for the choice of the excavation and production systems. We evaluated this aspect by calculating the *specific rates of lateral changes* for the bauxite thickness of neighboring boreholes. An example of this evaluation is presented in Figure 5. In the orebodies of our test sector, these specific rates of lateral changes are often very strong, and they may vary quickly in the different directions, making difficulties in the choice of the mining methods. Note that the boreholes beyond the range of influence were excluded from this evaluation. The entire productive sector has been evaluated in this way. The exploration is complete in this respect. That is, the lateral thickness changes do not exceed a technically determined value.

It is mathematically possible to aggregate all the discussed aspects into one *fuzzy completeness index of the exploration*, following the methodology of Luo and Dimitrakopoulos (2003) for their fuzzy mineral favorability index. This is a useful estimator for the stakeholder, but for the mining engineer, planning and starting the mining operations, it is more useful to evaluate and to compare all the discussed aspects separately. We recommend therefore the stepwise evaluation of each aspect after every exploration stage and making decisions after ranking them with

TABLE 2. MAIN STATISTICS OF SELECTED CHEMICAL COMPONENTS AT THE END OF EXPLORATION STAGES

Chemical components	Stage 12	Stage 8	Stage 6	Stage 4	Stage 2
SiO_2					
Mean (%)	5.30	5.70	5.70	5.50	5.40
Standard error of the mean (%)	±0.20	±0.20	±0.30	±0.30	±0.70
Standard deviation (%)	±2.10	±2.30	±2.30	±2.2	±2.40
Coefficient of variation (%)	39.00	40.00	41.00	40.00	45.00
Skewness	0.02	–0.05	–0.03	0.03	–0.16
Min (%)	1.30	1.30	1.30	1.30	1.30
Max (%)	9.90	9.50	9.50	9.50	9.40
Al_2O_3					
Mean (%)	51.20	51.60	51.90	51.90	52.80
Standard error of the mean (%)	±0.30	±0.40	±0.40	±0.50	±1.00
Standard deviation (%)	±3.60	±3.70	±3.50	±3.20	±3.70
Coefficient of variation (%)	7.00	7.00	7.00	6.00	7.00
Skewness	0.24	–0.09	0.45	0.88	–0.89
Min (%)	38.70	38.70	42.90	44.00	44.00
Max (%)	64.10	63.10	63.10	63.10	59.50
Fe_2O_3					
Mean (%)	24.60	24.60	24.30	24.30	25.30
Standard error of the mean (%)	±0.30	±0.40	±0.40	±0.60	±1.00
Standard deviation (%)	±3.80	±3.90	±3.50	±3.80	±3.50
Coefficient of variation (%)	16.00	16.00	15.00	16.00	14.00
Skewness	–0.89	0.65	–0.77	–0.94	0.38
Min (%)	10.30	11.10	11.10	11.10	20.10
Max (%)	36.70	36.70	32.60	32.60	32.60
CaO					
Mean (%)	1.00	0.90	0.89	0.76	0.60
Standard error of the mean (%)	±0.08	±0.09	±0.12	±0.19	±0.27
Standard deviation (%)	±0.83	±0.81	±0.85	±0.87	±0.55
Coefficient of variation (%)	83.00	90.00	96.00	114.00	91.00
Skewness	1.32	1.54	1.74	2.57	1.54
Min (%)	0.11	0.11	0.11	0.11	0.20
Max (%)	3.90	3.90	3.90	3.90	1.38

TABLE 3. PRIOR CATEGORIZATION OF SELECTED BOREHOLE SITES

Borehole	Exploration stages										
	1	2	3	4	5	6	7	8	9	10	11
H-2564	?	5	3	3	3	3	3	3	3	3	2
H-2557	5	5	5	5	5	5	5	5	5	5	1
H-2556	?	?	?	?	?	?	?	?	?	?	5
H-2555	?	?	?	?	?	?	?	?	?	?	5
H-2554	?	(1)	(1)	3	3	3	3	3	3	3	2
H-2553	?	(2)	(1)	3	3	3	3	3	3	3	2
H-2552	?	(3)	3	3	3	3	3	3	3	3	2
H-2551	?	(1)	(1)	1	1	1	1	1	1	1	3
H-2550	5	5	3	3	3	3	3	3	3	3	2
H-2549	?	5	5	5	5	5	5	3	3	3	4
H-2548	?	?	?	?	?	?	?	?	?	?	5
H-2547	?	5	3	3	3	3	3	3	3	3	3
H-2546	?	5	(1)	1	1	1	1	1	1	1	1
H-2545	?	5	(4)	(4)	(4)	(4)	(4)	(4)	4	4	n.a.
H-2544	?	?	5	?	?	?	?	5	3	2	n.a.
H-2543	5	5	5	?	?	?	?	?	?	5	n.a.
H-2542	5	5	5	?	?	?	?	?	?	4	n.a.
H-2541	?	?	?	?	?	?	?	?	5	2	n.a.
H-2540	?	?	?	?	?	?	?	?	?	2	n.a.
H-2539	?	?	?	?	?	?	?	?	?	4	n.a.

1. Site within the productive area.
2. Site on the border of the productive area.
3. Site within the possible area.
4. Site on the outer border of the possible area.
5. Site within unproductive area (clayey bauxite and bauxitic clay).

( ) Site categorized by extrapolation.
? Not categorized site, outside the ranges of influence.
Bold numbers: categories after drilling the corresponding borehole site.

TABLE 4. SUMMARY RESULTS OF PRIOR CATEGORIZATION OF THE FIRST SEVEN EXPLORATION STAGES

CATEGORIES	Productive	Possible	Unproductive	Sum of row
Productive	20	16	14	50
Possible	10	17	11	38
Unproductive	0	2	12	14
Sum of column	30	35	37	102

TABLE 5. SUMMARY RESULTS OF PRIOR CATEGORIZATION OF ALL 12 EXPLORATION STAGES

Categories	Productive	Possible	Unproductive	Sum of row
Productive	28	35	37	100
Possible	12	32	25	69
Unproductive	0	2	32	34
Sum of column	40	69	94	203

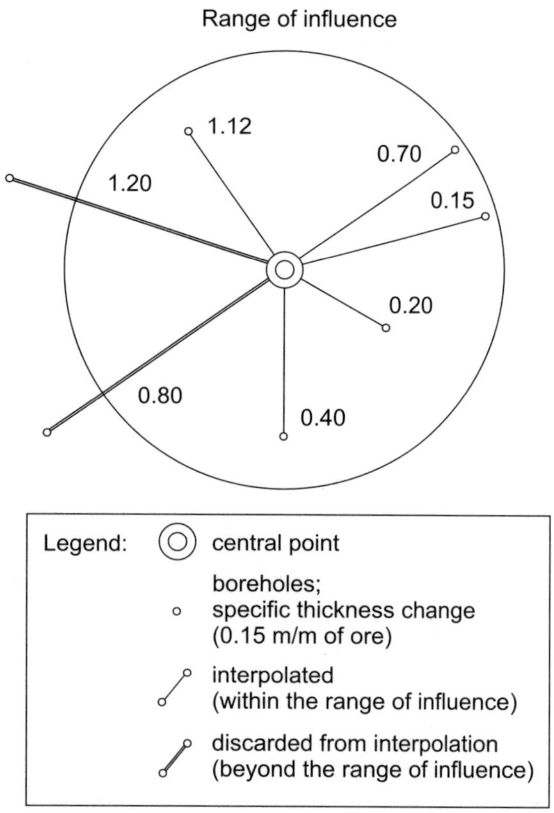

Figure 5. Evaluation of the specific rate of changes for ore thickness in an underground mine around a measured central point (borehole).

respect to both completeness (reliability) and the additional costs of the drilling of further boreholes.

Two general experiences can be formulated as follows. First, even in the case of the best planned and evaluated exploration, random effects (over- or underestimation of the given variable) cannot be excluded, mainly in the early stages of the exploration campaign. Second, a reliable deposit model is the precondition of any evaluation in this respect.

CONCLUSIONS

The completeness of a mineral exploration program can be evaluated by a joint application of the fuzzy set theory and prior probabilities. The establishment of appropriate computerized databases is indispensable for these tasks.

The method consists of the stepwise evaluation of successive exploration stages using information available at each stage.

According to our experiences, completeness of exploration is achieved at different stages of exploration regarding the different evaluated aspects. The criterion for completeness should be the decrease in variability or complete equalization of the given aspect.

REFERENCES CITED

Bárdossy, Gy., R.Szabó, J., and Varga, G., 2003, A new method of resource estimation for bauxite and other solid mineral deposits: BHM, Berg- und Hüttenmännische Monatshefte Leoben, v. 148, p. 57–64.

Bárdossy, Gy., and Fodor, J., 2004, Evaluation of uncertainties and risks in Geology: Berlin, Heidelberg, London, New York, Springer-Verlag, 221 p.

Goovaerts, P., 1997, Geostatistics for natural resources estimation: New York, Oxford University Press, 483 p.

Henley, S., 2000, Resources, reserves and reality: Earth Sciences Computer Applications, v. 15, no. 10, p. 1–2.

Luo, X., and Dimitrakopoulos, R., 2003, Data-driven fuzzy analysis in quantitative mineral resource assessment: Computers and Geosciences, v. 29, p. 3–13, doi: 10.1016/S0098-3004(02)00078-X.

Pannatier, Y., 1996, VARIOWIN: Software for spatial data analysis in 2D: New York, Springer-Verlag, ISBN 0-387-94679-9. 91 p.

SPSS Base 12.0 User's Guide, 2003: Chicago, SPSS, 537 p.

Tukey, W., 1977, Exploratory data analysis: Reading, Massachusetts, Addison-Wesley, 688 p.

Wellmer, F.W., 1989, Economic evaluations in exploration: Berlin, Heidelberg, London, New York, Springer-Verlag, 150 p.

Wood, R., and Curtis, A., 2004, Geological prior information and its applications in solving geoscientific problems, *in* Curtis, A, and Wood, R., eds., Geological prior information: Informing science and engineering: Geological Society [London] Special Publication 239, p. 1–14.

Yamamoto, K., 1999, Quantification of uncertainty in ore reserve estimation: Applications to Chapada copper deposit, Brazil: Natural Resources Research, v. 8, p. 153–163, doi: 10.1023/A:1021894703729.

Manuscript Accepted by the Society 18 June 2005

Printed in the USA

Geological Society of America
Special Paper 397
2006

The system integrator in distributed data systems: The OPeNDAP Data Connector (ODC)

Peter Cornillon
Graduate School of Oceanography, University of Rhode Island, Narragansett, Rhode Island 02882, USA

John Chamberlain
OPeNDAP, 165 Dean Knauss Drive, Narragansett, Rhode Island 02882, USA

Daniel Holloway
Graduate School of Oceanography, University of Rhode Island, Narragansett, Rhode Island 02882, USA

ABSTRACT

This paper investigates the evolution of the end-to-end data system and in particular what constitutes a data system in today's rapidly evolving environment of distributed data system components. Particular attention is devoted to the role of the data system integrator, a computer program or set of programs that may run on the user's computer or at a remote site. By integrating data discovery and delivery with the user's analysis environment, the data system integrator provides the user with seamless access to remote data archives from his or her visualization or analysis program. The data system integrator is the component that effectively defines the data system in an environment in which system components are being produced independently by a broad range of data and service providers. What makes the data system integrator possible is the adoption of a relatively small number of data access protocols, metadata descriptions and data set directories. At the same time, the lack of convergence on *one* data access protocol, *one* directory access protocol, and *one* metadata standard is what necessitates the data system integrator. The data system integrator provides the multiprotocol glue necessary to achieve interoperability over a broad range of service providers. The OPeNDAP Data Connector (ODC) is discussed in detail as an example of a data system integrator designed to run on the user's computer. The requirements of the various functional elements of the system integrator are described, and it is shown that the system integrator is by necessity a multiprotocol program.

Keywords: Web services, distributed data access, data access protocol.

Cornillon, P., Chamberlain, J., and Holloway, D., 2006, The system integrator in distributed data systems: The OPeNDAP Data Connector (ODC), *in* Sinha, A.K., ed., Geoinformatics: Data to Knowledge: Geological Society of America Special Paper 397, p. 225–234. doi: 10.1130/2006.2397(18). For permission to copy, contact editing@geosociety.org.

1. INTRODUCTION

With the development of interoperable data access protocols and architectures that use them, the notion of "the data system"[1] is undergoing a rapid change from one based on a monolithic centralized architecture to one that is diverse and highly distributed. This is true of data systems over a broad range of disciplines: the earth sciences, the space sciences, genetics, etc. We postulate that an end-to-end data system in this highly distributed environment is in fact defined by a data system integrator, a program (or group of programs) that interfaces the various system components in such a way that the user can move seamlessly from data discovery through delivery to analysis. The objective of this paper is to introduce the data system integrator. This is done in the context of the OPeNDAP Data Connector (ODC), a data system integrator based on the OPeNDAP[2] data access protocol. The ODC provides a great deal of functionality, but only those elements that are likely to be common to the general system integrator are discussed. We begin with an overview of the evolution of the "data system" because this evolution is indicative of trends that we believe will continue into the future and hence should be addressed in the design of system integrators.

Historically, data systems have been developed by a centrally managed group that took responsibility for all aspects of the system. None of these systems provided for the complete range of data system functionality—discovery to analysis—over a broad range of data providers and data types. The first generation of data systems involved a single computer that could only be accessed locally. The system consisted of the data, a search capability, and an ability to manipulate the data. Adding data to the system was tedious at best. Programs had to be written to ingest data from storage media. Then the ingested data had to be cataloged and stored in a format that the data system could readily access. All aspects of these systems were controlled by the system builder, and such systems did not scale well either from the perspective of added functionality or from the perspective of the data available.

The ability to network computers and in particular the advent of wide-area networks resulted in a substantial change in the approach taken to the development of data systems. The first significant system components to be built were online directory services designed to aid in the discovery of data. Initially, directory entries provided the user with a point of contact at the originating archive such as a phone number or postal address. As more data became available electronically, directory entries associated with these data were augmented with the electronic address of the data, generally the Uniform Resource Locator (URL) of the associated FTP site. To obtain data of interest, the user located the data set in the directory, then contacted the site holding the data and ordered them for either electronic or mail delivery. Once delivered, the user entered the data into an application program for analysis.

The next step in the evolution was to build systems that integrated the discovery and delivery functions. Such systems are often referred to as "one-stop shopping" systems. They provide for a seamless connection between the data discovery component and the data order component, but little beyond that. These systems do not provide for an end-to-end integrated solution to data access; they are missing the component that provides seamless access to a data analysis capability. The Earth Observing System Data Information System (EOSDIS)[3] is an example of such a system. EOSDIS provides access to a broad range of satellite-derived data products, but this access is limited to removable media or FTP transfers of files.

Because "one-stop shopping" systems are limited to data that can be ordered through the system, work has continued on the development of system components that provide for data discovery only, such as the Global Change Master Directory (GCMD)[4]. These systems provide pointers to a much broader range of data sets, but without the ability to order these data in a consistent fashion. The GCMD does, however, provide a simple protocol that allows for remote access to its data set list. It also includes fields for URLs as part of the information associated with each data set. Other cataloging efforts are under way as well. For example, the Thematic Realtime Environmental Distributed Data Services (THREDDS)[5] project of the University Corporation for Atmospheric Research (UCAR) provides for distributed hierarchical catalogs. The catalogs are hierarchical in the sense that they provide for general data set descriptions (the directory level) as well as for listings of the contents of large data sets (the inventory level). The directory level links seamlessly to the inventory level, which may in turn contain URLs pointing to data servers. THREDDS differs from the GCMD in that it provides for an integrated inventory and in that it is distributed. In the THREDDS vision, each data provider installs a THREDDS catalog, which may be accessed remotely either from an application or from another THREDDS catalog.

Independently of the order and discovery system components, several large projects are developing data access protocols with which delivery of a well-defined data stream to the user's application package is made possible. The oldest of these is the Open-source Project for a Network Data Access Protocol (OPeNDAP). In addition to providing a data access protocol, OPeNDAP provides software that makes local data accessible to remote locations regardless of local storage format, the server

[1]In this paper, the data system is confined to those components ranging from data discovery—finding the data of potential interest for a given problem—through analysis or visualization of these data. It does not include the sensors that collect the data or the preliminary processing performed on the data prior to their archival for public access.

[2]http://opendap.org

[3]http://spsosun.gsfc.nasa.gov/eosinfo/Welcome/index.html

[4]http://gcmd.nasa.gov/

[5]http://www.unidata.ucar.edu/projects/THREDDS/index.html

side. It also provides tools for transforming existing applications into OPeNDAP clients, the client side. More recently, the Open Geospatial Consortium (OGC) has developed an architecture addressing the same basic issue of data delivery, although there are some fundamental differences (addressed in section 2) between the architectures.

The evolution of data system development is clearly away from centrally designed, implemented, and maintained end-to-end systems toward the development of system components, with different groups developing and managing different system components. Given the lack of central authority, these components simply present a disjointed suite of services to the end user. The data system integrator brings order to this environment by interfacing the components in a structured fashion. It effectively defines the data system.

The data system integrator allows the user to perform all essential preanalysis activities—finding, browsing, subsetting, acquiring, and delivering to the user's analysis software data of interest. In each of these basic areas, there may exist specialized applications that can do the task more efficiently or that provide the user with greater functionality. The advantage of the system integrator is that it provides for a seamless interface between functional components, thus resulting in a substantial savings in time. The integrator also provides a uniform interface for functions that in stand-alone system components may be represented by several different programs and hence several different user interfaces.

The data system integrator may operate remotely from the user or may reside on the user's computer. Web-based data portals, such as that of the Council for the Central Laboratory of the Research Councils (CCLRC) (CCLRC, 2004), are examples of the former, while the ODC is an example of the latter. Specifically, the ODC is a Java-based program that is downloaded to and run on the user's computer. As such, it provides for more rapid response to a broad range of functions than the Web-based data portal, and it operates as a front end to a variety of application programs such as MATLAB and Excel.

The remainder of this paper addresses the ODC as an example of a data system integrator. The general characteristics of OPeNDAP, which lies at the core of the ODC, are also described. Because OPeNDAP is not heavily used in the geophysics community, but could be, its description is presented in a bit more detail than is needed to understand the ODC; this manuscript also serves to introduce OPeNDAP to the geophysics community. A detailed description of the OPeNDAP architecture is available in Cornillon et al. (2003).

2. THE OPeNDAP ARCHITECTURE

In 1994, the University of Rhode Island received NASA and NOAA funding to develop a data system that would facilitate access to oceanographic data over the network. The system developed with this funding was referred to as the Distributed Oceanographic Data System (DODS). Following this early work, additional funding was obtained from the National Science Foundation (NSF) and the National Oceanographic Partnership Program (NOPP) to continue development of the system. In the late 1990s, it became clear that the data access protocol being developed as part of the overall effort applied to nonoceanographic data as well as to oceanographic data. In 2000, in an effort to clarify the distinction between the data system and the discipline-neutral data access protocol at its core, the DODS effort was divided into the National Virtual Ocean Data System (NVODS) (the oceanographic part) and OPeNDAP, Inc. (focusing on the data access protocol). Specifically, OPeNDAP, Inc., is a nonprofit corporation formed to maintain, evolve, and promote a data access protocol that would facilitate network access to scientific data. In this paper, "OPeNDAP" is used to refer both to the software developed and maintained by OPeNDAP, Inc., and to the corporation, although the latter usage is rare.

The two fundamental objectives of the OPeNDAP infrastructure are (1) to isolate the format in which data are stored at the data provider's site from the format required by the user's application software, and (2) to allow users to access remotely held data directly from their analysis packages as though the data reside locally. An important characteristic of the data access protocol is that it provides for subsetting of the data at the server. This allows the user to efficiently access only those portions of very large data sets that are of interest.

The basic operations involved in an OPeNDAP access are described in the context of an example in which a MATLAB[6] user in Tallahassee, Florida, accesses a subset of a global shear wave tomography model output data set archived as a netCDF file in Boulder, Colorado[7] (Figs. 1 and 2)[8]. Data are requested from the user's application via the OPeNDAP data access protocol or DAP (Fig. 1). In this example, the application is MATLAB, but it could be any OPeNDAP-enabled application[9]. Furthermore, the request in this example was made from a command-line client, but the request could have been made programmatically from the client computer or via a graphical user interface. The server parses the request, extracts the subset of interest from the data set[10], converts it to the OPeNDAP data model, compresses it, and returns it to the client. On receipt, the client decompresses the

[6]http://www.mathworks.com/products/

[7]The data shown in this example, obtained via OPeNDAP using MATLAB, have been moved to the GEON site in San Diego, California. The URL requested was http://geon.unavco.org/opendap/nph-dods/data/tomography/bkS-VelPertOn2DegGrid.nc?vel[0:0][40:60][67:70]

[8]This example has been animated in a Flash video: http://www.opendap.org/movies/overview.html

[9]Supported analysis packages range from commercial products such as IDL, Microsoft Excel, or MATLAB to freeware such as ncBrowse (http://www.epic.noaa.gov/java/ncBrowse), GrADS (http://grads.iges.org/grads/index.html), Ferret (http://ferret.pmel.noaa.gov/Ferret), or IDV (http://www.unidata.ucar.edu/software/metapps/index.html), to programs developed by the user and linked with the OPeNDAP libraries.

[10]In this example, the data are assumed to be stored in netCDF, but many other Application Program Interfaces (APIs)/formats are supported, e.g., HDF, FITS, FreeForm, ESML, MATLAB, SQL via a JDBC server, GRIB, and BUFR.

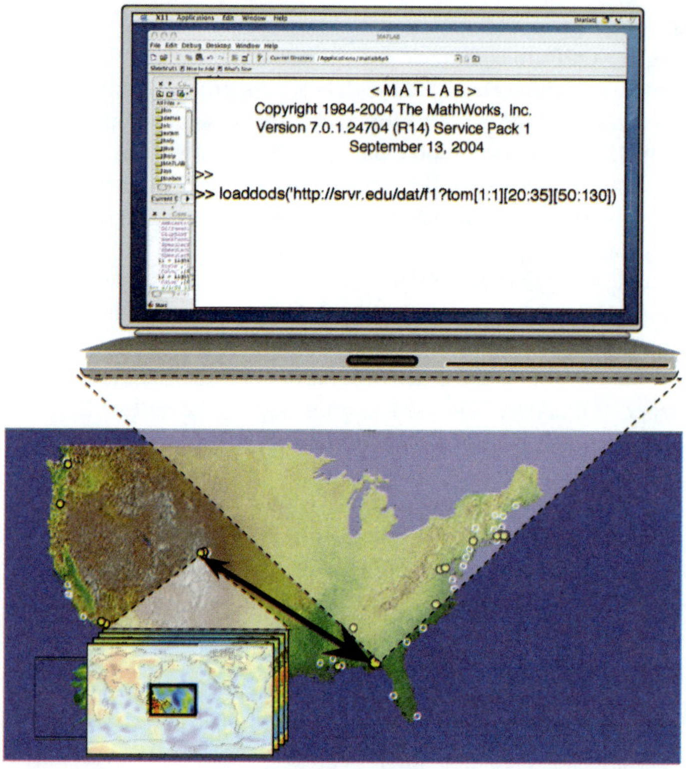

Figure 1. Shear wave data residing in an archive in Boulder, Colorado, are requested via OPeNDAP from a client running MATLAB in Tallahassee, Florida.

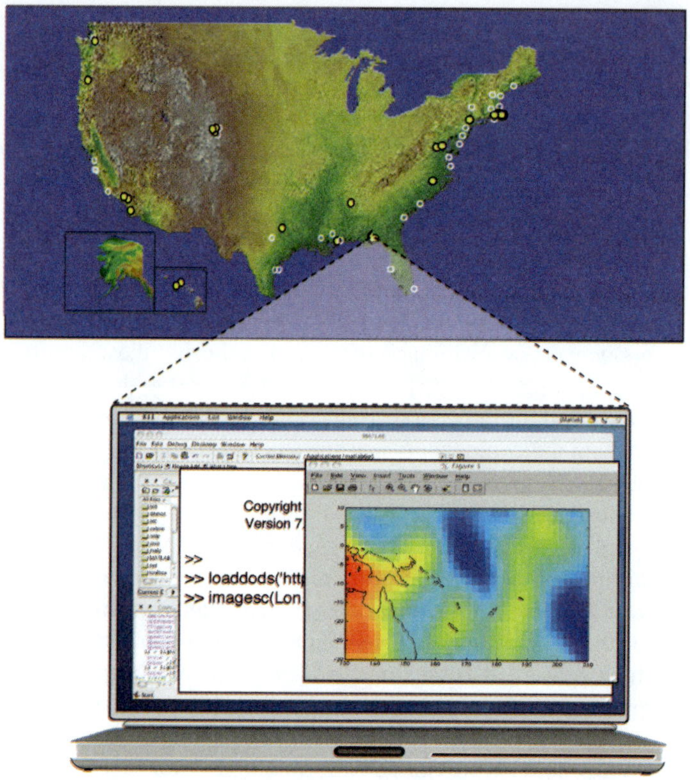

Figure 2. The retrieved data are plotted in MATLAB.

data, transforms the data from the OPeNDAP data model to the data model of the client application, and instantiates the data in the workspace of the analysis package. At that point, the data are available to the routines in the analysis package (Fig. 2).

A bit more detail: The DAP consists of a data model and its implementation. The DAP data model is made of a small number of low-level "atomic" data types (based on C++ data types) and a few higher-order "constructor" types, data types constructed from atomic data types, such as a sequence of integer triplets. The DAP also includes operators and an external data representation, a procedure for retrieving data and ancillary data from remote platforms, and an API consisting of OPeNDAP classes and data access calls designed to implement the protocol. The OPeNDAP core library has been implemented in both C++ and Java. As mentioned above, implementation of the DAP provides a means for the client application to specify simple projection and subsetting operations in a data request to the server. Because subsetting and projection operations are completely generic and because the data model is based on computer data types, the DAP is "discipline-neutral". This is one of the reasons why it is used in a number of different disciplines, e.g., oceanography, meteorology, land cover studies, and Sun-Earth connection research. It is not heavily used in the geophysics community. However, given the discipline neutrality of the data access protocol, there is nothing that precludes this community from making use of it.

3. THE OPeNDAP DATA CONNECTOR

The OPeNDAP Data Connector (ODC) is a data system integrator for OPeNDAP-accessible data sets. In addition to providing the data user with a broad range of preanalysis functions, it also provides the data provider with a convenient way to test his or her OPeNDAP server. In the remainder of this section, the details of the ODC are presented along with the rationale for the choices made.

3.1. ODC General Architecture

As discussed above, the OPeNDAP data access protocol is discipline-neutral. This means that semantic metadata (those metadata that describe the data, such as units and projections) are not mandated as part of the protocol, although the protocol does provide for the transmittal of whatever semantic metadata the data provider would like to associate with the data. The benefit of a discipline-neutral protocol is that it may be used in any setting requiring remote access to data, be it scientific, medical, industrial, business, etc. In addition, it facilitates the serving of data; the data provider need only be concerned about information related to the syntax of the data (the data type, array size, relationship between arrays, etc.), not their semantics, in order to serve them via the protocol. The cost of a discipline-neutral protocol is the lack of semantic consistency of the metadata together with a lack of organizational consistency of the data themselves. This makes writing semantically aware clients difficult at best. As with the OPeNDAP data access protocol, the ODC is based on a discipline-neutral model while at the same time providing for the use of semantic information when available. This allows the ODC to be used with any data set served via OPeNDAP but means that the basic functions that it provides—search, subsetting and plotting—will, in general, be performed based on the syntax of the data, as opposed to their semantics. This is discussed in more detail in section 3.2.

Figure 3 illustrates the architectural relationship between the ODC and other system components. Through the ODC, any data set served via OPeNDAP is accessible from a user's computer. The user can use the ODC to find sites hosting the data, retrieve the data, browse them, and moderate transfers to specialized storage and analysis programs. The ODC functions as middleware in the architecture. The key point here is that the middleware piece ties many disparate source formats to an equally disparate set of destination applications in a unified way.

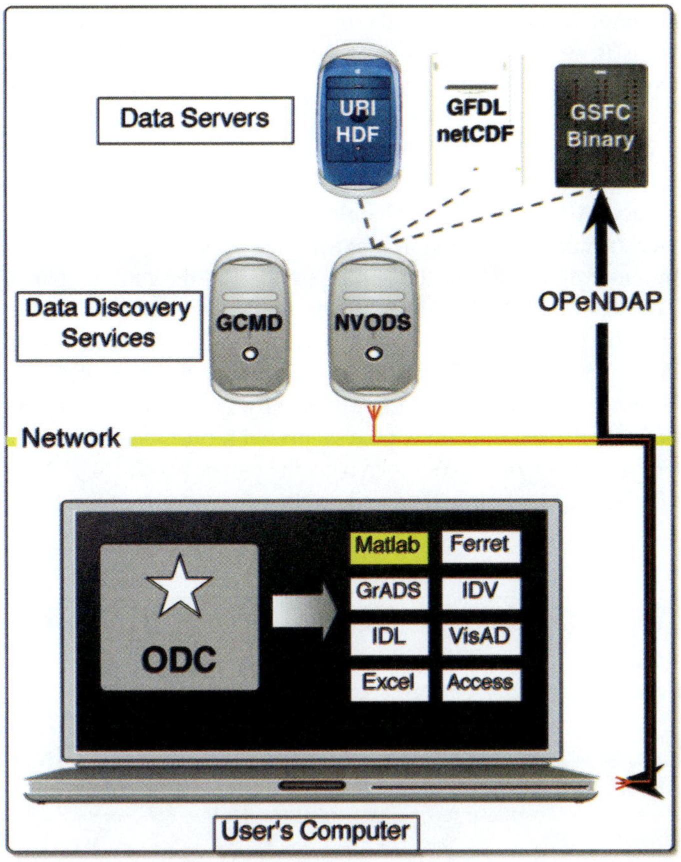

Figure 3. The architectural relationship of the ODC to other system components.

The ODC communicates with data discovery services via the protocol that they have established for access to their databases. Two data discovery services are indicated in Figure 3, the GCMD and the NVODS directory. (All system components shown in the figure—data discovery services, data servers, and the user's computer—are connected to the internet, but for clarity only those connections that are active are indicated.) In this example, the ODC is shown accessing the NVODS directory, the solid red line connecting the user's computer with the NVODS computer. (See section 3.2.1 for a description of data discovery within the ODC). A dashed line connecting a discovery service with a data archive is meant to indicate that there are data sets at the archive that are listed at the discovery service. These lines are dashed because, at present, discovery service databases are not automatically updated from the data servers, rather the process is performed manually. Work is in progress to automate this process for the NVODS directory using THREDDS catalogs.

The eight small rectangles shown on the user's computer represent OPeNDAP-enabled analysis packages. The user's computer may be running any OPeNDAP-enabled application, one or more of those shown in the figure as well as others not shown. In the example, MATLAB has been highlighted, indicating that it is the package being used. Because the ODC is also running on the user's computer, any communication between the ODC and an analysis package is performed locally; i.e., such communication is not performed over the network.

The thick black line labeled OPeNDAP shows the ODC accessing a data archive (a binary data set at NASA's Goddard Space Flight Center (GSFC) in this case) via the OPeNDAP data access protocol.

3.2. ODC Functionality: The Specifics

The functions performed by the ODC are arranged in five groups: search, retrieve, view, help, and feedback, each associated with a multipane panel in the connector's graphical user interface (GUI). The panels are accessed via clickable tabs at the top of the main panel. The tabs are arranged from left to right based on the order in which the user is expected to proceed, at least for the first three tabs—from search to view The user may move between tabs either by explicitly clicking the tab or by finalizing the steps in a given panel; i.e., operations in some of the panels will automatically close that panel and open the next one. Each of the functional groups consists of several functional elements. The functional groups associated with search, retrieve, and view are described in more detail below, and the relationships between the various panels as well as between the ODC and the user's application are shown in Figure 4. The help panel simply provides help with regard to use of the ODC, and the feedback panel provides a mechanism for the user either to send a free text e-mail message to the ODC design and maintenance team or to register a bug in the OPeNDAP project bug database. The following is not a complete description of the operation of the ODC, rather the focus is on the functionality of the connector from the perspective of a data system integrator.

3.2.1. Search Panel: Data Discovery

There are two ways in which the user may specify an OPeNDAP URL in the ODC: through one of the discovery services available in the search panel, the focus of this section, or by entering a URL in the space provided in the retrieve panel. The latter enables the ODC to be used with any OPeNDAP-enabled data set, not just those listed in the services supported in the search panel. In particular, this means that the ODC can be used with data that are served locally or with data that may not yet be listed with one of the supported services. Interestingly, Google or any other Web search engine can be used to search for OPeNDAP-accessible data and, in the past, we have found data sets that are not listed at a supported service in this way. Data discovery via a Web search engine is, however, much more tedious than using the services provided in the search panel.

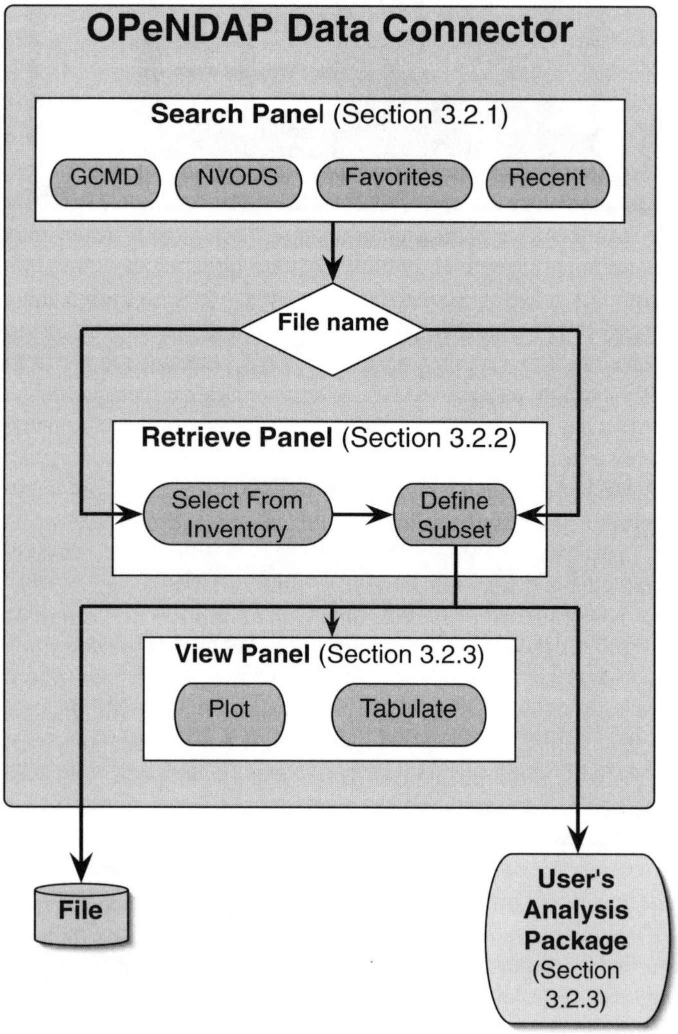

Figure 4. Basic functional elements of the ODC used to search, retrieve, and view data, and the relationship between the ODC and the user's application package.

The ODC's search capability features a pluggable programming interface that allows external search modules to be dynamically added. This was done to allow the ODC to make use of any data discovery service that provides access to OPeNDAP URLs. Additionally, this feature is useful for those organizations that use the ODC internally to search private databases.

The ODC is currently configured to make use of two external discovery services: the GCMD and the NVODS directory. There are two fundamental differences between these discovery services. One relates to the objects, OPeNDAP URLs, identified in the databases associated with the services. Specifically, the GCMD lists each data set separately while the NVODS directory provides a list of data sets as well as collections of data sets. The distinction between a data set and a collection of data sets is addressed in more detail in section 3.2.2. The second difference relates to the amount of metadata associated with the objects referenced in the associated databases. The GCMD provides extensive metadata designed primarily to facilitate finding data of interest. The NVODS directory is simply a list of titles of OPeNDAP-accessible data sets or of collections of data sets. The NVODS directory is more inclusive than the GCMD, but it is also more difficult to find a data set of interest using this service.

In addition to the two data discovery services, the ODC also allows the user to add data sets to a "favorites" list, and the ODC maintains a list of recently selected data sets. The user selects the service desired from a set of tabs at the top of the search panel. Additional tabs may be added as access to more discovery services are provided.

Once a data set or collection of data sets has been identified using the selected discovery service, the related OPeNDAP URL may be sent to the retrieve panel. This closes the search panel and opens the retrieve panel. It also performs the first operation on the URL that is performed when a data set is selected in the retrieve panel. This is discussed in more detail in section 3.2.2. The important point here is that although the steps to select a data set may differ from one discovery service to another, the action taken once a data set has been selected is the same.

The differences between GCMD and NVODS discovery services provide a good example of the role played by the system integrator as well as the way in which the user locates a data set of interest.

GCMD. Each data set associated with the GCMD contains, among other information, a well-defined name for each of the variables in the data set, the latitude and longitude range covered by the data, the data's temporal extent, and a text description of the data set[11]. The spatial and temporal information, referred to as "semantic search metadata," is often used for space-time

[11]See http://gcmd.gsfc.nasa.gov/User/difguide/difman.html for a description of the Directory Interchange Format (DIF), the metadata record used to describe data sets listed in the GCMD.

searches in discovery services. Any information associated with a given data set is accessible from the GCMD server located at http://gcmd.nasa.gov via an XML-driven query. The server's query interface supports a "free text" search and a predefined keyword search, both with an optional spatial-temporal qualifier. The ODC exposes this interface to the user with text entry boxes and lists showing all the available keywords. Definition of a spatial region of interest is performed graphically or in text boxes with a standardized map widget. Once a search has been specified, the information is passed to the GCMD, and the GCMD responds with a list of data set titles, each of which is associated with a URL. The user selects the data set of interest from this list and "sends" it to the retrieve panel.

NVODS directory. The NVODS directory is accessible as a static XML document maintained by OPeNDAP at http://xml.opendap.org/datasets/datasets.xml. The ODC retrieves this document and presents its contents as a searchable tree. The highest level of the tree is a list of data provider sites, and the lowest level lists either individual data sets or locations containing a group (or collection) of data sets. As with the GCMD, the user selects the entry of interest from this list and "sends" it to the retrieve panel.

3.2.2. Retrieve Panel: Building the Data Request URL

The local data catalog. The primary role of the retrieve panel is to help the user specify a subset of the selected data set: the subset that will be retrieved from the OPeNDAP server for display in the view panel or for analysis in the user's software application. The complication here, which the ODC solves, is that many data sets consist of a large number of files and there is no consistency in the way inventories for such multifile data sets are provided. In fact, the problem associated with inventories is one that has been largely ignored to date in the development of distributed data systems. The result is that several ad hoc mechanisms to represent inventories have been developed for OPeNDAP-accessible data sets. This means that the operations required to obtain a subset of the data set of interest depends on the data object identified in the discovery service and on the characteristics of the server itself. In order to better understand this, it is necessary to understand the OPeNDAP URL. The OPeNDAP URL consists of six primary parts: the TCP/IP application protocol, the machine name, the OPeNDAP server, the directory in which the data reside, the file name, and the constraint that is to be applied to the data set. Below is a fictitious OPeNDAP URL:

http://	data.umm.edu /	cgi-bin/nph-dods /	data/run_1/	tsuv.nc	?T[1:5], S[1:5]
protocol	machine name	server	directory	file name	constraint expression

The server parses the information following the server name to determine how to respond to the request; everything that follows the server name is effectively an argument of a request to the server. The DAP specifies the form of the constraint expression, if present, but it does not specify the structure of the directory and file name portion of the URL. Server developers have taken advantage of this to add specifications for specialized server operations in this portion of the URL.

One OPeNDAP server specialization is relevant here: the ability to provide a single entry point for a collection of the data holdings at the server. Two forms of this exist, one associated with servers supplied by the OPeNDAP project and one associated with the GrADS-Data Server (GDS)[12]. In both cases, the server recognizes a URL for which neither a file name nor constraint is present. In the case of OPeNDAP-supplied servers, when the URL ends with a directory name, such as "data/" in the example shown above, the response to the request is a list of the contents of the specified directory. This list, with one exception, is identical to the list returned by a Web server if the URL entered in a browser ends in "/". The exception is that all URLs in the list include the OPeNDAP server name followed by a subdirectory name, i.e., end in a "/", or a file name. We refer to this list or collection of data sets as a "catalog" to differentiate it from an inventory for a particular data set in that the list may in fact include URLs for a number of different data sets. In the case of the GDS, a request for a list of data accessible via the server is expressed by a URL for which the argument of the server is null; no directory name is supplied. A request based on such a URL returns a list, but the list in this case contains the titles of the data objects to which the list refers, as opposed to a list of file names and directory names; hence it is easier to read. The list also contains URLs for each OPeNDAP object on the list.

The capability offered by the specialization outlined above, either to traverse the local directory structure or to provide a list of OPeNDAP-accessible data sets, allows the data provider to effectively provide an inventory, albeit a coarse one, for multifile data sets in addition to a list of local data sets. For example, in the case of OPeNDAP-supplied servers, the directory identified might point to the top of a tree that includes a number of data sets, some of which are single-file data sets (in which case the associated URL would be a file name), and some of which are multifile data sets (in which case the associated URL might be a directory containing the files of interest or additional subdirectories).

The point here is that for the ODC to perform as a seamless integrator of system components, it must be capable of presenting these lists to the user in a similar form regardless of the form in which they are received from the server. This is done in the retrieve panel. When a data object has been selected from a discovery service in the search panel and sent to the retrieve panel or has been entered manually in the retrieve panel, the associated URL is parsed to determine if it is a file name or a local catalog, a URL ending in a "/". If the latter, the ODC requests the list from the server. If the former, the ODC requests the metadata that are formally associated with the data set as an OPeNDAP metadata object. This is discussed in more detail in the next section.

For those cases in which the URL points to a list of data objects, the user traverses the directory tree until file URLs are found. Selecting a file URL at this point will result in the ODC requesting the OPeNDAP metadata in precisely the same fashion

[12]http://grads.iges.org/grads/gds/

as was done when the URL passed in from the search panel was an OPeNDAP-accessible file.

Subset control. The next step in the process is to define a subset of interest in the selected file. The problem of standardized subsetting of structured data was one of the primary goals of the DAP at its inception 11 years ago. A graphical client program like the ODC solves the subsetting problem from the user's perspective because it presents an interface to an OPeNDAP server's subsetting capabilities. This interface is populated using the OPeNDAP-accessible metadata obtained as discussed in the previous section. These metadata consist of the syntactic metadata that describe the data type, array size if an array, etc., and the semantic metadata. They are requested by appending a suffix to the file name in the URL (Cornillon et al., 2003). The subsetting interface in the retrieve panel consists of the list of variables in the file and the attributes associated with these variables. For each variable, the interface also includes user-selectable fields with which the user may define a subset of the data; i.e., the user not only has a graphical visualization of the structure of the data, but can also use on-screen controls to specify which part of the data to retrieve.

The selection capabilities offered in the retrieve panel are discipline-neutral; i.e., there is nothing in the selection that depends explicitly on the semantics of the data. Selection of arrays is performed by row and column, as opposed to being based on the semantic meaning of the rows and columns, such as latitude and longitude or temperature and salinity, and selection of sequences (points and vectors in GISs) is performed based on the values of the variables in the sequence. In the latter, the selection may have semantic meaning if the variables are in readily recognized units, but there is nothing that requires this in the DAP. Having said this, we note that it would be straightforward to add a module to the ODC that would make use of semantic metadata that may be associated with a data set for the subsetting operation. For example, a module could be added that would recognize Cooperative Ocean-Atmosphere Research Data Standard (COARDS)–compliant data sets and make use of the associated metadata. The ODC could of course still work with other data sets; it would simply default to the discipline-neutral selection mechanism already in place.

3.2.3. Display and Analysis of the Subsetted Data

Once a subset of the data has been defined in the retrieve panel, the user may (1) save the data locally as a file, (2) display them in the view panel, or (3) pass them to an analysis package. It is important to stress that the user may return to any tab in the ODC at any time. So, if the user would like to view a subset of the data in the view panel, there is nothing that prevents the same data, or a modified subset, from later being imported directly into an analysis package. The focus of the remainder of this section is on the display capability of the ODC and the link between the ODC and external analysis programs.

View panel. When the user selects the **Output to** option in the retrieve panel, the ODC passes the data to the view panel and displays them either as text or as a table in the text or tabular frames, or it makes them accessible for plotting in the plotting frame. The plotter is designed to handle the following basic plot types:

- Pseudocolor—a two-dimensional color rendition of the data
- Vectors—if the data consist of at least two variables of the same data type and dimensionality, presumably two orthogonal components of a vector
- (X, Y)—an (x, y) plot either as a line or as points
- Histogram—a histogram of the data

In the plot frame, the user can easily select any of the plot types as well as which variables to plot.

The primary motivation for providing a plotting capability in the ODC is to help the user to understand the basic structure of the data as well as issues such as coverage and quality; i.e., it has been designed as a browse tool, not as an analysis tool. It is anticipated that once the subset of interest has been clearly identified, it will be moved directly to the user's analysis package.

Passing the data to an analysis package. The interaction between the ODC and the user's application is fundamentally different from that between ODC panels. In the latter, the data are pushed from one panel to the next. In the former, the data are pulled either from the ODC or directly from the OPeNDAP server.

Via the ODC. In some cases, rather than writing an interface to ingest an OPeNDAP data stream into an application, the ODC, which can read any OPeNDAP data stream, is used instead. This is true of clients such as Microsoft Excel, which require that the data be flattened prior to ingestion and for which an OPeNDAP client has not already been written. Using ordinary TCP/IP, the ODC can receive a request from another process for data. This allows data to be transferred directly from the ODC to other applications without the use of an intermediate file. The other application needs to have a plug-in or script that connects to the ODC and acquires the data. Currently plug-ins are available for Microsoft Access and Microsoft Excel. The protocol for the transfer has been deliberately made simple so that software developers can easily write such scripts or modules for their applications.

Directly from the originating OPeNDAP server. For applications for which a client already exists to ingest an OPeNDAP data stream, scripts are available to facilitate ingesting the data into the application's workspace. Such a script currently exists for MATLAB, and the intent is to develop them for other OPeNDAP-enabled applications. In the MATLAB case, the script requests the list of constrained URLs from the ODC. It then issues the request(s) to the OPeNDAP server(s) for the subset(s) of interest. The OPeNDAP client interns the returned data in the workspace of the application package.

4. DISCUSSION

The primary objective of the system integrator is to reduce the burden on the user of the preliminary operations in a task requiring data analysis, those operations that are only peripheral to the actual analysis of the data. As such, the emphasis in the design of an integrator is on the speed and convenience of labor-intensive steps. By assembling into a single interface access to the operations needed to find, subset, and acquire data, the system integrator dramatically reduces the level of effort required to accomplish these tasks. Although different system integrators will likely address in different ways the problems associated with integrating distributed resources to achieve the goal outlined above, the basic issues that they will all need to address will be similar. In the following, we discuss these issues from the perspective of the OPeNDAP Data Connector.

4.1. Communication Between Integrator Functional Elements

To begin with, the integrator must provide for seamless communication between the various steps that it supports. This means that it must pass data identifiers from one step to the next in the discovery/subsetting/retrieval process, or it must provide a simple mechanism for the user to do so. In the ODC, the data identifiers are OPeNDAP URLs, and they are automatically passed (pushed) from one panel to the next at certain points in the process. Furthermore, when a URL is received by a new panel, operations are automatically performed on it, and the results are used to configure the panel. The result is that, for simple data sets, one moves from finding the data to plotting them in a few minutes with a very small number of key clicks, as few as ten.

Key to the reduced interaction with the user are the default decisions made along the way. However, for this to be acceptable to the user, the user must be provided with the capability to override these defaults. The ODC has been designed and built with this in mind.

Another important characteristic of the integrator in this regard is the increased uniformity in the handling of URLs or data as one progresses through the process. In the first step in the ODC (the search panel), two discovery services are available, but by the next step (the retrieve panel), most of the idiosyncrasies have been addressed, and one mechanism to traverse inventories and another to define a data subset suffices. In the last panel (the view panel), there is complete homogeneity in how the data are handled. This means that regardless of how the data were located, they are dealt with in a consistent way in the view panel or in the step passing them to the user's application.

4.2. The Need to Deal with Multiple Access Mechanisms

Given the distributed responsibility in the development of data system components, there is a near certainty that different access mechanisms are going to be used to access different data types. In the example discussed in this paper, three different access mechanisms are used: one for the GCMD, one for the NVODS directory, and one to access the data themselves. The integrator must be able to deal with these different access mechanisms.

In the ODC, only one protocol is used to access data. This greatly simplified the later steps in the overall process as discussed above. However, one could imagine an integrator that handles different protocols to access the data themselves. This could result in a significant increase in accessible data, but also in the complexity of the integrator—both from the perspective the ODC developer and from the perspective of the user.

4.3. The Ability to Handle Different Types of Data Objects

The integrator must be able to conceptually deal with very different data objects, ranging from lists of data sets to data set inventories to the data themselves.

4.4. The Need to Provide for Expansion

The distributed responsibility in the development of system components also leads to competing services for various portions of the problem. It is therefore quite probable that new services will be developed that may be of interest to those using the system integrator. In order to accommodate this, the integrator should be designed so that additional services can easily be added. In the ODC, such a capability was demonstrated with those services associated with data discovery.

4.5. The Ability to Enter Information Locally

A fundamental concept of many OPeNDAP clients is that they have the ability to read data over the network via the DAP and that they have the ability to read data locally from their file system. In exactly the same way, we found that in addition to entering OPeNDAP URLs into the ODC via a discovery service, the ODC needed to offer the user the ability to manually enter URLs that point to OPeNDAP-accessible data. In this way, the ODC is not limited to a tool that will only work with data sets defined in the first (search) step of the process.

5. CONCLUSIONS

In this paper, we have investigated the concept of the data system integrator as the defining component of a modern end-to-end data system consisting of separately developed and maintained data system components. The end goal of an application acting as a data system integrator is to increase the productivity of the user. Adding new ways for the researcher to look at data is certainly useful, but the biggest payoff is making work more efficient so the scientist has more time to invent their own ways of looking at the data. The OPeNDAP architecture in conjunction with the ODC creates this efficiency by turning data acquisition into a point-and-click operation.

It is clear that as data system components on the Web increase, the number and capability of data system integrators is also going to increase. This paper has been presented with the intent of initiating a discussion with regard to the issues with which the data system integrator should be dealing.

ACKNOWLEDGMENTS

This research was performed with support from the National Aeronautics and Space Administration as part of the Research, Education, and Applications Solutions Network (REASoN) effort (NNG04GH13A) and with salary support to P. Cornillon from the State of Rhode Island and Providence Plantations. The authors would also like to thank Charles Meertens, Lola Olsen, and Robert Raskin for their many helpful comments made as part of the review process.

APPENDIX. ACRONYMS

API Application Program Interface
BUFR Binary Universal Format Representation
CCLRC Council for the Central Laboratory of the Research Councils
COARDS Cooperative Ocean-Atmosphere Research Data Standard
DAP Data Access Protocol
DIF Directory Interchange Format
DODS Distributed Oceanographic Data System
EOSDIS Earth Observing System Data Information System
ESML Earth System Markup Language
FITS Flexible Image (or Interchange) Transport System
FTP File Transport Protocol
GCMD Global Change Master Directory
GDS GrADS-Data Server
GIS Geographic Information System
GrADS Grid Analysis and Display System
GRIB Grid in Binary
GSFC Goddard Space Flight Center
GUI Graphical User Interface
HDF Hierarchical Data Format
IDL Interactive Display Language
IDV Integrated Data Viewer
JDBC Java Database Connectivity
NASA National Aeronautics and Space Administration
netCDF Network Common Data Format
NOAA National Oceanic and Atmospheric Administration
NOPP National Oceanographic Partnership Program
NSF National Science Foundation
NVODS National Virtual Ocean Data System
ODC OPeNDAP Data Connector
OGC Open Geospatial Consortium
OPeNDAP Open-source Project for a Network Data Access Protocol
REASoN Research, Education, and Applications Solutions Network
SQL Structured Query Language
TCP/IP Transmission Control Protocol/Internet Protocol
THREDDS Thematic Realtime Environmental Distributed Data Services
UCAR University Corporation for Atmospheric Research
URI University of Rhode Island
URL Uniform Resource Locator
XML Extensible Markup Language

REFERENCES CITED

CCLRC, 2004, CCLRC World Wide Web page: http://www.escience.clrc.ac.uk/web/projects/dataportal (October 2005).

Cornillon, P., Gallagher, J., and Sgouros, T., 2003, OPeNDAP: Accessing data in a distributed, heterogeneous environment: Data Science Journal, v. 2, p. 164–174.

MANUSCRIPT ACCEPTED BY THE SOCIETY 18 JUNE 2005

Geological Society of America
Special Paper 397
2006

Bird's-eye view of data mining in geosciences

Rahul Ramachandran
John Rushing
Xiang Li
Information Technology and Systems Center, University of Alabama in Huntsville, 301 Sparkman Drive, Huntsville, Alabama 35899, USA

Chandrika Kamath
Center for Applied Scientific Computing, Lawrence Livermore National Laboratory, 7000 East Avenue, Livermore, California 94551, USA

Helen Conover
Sara Graves
Information Technology and Systems Center, University of Alabama, 301 Sparkman Drive, Huntsville, Alabama 35899, USA

ABSTRACT

Data mining is presented in this paper as an example of geoinformatics, to illustrate how a collaborative effort between data-mining experts and geoscientists is essential for analyzing large volumes of data to extract knowledge. There is a wide variety of mining algorithms and tools available for both predictive and descriptive data mining that have been applied on a variety of geoscience problems. This paper provides a high-level overview of the data-mining process, several mining algorithms and tools, and their application to the geosciences. It also describes the geoscience data characteristics that make mining these data a challenge, and details some of the common data preparation operations that are required to make the geoscience data ready for mining. Three data-mining applications are presented in detail in this paper to demonstrate the usefulness of data mining to solve specific geoscience problems. Issues relevant in determining a successful data-mining application in geoscience are presented as lessons learned. Finally, the paper looks at challenges ahead by focusing on two relevant research areas for data mining and geosciences: mining streaming data in real time and exploiting the emerging cyberinfrastructure for mining.

Keywords: data mining, geospatial data, mining algorithms, cyberinfrastructure, mining applications.

1. INTRODUCTION

There is a growing recognition that Earth functions as a complex system and that existing information science infrastructure and science practices are inadequate to address the many complex science problems. To solve these complex problems requires new approaches to analyzing, modeling, and developing science databases. Geoinformatics encompasses diverse collaborative efforts between computer science and the geosciences to solve these complex scientific questions (Allison et al., 2002). Geoinformatics is thus a field where geoscientists and computer scientists join forces to leverage emerging information technologies

Ramachandran, R., Rushing, J., Li, X., Kamath, C., Conover, H., and Graves, S., 2006, Bird's-eye view of data mining in geosciences, *in* Sinha, A.K., ed., Geoinformatics: Data to Knowledge: Geological Society of America Special Paper 397, p. 235–247, doi: 10.1130/2006.2397(19). For permission to copy, contact editing@geosociety.org.

for integrated data management and analysis. Data mining of geospatial data is one such information technology, focusing on knowledge discovery and management.

Before providing an introductory overview of data mining, terms such as data, information, knowledge, and knowledge management and discovery need to be defined. The definitions used in this paper are based on those given by Singpurwalla (2003). "Data" is defined as something that is observable and therefore measurable; it is therefore factual. Knowledge is a statement about a hypothesis, and science is organized knowledge, i.e., a collection of one or more hypotheses in some logical manner. The testing of a hypothesis is often based on data, suitably organized. Information is a measure of uncertainty about a hypothesis, and the role of the data is to change the amount of information. Data by themselves cannot formulate a hypothesis; rather they support or negate a hypothesis. Finally, knowledge management and discovery is a systematic use of data to test a hypothesis, and data mining is one aspect of knowledge management and discovery.

The objective of this paper is to provide a high-level overview of data mining and its application in geosciences *as an excellent example of geoinformatics*. This paper will provide an introduction to the data-mining process, describe the geoscience data characteristics that make it challenging for mining, detail some of the common data preparation operations that are required to make the geoscience data ready for mining, and cover some popular mining algorithms. Selected mining applications in geosciences will be described to illustrate the many ways data mining can be useful. A review of some of the mining toolkits available to geoscientists will also be covered. Keys to a successful data-mining application will be covered in section 8. Finally, the paper will look ahead at two challenges relevant to data mining within the context of a comprehensive knowledge management and discovery system to solve complex geoscience problems. Additional information on spatial data mining can be found in Shekhar and Chawla (2002), Fotheringham et al. (2002), Miller and Han (2001), and Koperski et al. (1996).

2. DEFINITION OF DATA MINING

Data mining is the process of finding the valuable information nuggets hidden among the tons of data debris. Data mining, or knowledge discovery in databases (KDD), has been defined as "the non-trivial process of identifying valid, novel, potentially useful, and ultimately understandable patterns in data" (Fayyad et al., 1996). The volume of raw data being collected and stored by different earth science archives today defies even the partial manual examination by scientists. Data mining addresses this problem by bringing to bear techniques and algorithms from a multitude of disciplines to analyze and explore these massive data sets. Data mining is thus a multidisciplinary domain borrowing techniques from the fields of machine learning, statistics, databases, expert systems, pattern recognition, and data visualization. Data mining is an iterative, multistep process. The first step consists of defining the mining objective and selecting the appropriate data. The second step consists of data preparation and involves cleaning and preprocessing the data. The data preparation step can include feature extraction and data reduction operations. The data preparation is typically the most labor-intensive step within the mining process. The data preparation step is followed by the selection of the mining algorithm(s) and the application of the mining algorithm(s) on the data. The final step involves the analysis and interpretation of the mining result or the discovered patterns.

It is important to elucidate the subtle difference between scientific data analysis and data mining. Scientific data analysis is generally a hypothesis-driven process as compared to data mining, which is a data-driven process to discover knowledge. Data mining is thus a supplementary tool that can be used by scientists in conjunction with their current methods of data analysis. In fact, the usefulness of exploiting existing science analysis algorithms within the data-mining process will be discussed later.

3. CHARACTERISTICS OF GEOSCIENCE DATA

Geoscience data present unique challenges for mining because of their unique characteristics. Traditional statistical techniques work on the assumption of statistical independence and run into trouble with spatial data where the neighboring data points may be interrelated (spatial and temporal autocorrelation). Geoscience data are heterogeneous by nature, and this heterogeneity encompasses spatial variations, states of processing, and formats of storage. One of the first challenges in mining such data sets is dealing with the numerous data formats, which can range from simple file formats such as binary and ASCII to more complicated self-describing formats such as Hierarchical Data Format (HDF), Network Common Data Format (netCDF), and others. With the emergence of XML (Bray et al., 1998; Byron and Malhotra, 2001), there have been several interchange technologies designed and developed, such as the Earth Science Markup Language (ESML) (Ramachandran et al., 2004a), eXtensible Data Format (XDF) (Shaya et al., 2001), Geography Markup Language (GML) (Cuthbert and Lake, 2000; Cox et al., 2001), and eXtensible Scientific Interchange Language (XSIL) (Williams, 2000), to address this data format heterogeneity problem.

In addition to heterogeneous data formats, most geoscience data are generally stored as multidimensional arrays or as raster data. One of the biggest challenges with raster data is feature extraction, where the feature is a geoscience phenomenon such as a cyclone. Using individual data points as features is problematic and does not necessarily give the best results. A typical approach used with raster data is to extract features either by using some image-processing algorithms such as texture analysis or by using heuristic algorithms developed by geoscientists, storing the attributes of the feature as vector data and then applying mining algorithms on the vector data. Furthermore, features in geoscience data have an inherent spatial and temporal component that has to be properly represented in these vector data.

In addition to the raster data type, some of the geoscience data could also be a time series, representing a sequence of measurements. Time series data of multiple measurements can be difficult to analyze, especially if the measurements are made at a different time scales. Some geoscience data sets can have a very specific data structure where the structure itself embeds a semantic meaning. For example, Lightning Image Sensor (LIS) data (Christian et al., 2000) are structured as a tree hierarchy with each level a specialized instance of the previous level. Converting such formats into flat files for mining tends to result in sparse data and loss of information. Geoscience data used in mining applications can also be in various states of data processing, starting from uncalibrated raw data, to higher-level products that are the end result of science data-processing algorithms. These various processing levels require the data-mining expert to fully understand the data to utilize them successfully in a mining application. It is also rare in geoscientific mining applications that a single data set is used. Typically, multiple data sets from different sensors and models are used. This creates a challenging data fusion problem where the feature of interest has to be extracted from these multiple data sets and fused as a single vector data set for mining. It is uncommon that the data from different sensors are measured at the same resolution or rate. Consequently, the different spatial and temporal scales have to be taken into account while fusing these multiple data sets. Furthermore, these data sets typically have to be transformed from different native coordinate systems to a projected coordinate system.

4. DATA PREPARATION

Data preparation is the most labor-intensive step in the data-mining process. As stated in the previous section, the first step in geoscience data preparation involves framing the science problem into a mining problem, i.e., extracting the feature of interest from geoscience data and representing the feature as vector data with a set of attributes. For example, tropical cyclones could be a feature, and the set of attributes describing this feature can be size, wind speed, location, etc. Even after these vector data have been created, there may be several reasons to perform additional data preparation operations. The data may be incomplete, lacking the right attributes or attributes values. The data may also have noise due to errors and outliers, and may have inconsistencies. The data preparation step can involve many different operations. Data cleaning involves the removal of noise and correcting the inconsistencies in the data. This may entail filling in missing values, smoothing noisy data, and identifying and removing outliers. Data preparation can also involve data integration, and in geosciences this would primarily focus on combining data from multiple sources, such as different sensors observing the same phenomenon. In some cases, the data have to be normalized, and hence the values are transformed. In some mining applications, one has to remove the trend from the data, or "detrend" the data so that the results are not biased by the trend terms. Therefore, detrending becomes a necessary data preparation step before meaningful results can be obtained from mining. An example of detrending in geosciences is the removal of seasonality while mining climatology data sets (Tan et al., 2001). Data reduction is also an important operation in the data preparation step. It can involve subsetting the data over a spatial or a temporal region of interest, or aggregating the data to make them more manageable. An example of aggregation is the use of principal component analysis (PCA) on hyper- or multispectral satellite data to reduce the number of relevant spectral channels for analysis. Data reduction can also require dimensionality reduction by finding the best attributes for mining and discarding the irrelevant ones. The use of attribute selection algorithms in a geoscience example to find the optimal attributes for classification is described in section 6.2.

5. OVERVIEW OF POPULAR DATA-MINING ALGORITHMS

Data-mining algorithms can be categorized based on their purpose as predictive or descriptive mining algorithms (Dunham, 2003). Predictive algorithms make predictions about values of data using a set of known results, or "truth data." Classification, regression, time series analysis, etc., are examples of predictive mining algorithms. Descriptive mining algorithms are exploratory in nature and identify patterns or relationships in the data. Examples of descriptive algorithms include clustering, summarization, association rules, etc. Some of the commonly used predictive and descriptive mining algorithms used in geosciences will be described in this section. A more comprehensive description of these algorithms and their advantages and limitations can be found in Dunham (2003), Han and Kamber (2001), Duda et al. (2000), Witten and Frank (2000), and Mitchell (1997).

5.1. Predictive Algorithms

Data mining using classification algorithms is a multistep process that requires the user to identify the problem, i.e., select the target feature, create a truth data set consisting of a set of attributes that adequately captures the characteristics of the target feature, partition the truth set into training and testing sets, select an appropriate classification algorithm, train the classifier using the training set, test the classifier using the test data set, evaluate the classifier performance, and finally apply the classifier on data to detect the feature. Some of the classifiers commonly used in data mining are described below.

a. Decision Trees

Decision trees classify instances of data by sorting them as a tree from a root node to leaf nodes representing the classification of the instance. Each node in the tree specifies a test of one of the attributes, and the branches from the node correspond to the possible values for the attribute. Two commonly used decision trees are ID3 (Quinlan, 1986) and C4.5 (Quinlan, 1993). ID3 employs a top-down, greedy search through the space of possible decision trees. The tree is constructed by statistically evaluating all the attributes. A statistical property called "information gain" is used in the evaluation process. Information gain is a measure of how well a given attribute separates the training sample according to the target classification. The best attribute is selected and is used as the test at the root node of the tree. A descendant node is created for each of the possible values of the root node. The entire process is repeated using the training samples. C4.5 builds a decision tree from training samples, allowing overfitting. It then converts the tree into a set of rules where each rule is a path from the node to the tree, and pruning is applied to improve the accuracy by removing preconditions in a rule. Decision trees have been used in several geoscience applications. Friedl et al. (1999) examined methods to maximize classification accuracies using decision trees to map land cover from multitemporal AVHRR (Advanced Very High Resolution Radiometer) imagery at continental and global scales. Simard et al. (2000) used a decision tree classifier to extract thematic information on the tropical vegetation cover over the coastal region of Gabon from SAR image mosaics. Decision tree classifiers were used on multitemporal satellite data from the Enhanced Thematic Mapper Plus instrument to map 11 land cover types at the Delaware Water Gap National Recreation Area near Milford, Pennsylvania (de Colstoun et al., 2002). Theron et al. (2002) applied data-mining techniques such as decision trees to the paleo-oceanography problem of reconstructing the surface-water dynamics in the South China Sea during the last 130,000 years.

b. Bayes Classifier

Bayes classifiers utilize a probabilistic approach for classifying data. These classifiers are based on the assumption that attributes in the training examples are governed by probability distributions and that optimal decisions can be made by reasoning about these probabilities together. Bayes theorem is utilized to calculate the conditional probability of a data point belonging to a particular class. Bayesian classifiers allow prior knowledge (initial probability) to be combined with observed data to determine the conditional probability. When these initial probabilities are unavailable, assumptions have to be made regarding the underlying distribution. The Naïve Bayes classifier is based on the assumption that the attribute values are conditionally independent, and this assumption simplifies the conditional probability calculation by simply taking the products of independent probabilities for each attribute. Stark et al. (1995) used a Bayes classifier along with a neural network classifier to generate vegetation maps consisting of five different vegetation classes from Landsat and Spot images.

c. Neural Networks

An artificial neural network has been developed as a mathematical model of human cognition based on the assumptions that processing occurs at many simple elements called "neurons" and the signals are passed between the neurons over connection links (Fausett, 1994). These connection links have associated weights that impact the transmitted signal. Each neuron applies an activation function to the net input signals to determine the output. The back-propagation neural network (BPNN) is the most commonly used artificial neural network algorithm for data mining. The BPNN algorithm learns weights for a multilayer network, given a network with a fixed set of neurons and interconnections (Rumelhart et al., 1986a; 1986b). It attempts to minimize the squared error between the network output values and the target values of these outputs by finding the optimal weight values. The basic architecture of a BPNN consists of an input layer, an output layer, and one or more hidden layers of neurons. The training of the BPNN involves three phases: the feed forward of the input training pattern, the calculation and the back-propagation of the associated error, and the adjustments of the weights. During the feedforward phase, each input unit receives an input signal and broadcasts this signal to each of the hidden units in the next layer. Each hidden unit then computes its activation and sends its signal to the hidden unit in the next layer. Once the signal reaches the output unit, the output unit compares its activation signal to the target value to determine the associated error. This error is then propagated backward to distribute it to all the units in the previous layers. Once all the errors at each layer have been determined, the weights of all the layers are adjusted simultaneously. This process is repeated until the error falls below a specified threshold or the number of allowable iterations is exceeded. Once the network has been trained, the computed weights are used in a feedforward manner to classify data. Neural networks have been applied to several geoscience problems. A BPNN was used to predict corn yield from topographic features, vegetations, and texture indices (Serele et al., 2000). A neural network was applied on Landsat Thematic Mapper (TM) data for vegetation classification for the Cleveland National Forest (Carpenter et al., 1997).

d. Instance-Based Classifiers

Instance-based learning classifiers store expert-classified truth or training data, and when a new sample is presented to these classifiers, a set of similar instances are retrieved from memory to classify this new sample. K-Nearest Neighbor is an example of an instance-based learning classifier. The K-Nearest Neighbor algorithm assumes that all instances correspond to points in n-dimensional space and uses a distance measure to quantify similarity. When a classification is to be made for a new data sample, its distance to all the samples in the training set are determined, and the k samples closest to it are considered. The new data sample is then assigned to the class that contains the most data samples in the set of the k closest data samples. The K-Nearest Neighbor algorithm calculates the distances between instances based on all the attributes.

e. Support Vector Machines

Different classifiers use different methods to define the discriminate functions to minimize the classified error. If the discriminate function is too complex, it can overfit the training examples. However, if one limits the complexity of the discriminate function, one gets higher errors. The best function is the one that explains the data well and has small error. The "curse of dimensionality" from statistics states that the estimation problem increases as the dimensions of the feature space increase. However, statistical learning theory counters that learning can be made simpler if one uses a function of low complexity for discrimination. Support vector machines (SVMs) use this principle and project training samples into higher dimensions to use a simple discriminate function (Müller et al., 2001). SVMs attempt to avoid both overfitting and underfitting the training data and do this by minimizing an estimate of the overall risk. There have been several applications of SVMs for geoscience problems. SVMs have been used to classify crops in hyperspectral imagery (Camps-Valls et al., 2004), and for the retrieval of oceanic chlorophyll concentration from marine reflectance (Zhan et al., 2003). Azimi-Sadjadi et al. (2000) used SVMs for cloud classification on GOES 8 (Geostationary Operational Environmental Satellite) imagery data using the infrared (IR) channel only. Hermes et al. (1999) applied SVMs for land usage classification.

5.2. Descriptive Algorithms

Two commonly used descriptive mining operations are clustering and association rules. Clustering algorithms can be classified as hierarchical or partitional. Clustering algorithms are similar to classifiers in that they group the data. However, unlike classifiers, clustering algorithms do not require truth data for training. Clustering algorithms discover the groupings in the data based on similarities in the data attributes. These algorithms are often useful as an exploratory tool where the general characteristics of the data can be observed by the cluster results. These algorithms can identify outlier points that vary substantially from the rest of the data. They can also be used to summarize the data. The principle behind clustering algorithms is simple. The clustering algorithms use a similarity measure to check if a sample in the data is similar to the other samples in the data set. The samples that are similar are grouped together. Different similarity measures, such as Euclidean distance, cosine coefficient, correlation coefficient, and Canberra metric, can be used by the clustering algorithms. Using these measures, each data sample is compared to the others in the data. Unfortunately, the algorithm complexity of performing these comparisons is of the order N^2. To make clustering tractable for large data sets, some of the clustering algorithms use mathematical tricks such as randomly selecting cluster points, looking for furthest values, etc., to reduce the algorithmic complexity. Different clustering algorithms have been used in various geoscientific mining applications. Smyth et al. (1999) used a mixture model approach to identify the cluster structure in atmospheric pressure data. Gaffney et al. (2001) applied mixtures of regression models to general circulation model data to cluster extratropical cyclone trajectories.

a. Single-Link Clustering

The single-link clustering algorithm is also referred to as "nearest neighbor clustering" and is a hierarchical clustering algorithm where nested sets of clusters are created. It is based on the idea of connected vertices in a graph. Two clusters are merged if the minimum distance between two points or vertices is less than or equal to the threshold distance being considered. The result of hierarchical algorithms such as single-link is a tree data structure called a "dendogram." These algorithms are suited for applications that require generation of taxonomies.

b. K-Means

K-Means is an example of a partitional clustering algorithm where the desired number of clusters to partition the data is specified to the algorithm. Initial cluster means are randomly selected, and patterns are assigned to these closest cluster means. New cluster means are then calculated, and the patterns are reassigned to these new cluster means. This process is iterated until the convergence criterion is met. This criterion can be based upon squared error or the number of iterations. K-Means is one of the most widely used clustering algorithms and has been applied on several geoscience data sets. Funk et al. (2001) used K-means clustering on hyperspectral data to detect faint gas plume signatures. K-Means has been used for forest change mapping using Landsat data (Leckie et al., 2002), and in the spectral classification process to map canopy closure and leaf area index in Airborne Spectrographic Imager data (Hu et al., 1999). K-Means clustering classification has also been used to separate the pixels of images with differing scattering mechanisms to classify polar sea ice in the NASA scatterometer data (Remund et al., 1998). K-Means using a correlation coefficient as a similarity measure has also been used to divide ecosystem model output into disjoint regions in an automated fashion to identify regions with similar short-term and long-term climate characteristics (Steinbach et al., 2001, 2003).

c. DBSCAN

Density-based spatial clustering of applications with noise (DBSCAN) is a density-based clustering algorithm that groups clusters of arbitrary shape in spatial data sets with noise. DBSCAN generates clusters by checking a specified neighborhood around each data point, and if the data points within this neighborhood exceed the minimum points specified, then a new cluster is formed. This process is iterated until no new points can be added to any clusters. DBSCAN can generate clusters of arbitrary shape and can handle outliers.

d. Association Rules

Association rules have traditionally been used in business applications to find relationships present among attributes in large data sets. Association rules are designed to capture information about items that are frequently associated with each other. Let $A_1, A_2, \ldots, A_{m+n}$ be distinct items. Assume that when items $A_1 \ldots A_m$ occur together, items $A_{m+1} \ldots A_{m+n}$ also occur in the same instance. This information is represented in the form of an association rule as given below:

In the above equation, the "^" symbol represents the logical AND operator, and the "⇒" symbol represents implication.

Each association rule has a support value and a confidence value associated with it. The support for a rule is defined as the percentage of data that contain the preconditions of the rule. The confidence value for a rule is the minimum percentage of instances in which both the preconditions and conclusions of a rule are true. Rule-mining algorithms identify all of the association rules for a particular data set that meet a set of user-defined criteria. Since the total number of items of interest is large, it is not practical to enumerate all combinations of these items. Users typically specify minimum values for support and confidence of rules, and often specify constraints on the rule format. The association rules algorithm has been used in geosciences to find spatial association patterns between ecological regions of interest and climate variables using ecological model output (Tan et al., 2001).

6. APPLICATIONS OF MINING IN GEOSCIENCES

This section describes three examples of mining geoscience data to illustrate the many ways in which data mining can be useful. The first application is an example of the direct use of a classification algorithm on imagery data to detect and identify regions of interest. The second example is also a classification problem but uses a heuristic algorithm based on domain knowledge to extract features and a classifier to differentiate between the true features and false signatures. The third application is a current work in progress that uses a clustering algorithm to find patterns in the output of an ensemble of numerical models.

6.1. Cumulus Cloud Detection in GOES Satellite Imagery

Clouds are important modulators of Earth's radiative energy budget, and changes in the climate system can alter their spatial and temporal distribution. Changes in cloud distribution, in turn, impact the radiative energy budget, which then feeds back to the climate system. General circulation models (GCMs) are widely used to investigate climate change scenarios. Adequate representation of radiative interactions of clouds in GCMs is crucial for such studies. For cloud systems of large areal extent, the plane-parallel radiative transfer assumption is reasonable. However, for boundary layer cumulus cloud fields, in both long- and short-wave portions of the electromagnetic spectrum, three-dimensional radiative transfer is important. A better parameterization is required for cumulus cloud fields for the models. There are other areas that would benefit from improved understanding of boundary layer cumulus cloud field structure. Since shallow convection is strongly influenced by surface characteristics, areas of significant land use change are expected to show indications of local climate modification (Nair et al., 2000).

Satellite data are ideal to study the boundary layer cumulus cloud field characteristics. Analysis of large volumes of data is necessary to develop good statistical descriptions of cloud field structure. This requires the use of automated methods for identifying cumulus cloud fields in satellite imagery. In this case study, data-mining techniques were applied to detect boundary layer cumulus cloud fields in Geostationary Operational Environmental Satellite (GOES) images. In the data preparation step, the background is subtracted from the visible image. The visible background is determined by taking the minimum pixel value for the previous week for each time of day. This background pixel level is the intensity reflected from the ground in the absence of any cloud cover. The visible images are also quantized to improve the discriminatory power of statistical texture measures. A general rule of thumb, which was used in this case, is that the images should be quantized to the smallest number of levels where the texture boundaries are still clearly visible in the quantized images. For the visible GOES images and the two classes of interest, three levels were used. Next, the texture features were computed from the quantized images. Two texture measures, association rule (Rushing et al., 2001) and gray level run length (GLRL) (Chu et al., 1990) were applied to compute texture features. In the final step, a Naïve Bayes classifier was used to produce a classification for each pixel in the source image, based on its texture features, to generate a segmented image consisting of class label. An example GOES image, expert-labeled truth, and the segmented images produced by the mining process can be seen in Figure 1. The trained classifier was then used by the scientist to process through large volumes of images to develop good statistical descriptions of cloud field structure.

6.2. Mesocyclone Detection in NEXRAD Radar Data

Mesocyclones are rotating updraft/downdraft structures inside severe thunderstorms. Detection of mesocyclones is important for severe weather forecasts because over 90% of mesocyclones are accompanied by severe weather such as tornadoes or large hail. Experienced radar analysts can visually identify mesocyclone signatures from the Weather Surveillance Radar–1988 Doppler (WSR-88D) data. However, this manual identification process is tedious and time-consuming and can be overwhelming during a severe weather outbreak. Consequently, an automated mesocyclone detection algorithm (MDA) is necessary for assisting meteorologists in severe weather forecasts. The National Severe Storm Laboratory (NSSL) MDA is one such algorithm.

Li et al. (2004) conducted an investigative case study to detect and classify mesocyclone signatures from WSR-88D data using data-mining techniques. This research consisted of two components. The first component was to develop a feature detection algorithm to detect mesocyclone signatures from WSR-88D radar data. In addition to accuracy, the major requirement for the algorithm was computational speed; the algorithm had to be fast enough to efficiently and expeditiously process large volumes of archived WSR-88D data. The second component of this research was to identify "true" mesocyclone signatures using the detected algorithm output. A truth data set was available for classification purposes.

The MDA developed in this study combines domain knowledge and image-processing techniques and contains three steps:

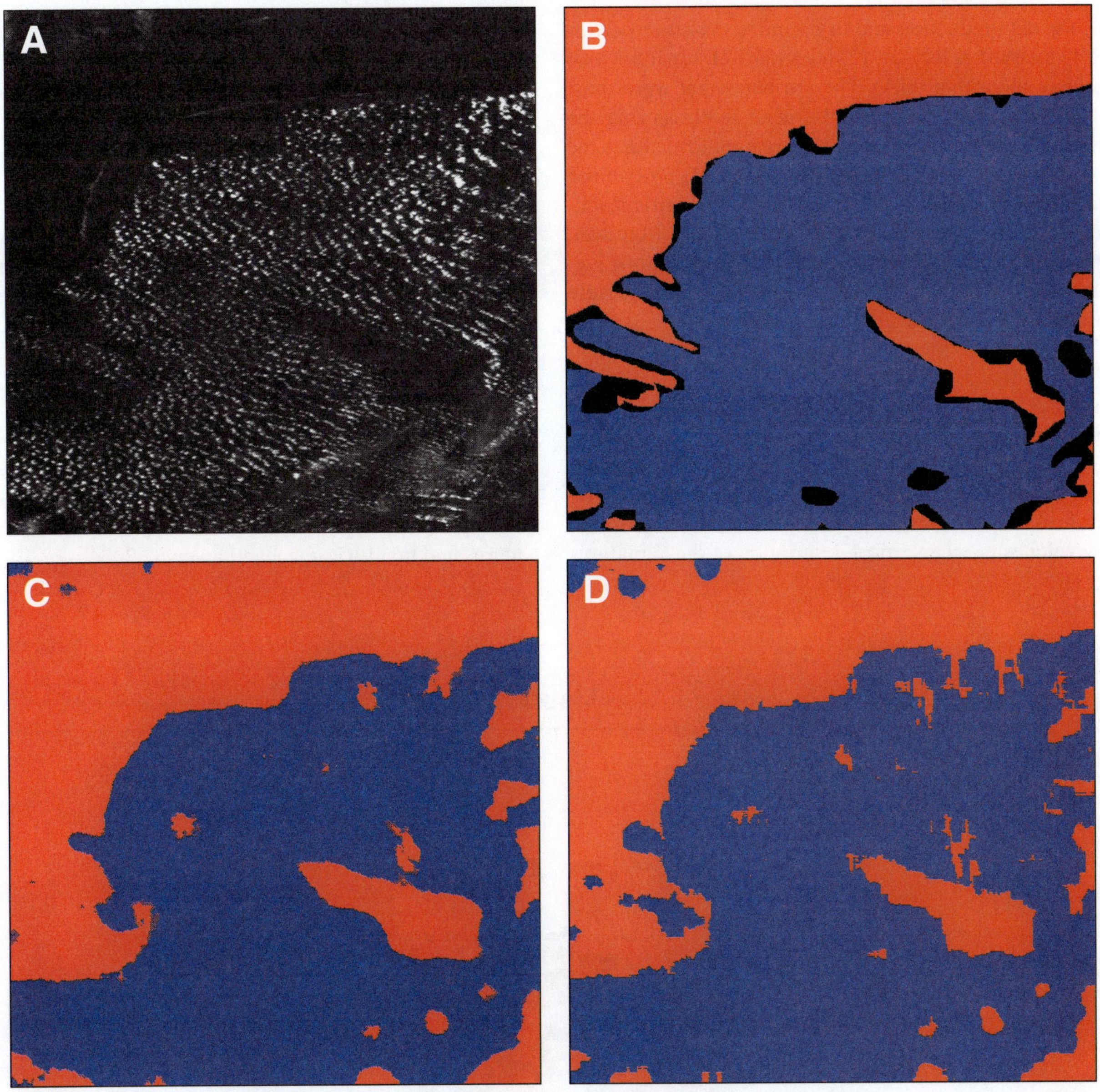

Figure 1. Results from cumulus cloud detection in GOES satellite imagery. A: Original image. B: Expert labeled truth image. C: Cloud mask using association rules. D. Cloud mask using gray level run length.

(1) identifying one-dimensional shear segments, (2) identifying two-dimensional shear regions in a scan sweep using a region-growing technique, and (3) collocating shear regions across elevation scans into three-dimensional mesocyclone signatures. A feature vector is calculated for each two-dimensional shear region with attributes including central azimuth angle, central range, central height, diameter, shear, maximum gate-to-gate velocity difference (GTGVD), rotational velocity difference (V_{rot}), and mesocyclone strength index. For a three-dimensional mesocyclone signature, a number of attributes such as base height, top height, core base height, and core top heights are calculated in addition to the mean values of the corresponding two-dimensional features. Each identified three-dimensional mesocyclone signature was labeled as either "true" or "false" based on the information from the truth data set. The three-dimensional signature data set was then randomly split into equal halves to be used as the training and testing data in the classification process. The resulting algorithm was evaluated by comparing it with the NSSL MDA. The comparisons showed that the results from the mining MDA were similar to those from the NSSL MDA in both the number of detected mesocyclones and the distributions of feature attributes (see Fig. 2).

Two classifiers, Naïve Bayes and support vector machines, were trained to weed out the false mesocyclone signatures. These false signatures may be due to noise or other artifacts. Three attribute selection methods were applied to the two classifiers for data reduction and to optimize classifier performances: (1) forward selection, (2) backward elimination, and (3) the genetic algorithm (GA). The attributes that were most frequently selected were considered crucial in classifying the mesocyclone signatures. Four metrics were used to evaluate the performance of each classifier on mesocyclones: (1) probability of detection (POD), (2) false alarm rate (FAR), (3) critical success index (CSI), and (4) Heidke skill score (HSS) (Marzban, 1998). Due to the heavy skewness of detected mesocyclone signatures toward the "false" class, the CSI is the proper measure in evaluating classifier performance.

The results from the classification show that although the CSI values for different attribute selection methods are very close, the attributes optimized for the classifiers varied from method to method. Also, the total number of attributes selected varied for different classifiers. The occurrence of an attribute in the attribute selection process was calculated. The results confirm that the most important attribute to classify the true mesocyclones are the vertical description of the signatures (base height, depth and height of maximum GTGVD) and the strength of the rotation (strength index and maximum V_{rot}). The secondary attributes include the maximum diameter and maximum reflectivity features. This is consistent with the domain knowledge.

6.3. Mining Ensemble Model Data

Ensemble modeling is a simulation suite using different initial conditions, models, and boundary conditions to generate a large volume of simulation output. Traditionally used in weather prediction, these simulation outputs are composited to generate probability forecasts. In an ongoing study (Wilhelmson et al., 2003), model simulations with different initial conditions were performed to investigate different cell interactions favorable for subsequent tornado formation. These simulations created a large parameter space of thunderstorm cell interaction and the consequent storm development. This search space is being mined for patterns and trends (Ramachandran et al., 2004b).

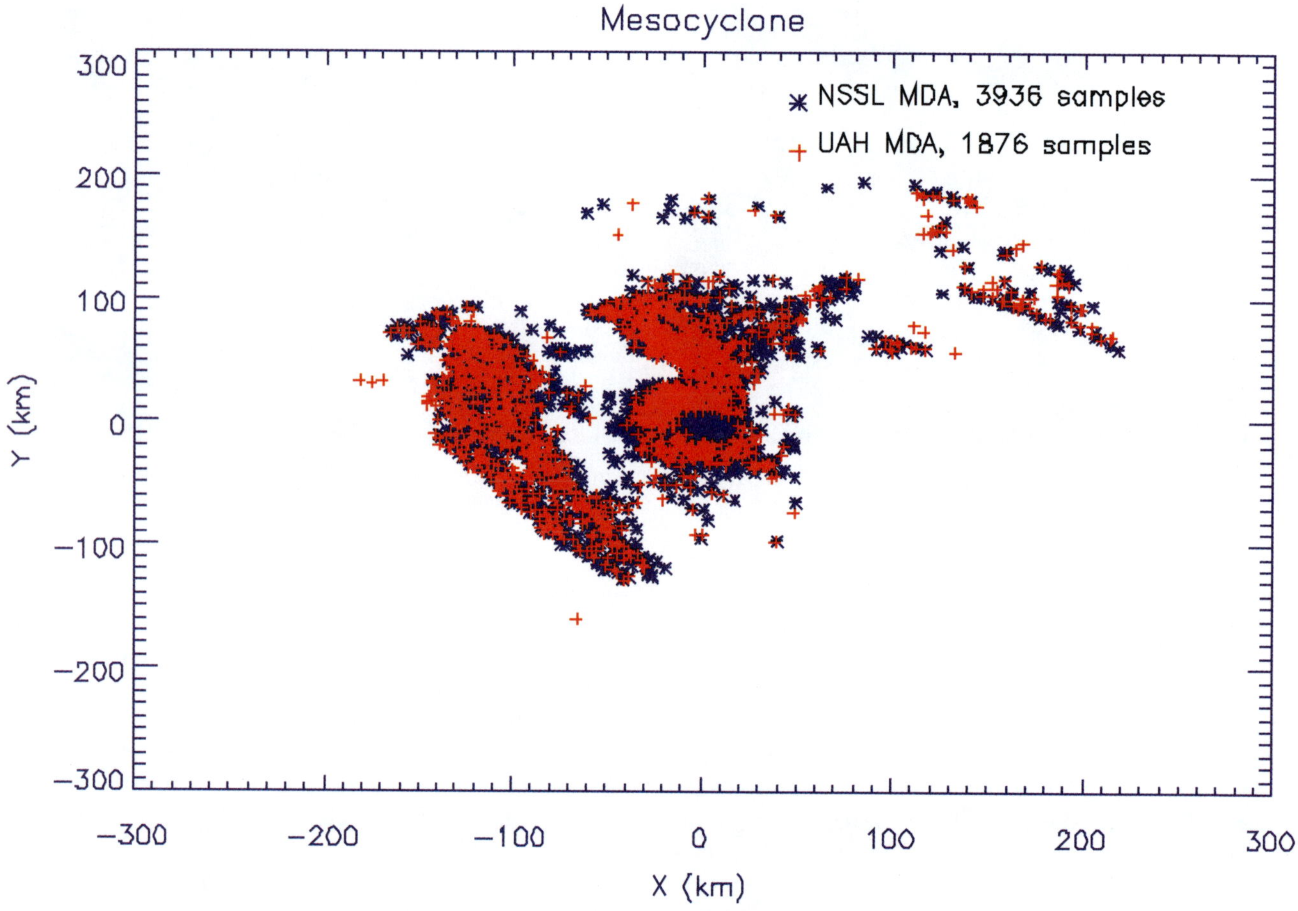

Figure 2. Spatial location distribution of the identified mesocyclone signatures detected from NSSL MDA and UAH MDA for one of the cases. NSSL—National Severe Storm Laboratory; UAH—University of Alabama, Huntsville; MDA—mesocyclone detection algorithm.

A total of 232 runs of the Weather Research and Forecasting (WRF) Model (http://www.wrf-model.org/index.php) with different initial conditions were performed. For each run, the first storm cell was placed at the center, and the location of a second storm cell was varied. The model runs were made with a 1 km by 1 km horizontal resolution and variable vertical resolution. Data were saved every ten time steps, i.e., once every minute of simulation time, up to 2.5 h. Each simulation run created 151 output files, with the size of each ~27 MB. A heuristic mesocyclone detection algorithm for the model data was developed based on domain knowledge. This detection algorithm was used to search the model output for each run to extract mesocyclones and generate a set of attributes for the detected mesocyclones. In addition to the detection algorithm, a mesocyclone tracking algorithm was developed to keep track of the temporal continuity of the detected mesocyclones. Examples of this detection and tracking algorithm can be seen in Figure 3 for two specific simulation runs. Using this detection and tracking algorithm, each model run was summarized by attributes such as the total number of mesocyclones detected for a model run, number of mesocyclones with duration greater than 30 min, number of mesocyclones with depth greater than 4 km, number of mesocyclones with vorticity greater than a specified threshold, and the number of mesocyclones with base lower than 2 km. The K-Means clustering algorithm was run on these summary data. The clustering algorithm discovered some interesting patterns. Certain runs grouped together in a particular cluster showed a tendency to have a larger number of mesocyclones with greater depth, high vorticity values, and lower base values but shorter durations. The runs belonging to this cluster tended to have the second storm cell location farther away from the first storm cell. The runs with the second storm cell closer to the first storm cell exhibited mesocyclones with longer durations. A new attribute was added to the model run summaries to indicate those mesocyclones that exhibited long durations, a certain minimum depth, a specified or higher vorticity, and a low base. The purpose of this attribute was to single out the significant mesocyclones in each of the runs. This attribute produced interesting cluster locations of runs that produced significant mesocyclones. These cluster locations corroborated the specific spatial orientation of the cells hypothesized in modeling literature.

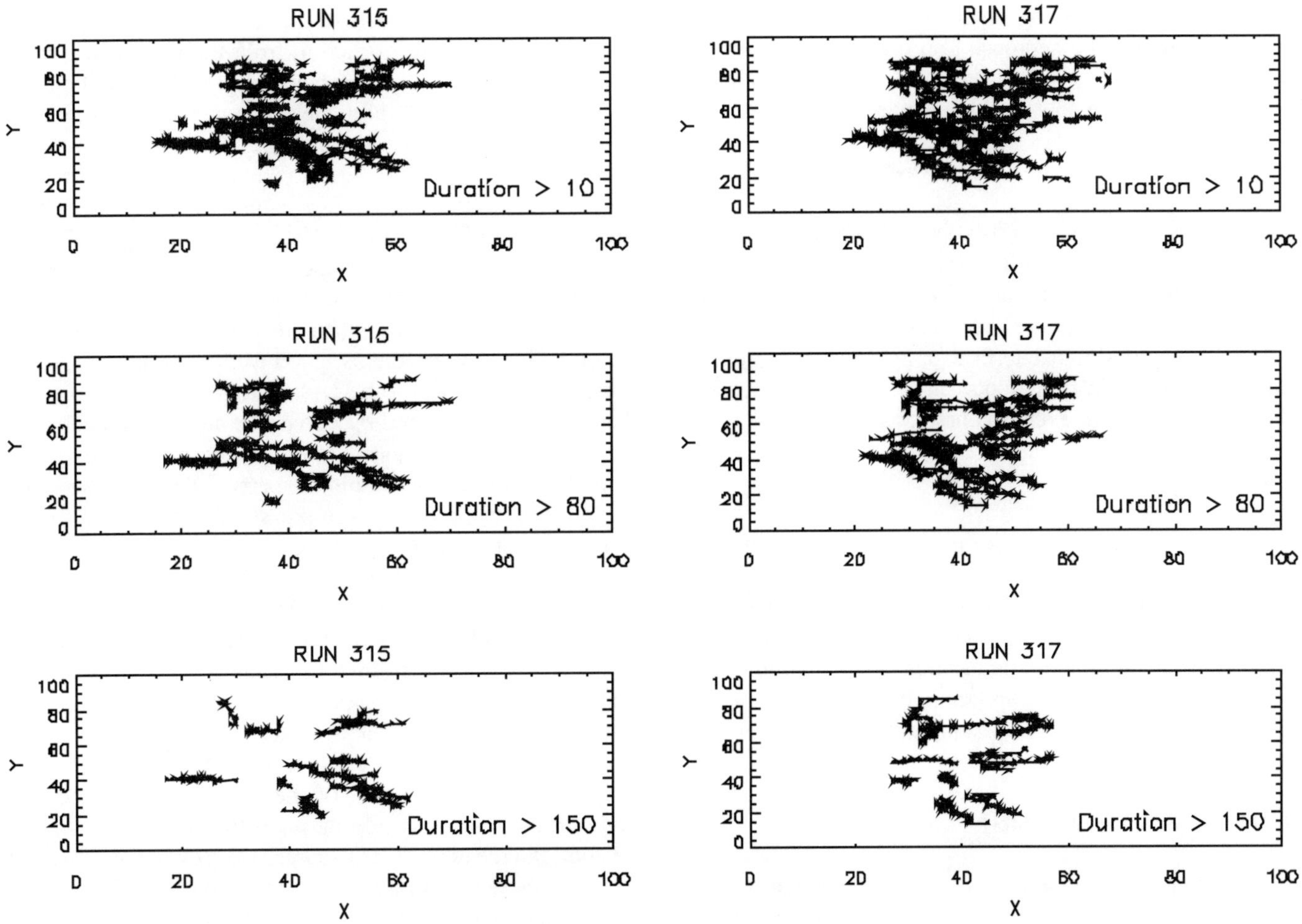

Figure 3. Plot summarizing the results from the mesocyclone tracking algorithm for two specific simulation runs.

7. OVERVIEW OF DATA-MINING TOOLS

There are several commercial and freeware data-mining toolkits, systems, and stand-alone algorithms that are available. Four toolkits that are fairly comprehensive and are freely available to the research community are described in this section.

7.1. Algorithm Development and Mining (ADaM) Toolkit

ADaM (http://datamining.itsc.uah.edu/adam/) is a data-mining toolkit designed for use with scientific and image data. It includes classification, clustering, feature selection, model validation, data cleaning, image processing, optimization, and association rule–mining capabilities. The system consists of a set of individual algorithms or components that can be used together to perform complex tasks. Components are packaged as stand-alone executables and as Python modules for easy scripting. The toolkit is implemented in C++.

Packaging the components in these forms facilitates the rapid development of scientific data-mining applications, while allowing for efficient implementation of performance-critical components. Care has been taken to ensure that the components of the system are as independent as possible, in order to make it possible to use subsets of the components as appropriate for given applications. The approach also makes it possible to use third-party components with ADaM. All components are available for Microsoft Windows and Linux platforms.

7.2. Waikato Environment for Knowledge Analysis (WEKA)

WEKA (http://www.cs.waikato.ac.nz/ml/weka/) is an open-source data-mining toolkit written in Java. It has tools for classification and regression, clustering, association rule mining, feature selection, data preprocessing, and visualization. The tools can be driven by a graphical user interface, or by direct calls from Java code. New machine learning tools can be added to the toolkit by deriving classes from the relevant base class.

7.3. Orange

Orange (http://magix.fri.uni-lj.si/orange/default.asp) is a component-based data-mining toolkit written in C++. Orange includes Python wrappers so that its components can be scripted together, and also graphical user interface objects for visual programming. Like all the toolkits mentioned here, Orange can be extended by adding new components. It includes classification and prediction, feature selection, visualization, and model validation capabilities.

7.4. Data to Knowledge (D2K)

D2K (http://alg.ncsa.uiuc.edu/tools/docs/d2k/manual/index.html) is a data-mining toolkit written in Java. It has a workflow-oriented visual programming environment in which components can be linked together in flexible ways. Data compatibility in the control flow is enforced by type checking using the class hierarchy. Each mining component exposes a set of input and output ports, and these ports are connected together to achieve the desired workflow. D2K includes classification and prediction, clustering, feature selection, model validation, data cleaning, and association rule–mining capabilities.

8. LESSONS LEARNED

The key issue that determines the success of a geoscientific data-mining application is the close collaborative relationship between the geoscientists and the mining experts. Just as analysis on high-dimensional data sets devoid of data mining has limitations, data mining devoid of geoscience domain expertise will not discover the valuable information nugget. The collaborative relationship between the two experts has to be symbiotic in order for the mining application to be successful. The geoscientist can assist the data-mining expert in several aspects of the mining process, such as providing domain knowledge to understand the problem and the data set being used, providing the truth data, helping during data preparation, helping phrase the science problem as a mining problem, and finally in the interpretation and analysis of the results. The data preparation step involves feature extraction from the geoscience data and the selection of the right attributes to represent and characterize this feature. In addition, these features have inherent spatial and temporal attributes that must be represented correctly. The geoscientist can provide heuristic algorithms based on domain knowledge for extracting these features, allowing the data-mining process to leverage the existing scientific analysis process and research. In addition, the geoscientist can help select most or all the right attributes that describe the feature and gather enough instances of the attributes for the feature of interest. For classification problems, good truth data that are representative of the features to be classified are essential. These truth data have to be created by domain experts who label samples of particular classes based on their experience. Geoscientists can also help the data-mining experts understand the scope of the mining problem. For example, some of the features of interest can exist as minority classes, which exist infrequently in the data. In such applications, standard methods of accuracy, where equal weight is given to identification of the feature class and the other classes, cannot be used for evaluation. Using such accuracy measures can give misleading accuracy scores since the samples of other classes in the data outnumber the feature class. With the help of the geoscientist, the data-mining expert can account for such issues and use other accuracy measures for a better representation of the performance of the mining algorithms. Finally, a mining algorithm will always generate results, and the geoscientist's expertise is essential in evaluating whether these results are meaningful and useful. The geoscientist can provide interpretations for the mining results and discovered patterns that a data-mining expert cannot provide. For example, a data-mining expert may not be aware of an existing hypothesis that a discovered pattern might corroborate.

9. CHALLENGES AHEAD

As information technology evolves, it directly affects data mining and creates new challenges open for active research. Two such areas of active research that are relevant to geosciences and specifically to geoinformatics are discussed here.

9.1. Mining Data Streams

Traditionally, data mining is a postarchiving operation that is performed after all the data have been collected. In such operations, the set of training samples is fixed, and a classifier is trained and evaluated using those samples. However, due to improvements in network capabilities, more and more sensors are being connected online and can provide users access to streams of data in real time. Mining streaming data is a challenge because it is not generally possible to store all of the samples produced during the lifetime of the stream. This means that the mining algorithms must be modified to make only a single pass over the data. Other complications can arise as well. For some phenomena of interest, the samples may be very unevenly distributed with respect to both the class and the range of observed sample values. Data streams with unevenly distributed samples can occur in a wide variety of applications. For example, severe weather phenomena may occur much more frequently during certain seasons than in others, and the observations relevant to detecting them can vary seasonally as well. Monitoring of streaming video presents similar challenges, such as changing lighting conditions and the influences of weather. Furthermore, the underlying process or phenomenon of interest may change over time. The mining algorithms must continuously adapt to these changes, generally referred to as "concept drift" (Widmer and Kubat, 1996). These challenges have led to the development of specialized mining algorithms for streaming data, including algorithms for classification and prediction, association rule mining, and clustering.

Supervised pattern classifiers learn phenomena of interest from labeled samples. The learning algorithms employed by the classifiers must be modified to accommodate data streams. One way to accomplish this is to learn incrementally from the samples. Incremental learning involves creating a model from an initial set of samples and then subsequently updating that model as new samples become available. There are now incremental learning algorithms for a variety of types of classifiers, including decision trees (Domingos and Hulten, 2000; Gehrke et al., 1999), neural networks (Chakraborty and Pal, 2003; Zhou et al., 2000; Parekh et al., 2000), and support vector machines (Rüping, 2001; Syed et al., 1999). Another approach is to use an ensemble of classifiers (Street and Kim, 2001). In this approach, the data stream is partitioned into groups of samples, each of which is used to train an independent classifier. The ensemble then classifies new patterns based on the responses of the individual classifiers, often by using a weighted voting approach. Ensemble classifiers have been shown to be effective in dealing with concept drift (Wang et al., 2003).

Clustering algorithms for streaming data are distinct from standard clustering algorithms in that they must make only a single pass over the data. The goal of the clustering process, grouping together patterns with similar characteristics, remains unchanged. This goal is achieved by using a clustering quality measure such as the K-median objective. The goal of the k-median objective is to identify k clusters such that the maximum radius of a cluster is minimized. Efficient one-pass clustering algorithms that meet this objective include a multilevel divide-and-conquer approach to that makes only a single pass through the data (O'Callaghan et al., 2002; Charikar et al., 2003). Data streams changing with time pose additional challenges, and algorithms have been developed to allow dynamic clustering at different time scales (Aggarwal et al., 2003).

Specialized stream-mining algorithms have also been incorporated into data-mining toolkits. One example is the Mining Alarming Incidents in Data Streams (MAIDS) extension to the D2K toolkit (Cai et al., 2004). MAIDS provides a common framework for stream data mining and has a wide variety of tools including data cube construction and query, frequent pattern mining, classification, clustering, and visualization.

9.2. Cyberinfrastructure Implications

Mining large complex problems in geosciences requires access to distributed data and to large computational resources. The Grid has been conceptualized as a new class of infrastructure built upon the World Wide Web (Foster, 2002) and addresses the broader notion of a cyberinfrastructure with "a coordinated resource sharing and problem solving approach between dynamic multi-institutional virtual organizations" (Foster et al., 2001). There are several Grid efforts under way in different domains of geosciences that aim to build this cyberinfrastructure required to accelerate scientific research. Geosciences Network (GEON) (http://www.geongrid.org/) in solid Earth, the Earth System Grid (ESG) in climate studies (https://www.earthsystemgrid.org/), the Network for Earthquake Engineering Simulation Grid (NEESgrid) in earthquake engineering (http://it.nees.org) and Linked Environments for Atmospheric Discovery (LEAD) in mesoscale meteorology (http://lead.ou.edu/) are some examples of Grid effort in geosciences. These large infrastructure projects are building the foundations to provide scientists direct access to distributed data and computational resources for data analysis and mining. The challenge lies in adapting the existing mining tools to exploit these resources, such as providing coarse-grain and fine-grain parallelization to make mining algorithms more computationally efficient. One of the goals of the LEAD project is to provide data mining in the context of Grid services, in order to fully exploit not only the readily available distributed data but also the computational resources. The ADaM toolkit is being redesigned as Grid service components for this project and will be available as a shared community resource for researchers and students.

Access to data and the previous analysis results in the cyberinfrastructure will facilitate multilevel mining and the generation of knowledge bases for geoscience. With the new cyberinfrastructure, one will have access not only to data but also to results generated by other mining and data analysis. This will allow researchers to incrementally build knowledge using these existing results as data for their own analysis and mining. This infrastructure providing seamless access to organized data and results will allow multilevel mining, or leveraging results from other mining and analysis, thus creating a knowledge management and discovery system capable of solving complex geoscience problems.

REFERENCES CITED

Aggarwal, C., Han, J., Wang, J., and Yu, P.S., 2003, A Framework for Clustering Evolving Data Streams, *in* International Conference on Very Large Data Bases (VLDB'03), Berlin, Germany.

Allison, M.L., Snyder, W.S., and Walker, J.D., 2002, Geoinformatics: A Nascent Revolution in the Earth Sciences: GSA Today, v. 12, no. 7, p. 17, doi: 10.1130/1052-5173(2002)012<0017:GANRIT>2.0.CO;2.

Azimi-Sadjadi, M.R., and Zekavat, S.A., 2000, Cloud classification using support vector machines, *in* Proceedings, Geoscience and Remote Sensing Symposium, IEEE International, Honolulu, Hawaii, USA, v. 2, p. 669–671.

Bray, T., Paoli, J., and Sperberg-McQueen, C.M., 1998, XML 1.0 Specification: W3C.

Byron, P., and Malhotra, A., 2001, XML Schema Part 2: Datatypes: World Wide Web Consortium.

Cai, Y.D., Clutter, D., Pape, G., Han, J., Welge, M., and Auvil, L., 2004, MAIDS: Mining Alarming Incidents from Data Streams, *in* 23rd ACM SIGMOD (International Conference on Management of Data), Paris, France.

Camps-Valls, G., Gomez-Chova, L., Calpe-Maravilla, J., Martin-Guerrero, J.D., Soria-Olivas, E., Alonso-Chorda, L., and Moreno, J., 2004, Robust Support Vector Method for Hyperspectral Data Classification and Knowledge Discovery: IEEE Transactions on Geoscience and Remote Sensing, v. 42, no. 7, p. 1530–1542, doi: 10.1109/TGRS.2004.827262.

Carpenter, G.A., Gjaja, M.N., Gopal, S., and Woodcock, C.E., 1997, ART neural networks for remote sensing: Vegetation classification from Landsat TM and terrain data: IEEE Transactions on Geoscience and Remote Sensing, v. 35, no. 2, p. 308–325, doi: 10.1109/36.563271.

Chakraborty, D., and Pal, N.R., 2003, A Novel Training Scheme for Multilayered Perceptrons to Realize Proper Generalization and Incremental Learning: IEEE Transactions on Neural Networks, v. 14, no. 1, p. 1–14, doi: 10.1109/TNN.2002.806953.

Charikar, M., O'Callaghan, L., and Panigrahy, R., 2003, Better streaming algorithms for clustering problems, *in* 35th ACM Symposium on Theory of Computing (STOC), San Diego, California, USA.

Christian, H., Blakeslee, R.J., Goodman, S., and Mach, D., 2000, Algorithm Theoretical Basis Document (ATBD) for the Lightning Imager Sensor (LIS): Huntsville, Alabama, Marshall Space Flight Center, NASA, p. 53.

Chu, A., Sehgal, C.M., and Greenleaf, J.F., 1990, Use of gray value distribution of run lengths for texture analysis: Pattern Recognition Letters, v. 11, no. 6, p. 415–420, doi: 10.1016/0167-8655(90)90112-F.

Cox, S., Cuthbert, A., Lake, R., and Martell, R., 2001, Geography Markup Language (GML) 2.0 OpenGIS Implementation Specification: OGC Document Number 01-029.

Cuthbert, A., and Lake, R., 2000, Geography Markup Language (GML) v1.0: OGC Document Number 00-029.

de Colstoun, E.B., Story, M.H., Thompson, C., Smith, T.G., and Irons, J.R., 2002, Vegetation mapping using multi-temporal ETM+ data and a decision tree classifier, *in* Proceedings, Geoscience and Remote Sensing Symposium, Toronto, Ontario, Canada, 2002 IEEE International, v. 5, p. 2890–2892.

Domingos, P., and Hulten, G., 2000, Mining High-Speed Data Streams, *in* Sixth ACM SIGKDD International Conference on Knowledge Discovery and Data Mining, Boston, Massachusetts, p. 71–80.

Duda, R.O., Hart, P.E., and Stork, D.G., 2000, Pattern Classification: Wiley-Interscience.

Dunham, M.H., 2003, Data Mining: Introduction and Advanced Topics: Pearson Education.

Fausett, L., 1994, Fundamentals of Neural Networks: Englewood Cliffs, New Jersey, Prentice Hall.

Fayyad, U.M., Piatetsky-Shapiro, G., and Smyth, P., 1996, From Data Mining to Knowledge Discovery: An Overview, *in* Fayyad, U.M., et al., eds., Advances in Knowledge Discovery and Data Mining: MIT Press.

Foster, I., 2002, The Grid: A New Infrastructure for 21st Century Science: Physics Today, v. 55.

Foster, I., Kesselman, C., and Tuecke, S., 2001, The Anatomy of the Grid: Enabling Scalable Virtual Organizations: International Journal of Supercomputer Applications, v. 15, no. 3.

Fotheringham, A.S., Brunsdon, C., and Charlton, M., 2002, Geographically Weighted Regression: The Analysis of Spatially Varying Relationships: John Wiley & Sons.

Friedl, M.A., Brodley, C.E., and Strahler, A.H., 1999, Maximizing land cover classification accuracies produced by decision trees at continental to global scales: IEEE Transactions on Geoscience and Remote Sensing, v. 37, no. 2, p. 969–977, doi: 10.1109/36.752215.

Funk, C.C., Theiler, J., Roberts, D.A., and Borel, C.C., 2001, Clustering to improve matched filter detection of weak gas plumes in hyperspectral thermal imagery: IEEE Transactions on Geoscience and Remote Sensing, v. 39, no. 7, p. 1410–1420, doi: 10.1109/36.934073.

Gaffney, S., Robertson, A., and Smyth, P., 2001, Clustering of Extra-Tropical Cyclone Trajectories using Mixtures of Regression Model, *in* Klamath, C., ed., Fourth Workshop on Mining Scientific Datasets in conjunction with 7th ACM SIGKDD International Conference on Knowledge Discovery and Data Mining, San Francisco, California, USA.

Gehrke, J., Ganti, V., and Ramakrishnana, R., 1999, BOAT—Optimistic Decision Tree Construction, *in* Proceedings of the ACM SIGMOD Conference on Management of Data, Philadelphia, Pennsylvania, USA.

Han, J., and Kamber, M., 2001, Data Mining: Concepts and Techniques: Morgan Kaufmann.

Hermes, L., Frieauff, D., Puzicha, J., and Buhmann, J.M., 1999, Support vector machines for land usage classification in Landsat TM imagery, *in* Proceedings, Geoscience and Remote Sensing Symposium, IGARSS '99, IEEE 1999 International, Hamburg, Germany, v. 1, p. 348–350.

Hu, B., Innanen, K., Miller, J., and Hollinger, A., 1999, Retrieval of leaf area index and canopy closure from high-spatial resolution CASI data over the BOREAS flux tower sites, *in* Proceedings, Geoscience and Remote Sensing Symposium, IGARSS '99, IEEE 1999 International, Hamburg, Germany, v. 2, p. 1084–1086.

Koperski, K., Adhikary, J., and Han, J., 1996, Spatial Data Mining: Progress and Challenges: Workshop on Research Issues on Data Mining and Knowledge Discovery (DMKD), Montreal, Canada.

Leckie, D.G., Walsworth, N., Dechka, J., and Wulder, M., 2002, An investigation of two date unsupervised classification in the context of a national program for Landsat based forest change mapping, *in* Proceedings, Geoscience and Remote Sensing Symposium, IGARSS '02, 2002 IEEE International, v. 3, p. 1307–1311.

Li, X., Ramachandran, R., Rushing, J., Graves, S., Kelleher, K., Lakshmivarahan, S., Douglas, K., and Jason, L., 2004, Mining NEXRAD Radar Data: An Investigative Study, Interactive Information and Processing Systems (IIPS), American Meteorological Society Annual Meeting, Seattle, Washington, USA.

Marzban, C., 1998, Scalar Measures of Performance in Rare-Event Situations: Weather and Forecasting, v. 13, p. 483–493.

Miller, H.J., and Han, J., 2001, Geographic Data Mining and Knowledge Discovery (1st edition): CRC Press.

Mitchell, T.M., 1997, Machine Learning: McGraw-Hill.

Müller, K.-R., Mika, S., Rätsch, G., Tsuda, K., and Schölkopf, B., 2001, An Introduction to Kernel-Based Learning Algorithms: IEEE Transactions on Neural Networks, v. 12, no. 2, p. 181–201, doi: 10.1109/72.914517.

Nair, U.S., Welch, R.M., Lawton, R.O., and Pielke, R.A., Sr., 2000, Impact of land use on Costa Rican regional climate, *in* Proceedings, Geoscience and Remote Sensing Symposium, IGARSS 2000, IEEE 2000 International, Honolulu, Hawaii, USA, v. 5, p. 1993–1995.

O'Callaghan, L., Mishra, N., Meyerson, A., Guha, S., and Motwani, R., 2002, Streaming-data algorithms for high quality clustering, *in* 8th International Conference on Data Engineering, p. 685–694.

Parekh, R., Yang, J., and Honavar, V., 2000, Constructive neural network learning algorithms for pattern classification: IEEE Transactions on Neural Networks, v. 11, no. 2, p. 436–451, doi: 10.1109/72.839013.

Quinlan, J.R., 1986, Induction of decision trees: Machine Learning, v. 1, p. 81–106.

Quinlan, J.R., 1993, C4.5: Programs for Machine Learning: San Mateo, California, Morgan Kaufmann.

Ramachandran, R., Graves, S., Conover, H., and Moe, K., 2004a, Earth Science Markup Language (ESML): A solution for scientific data-application interoperability problem: Computers and Geosciences, v. 30, no. 1, p. 117–124, doi: 10.1016/j.cageo.2003.08.010.

Ramachandran, R., Li, X., Movva, S., Rushing, J., Tanner, S., Graves, S., Jewett, B., and Wilhelmson, R., 2004b, Mining Ensemble Model Data, *in* Seventh Workshop on Mining Scientific and Engineering Datasets, Society for Industrial and Applied Mathematics International Conference on Data Mining, Lake Buena Vista, Florida.

Remund, Q.P., Long, D.G., and Drinkwater, M.R., 1998, Polar sea-ice classification using enhanced resolution NSCAT data, *in* Proceedings, Geoscience and Remote Sensing Symposium, IGARSS '98, 1998 IEEE International, Seattle, Washington, USA, v. 4, p. 1976–1978.

Rumelhart, D.E., Hinton, G.E., and Williams, R.J., 1986a, Learning representations by back propagating error: Nature, v. 323, p. 533–536.

Rumelhart, D.E., Hinton, G.E., and Williams, R.J., 1986b, Learning internal representations by error propagation: Parallel Distributed Processing, v. 1, p. 318–362.

Rüping, S., 2001, Incremental learning with support vector machines, *in* First IEEE International Conference on Data Mining (ICDM), San Jose, California, USA, p. 641–642.

Rushing, J.A., Ranganath, H.S., Hinke, T., and Graves, S.J., 2001, Using association rules as texture features: IEEE Transactions on Pattern Analysis and Machine Intelligence, v. 23, no. 8, p. 845–858, doi: 10.1109/34.946988.

Serele, C.Z., Gwyn, Q.H.J., Boisvert, J.B., Pattey, E., McLaughlin, N., and Daoust, G., 2000, Corn yield prediction with artificial neural network trained using airborne remote sensing and topographic data, *in* Proceedings, Geoscience and Remote Sensing Symposium, IGARSS 2000, IEEE 2000 International, Honolulu, Hawaii, v. 1, p. 384–386.

Shaya, E., Thomas, B., and Cheung, C., 2001, Specifics on a XML Data Format for Scientific Data, ASP Conference Series, Volume 238: Astronomical Data Analysis Software and Systems X: San Francisco, California, USA, p. 217.

Shekhar, S., and Chawla, S., 2002, Spatial Databases: A Tour: Prentice Hall.

Simard, M., Saatchi, S.S., and De Grandi, G., 2000, The use of decision tree and multiscale texture for classification of JERS-1 SAR data over tropical forest: IEEE Transactions on Geoscience and Remote Sensing, v. 38, no. 5, p. 2310–2321, doi: 10.1109/36.868888.

Singpurwalla, N.D., 2003, Knowledge management and information superiority (a taxonomy): Journal of Statistical Planning and Inference, v. 115, no. 2, p. 361–364, doi: 10.1016/S0378-3758(02)00164-7.

Smyth, P., Ide, K., and Ghil, M., 1999, Multiple regimes in northern hemisphere heights fields via mixture model clustering: Journal of Atmospheric Sciences, v. 56, p. 3704–3723, doi: 10.1175/1520-0469(1999)056<3704:MRINHH>2.0.CO;2.

Stark, E., Eltoft, T., and Braathen, B., 1995, Performance of vegetation classification methods using synthetic multispectral satellite data, *in* Proceedings, Geoscience and Remote Sensing Symposium, IGARSS '95, "Quantitative Remote Sensing for Science and Applications," IEEE International, Florence, Italy, v. 2, p. 1276–1278.

Steinbach, M., Klooster, S., Tan, P.-N., Potter, C., Kumar, V., and Torregrosa, A., 2001, Clustering earth science data: Goals, issues and results, *in* Kamath, C., ed., Fourth Workshop on Mining Scientific Datasets in conjunction with 7th ACM SIGKDD International Conference on Knowledge Discovery and Data Mining, San Francisco, California, USA.

Steinbach, M., Tan, P.-N., Kumar, V., Klooster, S., and Potter, C., 2003, Discovery of Climate Indices Using Clustering, *in* KDD-2003, The Ninth ACM SIGKDD International Conference on Knowledge Discovery and Data Mining, Washington, D.C., USA

Street, W.N., and Kim, Y.S., 2001, A streaming ensemble algorithm (SEA) for large-scale classification, *in* Seventh ACM SIGKDD International Conference on Knowledge Discovery and Data Mining, San Francisco, California, USA, p. 377–382.

Syed, N.A., Liu, H., and Sung, K.K., 1999, Handling concept drifts in incremental learning with support vector machines, *in* Fifth ACM SIGKDD International Conference on Knowledge Discovery and Data Mining, San Diego, California, USA p. 317–321.

Tan, P.-N., Steinbach, M., Kumar, V., Potter, C., Klooster, S., and Torregrosa, A., 2001, Finding Spatio-Temporal Patterns in Earth Science Data, *in* KDD Workshop on Temporal Data Mining.

Theron, R., Flores, J.A., Sierro, F.J., Pelejero, C., Grimalt, J., and Vaquero, M., 2002, Using data mining and visualization techniques for the reconstruction of ocean paleodynamics, *in* Proceedings, Geoscience and Remote Sensing Symposium, IGARSS '02, 2002 IEEE International, Toronto, Ontario, Canada, v. 4, p. 2382–2384.

Wang, H., Fan, W., Yu, P.S., and Han, J., 2003, Mining concept-drifting data streams using ensemble classifiers, *in* Ninth ACM SIGKDD International Conference on Knowledge Discovery and Data Mining, p. 226–235.

Widmer, G., and Kubat, M., 1996, Learning in the presence of concept drift: Machine Learning, v. 23, p. 69–101.

Wilhelmson, R., Droegemeier, K., Graves, S., Ramamurthy, M., Haidvogel, D., Jewett, B., Alameda, J., and Gannon, D., 2003, Modeling environment for atmospheric discovery, *in* Annual American Meteorological Society Meeting, Long Beach, California, USA.

Williams, R., 2000, XSIL: Java/XML for Scientific Data.

Witten, I.H., and Frank, E., 2000, Data Mining: Practical Machine Learning Tools and Techniques with Java Implementation: Morgan Kaufmann Publishers.

Zhan, H., Shi, P., and Chen, C., 2003, Retrieval of oceanic chlorophyll concentration using support vector machines: IEEE Transactions on Geoscience and Remote Sensing, v. 41, no. 12, p. 2947–2951, doi: 10.1109/TGRS.2003.819870.

Zhou, Z.H., Chen, S., and Chen, Z., 2000, FANNC: A fast adaptive neural network classifier: Knowledge and Information Systems, v. 2, no. 1, p. 115–129, doi: 10.1007/s101150050006.

Manuscript Accepted by the Society 18 June 2005

Printed in the USA

Geological Society of America
Special Paper 397
2006

A new GIS-driven geophysical database for the southwestern United States

Jesse B. Yoburn*
Matthew J. Fouch
J Ramón Arrowsmith
Department of Geological Sciences, Arizona State University, Tempe, Arizona 85287-1404, USA

G. Randy Keller
Department of Geological Sciences, University of Texas, El Paso, Texas, USA

ABSTRACT

Geologic and geophysical investigations of the southwestern United States have provided a broad range of noteworthy regional geological and geophysical data sets. These data are integral to understanding the rich and complex geologic history of the region, which has experienced major periods of subduction, uplift, and extension over the past 2 Ga. However, these data are not generally archived in a consistent, digital format that enables interdisciplinary analysis. This paper presents the results of the development of a geographic information system (GIS) of significant geophysical data sets for the Colorado Plateau and southern Basin and Range as part of the SWGEONET project (http://www.geoinformaticsnetwork.org/swgeonet/). The data system contains the following ten compilations: seismic S-wave tomography, receiver functions, shear wave splitting, stress fields, heat flow, thermal conductivity, heat production, complete Bouguer gravity anomaly, geodetic velocity field, and digital elevation models. We utilize the attributes and key findings from these individual data sets in this new infrastructure to better constrain relationships between data sets. To demonstrate the utility of examining these varying data in a common framework, we present four case studies that use GIS tools to join these data sets and provide new views of regional dynamics. These case studies provide estimates of regional variations in crust and/or lithospheric thickness, correlations between independent seismic data sets near the boundaries of the upper mantle transition zone, the potential extent of coupling between the crust and mantle, and possible sources of crustal deformational forces across the region.

Keywords: geospatial analysis, geophysics, crust, mantle, GIS.

*Now at: ExxonMobil Exploration Company, Houston, Texas, USA.

Yoburn, J.B., Fouch, M.J., Arrowsmith, J R., and Keller, G.R., 2006, A new GIS-driven geophysical database for the southwestern United States, *in* Sinha, A.K., ed., Geoinformatics: Data to Knowledge: Geological Society of America Special Paper 397, p. 249–267, doi: 10.1130/2006.2397(20). For permission to copy, contact editing@geosociety.org.

INTRODUCTION

For the past decade, there has been a critical need within the geological and geophysical communities for a next-generation data infrastructure to develop a new level of constraints on the geodynamics and tectonic evolution of major geologic systems on Earth. The Colorado Plateau–Basin and Range region in Arizona and New Mexico is an example of such a major tectonic regime (Fig. 1). A broad range of geophysical and geological data sets have been used to characterize the spatial and temporal aspects of the regional tectonic history. Many of these data sets are publicly available, enabling the development and testing of geodynamical models for the area in a comprehensive fashion. Of primary importance is the ability to perform combined analyses of significantly different geological and geophysical data sets in a common framework without the requirement of a deep understanding of high-level data analysis tools or software. Additionally, such a data system must be accessible to the scientific community, complete with the ability to download raw data, metadata, and supplemental information.

The goal of the SWGEONET initiative (http://www.geoinformaticsnetwork.org/swgeonet/) is to provide the necessary infrastructure for an online, interactive data analysis and mapping system with associated data and metadata stores. SWGEONET is a collaborative effort between Arizona State University (ASU) and the University of Texas, El Paso (UTEP) to establish an online geological and earth science data repository. The backbone of this resource utilizes geographic information system (GIS) tools, which enables data sources to be easily collected and organized in a centralized location and common geospatial framework, while serving the needs of a wide range of users. We utilize ArcGIS, a software package created by Environmental Systems Research Institute (ESRI—http://www.esri.com/), due to its ease of use, wide acceptance in the geologic community, and integrated Internet publishing capabilities. Although GIS tools have not traditionally been used for deep (i.e., subcrustal) Earth geophysical applications, nearly every geophysical data set can be incorporated into the GIS environment. Although many geophysical data sets exist that cover the broad Colorado Plateau–Basin and Range region, most are stored at different institutions, in a variety of data formats, and with a range of map projections. A primary goal in

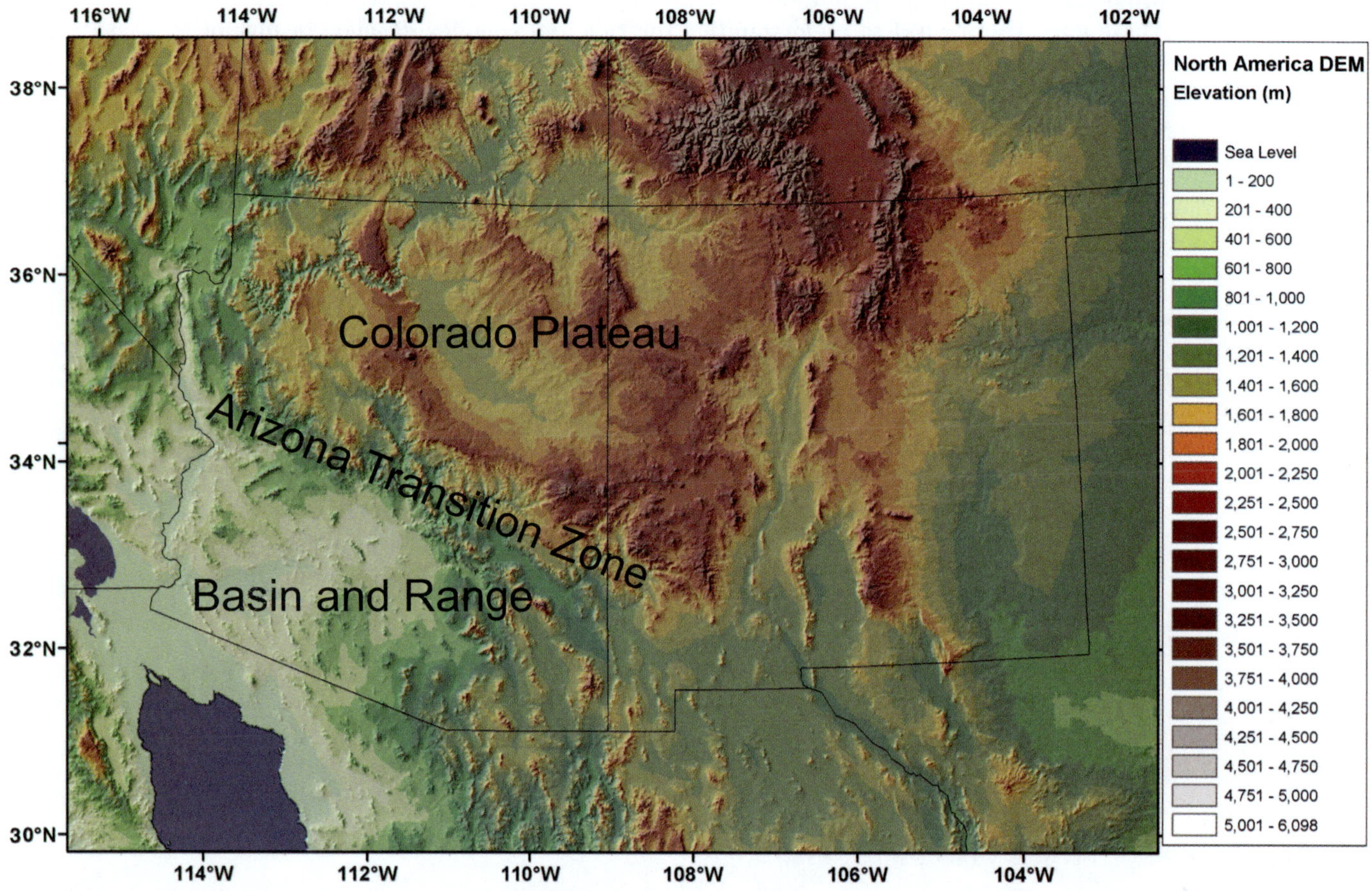

Figure 1. Study area showing the Colorado Plateau, Basin and Range, and Arizona transition zone. This figure was created by combining a semitransparent high-resolution digital elevation model (DEM) with a shaded relief map in ArcGIS.

this component of the SWGEONET project has been to centralize these data, establish an organizational scheme and framework for displaying these data, and develop innovative methods of data analysis using GIS as the primary analysis tool.

The centralized collection of these geophysical data, combined with the case studies described in this paper, can provide an improved synthesis of the tectonic history of the Colorado Plateau and Basin and Range regions by enabling a new approach to data integration. In this paper, we review the overall tectonic setting of the southern Colorado Plateau–Basin and Range region with special emphasis on processes causing regional surface deformation and structural anomalies in the subsurface. We then describe the geophysical data sets incorporated into the SWGEONET database and how each data set relates to aspects of the Colorado Plateau–Basin and Range region. Finally, we utilize GIS tools to combine pairs of these data sets and examine several case studies investigating regional tectonic processes. The results of these analyses provide new constraints on the relationship between a subset of geological and geophysical data sets previously unable to be compared in such a direct way.

TECTONIC SETTING OF THE COLORADO PLATEAU–BASIN AND RANGE REGION

The Colorado Plateau and southern Basin and Range province have experienced large-scale continental assembly, deformation, and eventual stability over the past ca. 2 Ga, and regional-scale uplift and deformation over the past ca. 100 Ma. The Laramide orogeny (80–40 Ma) and subhorizontal subduction of the Farallon plate led to magmatism, deformation, formation of high topography, and crustal thickening in the Colorado Plateau–Basin and Range region (Coney and Harms, 1984; Dickinson and Snyder, 1978). NW-oriented shear in the San Andreas system and a decrease in lateral support caused an eventual collapse of the topographically high region (Dickinson, 1989). E-W extension in the Basin and Range at ca. 15 Ma (Ward, 1991) and/or lithospheric delamination (e.g., Gao et al., 2004) caused further crustal thinning and extension along the Rio Grande rift in New Mexico. At ca. 8–10 Ma, crustal thinning waned (Eberly and Stanley, 1978) and the Colorado Plateau was uplifted, reversing the regional drainage from NE-directed (Laramide) to S-directed, and the Colorado River drainage basin was integrated from the southwestern Rocky Mountains to the Gulf of California.

A fundamental issue in western United States tectonics remains that the timing and duration of these regional events have not yet been resolved (Hendricks and Plescia, 1991). Key questions include the origin of support of the Colorado Plateau uplift, the state of isostatic compensation of the Colorado Plateau, and the extent of crust and mantle coupling during Basin and Range extension. Determining these relationships in a more integrated fashion can help us understand the larger picture of regional dynamics.

DATA SETS

Table 1 shows the range of data sets included in the SWGEONET geophysical database. These data sets were chosen for their degree of coverage of the study region, accessibility, and relevancy to the geophysical sciences. They can be divided into three main categories: point and continuous seismic data, gridded potential field data, and vector and miscellaneous data. In this section, we describe the basic characteristics of each of these data sets and refer the reader to both the references and our Web site (http://www.geoinformaticsnetwork.org/swgeonet/) for details. In raw form, the database contains nearly 500 MB of data and associated metadata.

Shear Wave Tomography

Seismic tomography provides information regarding the physical state of Earth's mantle and crust through imaging seismic wave speed variations. These variations generally are controlled by differences in temperature and/or composition—characteristics that affect material properties and thus the shear wave speed through a medium (e.g., Helffrich and Wood, 2001). High seismic wave speeds are commonly inferred to represent areas of

TABLE 1. DATASETS INCLUDED IN THE SWGEONET DATABASE

Dataset Name	Value	Units	# Data Points
Shear wave tomography	Velocity	km/sec	18,520 (20 depth slices)
Receiver functions	Depth	km	470
Shear wave splitting	Splitting time and fast polarization direction	sec and deg	58
Stress	Direction	deg	71
Heat flow	Areal flux	mW/m^2	589
Thermal conductivity	Conductivity	W/m·K	475
Heat production	Volumetric flux	mW/m^3	38
Gravity	Intensity	mgal	147,862
Geodetic velocity Field	Velocity and direction	mm/yr and deg from N	6
Aeromagnetic	Intensity	nT	1070877
Digital elevation model	Elevation	m	1 km resolution

cool, dense material, while low seismic wave speeds are generally inferred to represent areas of warm, less dense material (e.g., van der Lee, 2002).

The SWGEONET database includes the shear wave regional tomography model NA00 (van der Lee, 2002) calculated using the partitioned waveform inversion method (Nolet, 1990). Included in the database are 20 depth slices from NA00 approximately every 20–30 km transiting from the Moho to the base of the upper mantle. The model grid spacing is 0.25° in latitude and longitude, with 926 points covering Arizona and New Mexico within each model layer.

The NA00 shear wave model reveals several key features of mantle structure beneath the western United States (Fig. 2). First, a large, high velocity anomaly dips to the east from western North America. This N-S-trending feature ranges from ~3% faster than the surrounding rock at 50 km depth to ~1% faster at ~400 km depth and is interpreted to be the remnants of the subducted Farallon plate (van der Lee, 2002). Other seismic images of the region exhibit similar attributes. For example, Grand et al. (1997) produced independently derived high-resolution global tomography models for both P and S waves. Below the southern portion of the United States and extending into Mexico, they found a high velocity anomaly that dips to the east and extends to depths of nearly 2700 km that they also interpret to be the subducted Farallon slab.

Results from the LA RISTRA array, a linear array of 57 broadband seismometers extending from southeastern Utah to western Texas (Gok et al., 2003), provide higher-resolution estimates of regional seismic velocity structure. While not included in our current data set, their results indicate the presence of a

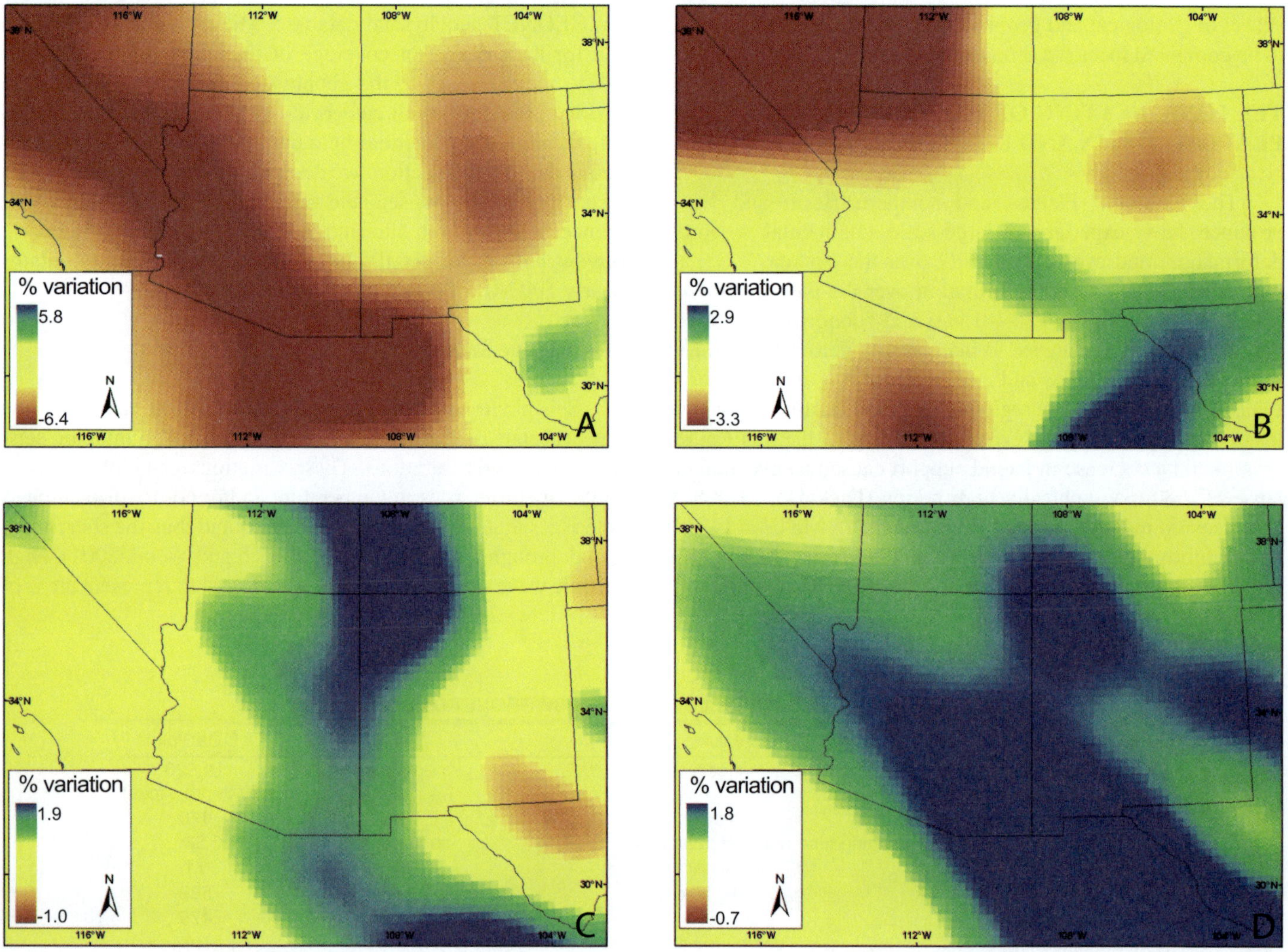

Figure 2. Shear wave velocity perturbations across the study area from the three-dimensional seismic velocity model NA00, shown in percent variation from the one-dimensional velocity model MC35 (van der Lee, 2002). Blue indicates higher than average seismic velocities, red indicates lower than average velocities at 150 km (A), 250 km (B), 419 km (C), and 669 km (D). Note extreme variation with depth for velocity anomalies across the region.

regional high shear wave velocity anomaly that dips toward the E to SE (Gao et al., 2004), though the geometry of LA RISTRA is such that the true dip of this anomaly cannot currently be resolved. Additionally, Gao et al. (2004) imaged a large low velocity anomaly centered beneath the Rio Grande rift. They infer that this low velocity anomaly may be the result of shallow mantle rising beneath the thinned lithosphere of the rift. West et al. (2004) used surface wave inversion of LA RISTRA waveforms to determine upper mantle seismic structure with good depth resolution of seismic wave speed variations. Their results confirm that the low seismic wave speed anomaly is shallow (~55–90 km) and likely does not originate from deep mantle upwelling. They also note a general thinning of the lithosphere from ~120–150 km beneath the Colorado Plateau to ~45–55 km beneath the Rio Grande rift.

Receiver Functions

Receiver functions provide a means of imaging first-order discontinuities in seismic velocities. In particular, they supply information regarding the depth to and sharpness of discontinuities, which are controlled by variations in temperature and/or composition (e.g., Ammon, 1991; Langston, 1977; Owens and Zandt, 1985). For most regional studies, the primary discontinuities of interest include the crust-mantle boundary (the Moho), a discontinuity usually considered to be due to compositional variations (e.g., Jarchow and Thompson, 1989). In addition, upper mantle discontinuities nominally found at 410 km and 660 km depth are generally ascribed to phase changes in olivine (Helffrich and Wood, 2001). Variations in the depth of these discontinuities can be due to thermal variations and/or compositional variations (e.g., Gilbert et al., 2003; Helffrich and Wood, 2001).

Receiver functions are commonly used to highlight converted seismic phases that originate from the existence of these discontinuities. The calculation of a receiver function typically involves deconvolving the vertical component of ground motion from the radial component (Ammon, 1991; Langston, 1977; Owens and Zandt, 1985). This operation effectively isolates the converted P-to-S phase ("Ps"), which arrives several seconds after the primary P arrival. The time difference between P and Ps is converted to a depth assuming an average one-dimensional reference seismic velocity model such as IASP91 (Kennett and Engdahl, 1991) or PREM (Dziewonski and Anderson, 1981). For regions with adequate coverage, receiver functions are typically stacked and spatially averaged to reduce noise and emphasize discontinuities (Dueker and Sheehan, 1997, 1998; Gilbert et al., 2003).

The receiver function data set in the SWGEONET database is derived from regional studies by Gilbert et al. (2003) and Gilbert and Sheehan (2004) that covered much of Nevada, Utah, and Colorado. While the data sets are located north of the overall SWGEONET coverage area, it still provides an important regional study that can be correlated with receiver function analyses currently under way along the southern Basin and Range–Colorado Plateau region from a new regional broadband seismometer array (Fouch et al., 2005).

These data sets provide depths relative to the Moho, 410 km, and 660 km discontinuities. They use a total of nearly 1000 receiver functions to determine the values shown in Figure 3. The data set includes grids with 190, 138, and 142 points for the Moho, 410 km, and 660 km discontinuities, respectively. Data points represent the location of common piercing points for statistical bins used to average the data. Each data point contains spatial reference information, average bin value of the depth to the discontinuity, and associated measurement error.

Variations in discontinuity structure are prevalent across the region (Fig. 3). They indicate a relatively thin crustal thickness of 25–30 km beneath the Basin and Range, thickening to 45–50 km beneath the Colorado Plateau region. Depth variations of the 410 km discontinuity are also evident (Fig. 3). In particular, slightly elevated depths of 400 km to 410 km are present in the western Basin and Range, and slightly depressed depths of 410 km to 420 km are prevalent along the Colorado Plateau. Similarly, variations in the 660 km discontinuity structure are present across the region. A small N-S–oriented relatively shallow discontinuity between 640 km and 655 km is evident beneath the Rocky Mountains in the center of the study area, while a relatively deeper 655 km to 670 km anomaly is apparent to the east and west beneath the Colorado Plateau and Basin and Range. Gilbert et al. (2003) concluded that there is not a strong correlation between the depths to the 410 km and 660 km discontinuities. This result suggests that a dynamic upwelling from mantle transition zone depths is likely not the cause of support of the high elevation in the western United States.

Shear Wave Splitting

Constraining the location and extent of deformation beneath the Colorado Plateau–Basin and Range region is an important component of unraveling the tectonic history of the region. A key method for inferring deformation in the lithosphere and sublithospheric mantle is via measurements of seismic anisotropy, or the dependence of seismic wave speeds on propagation and polarization direction. Seismic anisotropy in the upper mantle is likely generated by finite strain-induced lattice-preferred orientation (LPO) in mantle minerals, primarily olivine (e.g., Karato and Wu, 1993; Ribe, 1989; Silver and Chan, 1988). Seismic anisotropy is expressed through shear wave splitting, in which a shear wave passes through an anisotropic medium and splits into fast and slow quasi-shear waves with orthogonal polarizations (e.g., Silver, 1996; Savage, 1999). The orientation, strength, and coherence of the anisotropic layer can be measured for each split shear wave by calculating a fast polarization direction, ϕ, and a splitting delay time, δt.

The SWGEONET shear wave splitting data set contains 58 measurements of shear wave splitting (Fig. 4) and includes

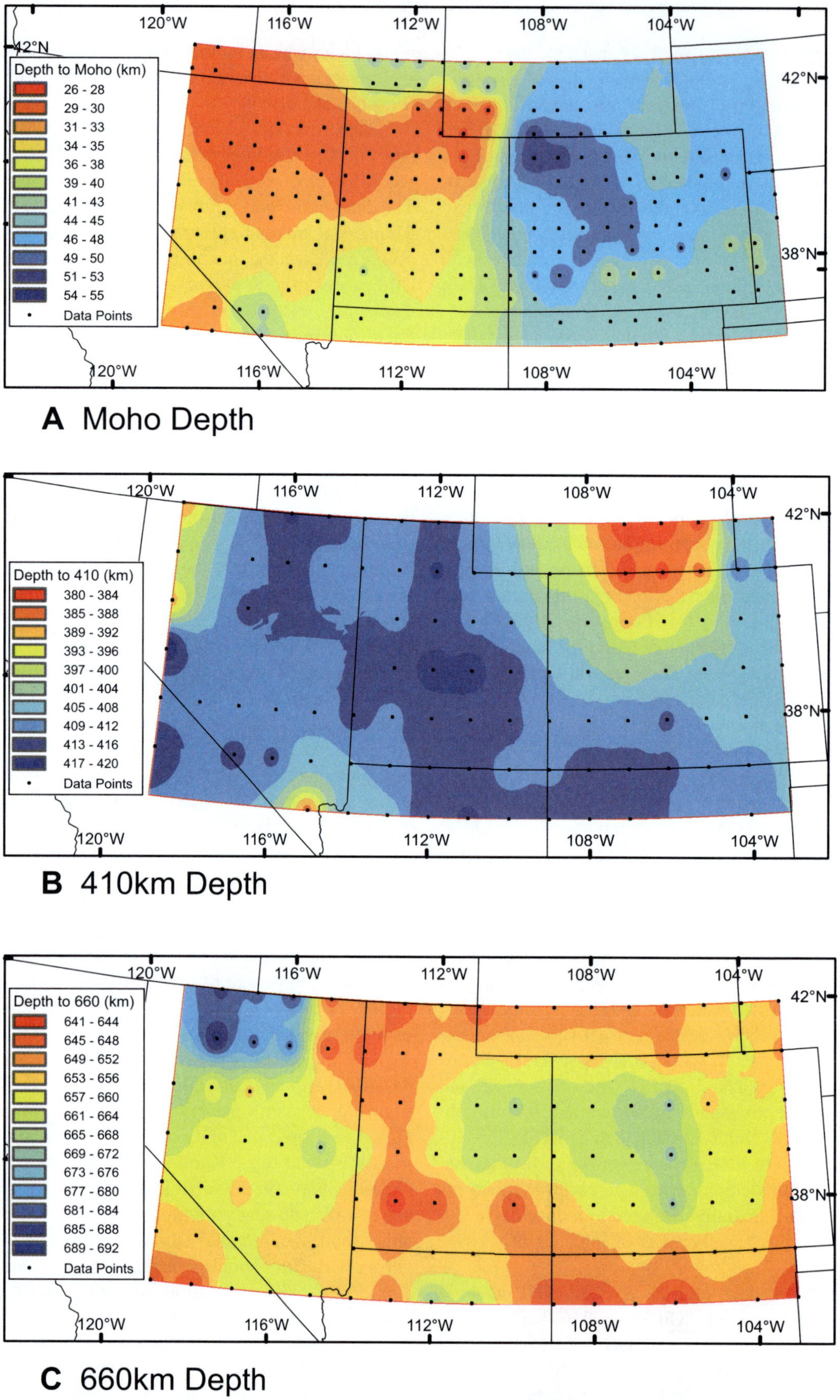

Figure 3. Results of receiver function analyses near the study region (Gilbert et al., 2003; Gilbert and Sheehan, 2004) for the Moho (A), 410 km depth (B), and 660 km depth (C). Values are shown as perturbations from estimated median depth. Region shown here is slightly north of the main study area.

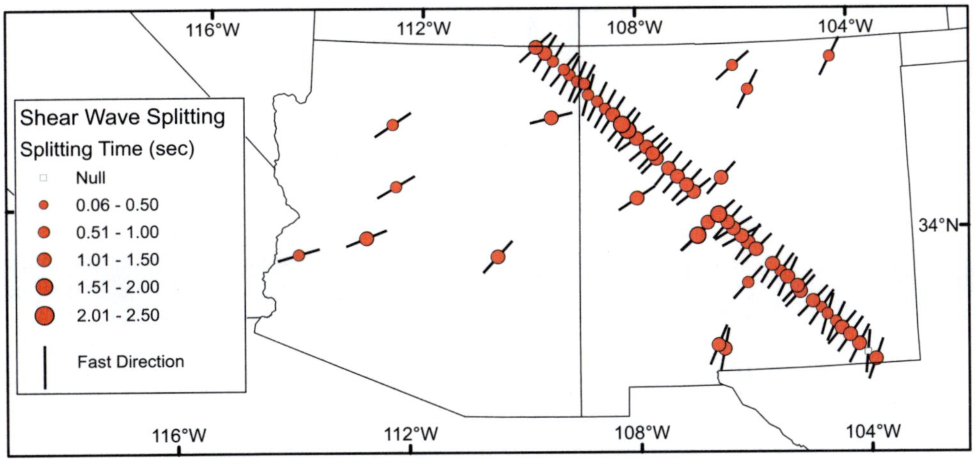

Figure 4. Shear wave splitting results for study area (see text for references). Azimuth of bar indicates fast polarization direction; size of circle denotes splitting time. Note general E-W to NE-SW fast direction trend across the western portion of the study area, and NE-SW to N-S trend across the Rio Grande rift region to the east.

results from Gok et al. (2003), Ruppert (1992), Sandvol and Ni (1994), and Sandvol et al. (1992). Attributes of these data include station name and location, shear wave splitting parameters (ϕ, δt), measurement error, and data source. Some stations have multiple measurements associated with them; in these cases, the database includes all measurements and the necessary metadata to distinguish between them.

Shear wave splitting across the SWGEONET study region is generally uniform (Fig. 4). The most prevalent trend is NE-SW fast polarization directions that are subparallel to North America absolute plate motion (Gripp and Gordon, 2002), particularly within the southernmost Colorado Plateau and southern Basin and Range (Ruppert, 1992; Sandvol and Ni, 1994). These fast directions are also parallel to regional Proterozoic terrane boundaries but are not parallel to current Basin and Range extension (Bird, 2002; Ruppert, 1992; Fouch et al., 2005). Splitting results from the LA RISTRA array also exhibit fast polarization directions oriented NE-SW for regions of the Rio Grande rift and northern Colorado Plateau (Gok et al., 2003). Conversely, in the southern Rio Grande rift, fast directions rotate to a more N-S direction and are nearly parallel to the rift axis (Sandvol et al., 1992). Splitting times range from ~1.0–2.0 s, indicating moderate amounts of anisotropy in the subsurface. These values are somewhat larger than the global average of 1.0 s (Silver, 1996).

Gok et al. (2003) concluded that mantle flow in a NE-SW direction is the likely cause of the bulk of the fast polarization directions across the region, possibly resulting from flow or counterflow initiated by the subduction of the Farallon plate. In related work, Silver and Holt (2002) conclude that the mantle flow is likely a result of both mantle density heterogeneity and decoupling of the overriding North American plate from the underlying mantle. Current analyses under way along the southern Basin and Range–Colorado Plateau region from a new broadband seismometer array will help elucidate the details of the source of shear wave splitting across the region to address this current debate (Fouch et al., 2005).

World Stress Map

The stress state within Earth's crust can be measured through observations of hydrofracturing, young geologic structures, earthquake focal mechanisms, and wellbore breakouts. The SWGEONET database contains 71 stress measurements that lie within the boundaries of Arizona and New Mexico and were collected from the World Stress Map project (Fig. 5) (Reinecker et al., 2003). In addition to the orientation of maximum horizontal compressive stress, each entry contains information regarding the location of the measurement, the method used to determine the stress field, measurement error, and a quality ranking (Reinecker et al., 2003).

The directions of maximum horizontal compression (S_{HMax}) in our study region are generally oriented N-S, but the limited number of data points does not allow for more than a generalized discussion of stress variations. Across the Rio Grande rift, S_{HMax} directions show a general N-S trend, while to the west in the southern Colorado Plateau and Basin and Range regions, S_{HMax} data exhibit both NW-SE and NE-SW trends. Across the Rio Grande rift, the direction of S_{HMax} is oriented orthogonally to the regional E-W-oriented extension direction.

Heat Flow, Thermal Conductivity, Heat Production

Heat flow, thermal conductivity, and heat production measurements provide information about local thermal anomalies, as well as the thermal state of the lithosphere and the deeper mantle. Local thermal anomalies can be important indicators of the tectonic evolution of a region. Heat sources within Earth may be radiogenic in nature, or from physical thermal anomalies such as magma bodies. Heat flow values are calculated using Fourier's Law ($q = k \times dt/dz$, where q is heat flow, k is thermal conductivity, and dt/dz is thermal gradient). The thermal conductivity of a rock is a measure of its ability to transfer heat and is generally dependent upon composition and density. Heat production is a

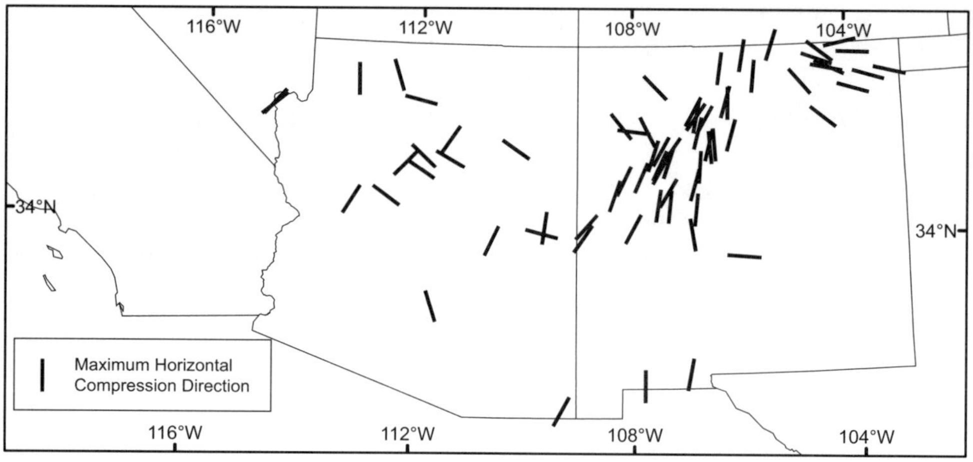

Figure 5. Orientations of maximum horizontal compressive stress direction (S_{Hmax}) measurements (Reinecker et al., 2003) (black lines). Note N-S trend across the Rio Grande rift region, and lack of a coherent trend across the southern Colorado Plateau in Arizona.

measure of the heat internally generated in a rock as the result of the decay of its radioactive constituents.

The SWGEONET heat data set is derived from Pollack et al. (1993) and references therein. The heat flow data set consists of 589 data points within Arizona and New Mexico, each with station name, location, and heat flow value (Fig. 6). Similar attributes are available for the thermal conductivity data set, which contains 475 points within Arizona and New Mexico (Fig. 7). The heat production data set only contains 38 individual data points, many of which are located in the southern regions of the states (Fig. 8).

A broad region of relatively low heat flow (40–50 mW/m^2) covers northern Arizona, with relatively high values of greater than 100 mW/m^2 in southern portions of the state. A N-S-trending high heat flow anomaly (75–200 mW/m^2) exists in central New Mexico. Thermal conductivity observations are similar to heat flow observations, with high anomalies (3.3–6.5 W/m·K) beneath the Rio Grande rift and in southern Arizona. Low anomalies (0.5–2.0 W/m·K) are observed in northern Arizona and portions of New Mexico. Heat production values are relatively high (2.5–5.0 mW/m^3) in southern Arizona and a portion of southern New Mexico, and relatively low (0.03–1.25 mW/m^3) in northern Arizona and New Mexico.

Assuming that heat flow is controlled primarily by a lithospheric thermal boundary layer, the measurements in our study region provide reasonable estimates of lithospheric thickness variations (Pollack et al., 1993). Higher heat flow values generally suggest a thinner lithosphere, while lower heat flow values suggest a thicker lithosphere, causing the heat source (mantle) to be farther away from the surface. Alternatively, regional crustal thickness differences may be the source of the heat flow variations. The low heat flow anomaly over northern Arizona corresponds with the thick crustal root of the Colorado Plateau (McQuarrie and Chase, 2000), while the higher values in the southern region of the state are consistent with the estimates for a relatively thin crust in that area (e.g., Frassetto et al., 2004). In New Mexico, Keller and Baldridge (1999) estimate a thin (25–30 km) crust beneath the southern Rio Grande rift and shallow mantle flow, yielding higher than average heat flow measurements. These conclusions are consistent with the high heat flow anomaly of 100–180 mW/m^2 across the Rio Grande rift region.

Gravity

Gravity anomaly data provide information on crustal thickness variations, local density anomalies, and depth to bedrock. In order to obtain complete Bouguer anomaly (CBA) values, calculations are made to account for the effects of latitude, elevation, and nearby topographic effects (e.g., Keller et al., 2005). The resulting CBA values indicate the variation in observed gravity from expected gravity based on elevation and latitude. Local increases in CBA indicate larger than expected density anomalies, while local decreases in CBA indicate smaller than expected density anomalies.

The SWGEONET gravity data set contains 147,862 CBA data values in a 15° by 8° region covering Arizona and New Mexico (Fig. 9). The data are part of the University of Texas, El Paso, Pan-American Center for Earth and Environmental Studies (PACES) gravity database that was created in cooperation with the U.S. Geological Survey, the National Geospatial-Intelligence Agency, and the National Oceanic and Atmospheric Administration (NOAA) (http://paces.geo.utep.edu/; Keller et al., 2005). The data exhibit high CBA values across southwestern Arizona and regions of eastern New Mexico. A weak N-S-trending linear anomaly exists across the Rio Grande rift in central New Mexico. Keller et al. (1990) showed that the dominant long-wavelength gravity anomalies in the southern Rio Grande rift region are likely the result of thinner crust resulting from rift-related extension. Regional gravity trends across Arizona can likely be explained by the thicker crust beneath the Colorado Plateau and thinner crust beneath the Basin and Range province (Sumner, 1989). Smaller anomalies within the larger general trends result from features such as sedimentary basins that produce gravity lows (Cordell, 1982; Keller and Baldridge,

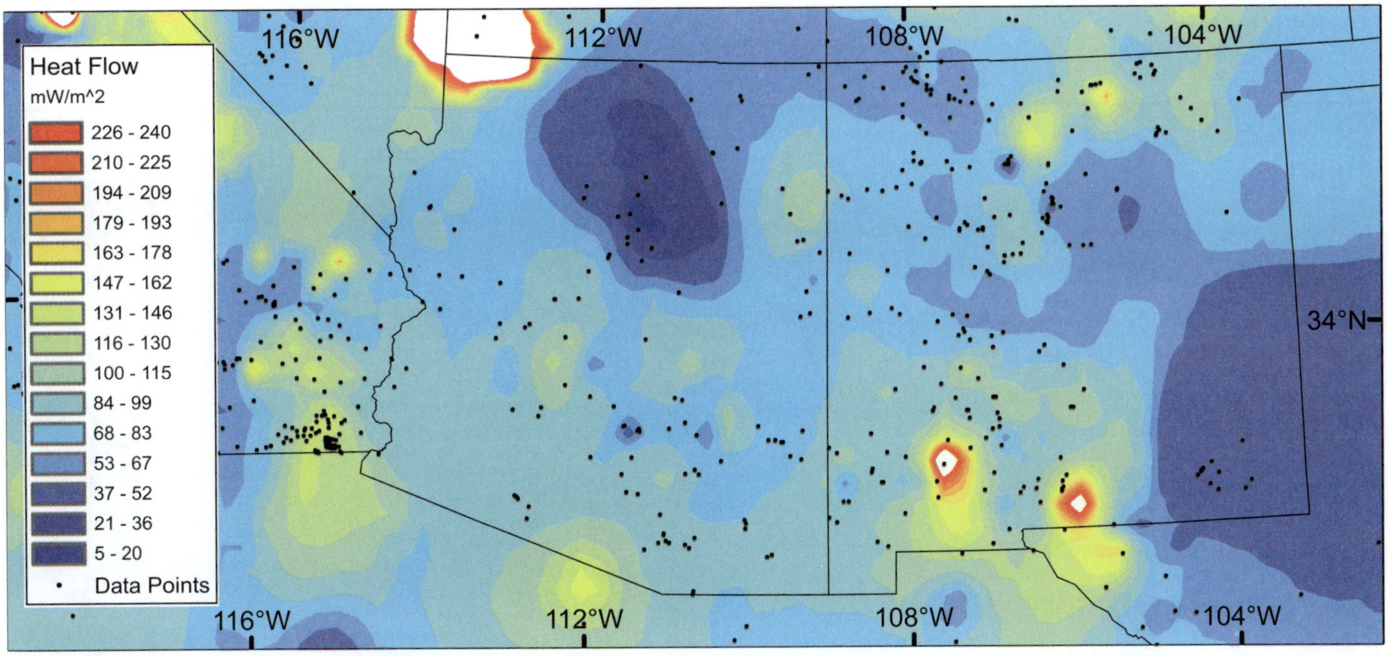

Figure 6. Inverse distance weighted (IDW) grid of heat flow measurements compiled by Pollack (1993). Note the broad high heat flow anomalies across the Rio Grande rift region, and the low anomaly across the southernmost Colorado Plateau region. These anomalies are inferred to be due to a thin crust and/or lithosphere, and thicker crust and/or lithosphere, respectively (Keller and Baldridge, 1999; McQuarrie and Chase, 2000).

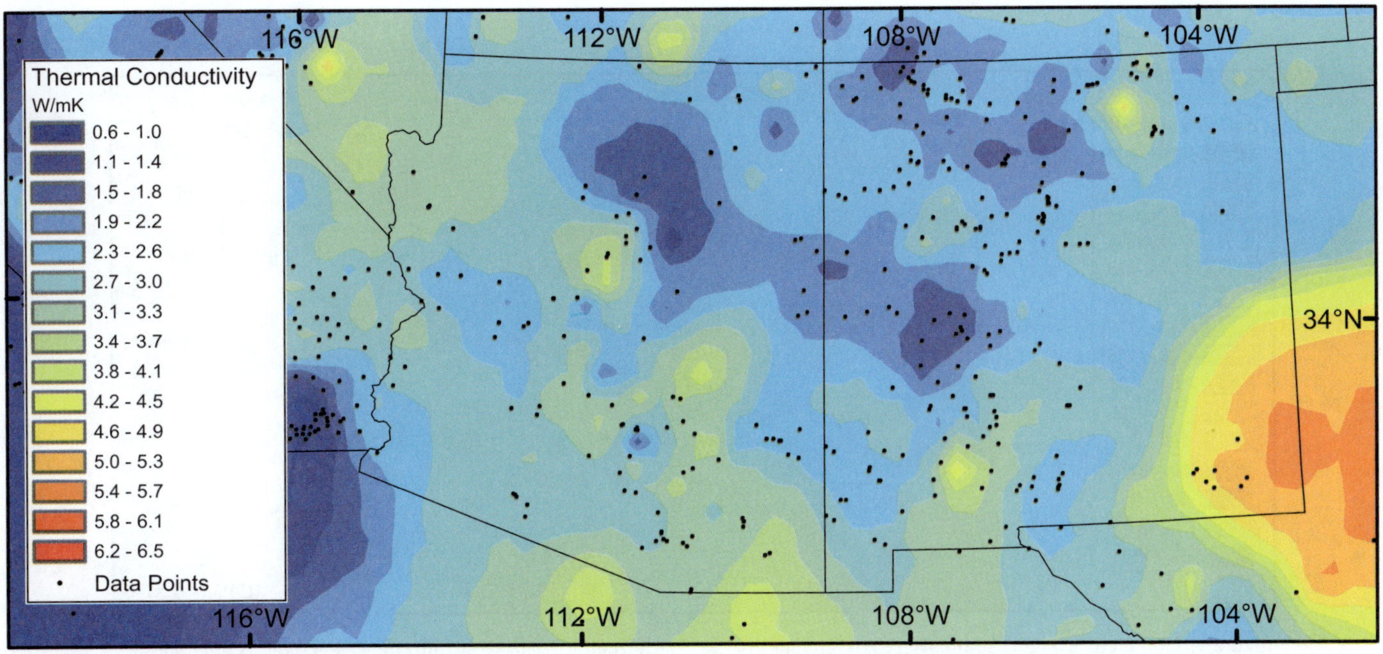

Figure 7. Inverse distance weighted (IDW) grid of thermal conductivity measurements compiled by Pollack (1993). Note the high anomalies beneath the Rio Grande rift and in southern Arizona. Low anomalies are visible in the northern portion of the study region. The direct relationship between heat flow and thermal conductivity suggests that regions of higher thermal conductivity (Basin and Range, Rio Grande rift) indicate thinner crust and/or lithosphere, and regions of lower thermal conductivity (Colorado Plateau) indicate thicker crust and/or lithosphere (Keller and Baldridge, 1999; McQuarrie and Chase, 2000).

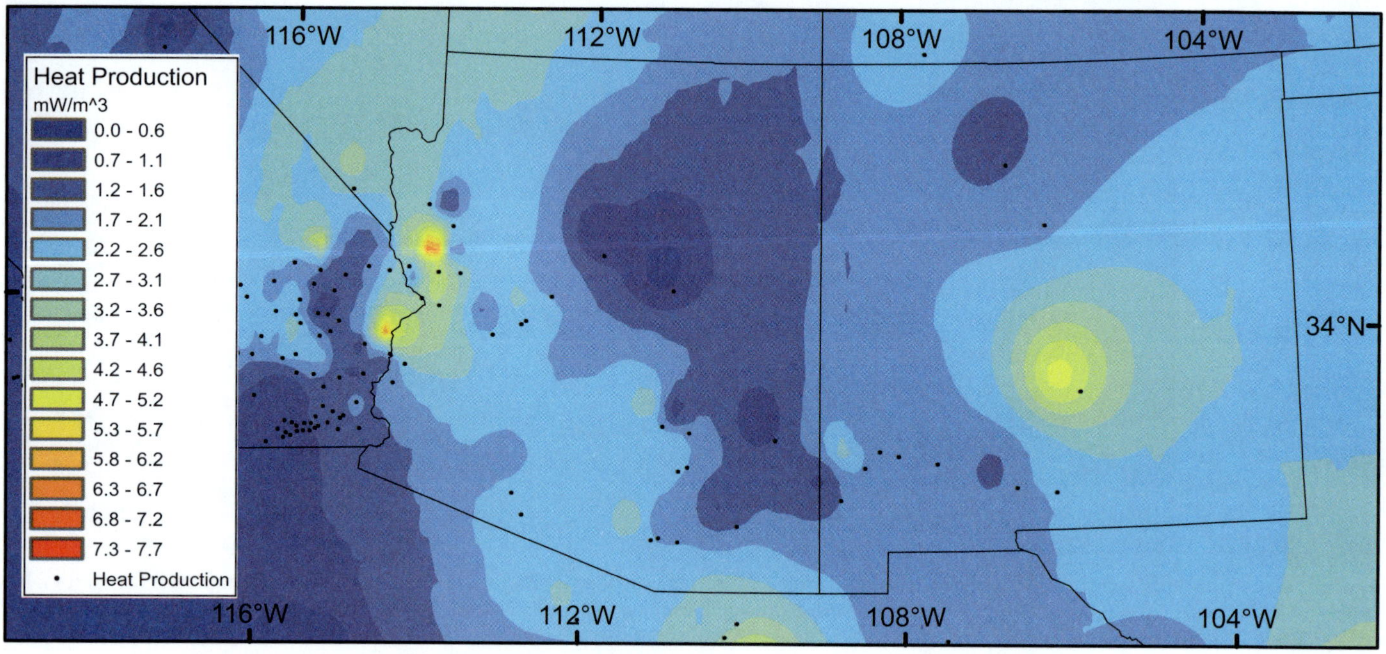

Figure 8. Inverse distance weighted (IDW) grid of heat production measurements compiled by Pollack (1993). Note high heat production anomaly in southern Arizona and the eastern regions, and low anomaly across much of the northern regions.

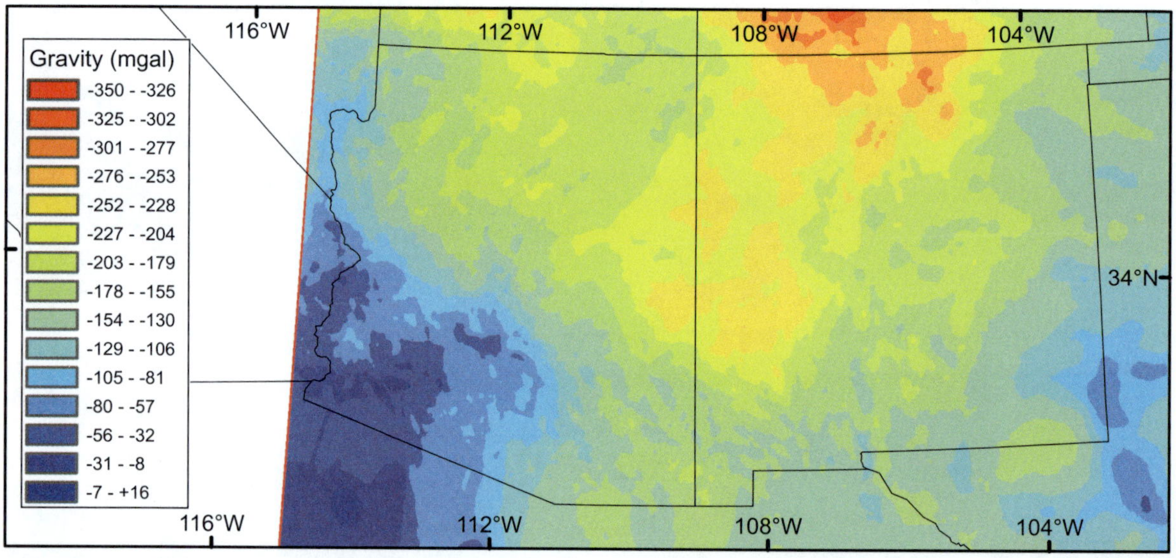

Figure 9. Complete Bouguer anomaly (CBA) values for the study region (data from http://paces.geo.utep.edu/). Higher values shown in blue; lower values shown in red. Lower gravity values imply a regional density deficit, such as the Colorado Plateau shown in red/orange, suggesting either a lack of isostatic compensation or variations in crustal thickness or density.

1999; Seeley and Keller, 2003) and metamorphic core complexes that produce gravity highs (Sumner, 1989).

Geodetic Velocity Field

Relative plate motions can be determined through the use of high-resolution global positioning systems (GPSs) and Very Long Baseline Interferometry (VLBI). The SWGEONET Geodetic Velocity Field (GVF) data set contains high-precision GPS-derived plate motions that provide important information regarding the deformation of western North America.

The GVF data set contains only four data points within Arizona and New Mexico, as much of the measurement effort is currently focused on the San Andreas fault system (Fig. 10) (Bennett et al., 1998). Direction and speed of surface motion are in vector form, with spatial reference information for the location of the observation point and associated error for the velocity value. Uncertainties for site velocities are on the order of 1–2 mm/yr.

The direction of GVF vectors for our study region exhibit an approximately SW-directed plate motion assuming the ITRF2000 reference frame with no net rotation (http://www.iers.org/iers/pc/itrs/). An ~2 mm/yr velocity gradient exists across the region. To the west, the influence of plate-boundary-parallel flow along the San Andreas fault system abruptly changes the direction and rate of surface motion.

CASE STUDIES: EXAMPLES OF INTEGRATED DATA ANALYSIS USING GIS

The data sets included in the SWGEONET database individually provide important constraints on geophysical structure across the Colorado Plateau–Basin and Range region. A primary goal of this study, however, is to provide a mechanism by which to integrate these data sets in a more quantitative form. Below we present four examples of analyses using GIS to combine these data sets as an integrative tool for further investigation of this complex region.

Case Study #1: Gravity Versus Elevation

The goal of this case study was to examine density anomalies and isostatic compensation beneath the Colorado Plateau and the Basin and Range province of Arizona and New Mexico. Isostatic compensation is an important tectonic phenomenon in the Colorado Plateau–Basin and Range region (McQuarrie and Chase, 2000), yet it is integral to determining the mechanism of the Colorado Plateau uplift without significant lateral deformation across the region. During this study, we derived a simple method for identifying regions that may be isostatically overcompensated or undercompensated.

We first compared the complete Bouguer gravity-anomaly (CBA) data set to the USGS HYDRO1k digital elevation model (DEM) (http://edcdaac.usgs.gov/gtopo30/hydro/na_dem.asp). We gridded the CBA data set using an inverse distance weighted (IDW) algorithm with an original resolution of 3.7 km/pixel, and a DEM resolution of 100 m/pixel. To provide consistent information for the analysis, we then resampled each data set to 1 km/pixel resolution. The results of this analysis show an overall linear trend of increasing elevation with decreasing gravity values (Fig. 11). Such a trend is expected, as CBA is calculated to vary as a linear function of elevation. This first-order linear trend suggests that most of the region is isostatically compensated. Populations of outliers beyond this first-order linear trend, however, demonstrate higher-order complexity in the region. Using GIS, we extracted the geographic locations of the outlier populations and compared them to regions of positive and negative relief for the Arizona and New Mexico region (Fig. 12). To first order, we

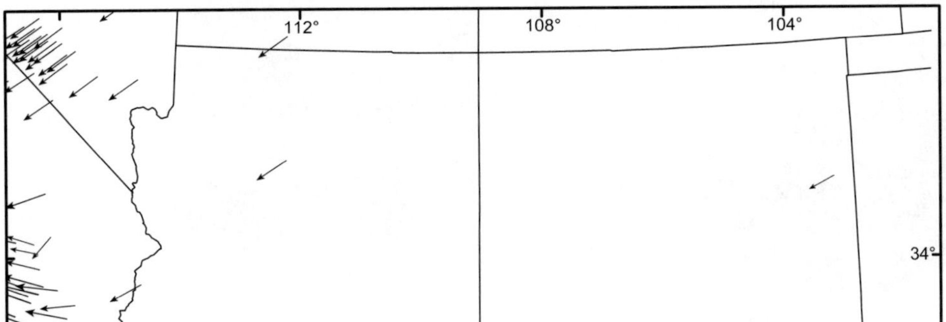

Figure 10. Absolute plate motion of the study area relative to a stable point in North America, in mm/yr using high-precision GPS measurements. Data from http://cfa-www.harvard.edu/space_geodesy/WUSC/; GPS site velocities for the SWGEONET region from http://sps.unavco.org/crustal_motion/dxdt/model/. Data are plotted using the ITRF2000 reference frame with no net rotation. Uncertainties for site velocities are on the order of 1–2 mm/yr. Note the increase in velocity in the southwestern portion of the study area showing a SW-directed flow of the Colorado Plateau and Basin and Range with ~2 mm/yr velocity gradient across the region. To the west, the influence of plate-boundary-parallel motion along the San Andreas fault system abruptly changes the direction and rate of motion.

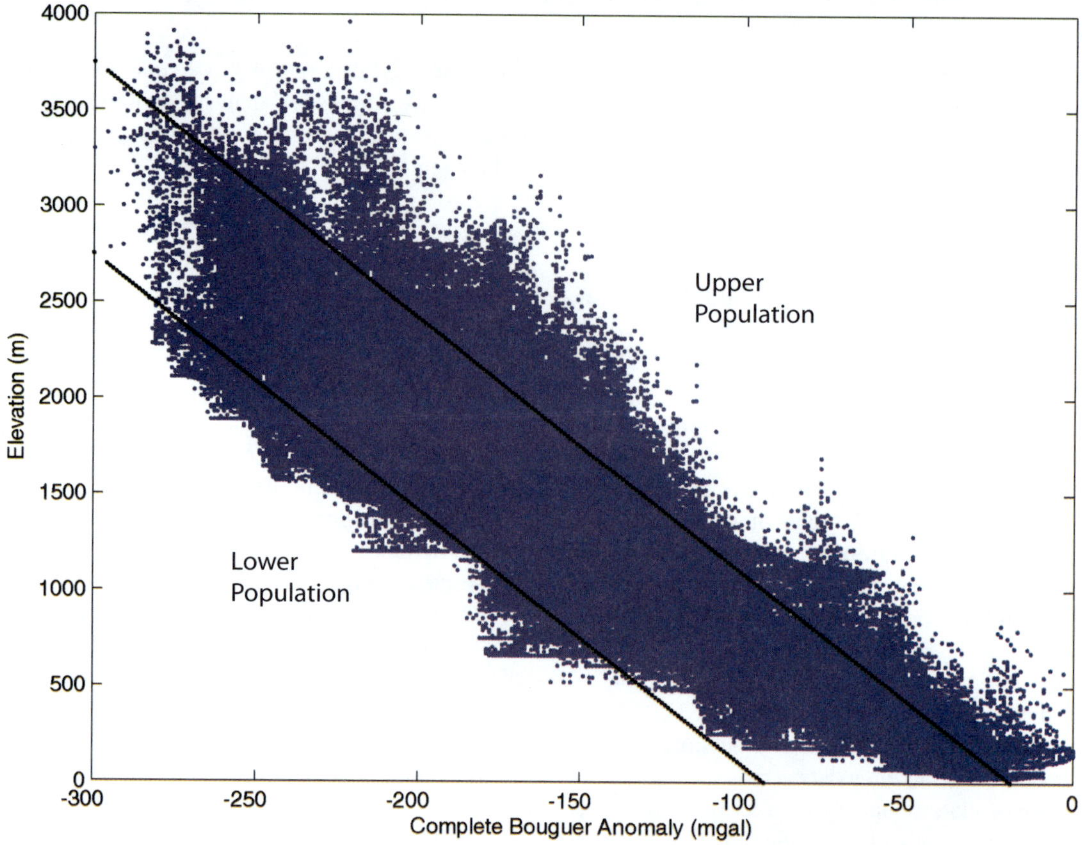

Figure 11. Complete Bouguer anomaly (CBA) versus elevation for the region. The first-order linear trend of higher elevations with lower CBA values is consistent with long-wavelength isostatic compensation across the region. The upper and lower outlier populations that deviate from this general trend have regional correlations and are shown in Figures 12 and 13.

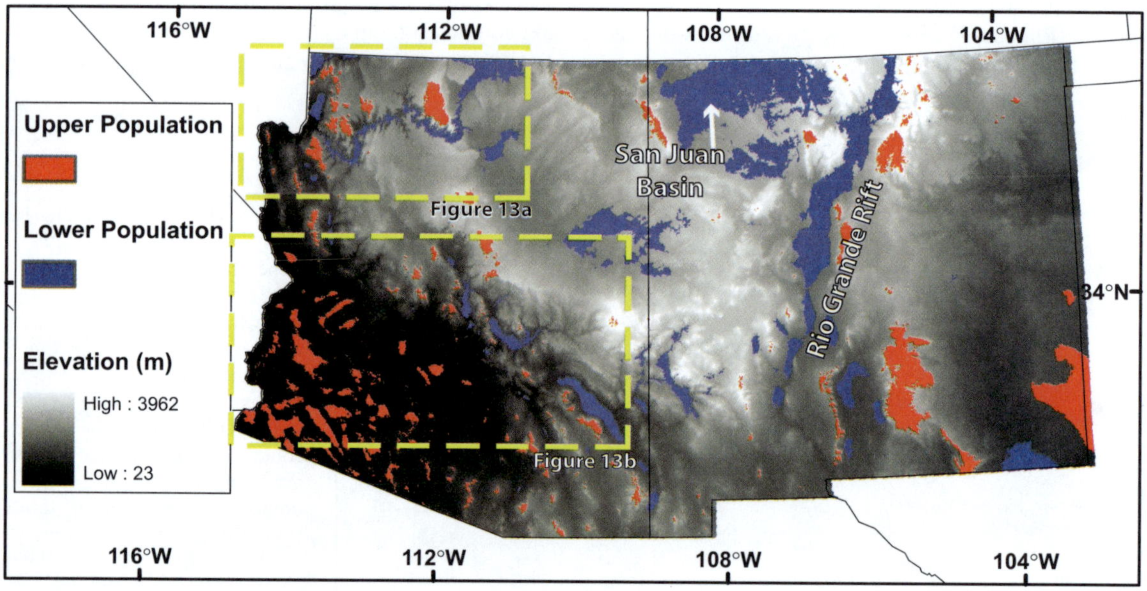

Figure 12. Upper and lower outlier populations from Figure 11. Red regions (upper population) generally correspond to regions of strong positive relief (i.e., Kaibab uplift and metamorphic core complexes—see Fig. 13), and blue regions (lower population) generally correspond to regions of strong negative relief (i.e., San Juan Basin, Rio Grande rift, and Grand Canyon—see Fig. 13).

found an upper outlier population corresponding to regions of strong positive relief, and a lower outlier population corresponding to regions of strong negative relief.

Chase et al. (2002) concluded that at 0.25° (~30 km) resolution, the Colorado Plateau and much of western North America is isostatically compensated. However, at a higher resolution of 1 km/pixel, our analysis identified several areas that are not consistent with this conclusion, and demonstrates that short-wavelength compensation is not ubiquitous across the region.

Some regions of negative relief, such as the Grand Canyon, the San Juan Basin, and the Rio Grande rift, may correspond to regions of overcompensation (Figs. 12 and 13A). For some regions such as the Grand Canyon, this result might be due to terrain corrections in the CBA calculation that cannot fully account for first-order discontinuous topographic changes (i.e., the steep walls of the canyon). Variations for other regions with broader negative relief, such as the San Juan Basin and the Rio Grande rift, are more likely due to thick low-density sediments in these low-lying regions. Another possible interpretation is that crustal density is lower or crustal thickness is higher than assumed, though this possibility is less likely for these regions.

Some areas of strong positive relief, such as the Kaibab uplift in northern Arizona and the metamorphic core complexes in southern Arizona, are located within undercompensated regions (Figs. 13A and 13B) that cannot be explained solely by inaccurate terrain corrections (e.g., Reynolds and Spencer, 1985). One possible explanation for this observation is that this region contains a higher crustal density or thinner crust. Reynolds and Spencer (1985) suggest that large-scale detachment faults delivered metamorphic rocks beneath these core complexes during the Mesozoic and Tertiary, providing the source for high densities across the region.

The inference of regional crustal density variations suggests large-scale variations in rock composition beneath these tectonic regimes. A higher-density crust could suggest major magmatic modification such as underplating in the history of the region, while a lower-density crust might suggest more sedimentary rock deposition or emplacement of felsic batholiths. While these issues currently have nonunique solutions, the addition of data sets in an integrated framework will improve our understanding of these correlations in the future.

Case Study #2: Receiver Functions Versus Tomography

In this case study, we examine the regional consistency between receiver functions and seismic tomography images to evaluate the correlation between these data sets near upper mantle seismic discontinuities. We utilized receiver function data from Gilbert et al. (2003) at depths of 410 km and 660 km, and data from depth slices of 419 km and 669 km from the three-dimensional seismic velocity model NA00 (van der Lee, 2002). We assume that both seismic velocity perturbations and upper mantle discontinuity depth differences are primarily determined by thermal variations (e.g., Helffrich and Wood, 2001). Temperature increases tend to reduce seismic velocities, while temperature reductions tend to increase seismic velocities. Temperature increases near the 660 km discontinuity would cause it to elevate, while temperature decreases would cause it to depress. Conversely, temperature increases near the 410 km discontinuity would cause it to depress, while temperature decreases would cause it to elevate. Reduced mantle temperatures therefore result in a thicker upper mantle transition zone, while elevated mantle temperatures result in a thinner upper mantle transition zone.

To evaluate possible correlations in mantle thermal structure from these independent seismic data sets, we calculated the ratio of upper mantle discontinuity depth variation to seismic velocity variation at corresponding nearest points. We define a ratio R as

$$R = \Delta D \times \Delta V_S$$

where ΔD is the variation in discontinuity depth relative to the median value, and ΔV_S is the variation in S-wave velocity of NA00 relative to the MC35 one-dimensional seismic velocity model (van der Lee, 2002).

Near 410 km, the combination of a shallower relative discontinuity depth (negative ΔD) and a higher relative S-wave velocity (positive ΔV_S) would produce a negative R value, suggesting a temperature decrease. Similarly, a deeper relative discontinuity depth (positive ΔD) and a lower relative S-wave velocity (negative ΔV_S) would produce a negative R value, suggesting a temperature increase. For the 410 discontinuity, a negative R value therefore implies a change in thermal properties consistent with both data sets. Conversely, near the 660 km discontinuity, a shallower relative depth (negative ΔD) and a slower relative S-wave velocity (negative ΔV_S) would produce a positive R value, suggesting an increase in temperature. A deeper relative discontinuity depth (positive ΔD) and a faster relative S-wave velocity (positive ΔV_S) would suggest a temperature reduction. For the 660 discontinuity, a positive R value therefore implies a change in thermal properties consistent with both data sets.

Near 410 km, positive R values are generally present beneath much of the region, particularly the study area (Fig. 14A). Our analysis therefore suggests no regional thermal correlation beneath the study area. Conversely, near 660 km, negative R values are present beneath the study area and therefore do exhibit a clear regional thermal correlation (Fig. 14B).

Interestingly, our results are not consistent with the interpretation of cool thermal anomalies due to Farallon slab subduction (e.g., van der Lee, 2002) or regional lithospheric delamination (e.g., Gao et al., 2004). While the observed variations in mantle structure could be due to regional compositional differences, the effects of these variations are less well constrained and are therefore beyond the scope of the current work. The first-order analyses performed here should be more fully evaluated using improved techniques that more fully examine the possible three-dimensional regional model space. We also note that the bulk of the study region is near the southern edge of the current data set,

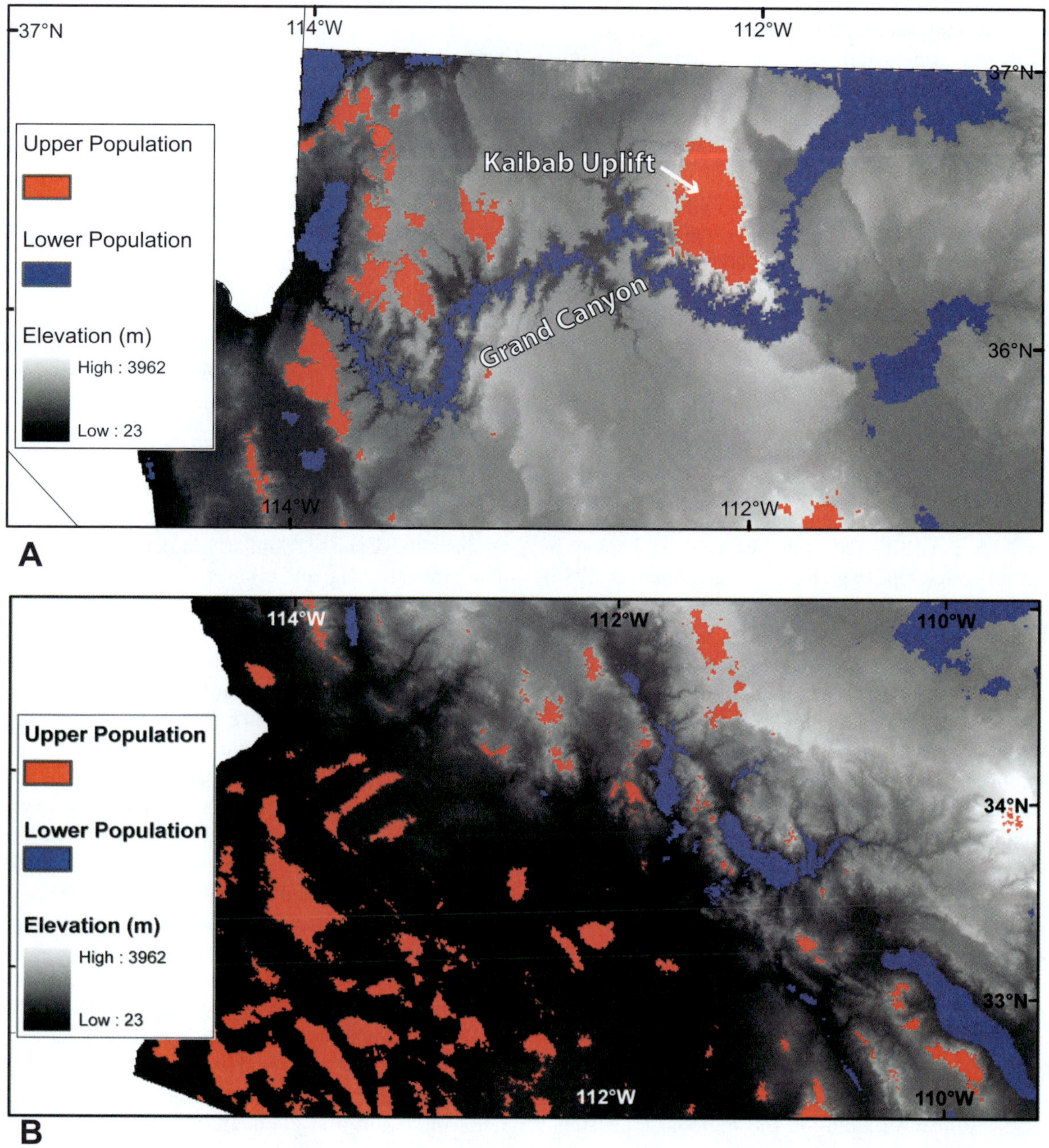

Figure 13. Detail of Figure 12 showing the close relationship between the outlier populations and areas of strong relief. A: Red regions (upper outliers, Fig. 12) correspond closely with the Kaibab uplift (strong positive relief); blue regions (lower outliers, Fig. 12) correspond closely with the Grand Canyon (strong negative relief). B: Red regions (upper outliers, Fig. 12) correspond closely with metamorphic core complexes of southwestern Arizona.

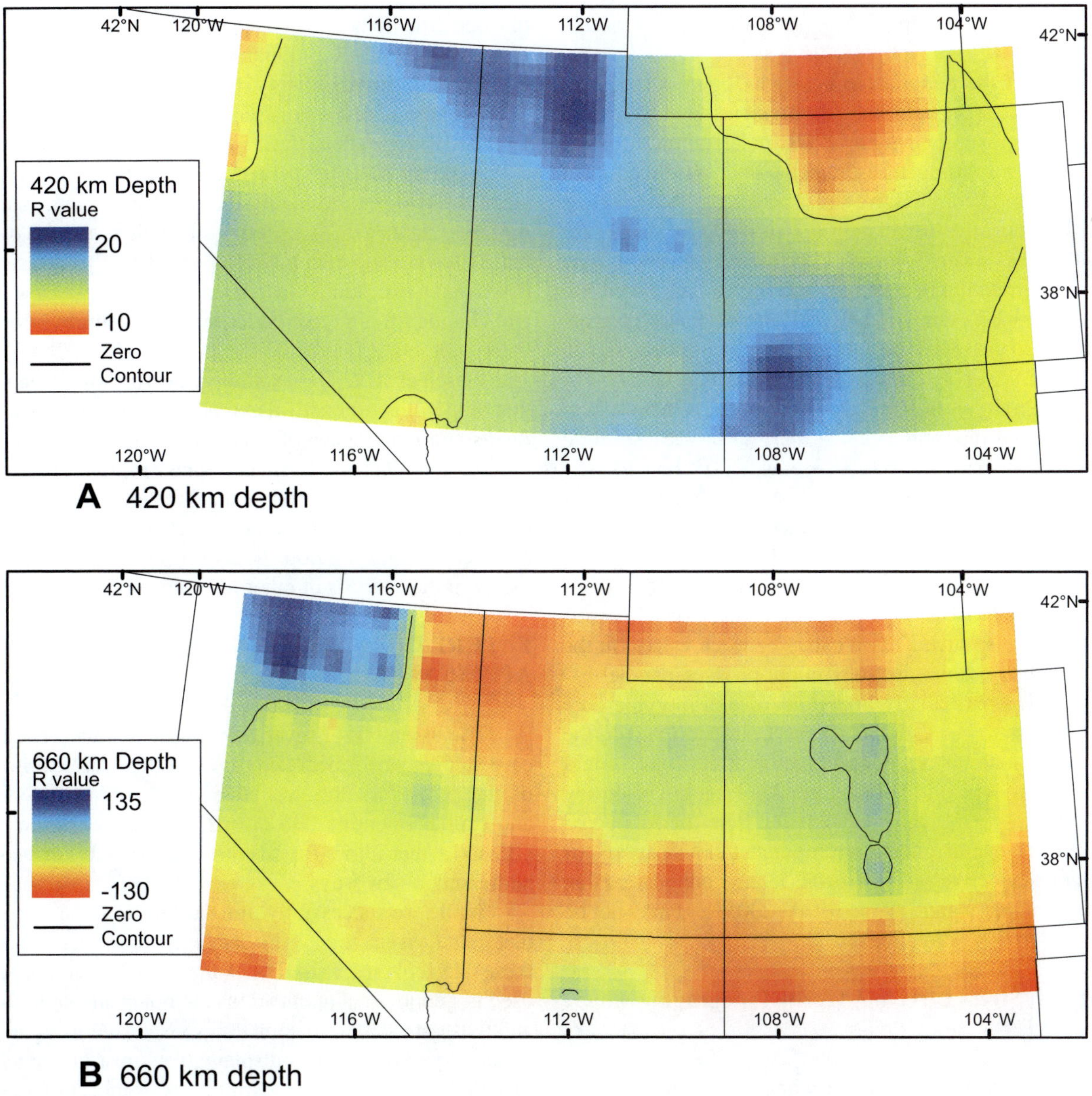

Figure 14. Relative receiver function variations multiplied by relative shear wave velocity values (R—see text for details). Data from van der Lee (2002) and Gilbert et al. (2003). A: 410 km region. Positive R values across much of the region suggest a lack of regional correlation between receiver function and tomography data sets. B: 660 km region. Negative R values across much of the region suggest a lack of regional correlation between receiver function and tomography data sets.

and that new analyses in the heart of the study region will provide better constraints on this structure (Fouch et al., 2005).

Case Study #3: Crustal Stress Versus Seismic Anisotropy

Here we examine correlations between observations of crustal stress and seismic anisotropy to evaluate the extent of mechanical coupling between the crust and mantle during deformation. Seismic anisotropy can be observed via examination of the birefringence, or splitting, of seismic shear waves. It is generally caused by deformation in the crust and/or mantle, providing a good marker for mapping strain and/or strain history. In the crust, seismic anisotropy can be caused by cracks, layering, and mineral alignment in the brittle upper crust, or flow, layering, and mineral alignment in the ductile lower crust. Similarly, mantle anisotropy likely develops from deformation in the lithosphere, flow in the sublithospheric mantle, or a combination of both (e.g., Savage, 1999, 2002; Silver, 1996). For a scenario in which the crust and mantle are fully coupled, the direction of maximum compressional stress in the crust (S_{HMax}) and shear wave splitting fast directions (FD) in the mantle should be parallel. Conversely, for scenarios where the crust and mantle are fully decoupled, S_{HMax} and FD will not necessarily correlate.

We examined the relationship between 36 stress measurements and 30 shear wave splitting measurements within Arizona and New Mexico. We compared the fast polarization direction of the shear wave splitting measurements and the direction of maximum horizontal compressive stress across the study region. Results show a negative correlation between FD and S_{HMax} for regions within the lower Colorado Plateau and transition zone and a positive correlation between FD and S_{HMax} in the Basin and Range region (Fig. 15). For the southern Colorado Plateau region, azimuths between the two data sets are generally orthogonal, and they are ~20°–25° relative to one other in the Rio Grande rift region. The general consensus is that flow in the asthenosphere is primarily responsible for the bulk of these shear wave splitting measurements (i.e., Savage, 2002; Gok et al., 2003; Fouch et al., 2005). This conclusion is generally consistent with the results of this analysis, which demonstrate a limited correlation between the S_{HMax} and SWS (shear wave splitting) data sets.

However, due to the small number of measurements, a complete analysis is not yet achievable for this area. Crustal anisotropy could also cause such a discrepancy between values; in the Basin and Range, crustal anisotropy might produce parallel FD and S_{HMax}, perhaps particularly in regions of metamorphic core complexes (Frassetto et al., 2004). In the Colorado Plateau, the thicker crust and long history of deformation of the region could produce more variability in splitting results. Mantle anisotropy beneath the region could be a result of flow from the subducted Farallon slab, a general plate-motion-parallel flow pattern, or flow parallel to the strike of Proterozoic terrane boundaries that extend into the upper mantle (Fouch et al., 2005) and may not be directly related to S_{HMax} directions.

Case Study #4: Stress Direction Versus Topographic Slope Direction

Here we address the question of whether local topographically driven forces or larger-scale tectonic stresses govern crustal deformation in the region. To accomplish this goal, we examined correlations between local downdip slope directions using digital elevation models (DEMs) and the direction of maximum horizontal compression from crustal stress measurements. Flesch et al. (2001) use a similar method to evaluate the source of tectonic stresses in the Tibetan Plateau.

We obtained stress directions from the SWGEONET World Stress Map data set (Reinecker et al., 2003), and slope vectors from the 1 km resolution DEM (USGS, 2003: http://edcdaac.usgs.gov/gtopo30/hydro/namerica.asp) of the study region (Fig. 16). We correlated the azimuth of the maximum horizontal compression (S_{Hmax}) from the stress values with the azimuth of the downdip direction of slope from 589 locations. For S_{HMax} values ranging from N and NE (and therefore S to SW), there appears to be no correlation with slope direction. For S_{HMax} values oriented approximately NW-SE, there are a greater population of slope directions oriented approximately NW-SE (i.e., approximately parallel to S_{HMax}) than directions oriented approximately NE-SW (i.e., approximately orthogonal to S_{HMax}).

Because regional horst and graben features reflect E-W Rio Grande rift extension, most mountain ranges trend N-S to NE-SW, yielding a higher percentage of E-W to SE-NW downdip slope directions surrounding the Rio Grande rift. A majority of the stress data (~75%) are from this region, and the data may be biased toward showing a N-S–oriented S_{HMax} slope trend for the Rio Grande rift. Additional data in tectonic regimes outside of the Rio Grande rift are required to provide a definitive correlation of these data sets.

Flesch et al. (2001) examined the Tibetan Plateau and found that slope direction and magnitude greatly influence stress azimuths and magnitudes. Similar calculations could be applied to the Colorado Plateau–Basin and Range region if additional stress data were available. Though the Tibetan Plateau is fundamentally different in that it is a large-scale contractional regime, any relationship between slope direction and magnitude with stress is useful in the analysis of other tectonic regimes.

FUTURE DIRECTIONS IN GIS SCIENTIFIC APPLICATIONS

The assembly of several geophysical data sets and the four case studies performed using GIS in this study showcase some of the powerful tools available for the SWGEONET project and similar scientific GIS efforts. We have described a suite of GIS tools that allow for scientifically valuable questions to be addressed in new ways.

While we successfully utilized GIS tools to import, manipulate, and visualize several data sets, drawbacks to employing these GIS tools still exist. For example, some symbols commonly used in geophysical applications are not available in this system, and thus data sets were sometimes visualized using nonstandard means. Symbols used to display stress measurements were not included as part of the standard library of tools provided with the GIS software, so several symbols had to be combined to create the necessary output. Furthermore, geodetic velocities are typically presented as vectors with error ellipses at their heads. We were unable to implement this symbology standard in ArcGIS. Tools such as Generic Mapping Tools (GMT) (Wessel and Smith, 1998) often provide these functions, but are not always fully compatible with propriety data formats used in GIS. One solution would be to either provide support for these functions through a new version of the GIS software or a third-party add-on, or integrate the tools and symbols of GMT into the GIS software.

An improved GIS system for the SWGEONET project would include, in an intuitive and highly configurable manner, all necessary geological and geophysical symbols used to describe data. Additionally, in the current configuration, basic mathematical and conditional operations between point and raster data sets can be performed, but comparisons between these data types are difficult. The ability to easily compare and perform mathematical

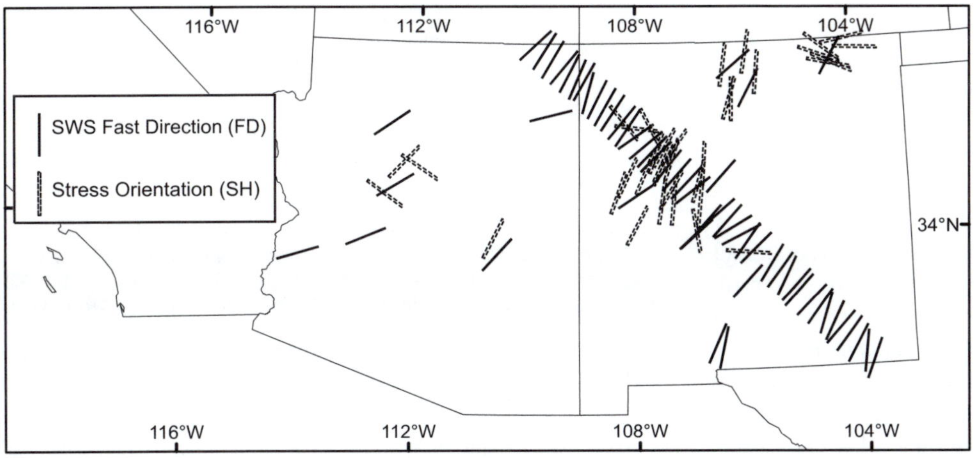

Figure 15. Shear wave splitting fast directions (black lines) and directions of maximum horizontal compression (blue lines) from the SWS (shear wave splitting) data set (Fig. 4) and stress data set (Fig. 5). Note the low azimuthal correlation near the Rio Grande rift, and nearly orthogonal orientations on the Colorado Plateau. Basin and Range values show some correlation, but the low number of measurement points prevents further conclusions.

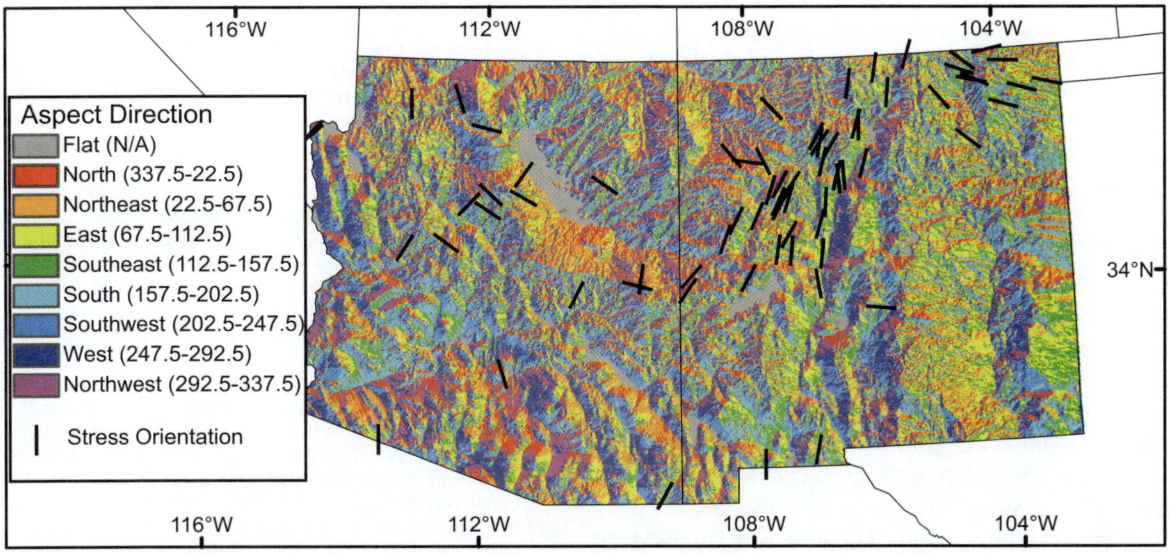

Figure 16. Aspect map of SWGEONET region showing slope orientations (color-coordinated) from digital elevation model with stress orientations (S_{HMax}) plotted as black lines (Fig. 5). Note general E-W slope orientations, primarily due to the N-S–trending horst and graben tectonic features from regional extension.

operations between point and raster data sets would therefore be an important improvement.

An improved system might also have the ability to export selected data as a module to be given to another party. This module could contain only raw data for a given region for all layers, corrected and plotted data for the region, or a combination thereof. The module would require no additional software and a minimum of hardware. It could also be used as part of a teaching module, without having to worry about extraneous data not required for the lesson. On a larger scale, the ability to enable users to import their own data into the online distribution system would significantly improve the total volume of data for a given project, while diverting resources needed to maintain the data sets to the data providers. As hardware resources increased, data sets outside of the geophysical sciences could be added, such as data from geochemical, petrological, and mineral physics studies.

Several other projects have attempted to develop similar types of databases, often with a high degree of success. For example, the Planetary Interactive GIS-on-the-Web Analyzable Database (PIGWAD) (http://webgis.wr.usgs.gov/) is an online interactive database used for the analysis of planetary map data sets. Users can add and remove data layers to suit their needs, export data, and retrieve metadata as needed. Like the SWGEONET database, the PIGWAD database contains many forms of data, including point, line, polygon, and raster (gridded) data sets. Scientific comparisons can also be done via an Internet interface, much like the SWGEONET project.

As another example, the burgeoning GEON project (http://www.geongrid.org/) aims to achieve a larger-scale goal of an earth-science-based multidisciplinary database. The GEON project provides the framework necessary for projects such as the SWGEONET initiative, in terms of both software and hardware needs. It also spearheads the effort to combine multiple genres of earth science in new and technologically advanced ways. A partner effort of the GEON project is NAVDAT (North American Volcanic and Intrusive Rock Database; http://navdat.geongrid.org/), which enables users to utilize a wide range of published chemical and age data for Late Cretaceous to Holocene igneous rocks in western North America. Products include interactive regional maps and visualizations of eruption age and chemical variations. This utility is an excellent example of a multidisciplinary approach to understanding earth science problems.

The combination of resources to enable progress of scientific research, visualization, modeling, and analysis will help further the exploration and understanding of large, complex Earth system dynamics. This paper presents a prototype of the vision represented for the earth science community by GEON. Incremental steps such as the one presented here are the beginning of a new path that taps the tools of cyberinfrastructure to lead us to an improved understanding of Earth systems.

CONCLUSIONS

The SWGEONET database provides a new framework for geophysical studies of the southwestern United States. We have incorporated a broad range of geophysical data sets into the GIS framework and have provided it to the public in an online, interactive data analysis and management system. We demonstrated through four case studies that these data sets can be interactively analyzed to provide new constraints on the dynamics of the Colorado Plateau–Basin and Range region. Features from similar data systems in other fields can be integrated to provide a new level of functionality with this system. The SWGEONET database is a significant step toward GEON, an integrated data system that hopes to provide a new level of transdisciplinary research in the earth sciences.

ACKNOWLEDGMENTS

We thank Lela Prashad, Jeff Conner, Steve Holloway, Paul Ivanich, and Mark Stevens for their technical expertise and assistance in producing this manuscript, the ASU GIS lab for GIS software expertise, and National Science Foundation awards ITR-0112960 and EAR-0225543. Additionally, we appreciate the assistance from those individuals who provided their data sets for this effort. Individual providers can be found on the SWGEONET "Data" pages at http://www.geoinformaticsnetwork.org/swgeonet/.

REFERENCES CITED

Ammon, C.J., 1991, The isolation of receiver effects from teleseismic P waveforms: Bulletin of the Seismological Society of America, v. 81, p. 2504–2510.

Bennett, R.A., Wernicke, B.P., and Davis, J.L., 1998, Continuous GPS measurements of contemporary deformation across the northern Basin and Range province: Geophysical Research Letters, v. 25, no. 4, p. 563–566, doi: 10.1029/98GL00128.

Bird, P., 2002, Stress direction history of the western United States and Mexico since 85 Ma: Tectonics, v. 21, no. 3, doi: 10.1029/2001TC001319.

Chase, C.G., Libarkin, J.A., and Sussman, A.J., 2002, Colorado Plateau: Geoid and means of isostatic support: International Geology Review, v. 44, no. 7, p. 575–587.

Coney, P.J., and Harms, T.A., 1984, Cordilleran metamorphic core complexes: Cenozoic extensional relics of Mesozoic compression: Geology, v. 12, p. 550–554, doi: 10.1130/0091-7613(1984)12<550:CMCCCE>2.0.CO;2.

Cordell, L., 1982, Extension in the Rio Grande rift: Journal of Geophysical Research, v. 87, no. B10, p. 8561–8569.

Dickinson, W.R., 1989, Tectonic setting of Arizona through geologic time, *in* Jenney, P., and Reynolds, S., eds., Geologic evolution of Arizona: Arizona Geological Society Digest: Tucson, Arizona, Arizona Geological Society, p. 1–16.

Dickinson, W.R., and Snyder, W.S., 1978, Plate tectonics of the Laramide orogeny, *in* Matthews, V., III, ed., Laramide folding associated with basement block faulting in the western United States: Geological Society of America Memoir 141, p. 355–366.

Dueker, K.G., and Sheehan, A.F., 1997, Mantle discontinuity structure from midpoint stacks of converted P to S waves across the Yellowstone hotspot track: Journal of Geophysical Research, v. 102, no. B4, p. 8313–8327, doi: 10.1029/96JB03857.

Dueker, K.G., and Sheehan, A.F., 1998, Mantle discontinuity structure beneath the Colorado Rocky Mountains and High Plains: Journal of Geophysical Research, v. 103, no. B4, p. 7153–7169, doi: 10.1029/97JB03509.

Dziewonski, A.M., and Anderson, D.L., 1981, Preliminary reference Earth model: Physics of the Earth and Planetary Interiors, v. 25, no. 4, p. 297–356, doi: 10.1016/0031-9201(81)90046-7.

Eberly, L.D., and Stanley, T.B., Jr., 1978, Cenozoic stratigraphy and geologic history of southwestern Arizona: Geological Society of America Bulletin, v. 89, p. 921–940, doi: 10.1130/0016-7606(1978)89<921:CSAGHO>2.0.CO;2.

Flesch, L.M., Haines, A.J., and Holt, W.E., 2001, Dynamics of the India-Eurasia collision zone: Journal of Geophysical Research, v. 106, no. B8, p. 16,435–16,460, doi: 10.1029/2001JB000208.

Fouch, M., Gilbert, H., Frassetto, A., Yoburn, J., Zandt, G., Beck, S., Owens, T., and Garnero, E., 2005, Crust and uppermost mantle structure beneath the southern Basin and Range/Colorado Plateau: Results from the COARSE broadband seismometer array and goals for regional USArray studies: EarthScope National Meeting, Santa Ana Pueblo, New Mexico.

Frassetto, A., Gilbert, H., Zandt, G., Owens, T.J., Fouch, M.J., Beck, S.L., and Garnero, E., 2004, Crust and upper mantle structure of the southern Basin and Range and Colorado Plateau of Arizona: Project COARSE: Eos (Transactions, American Geophysical Union), v. 85, no. 47, Fall Meeting Supplement, Abstract T51C-0464.

Gao, W., Grand, S.P., Baldridge, W.S., Wilson, D., West, M., Ni, J.F., and Aster, R., 2004, Upper mantle convection beneath the central Rio Grande rift imaged by P and S wave tomography: Journal of Geophysical Research, v. 109, no. B03, B03305, doi: 10.1029/2003JB002743.

Gilbert, H.J., and Sheehan, A.F., 2004, Images of crustal variations in the intermountain west: Journal of Geophysical Research, v. 109, no. B3, B03306, doi: 10.1029/2003JB002730.

Gilbert, H.J., Sheehan, A.F., Dueker, K.G., and Molnar, P., 2003, Receiver functions in the western United States, with implications for upper mantle structure and dynamics: Journal of Geophysical Research, v. 108, no. B5, 2229, doi: 10.1029/2001JB001194.

Gok, R., Ni, J.F., West, M., Sandvol, E., Wilson, D., Aster, R., Baldridge, W.S., Grand, S., Gao, W., Tillmann, F., and Semken, S., 2003, Shear wave splitting and mantle flow beneath LA RISTRA: Geophysical Research Letters, v. 30, no. 12, 1614.

Grand, S.P., van der Hilst, R.D., and Widiyantoro, S., 1997, Global seismic tomography: A snapshot of convection in the earth: GSA Today, v. 7, no. 4, p. 1–7.

Gripp, A.E., and Gordon, R.G., 2002, Young tracks of hotspots and current plate velocities: Geophysical Journal International, v. 150, no. 2, p. 321–361, doi: 10.1046/j.1365-246X.2002.01627.x.

Helffrich, G.R., and Wood, B.J., 2001, The Earth's mantle: Nature, v. 412, no. 6846, p. 501–507, doi: 10.1038/35087500.

Hendricks, J.D., and Plescia, J.B., 1991, A review of the regional geophysics of the Arizona transition zone: Journal of Geophysical Research, v. 96, no. B7, p. 12,351–12,373.

Jarchow, C.M., and Thompson, G.A., 1989, The nature of the Mohorovicic discontinuity: Annual Review of Earth and Planetary Science, v. 17, p. 475–506, doi: 10.1146/annurev.ea.17.050189.002355.

Karato, S.I., and Wu, P., 1993, Rheology of the upper mantle: A synthesis: Science, v. 260, no. 5109, p. 771–778.

Keller, G.R., and Baldridge, W.S., 1999, The Rio Grande rift: A geological and geophysical overview: Rocky Mountain Geology, v. 34, no. 1, p. 121–130, doi: 10.2113/34.1.121.

Keller, G.R., Morgan, P., and Seager, W.R., 1990, Crustal structure, gravity anomalies and heat flow in the southern Rio Grande rift and their relationship to extensional tectonics: Tectonophysics, v. 174, p. 21–37, doi: 10.1016/0040-1951(90)90382-I.

Kennett, B.L.N., and Engdahl, E.R., 1991, Traveltimes for global earthquake locations and phase identification: Geophysical Journal International, v. 105, p. 429–465.

Langston, C.A., 1977, Effect of planar dipping structure on source and receiver responses for constant ray parameter: Bulletin of the Seismological Society of America, v. 67, p. 1029–1050.

McQuarrie, N., and Chase, C.G., 2000, Raising the Colorado Plateau: Geology, v. 28, p. 91–94, doi: 10.1130/0091-7613(2000)028<0091:RTCP>2.3.CO;2.

Nolet, G., 1990, Partitioned waveform inversion and two-dimensional structure under the network of autonomously recording seismographs: Journal of Geophysical Research, v. 95, p. 8499–8512.

Owens, T.J., and Zandt, G., 1985, The response of the continental crust-mantle boundary observed on broad-band teleseismic receiver functions: Geophysical Research Letters, v. 12, p. 705–708.

Pollack, H.N., Hurter, S.J., and Johnson, J.R., 1993, Heat-flow from the Earth's interior: Analysis of the global data set: Reviews of Geophysics, v. 31, no. 3, p. 267–280, doi: 10.1029/93RG01249.

Reinecker, J., Heidbach, O., and Müller, B., 2003, The 2003 release of the World Stress Map: http://www.world-stress-map.org (October 2005).

Reynolds, S.J., and Spencer, J.E., 1985, Evidence for large-scale transport of the Bullard detachment fault, west-central Arizona: Geology, v. 13, p. 353–356, doi: 10.1130/0091-7613(1985)13<353:EFLTOT>2.0.CO;2.

Ribe, N.M., 1989, Seismic anisotropy and mantle flow: Journal of Geophysical Research, v. 94, p. 4213–4223.

Ruppert, S., 1992, Tectonics of western North America: A teleseismic view [Ph. D. thesis]: Stanford University, 216 p.

Sandvol, E., and Ni, J., 1994, Mapping seismic azimuthal anisotropy in the United States from LRSM short period data: Eos (Transactions, American Geophysical Union) Supplement, v. 75, no. 44, p. F481.

Sandvol, E., Ni, J., Ozalaybey, S., and Schlue, J., 1992, Shear-wave splitting in the Rio Grande rift: Geophysical Research Letters, v. 19, p. 2337–2340.

Savage, M.K., 1999, Seismic anisotropy and mantle deformation: What have we learned from shear wave splitting?: Reviews of Geophysics, v. 37, no. 1, p. 65–106, doi: 10.1029/98RG02075.

Savage, M.K., 2002, Seismic anisotropy and mantle deformation in the western United States and southwestern Canada: International Geology Review, v. 44, no. 10, p. 913–937.

Seeley, J.M., and Keller, G.R., 2003, Delineation of subsurface Proterozoic Unkar and Chuar Group sedimentary basins in northern Arizona using gravity and magnetics: Implications for hydrocarbon source potential: American Association of Petroleum Geologists Bulletin, v. 87, no. 8, p. 1299–1321, doi: 10.1306/031903200198.

Silver, P.G., 1996, Seismic anisotropy beneath the continents: Probing the depths of geology: Annual Review of Earth and Planetary Science, v. 24, p. 385–432, doi: 10.1146/annurev.earth.24.1.385.

Silver, P.G., and Holt, W.E., 2002, The mantle flow field beneath western North America: Science, v. 295, no. 5557, p. 1054–1057, doi: 10.1126/science.1066878.

Silver, P., and Chan, W.W., 1988, Implications for continental structure and evolution from seismic anisotropy: Nature, v. 335, p. 34–39, doi: 10.1038/335034a0.

Sumner, J.S., 1989, Regional geophysics of Arizona, *in* Jenney, P., and Reynolds, S.J., eds., Geologic evolution of Arizona: Arizona Geological Society Digest: Tucson, Arizona, Arizona Geological Society, p. 717–739.

van der Lee, S., 2002, High-resolution estimates of lithospheric thickness from Missouri to Massachusetts, USA: Earth and Planetary Science Letters, v. 203, no. 1, p. 15–23, doi: 10.1016/S0012-821X(02)00846-4.

Ward, P.L., 1991, On plate tectonics and the geologic evolution of southwestern North America: Journal of Geophysical Research, v. 96, no. B7, p. 12,479–12,496.

Wessel, P., and Smith, W.H.F., 1998, New, improved version of the Generic Mapping Tools released: Eos (Transactions, American Geophysical Union), v. 79, p. 579.

West, M., Ni, J., Baldridge, W.S., Wilson, D., Aster, R., Gao, W., and Grand, S., 2004, Crust and upper mantle shear wave structure of the southwest United States: Implications for rifting and support for high elevation: Journal of Geophysical Research, v. 109, B03309, doi: 10.1029/2003JB002575.

Manuscript Accepted by the Society 18 June 2005

Printed in the USA

Geological Society of America
Special Paper 397
2006

Designing a modular architecture for the structural geology ontology

Hassan A. Babaie
Department of Geosciences, Georgia State University, Atlanta, Georgia 30303, USA

John S. Oldow
Department of Geological Sciences, University of Idaho, Moscow, Idaho 83844-3022, USA

Abbed Babaei
Department of Biology, Geology & Environmental Science, Cleveland State University, Cleveland, Ohio 44115-2214, USA

Hans G. Avé Lallemant
Department of Earth Science, Rice University, 6100 Main Street, Houston, Texas 77005-1892, USA

A. John Watkinson
Department of Geology, P.O. Box 642812, Washington State University, Pullman, Washington 99164-2812, USA

ABSTRACT

Ontologies constitute one of the most important and enabling components of the semantic Web, allowing geoscientists to explicitly and formally model their knowledge base for sharing and reuse over a global network. The geoscience community, realizing the need for such knowledge-enhancing technologies, has started developing a series of geo-ontologies for specific fields. We analyze the problems of designing geo-ontologies, emphasizing issues related to conceptual modeling and system architecture, and present the preliminary conceptual model of part of the structural geology ontology (StructuralGeoOntology). We discuss the ontology development process and identify a set of useful steps and activities that enhance and facilitate the development of any geo-ontology.

Developing ontologies for any field in the geosciences becomes complex if all the concepts and their relationships in the field are included in a single, large ontology. To reduce complexity, we design the StructuralGeoOntology with a modular architecture involving multiple ontologies. This component-based ontology merges several homogeneous subontologies from allied subdisciplines in structural geology, and integrates ontologies from other fields. Each of the shared subontologies will have its corresponding relational database. The modular architecture leads to simplified and more efficient modeling, development, integration, maintenance, reasoning, and future extensibility for the ontologies and databases.

Keywords: domain ontology, structural geology, modular architecture, UML.

Babaie, H.A., Oldow, J.S., Babaei, A., Avé Lallemant, H.G., and Watkinson, A.J., 2006, Designing a modular architecture for the structural geology ontology, *in* Sinha, A.K., ed., Geoinformatics: Data to Knowledge: Geological Society of America Special Paper 397, p. 269–282, doi: 10.1130/2006.2397(21).

INTRODUCTION

Knowledge acquisition and advancement in any large geoscience domain typically are done through the work of the members specialized in different subdisciplines within the domain. The intradomain, cross-discipline interaction of these community members commonly leads to new discoveries and enhancement of the domain knowledge base. For example, discoveries made in fracture mechanics are shared by structural geologists studying fractures in the field. Traditionally, structural geologists make inferences about new information derived from newly acquired data by comparing it against existing information and applying their current knowledge. This process leads to a new and, ideally, better domain knowledge. For example, when new data about the orientation of foliation and lineation in a ductile shear zone are applied to determine kinematic information about the shear zone, the pattern in the data (new information) is compared to that in other shear zones or to the domain's knowledge base. This information processing requires knowledge, which traditionally resides in the mind of the domain experts, in this case structural geologists.

With the advent of the Internet, domain experts became overwhelmed with the rapidly growing size of new, available data and information. In this environment, it has become progressively more difficult for experts to sieve through huge volumes of information and to effectively decipher the patterns and implications of new information. Although a larger knowledge base may lead to better predictions and inferences, the sheer volume of data makes it difficult for humans to readily and effectively integrate the information.

Today, Web pages are designed for human consumption and typically contain presentation information (HTML) but no annotation (markup) about the content of the page. In the last few years, Extensible Markup Language (XML) [3], Resource Description Framework (RDF) [4], and RDF schema (RDFS) [5] have been used to develop markup languages and schemas to make the Web more efficient. These languages allow specification of information but not its meaning. XML, for example, is good for data format and structure but does not formally convey the semantics of the data. These and other languages (e.g., the OWL Web ontology language, [1]) are the backbone of the coming, machine-readable semantic Web (Berners-Lee, 1999), which will be an information and automated services environment in which machines can understand the contents of the pages in terms of information and meaning. The semantic Web will be achieved, between remote applications and software agents, through the presence of distributed, congruent ontologies available on the network, and with the help of rules of inference and application logic. With the semantic Web, we are entering a new stage of computing in which computers will be the point of entry to a global network of information exchange (Fensel, 2004). In this network, presentation, visualization, processing, integration, storage, and interchange of large amounts of data, information, and even knowledge can be achieved through machines and with the help of special, integrated technologies such as ontologies. Ontologies help scientists in a specific domain to explicitly define their common concepts and apply them for sharing, organizing, and reusing their large knowledge bases. Although more efficient in dealing with large volumes of information, the ontological information systems are not expected to perform better than the human experts. These so-called *ontology-driven* information systems, which abstract the domain knowledge base, allow improved communication between the human experts, through common vocabularies and better software reuse and system interoperability (Kishore et al., 2004). Ontologies will help the human experts to more effectively deal with their growing knowledge base.

Knowledge acquired through experience cannot be efficiently reused if it is not formally shared (Guarino and Welty, 2000; Welty and Guarino, 2001). Scientific books and journal articles traditionally collect and archive data, information, and knowledge in an unstructured way, and present them without a consensus or agreement on structure, content, or format. Formal knowledge management through ontologies involves systematic (mainly automated) knowledge acquisition, collection, analysis, organization, structure, representation, maintenance, and refinement (Benjamins et al., 1999). The importance and usefulness of ontologies in knowledge sharing and reuse are being increasingly recognized by scientific communities. In this paper, we introduce ontologies and their design, and discuss their benefits and application in sharing and reuse of information and knowledge in structural geology. We present the design of the structural geology ontology (StructuralGeoOntology) based on a modular system architecture (Rector, 2003), which involves multiple ontologies.

In this paper, "class," as in object-oriented languages such as Java, maps to complex geological concepts (e.g., Fault class), tables in a relational database, and XML elements. A "package" (e.g., Fracture package) is a collection of classes (e.g., Fracture, Fault, Segment) and corresponds to a relational database or Java or XML package (Babaie and Babaei, 2005a, 2005b). "Slots" correspond to attributes, properties, or relations of objects, fields in a relational database table, and XML elements. The naming scheme for class, slot, and package is as follows. We capitalize the first letter of a class name if it is a simple word, e.g., Fold, or Fault. In this paper, only the class and slot names are given in the Courier New font. If a class's name is compound, i.e., made of more than one word, we capitalize the first letter of each word and separate them with an underscore, e.g., Slaty_Cleavage or Strike_Slip_Fault. This naming scheme is consistent with the Protégé ontology development tool's best naming practice, which we intend to use for editing and implementing the StructuralGeoOntology. Slots' names are written in lower case, e.g., length and sense for the Fault concept (class). If a slot's name is made of more than one word we use lower case for all words and separate them with an underscore, e.g., net_slip, axial_plane. This way we can identify slots from classes, which start with capital letters.

ONTOLOGY

Ontology provides a shared and common understanding of a domain (e.g., structural geology) that can be communicated across different agents (people and software) in knowledge sharing and reuse. Ontology is a structured depiction or model of accepted and known facts in a specific field. The popular definition of ontology is given by Gruber (1993) as "formal, explicit specification of a shared conceptualization." This specification is centered on the identification of the relevant concepts within a domain, and their explicit definition and relations, so that they can become formalized and machine-readable. Guarino (1998) defines conceptualization as a set of relations defined in a domain space between a selected set of objects, which for structural geology could include rock, fault, and fold.

Ontologies must objectively capture consensual knowledge across the domain community, and not represent an individual's perception of that world or phenomenon (Gruber, 1993, 1995; Benjamins et al., 1999; Gruninger and Lee, 2002; Fensel, 2004). This means that the specification of the concepts and their relations must characterize the way knowledge is expressed in a specific domain. Successful ontologies are developed through a collaborative work of domain experts, system designers, and software developers. Ontologies, in their simplest form, include a controlled vocabulary (catalog), a list of terms and their meaning in a natural language (glossary), and terms and their relations (thesaurus). In more complex structured forms (Lassila and McGuinness, 2001; McGuinness, 2003), ontologies include terms, relations, and axioms to constrain the possible interpretation of terminology, and to provide formal but real-world semantics for contained information (Gruber, 1993, 1995).

Ontologies allow reuse of knowledge by employing a common vocabulary across heterogeneous software platforms (Guarino, 1998; Gomez-Perez et al., 2004). Reuse is possible through the inclusion mechanism, which allows one ontology to be used within another if they are available within the same node (i.e., server) or network environment (Maedche et al., 2003; [7]). The "included" ontologies (FractureGeoOntology, FoldGeoOntology) can be "restricted" within the "including" ontology (StructuralGeoOntology) by redefining the terms used in common and by making new axioms and assumptions.

Components of Ontology

Ontologies have five components: (1) classes, (2) relations, (3) functions, (4) axioms, and (5) instances (Gruber, 1993; Gomez-Perez et al., 2004). Domain concepts are represented by classes, which in structural geology include a wide range of processes and products. Representation of processes in geological ontologies is complicated due to the fact that many unknown processes can be involved in the formation of a geological object. In this paper, for the sake of simplicity, we deal only with the Fold, Fault, Foliation, and Lineation concepts and their relations and do not cover concepts that represent processes such as folding and deformation.

These classes are arranged in a taxonomic hierarchy where classes inherit the attributes of the classes that are higher in the hierarchy [8]. Each class has properties or slots, which describe various features and attributes of that concept. The restrictions on the slots are called *facets*, which define the allowed values for the slots. In our structural geology domain, each foliation concept (class) and its subclasses, for example, have an attitude property (slot), which can be defined either by the strike/dip or dip-amount/dip direction facets, but not by trend and plunge, which is solely used for linear features. Instances are defined when the slots of a class are filled by values. For example, the attitude of part of the San Andreas fault (an instance of the Fault concept) in a specific study area can be represented by any of the following alternative facets: N10W, 82NE (quadrant strike, dip), 350, 82NE (azimuth strike, dip), 82, 080 (dip amount, dip direction in azimuth), or 82, N80E (dip direction in quadrant). Subclasses represent specialization of the superclasses higher in the hierarchy. For example, the superclass Foliation includes the Slaty_Cleavage and Schistosity subclasses. Relations (properties) define the type of association between concepts (classes). Relations are usually binary (i.e., relate two classes) and have a domain and range (Gomez-Perez et al., 2004). For example, an axial planar foliation "intersects" the folded layer along the fold axis, which is a "Line". In this case, Foliation is the domain, and Folded_Layer is the range for the "intersects" relation.

DESIGN AND DEVELOPMENT OF ONTOLOGIES

The set of activities that lead to the development of an ontology is referred to as *ontological engineering* and includes the ontology development process, life cycle, methodologies, tools, and languages used to build them (Gomez-Perez et al., 2004). The vocabularies in the ontology describe domain concepts and their relationships. Gruber and Olsen (1994) defined ontological commitment as agreement amongst users to implement the shared vocabulary in a coherent and consistent way. Guarino (1998) defined ontological commitment as a function that links the vocabulary of the ontology with a conceptualization (e.g., structural geology). This means that when we commit to an ontology, we are making it clear how each term in the vocabulary maps to its intended meaning in the domain. For example, in the structural geology domain, the term "fracture" would include joint, shear fracture, fault, and fault zone. If several Web sites commit to the same ontology, then the software agents can retrieve and combine information from all these sites and present the coherent information in response to a query, for example, as an input to other applications.

In the following sections, we discuss the steps required to build the so-called *domain ontologies*, which are reusable in a specific domain such as structural geology or petrology. There is no unique or correct way of modeling a domain or, for that matter, its ontology. Building an ontology starts with describing the conceptualization of a given domain in a natural language (e.g., English). In these sentences that describe the domain, concepts are most likely the nouns and the relationships are the verbs. Design of an ontology varies depending on the scope, purpose, and application of the ontology. Our ontology must meet the projected task, reflect the real world, and be intuitive, extensible, and maintainable [8].

Identify the Domain and Scope of the Ontology

Building an ontology is commonly conducted in an iterative and incremental fashion [8]. The product of each iteration (version) requires evaluation and validation for accuracy and consistency and must be debugged by use in scenario-based applications. Ontological engineering starts with the definition of the domain and scope of the ontology, where we clearly identify the domain for which we are developing the ontology and define the use of the ontology. For this purpose, we need to determine the questions that we intend to answer by the ontology, and identify the people who use and maintain the ontology.

Our StructuralGeoOntology domain is intended for information interchange between structural geologists and software agents on Web sites that commit to the same underlying ontology. We intend to retrieve and integrate structural data and information from subontologies and their underlying databases and other Web sites. Assuming that the structural geology community will successfully use the StructuralGeoOntology, relevant, structured data and information can be retrieved from many sites that store a variety of structural information about a specific area. For example, we may be studying an exposure of cataclasite along the San Andreas fault and become interested to retrieve other structural, geochronological, geochemical, and isotopic data and information about the same exposure, from other Web sites on the network. This would work readily if all or most structural geologists commit to the same underlying ontology, which is not feasible. If subontologies in a specific field (e.g., structural geology) have different structures, mapping should be done to make them compatible and reusable. The subontologies can be used in a geological application through XML's import and inclusion mechanisms.

The scope of our ontology can be identified better if we sketch a list of *motivating scenarios* and *competency questions* that knowledge, based on the ontology, is intended to address (Gruninger and Fox, 1995). The motivating scenario is a detailed narrative about a situation in the domain, focusing on selected problems and tasks that need to be addressed. The motivating scenario can be used to identify the architectural requirements of the ontology, such as the mechanism to allow integration of GPS or GIS data into the system. Other architectural issues include the question of the compatibility of the modeling language (e.g., UML, Unified Modeling Language) with the implementation language (e.g., XML, RDF, OWL), or whether or not the implementation language includes a mechanism for representing constraints for the concepts and their relationships.

Some competency questions relevant to the structural geology ontology are as follows: How do we find the sense of shear of a fault from associated riedel fractures? How do folds and lineations relate to each other? How do fault segments link? How does a C-foliation relate to the kinematics in a ductile shear zone? The list of the competency questions will lead to other questions such as, Is there enough information in the ontology to answer the questions? Do answers need to be depicted in special ways (e.g., on stereonet, rose, or *P-T-t* diagrams)?

Judging from the list of the few competency questions above, it is clear that the StructuralGeoOntology will include information about many concepts such as shear zone, fault, fold, and lineation, and their relationships. In the earlier versions of the ontology, a limited number of domain concepts, which occur in the competency questions, is selected based on the scope and purpose of the ontology (Sowa, 1999). For example, earlier versions of the StructuralGeoOntology may focus only on a small set of spatial and geometric issues, and not the temporal or process-oriented aspects of the domain. Temporal and process subontologies can be built and integrated with the object ontologies at a later time.

The next step after identifying the competency questions is to analyze how to answer them, and explicitly state the assumptions needed for answering. These issues will affect the scope of the ontology. For example, to answer the question, Is the San Andreas fault active? we first need to answer another question: When is a fault considered active? A fault can be active if it is seismic or if it is creeping. Therefore, we must identify the Fault, Creep, and Earthquake concepts, and relate them to one another. We also need to make an assumption on the meaning of "active fault." Does "active" mean seismically active or does its meaning also include aseismic slip by creep. Concept definitions and assumptions must be acceptable and agreeable by designers and a reasonable size of domain experts and users. Meaning must be given mainly for the purpose of communication and must not be based on a specific language.

During the design, we must consider reusing existing ontologies if they exist. The StructuralGeoOntology probably will use information from other fields such as experimental deformation, field structural geology, rock and earthquake mechanics, paleoseismology, and microstructural analysis. Members of a few of these fields are already developing their domain ontologies (seismology, petrology, fracture mechanics). In addition to these geological ontologies, we should anticipate inclusion of general-purpose ontologies from free ontology libraries such as Ontolingua [9] or DAML [10] available on the Web. Moreover, the StructuralGeoOntology cannot be a single ontology, or maintenance will be very complicated and expensive; it will be broken into several subontologies, which will be integrated and shared (see below) [11]. Maintenance of ontologies is a main problem in knowledge management [12]. Thus, our design should accommodate extension, importation, inclusion, and redefinition of existing ontologies such as SWEET (Raskin and Pen, 2005) and the ontology for engineering mathematics [13] for units of measurements, time, and length.

Building the Taxonomic Hierarchy

For efficient reuse and componential development, we include related concepts in a group, which translate to package in UML

(Figs. 1 and 2). Within the modular architecture of the StructuralGeoOntology system, the list of the concepts identified from the competency questions will lead to the building of multiple ontologies for the structural geology domain. For example, we may need to make ontologies for fracture (FractureGeoOntology), foliation (FoliationGeoOntology), and fold (FoldGeoOntology); some of these packages are shown on Figure 1. The merging of the FractureGeoOntology, FoldGeoOntology, and other related ontologies will lead to a unified, single, shared ontology: StructuralGeoOntology (Pinto et al., 1999; Maedche et al., 2003; Klein, 2001). The StructuralGeoOntology will extend the SWEET ontology (Raskin and Pen, 2005) for higher-level concepts. It will also integrate ontologies from subjects other than geology, such as units of measure (Gruber, 1993; [13]). Integration is done through aggregation, combination, modification by extension, specialization, and adaptation (Pinto et al., 1999).

The component-based architecture of the StructuralGeoOntology requires identification of the modules (subontologies; packages in UML) that need to merge for developing the encompassing ontology. The packaging scheme during the early stages of design of the conceptual model is very important and plays a crucial role in the design (e.g., specification, class hierarchies) and maintenance of any database based on the UML model. The modular package architecture will help us avoid building a massive, single ontology and database to process and contain voluminous data for all the concepts in the intended ontology. The idea is that each package leads to one database, which can be integrated with other databases that are based on other packages. Another benefit of the package architecture is that it will allow mapping to Java packages and XML namespaces. Figure 2, a UML package diagram, shows a selected (partially shown), interdependent group of components (packages), each containing a set of selected related concepts. Each of the application ontologies (i.e., package) will map to its own database. For example, FractureGeoDB and FoldGeoDB will store data for the FractureGeoOntology and FoldGeoOntology, respectively (Fig. 1).

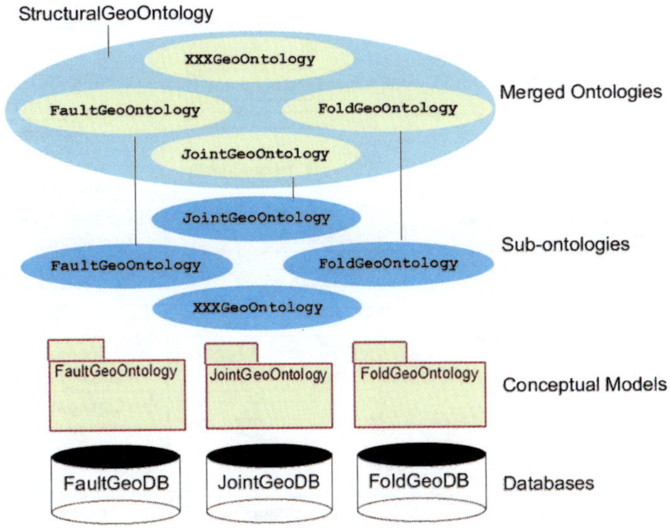

Figure 1. Modular system architecture of part of the StructuralGeoOntology. Only a few packages are shown. The large ellipse represents the shared StructuralGeoOntology, which contains several subontologies (e.g., FaultGeoOntology), shown with small ellipses, that are mapped to external subontologies. Each subontology maps to a UML package representing the conceptual model, shown with the folder icon (see Fig. 2). The UML package maps to a Java package (not shown) and/or a relational database (e.g., FaultGeoDB).

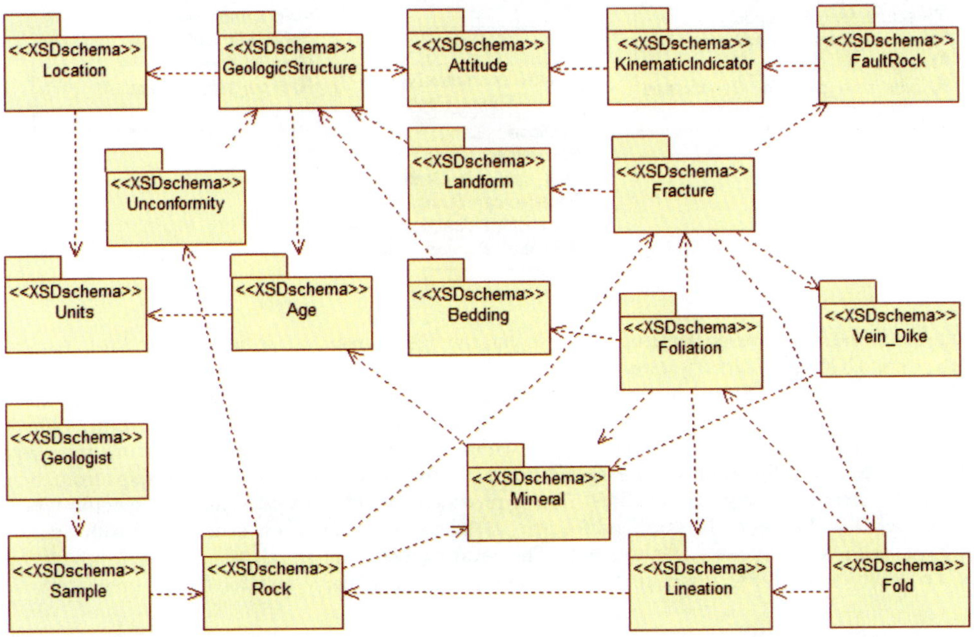

Figure 2. UML package diagram depicting some of the subpackages (shown with folder icons) in the StructuralGeoOntology. Each package contains several related concepts (i.e., classes; see Fig. 3). Package *stereotypes* are given between pairs of angular brackets (e.g., <<XSDSchema>>), for proper mapping to the XML schema. Package names are given in the middle of the folder icon (e.g., Fracture). Each package will map to a single subontology (e.g., FractureGeoOntology) and a database (FractureGeoDB). The dashed arrows indicate dependency of the packages, meaning that a package (e.g., Rock) will *include* other packages that it is pointing to (e.g., Mineral, Unconformity). If package A imports/includes package B, which itself imports package C, then package A does not have to import/include package C.

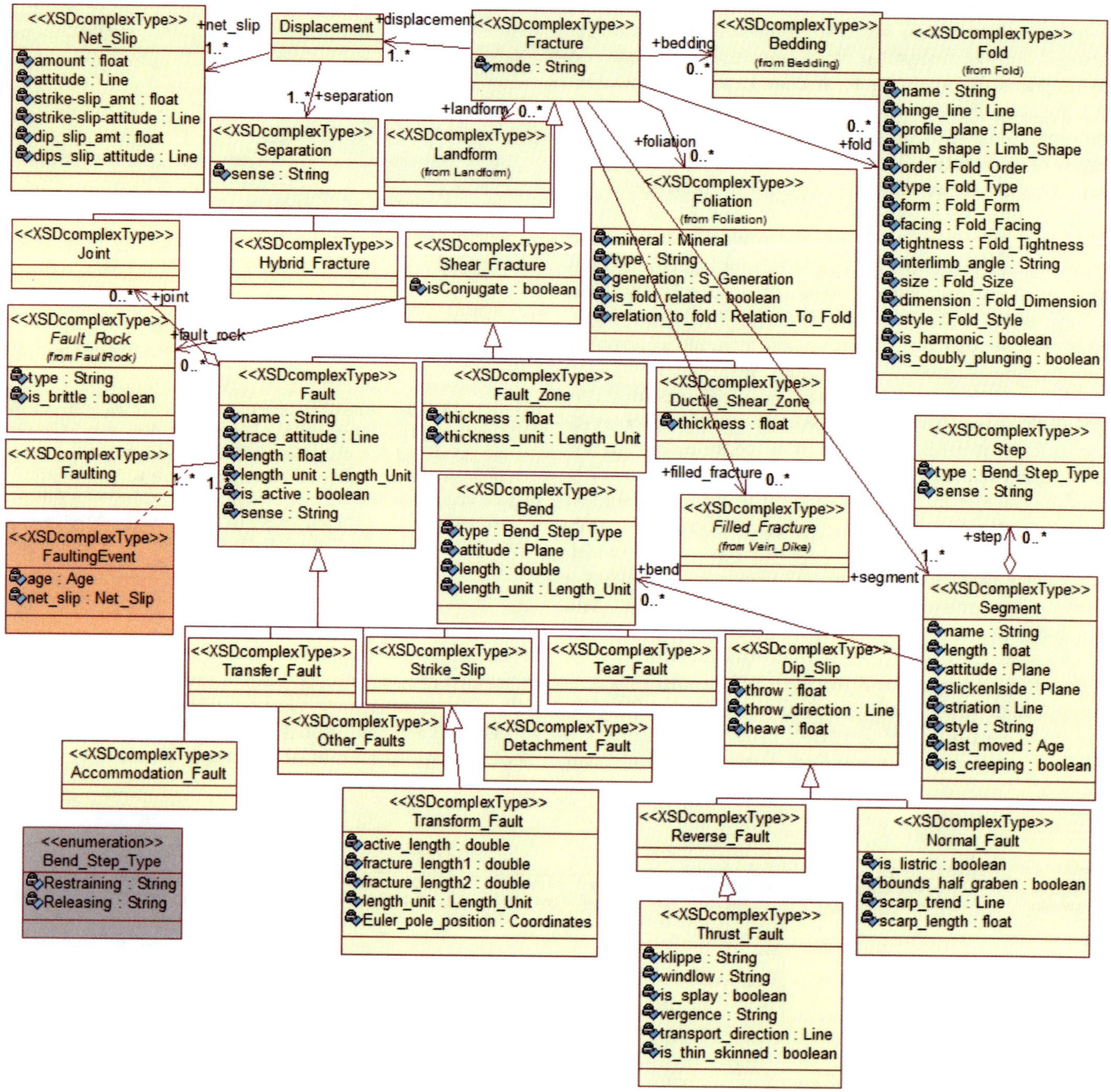

Figure 3. UML class diagram depicting the classes in the Fracture package of Figure 2, and their relationships, cardinalities, and role names. Each class represents a concept. Classes have a <<XSDComplexType>> *stereotype*, given in the top compartment, for proper mapping to elements in the XML schema. The name of each class also is given in the top compartment (e.g., Fault). The middle compartment lists the class attributes and their type, e.g., trace_attitude:Line, which means that the attribute "trace_attitude" of the Fault class is of type "Line," which is *included* into this package from another package (i.e., Attitude). Classes pointing up with open arrows (e.g., Transform_Fault) represent subclasses of their immediate parent (superclass; e.g., Strike_Slip; see text for explanation), inheriting all attributes and relationships from their superclass. The classes with the <<enumeration>> stereotype (e.g., Bend_Step_Type) provide a list of allowed values for specific types. Simple arrows represent associations between classes (e.g., between Fracture and Segment). The end of each arrow is decorated with a cardinality number (e.g., 1..*, which means one or more) and a role name (e.g., +segment). The FaultingEvent class is an association class between Faulting and Fault classes, allowing an M:N (i.e., many-to-many) relationship. Open diamonds represent "parts" relationship (aggregation).

The ontology-based system will share all these subontologies and databases (Fig. 1). Notice that these specialized databases are more useful for the community members than a single, extremely large, and probably impossible-to-build database covering every concept in structural geology.

The modular architecture (Wache et al., 2001; Stuckenschmidt and Klein, 2003) of our StructuralGeoOntology makes it much easier to accommodate inevitable change over time [14]. These changes can occur (1) in the domain, i.e., through new discoveries in the structural geology field; (2) in conceptualization, either due to changes in our view of the world and usage perspective or because of emergence of new tasks; (3) in explicit specification of the ontology (Noy and Klein, 2004).

Taxonomic Relations

The next step is to organize the knowledge to be represented and the ontological commitments for each of the merging ontologies at design time. We should identify synonyms for the same concept and only have one concept defined in the ontology. For example, if we believe that "class 1B fold" and "parallel fold" are synonyms, we should apply the most common usage, and document and allow using the synonyms. Our choice must be based on the consensus within the structural geology community. We need to find out if a class representing a concept is a subclass of another class (concept), or is at the same level in the hierarchy. There are four types of taxonomic relations: *subclass-of*, *disjoint-decomposition*, *exhaustive decomposition*, and *partition* (e.g., Gomez-Perez et al., 2004). The choice of these relations is very important for ontology design.

A concept C1 is a subclass of concept C2 if and only if every instance of C1 is also an instance of C2. The class hierarchy represents an "is-a" relation, i.e., the instance of every subclass is also an instance of its superclass. For example, every instance of Normal_Fault is an instance of the Fault class, or an instance of the C_Foliation class is an instance of the Foliation class, or an instance of the Boudin class is also an instance of the Lineation class. In other words, C_Foliation is a "kind of" Foliation, or a Normal_Fault is a kind of Fault. Care must be taken to avoid cyclic class hierarchies. This can happen when we define B to be a subclass of A, and at the same time B is the superclass of A. If B (e.g., Dip_Slip_Fault) is a subclass of A (e.g., Fault), and C (e.g., Normal_Fault) is a subclass of B, then C "is-a" subclass of A, i.e., a normal fault is a fault, which correctly reflects the domain's knowledge base. This represents the *transitive* property of the class hierarchy (i.e., if C = B and B = A, then C = A).

Siblings, i.e., classes that are in the same level within the hierarchy, must be at the same level of generality. For example, C_Foliation and Schistosity are siblings under the same Foliation parent. Contrary to many programming languages (e.g., Java), ontology languages allow *multiple inheritance*, meaning that a class can be a subclass of two different direct superclasses. For example, a Mylonite is both a Fault_Rock and a Metamorphic_Rock. Notice that subclasses, in addition to inheriting properties from their superclasses, define additional properties (slots), or have restrictions that are different from those of their superclass, or participate in relationships that their parents don't. If a subclass does not have additional properties, we may consider eliminating it from the hierarchy.

A disjoint-decomposition of concept C is a set of subclasses that do not have common instances (hence disjoint), and there may be instances of C that are not instances of any of the subclasses. This means that subclassing is not exhaustive or complete. For example, the superclass Lineation may be subclassed only by Boudin, Slickenline, and Stretching_Lineation. Even though these subclasses are disjoint, i.e., no lineation can be a boudin at the same time of being a stretching lineation, or slickenside, there are other lineations (e.g., intersection lineation) that are instances of Lineation and are not instances of any of these subclasses.

An exhaustive-decomposition of a concept C is a set of subclasses that represent all instances of their superclass but are not disjoint (i.e., have common instances); there is no instance of the superclass that is not an instance of at least one of the subclasses. In this case, subclassing is exhaustive and complete. For example, the Foliation concept can exhaustively (i.e., completely) be decomposed into many concepts (classes) such as Slaty_Cleavage, Gneissic_Banding, Schistosity, C_Foliation, and a few other types of foliation. In this case, every instance of the Foliation concept is represented at least by one of the subclasses of Foliation. The specific subclass types of foliation do not have to be disjoint, meaning that they can have a common instance.

A partition of a concept C is a set of subclasses that are instances of the superclass but do not share a common instance. For example, the Geologic_Structure concept (Fig. 4) can be partitioned into the Planar_Structure, Linear_Structure, and Complex_Structure subclasses. Every instance of Geologic_Structure is either Planar_Structure, Linear_Structure, or Complex_Structure, but an instance of Planar_Structure cannot be an instance of Linear_Structure at the same time. Notice that partitioning geologic structures into planar, linear, and complex types is only good if all geologic structures can be members of these three subclasses. Deciding whether two or more classes are disjoint may not always be obvious. For example, in our structural geology domain, we need to decide whether brittle and ductile deformations are disjoint. Notice that ductile deformation, at a microscopic scale, can be achieved through microfracturing (cataclastic flow). Rocks deformed through semibrittle deformation have components of both brittle and plastic deformation. We have to think about questions such as: Are Ductile and Cataclasis concepts disjoint?

It is important to have these hierarchical relationships in mind when relating concepts during the design stage. The hierarchies must be based on the domain theory that the ontology is representing. In other words, the definition of domain concepts and the relationship between these terms should lead to the type of hierarchy in the ontology. The process of term discovery and building the taxonomic hierarchy is iterative and incremental,

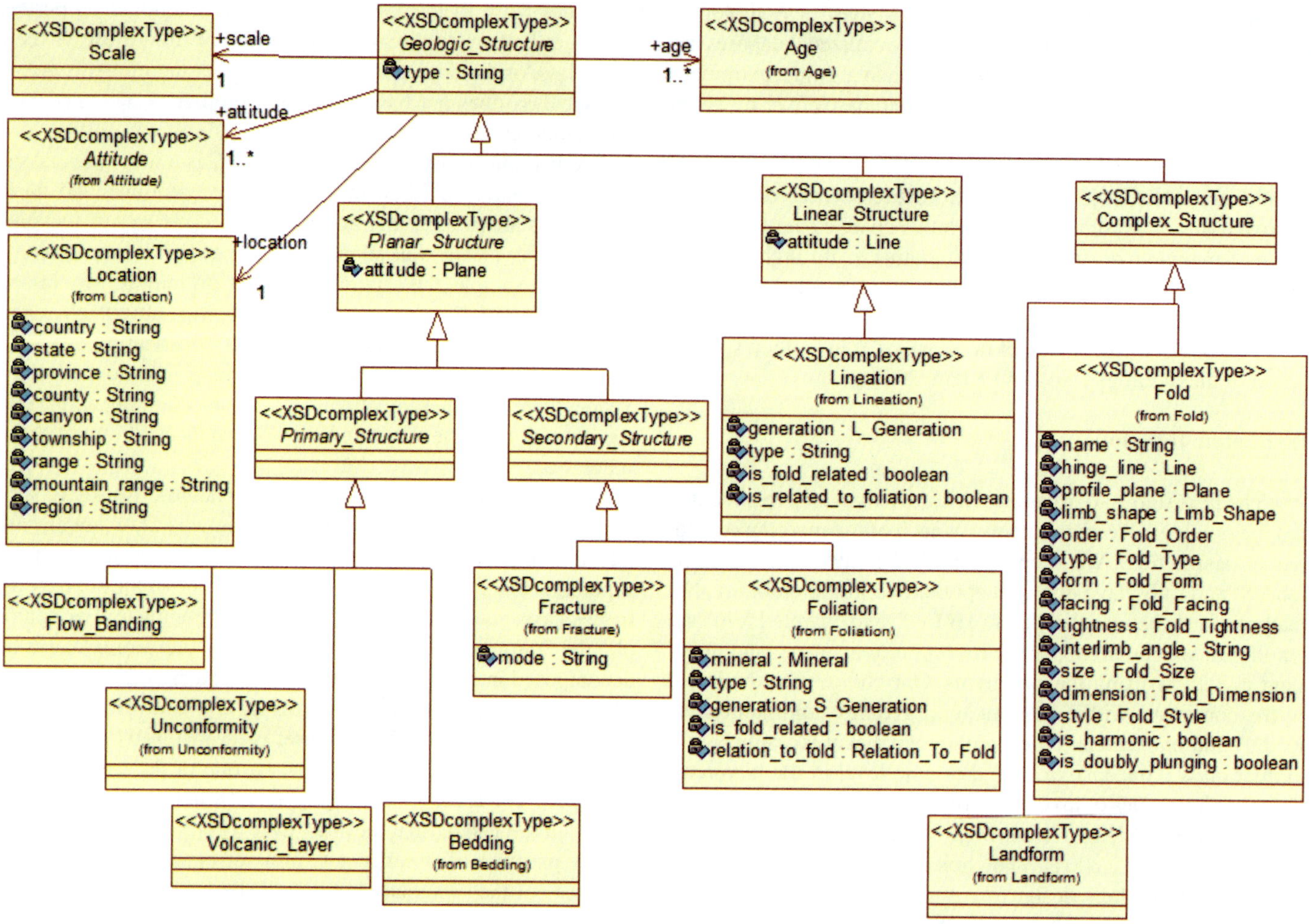

Figure 4. Class diagram of the GeologicStructure package. Notice that the subclasses such as Foliation, Fracture, and Fold inherit the attributes and relationships from their immediate and other superclasses. For example, every Foliation object will reference a scale, age, attitude, and location in addition to the attributes defined in its own class.

meaning that we start with a few concepts and gradually add to the complexity of the hierarchy. Ontologies may have up to a dozen levels of hierarchy. If there is only one subclass, the hierarchy may need to be revised by adding more levels. It there are too many levels, we may want to reconsider the number of levels.

Whether or not something should be modeled as property (slot) or class depends on the scope of our model and the projected task. For example, we should ask: Do we want to have C_Foliation and S_Foliation as two separate subclasses of Foliation, or just have a Foliation concept and then have a slot that defines the type of the foliation (C or S)? If the C_Foliation and S_Foliation concepts are very different and have different tectonic and kinematic implications, with their own properties, then we must represent them as separate classes. Deciding whether a particular concept must be modeled as a class or as an instance depends on the level of granularity [8]. The most specific concepts (i.e., particulars) are commonly modeled as individual instances. Classes that cannot have instances are labeled as *abstract*. An example of an abstract concept is Geologic_Structure (Fig. 4). We measure fault, foliation, and lineation, which are specific (concrete) types of this abstract concept. We create the abstract concepts only for better hierarchical organization, inheritance, and polymorphism in the application logic.

The class hierarchy can be developed in three different ways [8]: top-down, bottom-up, and a combination. None is better than others, and the choice depends on our view of the world (domain). Top-down means that we start at the top of the hierarchy from the most general concepts such as Geologic_Structure (Fig. 4) and come down the hierarchy to concepts such as Planar_Structure and Linear_Structure, and further down to Foliation and Fault under Planar_Structure, and Lineation under Linear_Structure, and even further down to C_Foliation and Schistosity under Foliation, and Slickenline and Boudin under Linear_Structure. Bottom-up starts, for example, from

Segment and goes up the hierarchy tree to Strike_Slip_Fault, to Fault, to Secondary_Structure, to Planar_Structure, and ultimately to Geologic_Structure. We have found the combination process, which uses both the top-down and bottom-up approaches, to be more practical for building the hierarchy.

Defining Facet, Domain, and Range for Slots

After identifying and defining the concepts (i.e., classes) and their hierarchy, we need to define the properties or slots for each concept, otherwise we will not be able to answer the competency questions listed above. For example, a Fault class will have many properties including name, attitude, sense, net_slip, length, and type, and will relate to other classes such as Fault_Rock, Faulting, and Segment. Notice that in UML, concepts map to classes, and slots map to attributes. Figure 3 shows the preliminary class hierarchy for the Fracture package that includes joint and shear fracture (including fault), and vein (under the Filled_ Fracture class).

For case-sensitive systems such as the Protégé ontology development tool, we cannot give the same name to a class and its slot even if the two same names are written in different cases (e.g., Fault and fault). Some properties have a simple type such as type and sense, which could be of type String and have values such as "strike-slip" or "left-lateral". The type for the "attitude" property (inherited from the Planar_Structure class) is complex and includes strike and dip, or dip-direction and dip-amount for planar structures, and trend and plunge, or pitch for linear structures; represented in both azimuth and quadrant formats (Fig. 4; see Babaie and Babaei, 2005a). This means attitude (as a concept) can be a class of its own instead of being a property. This is a better strategy because planar features such as bedding and fold axial plane, and linear features such as lineation and fold axis, also have their own "attitude." This allows reuse of concept and efficient maintenance of code [14].

Object properties can be of intrinsic, extrinsic, parts, and relationship types [8]. Intrinsic properties belong to the concepts, such as the attitude of a fold axis or a fault. Extrinsic properties are external properties of a class, such as the name of a fault or the area where the bend of a fault occurs. Parts represent concrete or abstract pieces (i.e., parts) of a whole. For example, steps, bends, and pull-apart basins are parts of a fault segment, which itself is a part of a fault (Fig. 3). In other words, faults have several segments, each with its own bends, steps, and pull-apart basins. Relationship properties represent relations to other individuals of other classes. For example, "Fault intersects Dike", and "Dike is_intersected_by Fault." The relations "intersects" and its inverse property "is_intersected_by" relate the Fault and Dike classes. Subclasses inherit properties (slots) from their superclass. This means that the Thrust_Fault class that inherits some of its properties from the Fault and other superclasses, has other properties that are unique to itself. We should attach slots to the most general class for code efficiency and the polymorphic behavior in object-oriented languages.

Each slot can have a specific permitted value type represented by its facet. For example, the sense slot of the Fault class can be of type String (Fig. 3). Common types include Number, Boolean, Enumerated, and Instance. Numbers could be Float or Integer. Non-whole numbers of a class, such as the amount of a fault's strike-slip component or its net slip, should be represented with the Float type. Boolean type can only take a yes or no, or true or false value. For example, a Fault class may have a slot called is_active, which can take a yes or no value (Fig. 3). Enumerated slots allow one of several listed value types. For example, the Stress_unit could be either atm, bar, MPa, Pa, kbar, or GPa; or a Bend of a Segment of a Fault could have either a restraining or releasing type (Fig. 3) Slots with the Instance type allow definition of relationships between individuals and must define a list of allowed classes. For example, a Fault class has a displacement slot that relates the Fault class to the Displacement class (Fig. 3). The subtle difference between displacement, slip, separation, etc., must be taken care of in one or more classes.

The permitted numbers a slot can have define the slot's cardinality. A fault can have one sense at a specific time and place. However, a fault can have more than one sense at different segments along the fault (e.g., along restraining and releasing bends). If we decide that a Fault class can only have one name slot, then the name's cardinality is one. However, a Fault class can have multiple segment slots (i.e., cardinality is many) (Fig. 3). We can define the minimum and maximum cardinalities for more precision. For example, a cylindrical fold can have a minimum and maximum of one axis and hingeline (Fig. 5). Notice that Cylindrical_Fold has an infinite number of axes, all with the same orientation, whereas Noncylindrical_Fold may have a minimum of zero axes, and a maximum of many (i.e., countable infinite number) hinge lines (inherited from the Fold class). The class hierarchy of the Fold package is shown in Figure 5, and the XML schema based on this class diagram is shown in Figure 6.

When a slot is of the Instance type, the allowed class is called a *range* of the slot. In other words, the range defines the values the property can take. The "deforms" slot (as in "Fold deforms Bedding") can have multiple values of Bedding (a fold often deforms many bedding planes), and values are instances of the Bedding class. Thus, the "deforms" slot is of type Instance, and Bedding is its allowed class type. The Bedding class is the range for the "deforms" slot. The domain of a property (slot) is the class of objects that the property can be applied to, and it does not need to be specified. For example, the domain of the has_net_slip property is the Shear_Fracture class (the range is the Net_Slip class), and the domain for the "deforms" slot is the Fold class (the range could also be Dike, or Bedding). It is always best to find the most general classes (i.e., superclasses), such as Fracture in Figure 3, when assigning the domain and range for a slot [8].

After classes and their properties (slots), with facets, cardinalities, domains, and ranges, are defined, we must fill the values of the slots with permitted values to create *instances* or

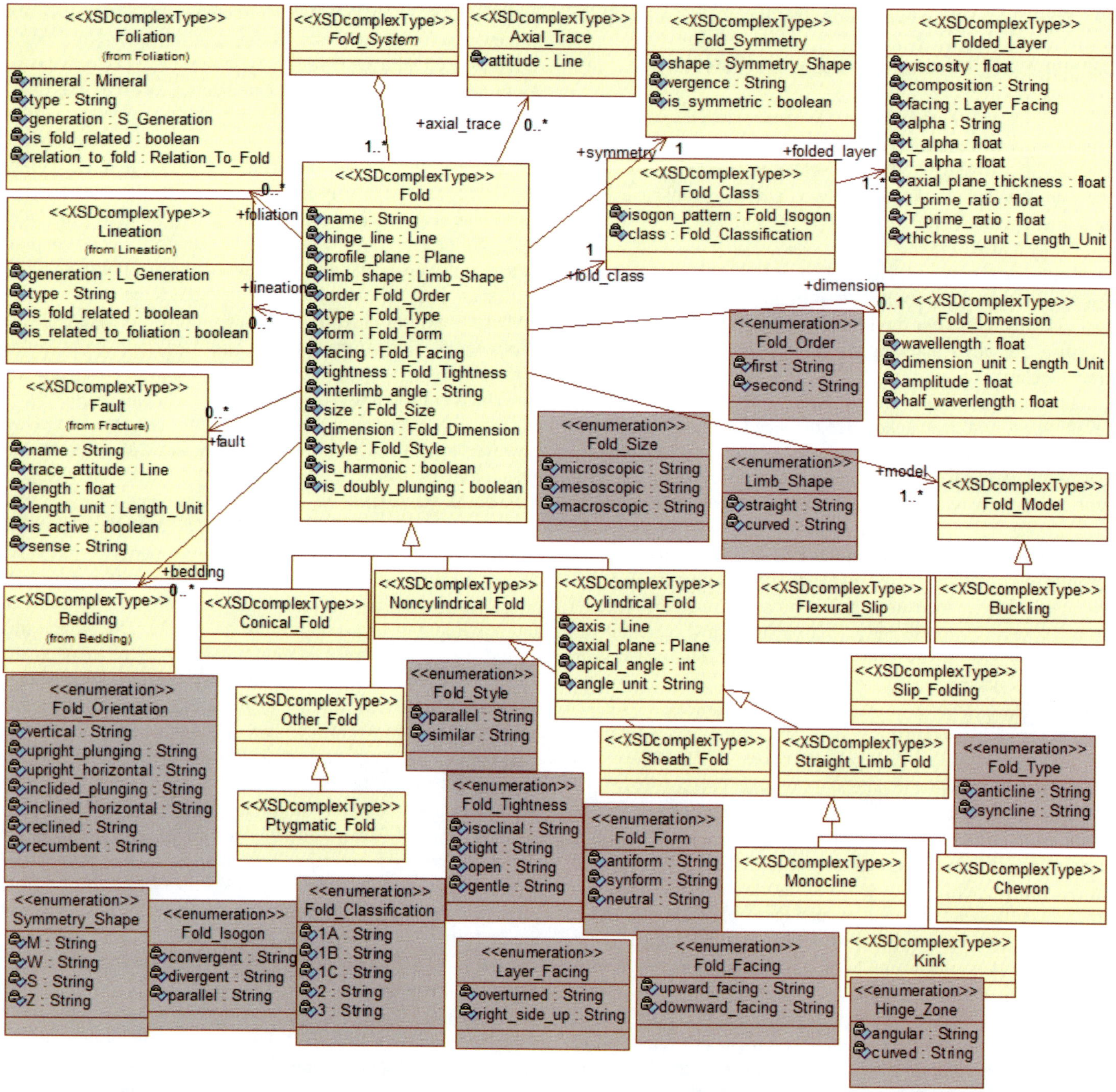

Figure 5. UML class diagram of the Fold package. Notice how the Fold class is referencing classes (e.g., Foliation, Lineation) that are in other packages. This shows that the package depends on the other packages that contain these outside classes (decorated with the "from …" clause in the top compartment). The Foliation, Lineation, and Fault packages (i.e., ontologies) must be included in the Fold ontology. See Figure 3 for symbols. For example, the Fold class references the Foliation class with the foliation reference (role name).

```
<xs:element name="Fold" type="Fold" substitutionGroup="Complex_Structure"/>
 <xs:complexType name="Fold">
  <xs:complexContent>
   <xs:extension base="Complex_Structure">
    <xs:sequence>
      <xs:element name="name" type="xs:string"/>
      <xs:element name="hinge_line" type="Line"/>
      <xs:element name="profile_plane" type="Plane"/>
      <xs:element name="limb_shape" type="Limb_Shape"/>
      <xs:element name="order" type="Fold_Order"/>
      <xs:element name="type" type="Fold_Type"/>
      <xs:element name="form" type="Fold_Form"/>
      <xs:element name="facing" type="Fold_Facing"/>
      <xs:element name="tightness" type="Fold_Tightness"/>
      <xs:element name="interlimb_angle" type="xs:string"/>
      <xs:element name="size" type="Fold_Size"/>
      <xs:element name="dimension" type="Fold_Dimension"/>
      <xs:element name="style" type="Fold_Style"/>
      <xs:element name="is_harmonic" type="xs:boolean"/>
      <xs:element name="is_doubly_plunging" type="xs:boolean"/>
      <xs:element name="symmetry" type="Fold_Symmetry"/>
      <xs:element name="fold_class" type="Fold_Class"/>
      <xs:element name="foliation" type="Foliation" minOccurs="0"
maxOccurs="unbounded"/>
      <xs:element name="fault" type="Fault" minOccurs="0" maxOccurs="unbounded"/>
      <xs:element name="axial_trace" type="Axial_Trace" minOccurs="0"
         maxOccurs="unbounded"/>
      <xs:element name="bedding" type="Bedding" minOccurs="0"
maxOccurs="unbounded"/>
      <xs:element name="planar_structure" type="Planar_Structure"
         maxOccurs="unbounded"/>
      <xs:element name="bedding" type="Bedding" minOccurs="0"
maxOccurs="unbounded"/>
      <xs:element name="lineation" type="Lineation" minOccurs="0"
         maxOccurs="unbounded"/>
      <xs:element name="model" type="Fold_Model" maxOccurs="unbounded"/>
      <xs:element name="dimension" type="Fold_Dimension" minOccurs="0"/>
    </xs:sequence>
   </xs:extension>
  </xs:complexContent>
 </xs:complexType>
```

Figure 6. XML schema code fragment for the Fold XML Schema element, mapped from the Fold class of the Fold package (see Fig. 5). Notice that the Fold class has inherited many attributes from the Complex_Structure and Geologic_Structure superclasses. The attributes are mapped into XML elements, and the cardinalities are given by the maxOccurs and minOccurs facets. Inheritance is shown with the <xs:extension base = "Complex_Structure"> statement, indicating that a Fold "is_a" Complex_Structure.

individuals (*particulars*) of the classes. For example, for the Fault class (Fig. 3), with slots name, type, sense, attitude, and net_slip, we can have the following instance: name: "San Andreas"; type: strike-slip; sense: right-lateral; attitude (strike, dip): 330, 90; net_slip (trend, plunge): 330, 05 (numbers are not real!).

PRINCIPLES OF ONTOLOGY DESIGN

Gruber (1993, 1995) identified the following principles in the design of ontologies: clarity and objectivity, completeness, coherence, extensibility, minimal encoding bias, and minimal ontological commitment. Clarity means that the ontology should give intended meaning to terms by objective definitions, and be documented with a natural language such as English. Ontology must minimize the ambiguity in the meaning and relationship of concepts by specifying the underlying assumptions. Completeness means that necessary *and* sufficient definitions are preferred over partial definitions given by necessary *or* sufficient definition.

Coherence means the ontology should lead to inferences that are consistent with the definitions. Maximum extensibility means we should be able to add new general or specialized terms without a need to revise the existing definitions. Scalability and reusability are important aspects of ontological design [14]. Given the difficulty and the cost (monetary and time-wise) of building ontologies, we should always preplan the scaling of our ontology and use existing ontologies. Ontologies must accommodate changes and evolution of the domain knowledge base over time (Noy and Klein, 2004; Stuckenschmidt and Klein, 2003). Minimal encoding bias means that the conceptualization must be done at the knowledge level and not be based on a specific encoding. We must represent the domain knowledge independent of any language and application so that autonomous software agents can access and use the ontology.

Minimal commitment means we only need to specify terms that are essential to the communication of knowledge consistent with the most general theory. Other design principles (Gomez-Perez et al., 2004) include ontological distinction, meaning that classes should be disjoint; diversification of hierarchies through a multiple-inheritance mechanism; modularity to minimize coupling between modules; minimization of the syntactic distance between sibling concepts by using the same class structure; standardization of naming.

Ontology construction process refers to all activities that are performed when building ontologies, such as management, development, and support activities (Gomez-Perez et al., 2004). The ontology development process identifies the activities but does not specify the order in which the activities should be performed. The ontology life cycle, however, identifies the time period (stage) in which each activity is performed.

The ontology management activities include scheduling, to identify the tasks, their arrangement, time, and resources; control, to check if scheduled tasks are completed as expected; and quality assurance, to see if the quality of products (ontology and its documentation) is satisfactory. The ontology development activities include pre-, syn-, and postdevelopment activities. The predevelopment activities (Gomez-Perez et al., 2004) deal with the type of platform where the ontology will be used, the applications where the ontology will be integrated, and the feasibility study, finding out if the ontology could be built or if it is suitable to be built. Syndevelopment activities consider the reason why the ontology is being built, and discover the intended uses and end users. This is the stage in which the conceptualization and formalization activities are done, and the ontology is implemented into an ontology language. The postdevelopment activities include maintenance involving updating and correction of the ontology. The ontology support activities include a series of parallel tasks that help to develop the ontology. These include knowledge acquisition, evaluation (of software environment), integration (reusing existing heterogeneous ontologies), merging (building a unique ontology from original homogenous ontologies), alignment (establishing mapping between ontologies without merging them), documentation, and configuration (recording all the changes; versioning) [6]

Fernandez-Lopez et. al. (1997) proposed a methodology (METHONTOLOGY) based on the software development process. They define an ontology life cycle in which knowledge acquisition, integration, and evaluation are performed more during conceptualization, and their role decreases during formalization and implementation. METHONTOLOGY includes an ordered sequence of tasks as follows (Gomez-Perez et al., 2004): (1) Build a glossary of relevant terms with their synonyms, acronyms, description, and type. (2) Build the concept taxonomic hierarchy, applying either top-down, bottom-up, or a combination approach. During this activity, we should identify disjoint concepts and identify four taxonomic relations: subclass-of, disjoint-decomposition, exhaustive-decomposition, and partition, as described above. Evaluate the hierarchy for possible errors. (3) Build the binary relation between the concepts and draw the relation diagrams. (4) Build the concept dictionary by tabulating domain concepts, class attributes, instance attributes, and relations. Instance attributes describe concept instances, and their value is different for each instance of a concept. Class attributes, on the other hand, which describe concepts (classes), are not inherited by subclasses or instances. (5) Produce the binary relation table of the relations included in the concept dictionary, by providing the name of the relation, names of the source and target concepts, cardinality, inverse relation, and mathematical property (e.g., symmetrical, transitive). (6) Produce the instance attribute table from the concept dictionary and provide the following for each instance attribute: name, concept, value type, measurement unit, precision and range of values, default values if any, minimum and maximum cardinality, other instance and class attributes and constants used to infer values of the instance attribute, other attributes that can be inferred using values of this attribute, and formulae or rules that allow inferring values of the attribute. Identify other ontologies for types for inclusion in the ontology (e.g., for measurement unit). (7) Produce a class attribute table by defining the same information as in the instance attribute table. (8) Define constants by giving their definition or name, value, value type, measurement unit, and the attribute that can be inferred from the constant. (9) Define formal axioms, which are logical expressions that are always true and used to specify constraints in the ontology. For each axiom, provide the name, natural language (NL) description, logical expression applying first-order logic (FOL), the concepts, the attributes and binary relations to which the axiom refers, and the variables used. (10) Produce a table of rules, needed in the ontology to infer knowledge, by providing its name, NL description, formal expression of the rule, the concepts, the attributes and relations that the rule refers, and the variables used in the expression. (11) Define relevant instances from the concept dictionary and define the name of the concept it belongs to, and attribute values if known.

Implementation of an ontology requires the application of one or more languages. Ontology languages must be intuitive to the human user and use a frame-like look and feel. They must have well-defined formal semantics with established reasoning properties in terms of completeness, correctness, and efficiency (Fensel,

2004). Moreover, they must be related to existing Web languages such as XML [3] and RDF [4] to ensure interoperability.

Issues related to the choice of ontological languages include the nature of the language; expressiveness of the language; inference and reasoning mechanisms in the language; whether or not the ontology development tool (e.g., Protégé) supports the language; support of the language for exchange between applications; integration of the language in other languages of knowledge representation on the Web, such as XML or HTML; existence of translators to transform the ontology implemented by the language to another language; whether or not translation is done with minimum loss of information. The languages to build ontologies include RDF, RDFS [5], DAML+OIL, and the latest Web ontology language OWL [1], which has evolved from DAML+OIL. Most of these languages are based on XML. OWL uses RDF, RDFS, and XML schema metadata and data types and is supported by many recent ontology editors (e.g., Protégé or SemanticWorks). Building on these languages, OWL allows specification of classes, properties, and subproperties in a much more useful and powerful way compared to its underlying technologies. The conceptual model can best be developed with UML.

SUMMARY

We presented ontologies and identified their value in the interchange, sharing, and reuse of information and knowledge in structural geology. We focused on the design issues of building class taxonomic hierarchies for a selected subset of structural geological concepts for the StructuralGeoOntology. This domain ontology, when completed, will represent major aspects of the structural geology domain. We presented the UML class diagrams depicting our conceptual models of Fracture, Fold, and Foliation concepts in the structural geology domain.

We described the major steps required in the development of ontologies and discussed the main pre- and syndevelopment issues of building the structural geology ontology. We emphasized on the need for a component-based architecture for ontologies. Our UML package diagram, representing this modular architecture, is designed to maximize ontology reuse and facilitate maintenance and inevitable ontology change to accommodate evolution of a structural geology knowledge base through time. The UML conceptual model presented in this paper is a first step to develop the StructuralGeoOntology and will be used to build and implement the ontology applying the Protégé ontology editor.

ACKNOWLEDGMENTS

Identification of the competency questions partly listed in this paper was carried out through the collaborative effort of many structural geologists who gathered in McCall, Idaho, in 2003 under the sponsorship of the National Science Foundation (EAR-0334367 to J.S. Oldow), and in Moscow, Idaho, during 2004 under auspices of GEON, the Cyberinfrastructure for the Geoscience (http://www.geongrid.org), supported by the National Science Foundation (EAR-0225421 to J.S. Oldow). We thank Atilla Aydin and a second, anonymous reviewer for their critical reviews which improved the quality of the paper.

REFERENCES CITED

Babaie, H.A., and Babaei, A., 2005a, Modeling Geological Objects with the XML Schema, in Babaie, H.A., and Ramachandran, R., eds., Applications of XML in Geosciences: Computers & Geosciences Journal Special Issue, v. 31, p. 1135–1150.

Babaie, H.A., and Babaei, A., 2005b, Developing the Earthquake Markup Language and Database with UML and XML Schema, in Babaie, H.A., and Ramachandran, R., eds., Applications of XML in Geosciences: Computers & Geosciences Journal Special Issue, v. 31, p. 1175–1200.

Benjamins, V.R., Fensel, D., Decker, S., and Gomez-Perez, A., 1999, $(KA)^2$: Building ontologies for the Internet, a mid term report: International Journal of Human-Computer Studies, v. 51, p. 687–712, doi: 10.1006/ijhc.1999.0275.

Berners-Lee, T., 1999, Weaving the Web: The original design and ultimate destiny of the World Wide Web by its inventor: New York, Harper Collins Publishers, 241 p.

Fensel, D., 2004, Ontologies: A silver bullet for knowledge management and electronic commerce (2nd edition): Berlin, Springer-Verlag, 162 p.

Fernandez-Lopez, M., Gomez-Perez, A., and Juristo, N., 1997, METHONTOLOGY: From ontological art towards ontological engineering: Stanford, California, Stanford University, Spring Symposium on Ontological Engineering of AAAI, p. 33–40.

Gomez-Perez, A., Fernandez-Lopez, M., and Corcho, O., 2004, Ontological Engineering—With examples from the areas of knowledge management, e-commerce and the semantic Web: London, Springer-Verlag, 401 p.

Gruber, T.R., 1993, A translation approach to portable ontology specification: Knowledge Acquisition, v. 5, p. 199–220, doi: 10.1006/knac.1993.1008.

Gruber, T.R., 1995, Toward principles for the design of ontologies for knowledge sharing: International Journal of Human-Computer Studies, v. 43, p. 907–928. Also available at http://citeseer.ist.psu.edu/gruber93toward.html.

Gruber, T.R., and Olsen, G., 1994, An ontology for engineering mathematics, *in* Doyle, J., Sandewall, E., and Torasso, P., eds., Principles of Knowledge Representation and Reasoning: Proceedings of the Fourth International Conference (Kr'94), Bonn, Germany, 1994, Morgan Kaufmann Publishers, San Francisco, California, p. 258–269.

Gruninger, M., and Fox, M.S., 1995, Methodology for the design and evaluation of ontologies, *in* Skuce, D., ed., International Joint Conference on Artificial Intelligence (IJCAI '95), Montreal, Canada, August 20–25, 1995, Workshop on Basic Ontological Issues in Knowledge Sharing, p. 6.1–6.10.

Gruninger, M., and Lee, J., 2002, Ontology application and design: Communications of the Association for Computing Machinery (ACM), v. 45, no. 2, p. 39–41, doi: 10.1145/503124.503146.

Guarino, N., 1998, Formal ontology and information system, *in* Guarino, N., ed., Formal ontology in information system: Proceedings of FOIS '98, Trento, Italy, 6–8 June 1998: Amsterdam, Netherlands, IOS Press, p. 3–15.

Guarino, N., and Welty, C., 2000, A formal ontology of properties, *in* Dieng, R., ed., Proceedings of the 12th International Conference in Knowledge Engineering and Knowledge Management (EKAW '00), Juan-Les-Pins, France, Lecture Notes in Artificial Intelligence LNAI 1937: Berlin, Germany, Springer-Verlag, p. 97–112.

Kishore, R., Zhang, H., and Ramesh, R., 2004, A helix-spindle model for ontological engineering: Communications of the American Computing Machinery (ACM), v. 47, no. 2, p. 69–75, doi: 10.1145/966389.966393.

Klein, M., 2001, Combining and relating ontologies: An analysis of problems and solutions: International Joint Conference on Artificial Intelligence (IJCAI '01) Workshop on Ontologies and Information Sharing, August 4–10, 2001, Seattle, Washington.

Lassila, O., and McGuinness, D., 2001, The role of frame-based representation on the semantic Web: Stanford, California, Stanford University, Knowledge Systems Laboratory, Technical Report KSL-01-02, 10 p.

Maedche, A., Motik, B., Stojanovic, L., Studer, R., and Volz, R., 2003, Ontologies for enterprise knowledge management: IEEE Intelligent Systems, v. 18, p. 26–33, doi: 10.1109/MIS.2003.1193654.

McGuinness, D., 2003, Ontologies come of age, *in* Fensel, D., et al., eds., Spinning the semantic Web: Cambridge, Massachusetts, MIT Press, p. 171–194.

Noy, N.F., and Klein, M., 2004, Ontology evolution: Not the same as schema evolution: Knowledge and Information Systems, v. 6, p. 428–440.

Pinto, S., Gomez-Perez, A., and Martins, J.P., 1999, Some issues on ontology integration, *in* Benjamins, R.B., ed., International Joint Conference on Artificial Intelligence (IJCAI '99) Workshop on Ontology and Problem Solving Methods: Lessons Learned and Future Trends: Stockholm, Sweden, Amsterdam, Netherlands, Central Europe Workshop Proceedings, v. 18, p. 7.1–7.12.

Raskin, R., and Pen, M.J., 2005, Knowledge Representation in the Semantic Web for Earth and Environmental Terminology (SWEET), *in* Babaie, H.A., and Ramachandran, R., Applications of XML in Geosciences: Computers & Geosciences Journal Special Issue, v. 31, p. 1119–1125.

Rector, A.L. 2003, Modularization of domain ontologies implemented in description logics and related formalisms including OWL, *in* Proceedings of the International Conference on Knowledge Capture 2003, Sanibel Island, Florida, 2003: American Computing Machinery (ACM), p. 121–128.

Sowa, J.F., 1999, Knowledge representation: Logical, philosophical, and computational foundations: Pacific Grove, California, Brooks Cole Publishing, 594 p.

Stuckenschmidt, H., and Klein, M., 2003, Integrity and change in modular ontologies, *in* Proceedings of the International Joint Conference on Artificial Intelligence (IJCAI '03), Acapulco, Mexico, August 9–15, 2003: San Mateo, California, Morgan Kaufmann, p. 900–905.

Wache, H., Vogele, T., Vkisser, U., Stuchenschmidt, H., Schuster, G., Neuman, H., and Hubner, S., 2001, Ontology-based integration of information—A survey of existing approaches, *in* Stuckenschmidt, H., ed., International Joint Conference on Artificial Intelligence (IJCAI '01) Workshop on Ontologies and Information Sharing, p. 108–117.

Welty, C., and Guarino, N., 2001, Supporting ontological analysis of taxonomic relationships: Data and Knowledge Engineering, v. 39, p. 51–74, doi: 10.1016/S0169-023X(01)00030-1.

INTERNET REFERENCES

[1] Dean, M., and Schreiber, G., 2003, OWL ontology language reference: W3C working draft: http://www.w3.org/TR/owl-ref/.

[2] Cyberinfrastructure for research and education in structural geology, preliminary report (2003): http://www.sci.uidaho.edu/cyber/http://www.w3.org/TR/owl-ref/.

[3] Bray, T., Paoli, J., Sperberg-McQueen, C.M., and Maler, E., 2000, Extensible Markup Language (XML) 1.0: W3C recommendation: http://www.w3.org/TR/REC-xml.

[4] Lassila, O., and Swick, R., 1999, Resource Description Framework (RDF): W3C proposed recommendation, January 1999: http://www.w3c.org/TR/WD-rdf-syntax.

[5] Brickley, D., and Guha, R., 2000, Resource Description Framework (RDF) schema specification 1.0: W3C candidate recommendation, http://www.w3.org/RDF/.

[6] Sowa, J.G., Building, sharing, and merging ontologies, 2001: http://users.bestweb.net/~sowa/ontology/ontoshar.htm.

[7] Farquhar, A., Fikes, R., Pratt, W., and Rice, J., 1995, Collaborative ontology construction for information integration: http://citeseer.ist.psu.edu/farquhar95collaborative.html.

[8] Noy, N.F., and McGuinness, D., Ontology development 101: http://www.ksl.stanford.edu/people/dlm/papers/ontology101/ontology101-noy-mcguinness.html.

[9] Ontolingua ontology library: http://www.ksl.stanford.edu/software/ontolingua/.

[10] DAML ontology library, 2004: http://www.daml.org/ontologies/.

[11] Klein, M., and Noy, N.F., 2003, A component-based framework for ontology evolution: http://citeseer.ist.psu.edu/article/klein03componentbased.html.

[12] Sindt, T., Formal operations for ontology evolution, 2003: http://citeseer.ist.psu.edu/683180.html.

[13] Gruber, T.R., and Olsen, G., 1994, An ontology for engineering mathematics: http://ksl-web.stanford.edu/knowledge-sharing/papers/engmath.html.

[14] Jarrar, M., and Meersman, R., 2002, Scalability and knowledge reusability in ontology modeling: http://citeseer.ist.psu.edu/jarrar02scalability.html.

Manuscript Accepted by the Society 18 June 2005

Printed in the USA